TAKING SIDES

Clashing Views in

Sustainability

SECOND EDITION

TAKING SIDES

Clashing Views in

Sustainability

SECOND EDITION

Selected, Edited, and with Introductions by

Robert W. Taylor
Montclair State University

The McGraw·Hill Companies

TAKING SIDES: CLASHING VIEWS IN SUSTAINABILITY, SECOND EDITION

Published by McGraw-Hill, a business unit of The McGraw-Hill Companies, Inc., 1221 Avenue of the Americas, New York, NY 10020.
Taking Sides is published by the **Contemporary Learning Series** group within the McGraw-Hill Higher Education division.

1 2 3 4 5 6 7 8 9 0 DOC/DOC 1 0 9 8 7 6 5 4 3

ISBN: 978-0-07-351453-6
MHID: 0-07-351453-5
ISSN: 2161-7651 (print)
ISSN: 2161-766X (online)

Acquisitions Editor: *Joan McNamara*
Marketing Director: *Adam Kloza*
Marketing Manager: *Nathan Edwards*
Senior Developmental Editor: *Jade Benedict*
Project Manager: *Jessica Portz*
Buyer: *Jennifer Pickel*
Cover Designer: *Studio Montage, St. Louis, MO*
Content Licensing Specialist: *Rita Hingtgen*
Cover Image: © *F1 online digitale Bildagentur GmbH/Alamy*
Media Project Manager: *Sridevi Palani*

Compositor: MPS Limited

www.mhhe.com

Editors/Academic Advisory Board

Members of the Academic Advisory Board are instrumental in the final selection of articles for each edition of TAKING SIDES. Their review of articles for content, level, and appropriateness provides critical direction to the editors and staff. We think that you will find their careful consideration well reflected in this volume.

TAKING SIDES: Clashing Views in SUSTAINABILITY

Second Edition

EDITOR

Robert W. Taylor
Montclair State University

ACADEMIC ADVISORY BOARD MEMBERS

Editors/Academic Advisory Board continued

Editors/Academic Advisory Board continued

Preface

There are a few ideas that have the capacity to alter human history. Perhaps sustainability is one of them. It has sprung into mainstream thought relatively recently, from the 1970s and 1980s, although its antecedents can be traced back to the late eighteenth and early nineteenth centuries. Today, it has moved from an abstract idea to practical applications in government offices and corporate boardrooms. Essentially it is defined by its characteristics rather than through its preciseness, and generally those characteristics are accepted societal goals, regardless of political disposition or ideological perspective. The notion that sustainability is related to human survivability, and the innate belief that this survivability is best achieved through the protection of its earthly ecosystem, has entered the mainstream. But, the question of what is the best way to achieve sustainability is open to debate. All decisions, particularly ones that affect both the environment and society, depend on a system of trade-offs. Humans will always alter the environment. But the question of whether this alteration is good or bad is subject to discussion, and the issue of whom is it good for, is also subject to debate. So, we have a series of conflicting viewpoints. What is it that we want to sustain? What is the best way to achieve sustainability? How much of our natural environment should be changed to meet the "needs" of humans? How do the needs of humans conflict with the survivability of other species? This automatically leads to conflicting viewpoints and a clash of opinions.

This collection of articles is built around 20 key issues which are significant to sustainability. The main technique—to present an issue as a pair of opposing essays—is designed to encourage critical thinking. It is well recognized that students tend to be biased toward positions that they hold, but the objective of this book is to make sure that students have access to views that are well presented and well thought out so that they can challenge themselves. While critical thinking is the art of analyzing and evaluating thinking with a view to improving it, it is above all a form of self-corrective thinking. A major challenge facing students in this information-laden world is not how to obtain information but how to evaluate it. The digital world has radically altered the quantity of knowledge available to us. Today, our main concern is not so much how to access information but how to evaluate the tsunami of conflicting ideas that overwhelms our daily information system.

Since sustainability is an ambiguous and controversial subject, the *Introduction* in this book provides a necessary synopsis that provides a basic understanding of sustainability and its many definitions and approaches. The remaining 20 issues—consisting of 40 YES–NO selections—are based upon key questions related to some aspect of sustainability. They are grouped together in five units—*Principles and Overview*; *Global Issues*; *Policy*; *Natural Resources*; and *Energy, Business, and Society*. The *introduction* to each issue has been expanded (from other Taking Sides books) to include more information and a deeper understanding of the various viewpoints associated with the issue. Each *issue introduction* includes a boxed *Learning Outcomes*, which lists the specific information that students should take out of each issue. There is a section, *Exploring the Issue*, which presents both critical questions related to the issue for the student to reflect on, and a section of *Is There a Common Ground*, which seeks to find whether the contrasting viewpoints can be reconciled. And finally, *Internet References* provide sources for further information on the topic and specific articles that had been alluded to in the *essay introduction*.

An added feature to *Taking Sides: Clashing Views in Sustainability* is a website, "i create Sustainability," dedicated to the issues of sustainability addressed in the book and administered by Mc-Graw Hill. The website can be found at www.icreate-sustainability.org and offers students an interactive capability. Now students can go directly to the website and post questions or viewpoints that will be answered. This is a new and exciting feature for the Taking Sides series.

Also available is a general guidebook, *Using Taking Sides in the Classroom*, which offers suggestions for adapting the pro–con approach in any classroom setting. An online version of *Using Taking Sides in the Classroom* and a correspondence service for Taking Sides adopters can be found at www.mhhe.com/cls/usingts. And finally, if you are interested in seeing the table of contents for any of the other titles in the Taking Sides series, please visit the Taking Sides website at http://mhhe.com/cls/takingsides/.

Robert W. Taylor
Montclair State University

The Educational Experience of Disciplinary Controversy*

Brent D. Slife
Brigham Young University

As a long-time user of the *Taking Sides* books, I have seen first-hand their educational impact on students. A student we will call "Brittany" is a prime example. Until her role in a *Taking Sides*' panel discussion, she had not participated once in class discussions. It is probably fair to say that she was sleepwalking through the course. However, once she was assigned to a "side" of the panel discussion, she vigorously pitched in "to do battle," as she put it, with the opposing team. She described a "kind of energy" as she and the rest of her team prepared for the upcoming debate. In fact, she found herself and her teammates "talking trash" good-naturedly with the opposing team before the actual discussion, despite her usual reserve. Because she wanted to win, she "drilled down" and even did extra research.

The panel discussion itself, she reported, was exhilarating, but what I noticed afterward was probably the most intriguing. Not only did she participate in class more frequently, taking more risks in class discussions because she knew her teammates would support her, she also found herself having a position from which to see other positions in the discipline. Somehow, as she explained, her advocating a particular position on the panel, even though she knew I had arbitrarily assigned it, gave her a stake in other discussions and a perspective from which to contribute to them. Brittany's experience nicely illustrates the unique educational impact of the *Taking Sides* Series.

Taking Sides is designed quite intentionally to shore up some of the weaknesses of many contemporary educational settings. The unique energy that Brittany experienced is a result of *Taking Sides* specific focus on the controversial side of academic disciplines. For several good reasons instructors and textbooks have traditionally focused almost exclusively on the more factual or settled aspects of their disciplines. This focus has led, in turn, to educational strategies that can rob the subject matter of its vitality.

Taking Sides, on the other hand, is uniquely structured to highlight the more issue-oriented aspects of a discipline, allowing students to care about and even invest in the subject matter as did Brittany. Involvement can spur a deeper understanding of the topic and help students to appreciate how knowledge advancement is sometimes driven by passionate positions. This focus has led, in turn, to educational strategies that can rob the subject matter of its vitality.

*The full text of this essay and references are available online at: http://highered.mcgraw-hill.com/sites/0076667771/information_center_view0/

Including the Controversial

A case could be made that a complete understanding of any discipline includes its controversies. Controversies may not be considered "knowledge" per se, depending on the discipline, but there is surely no doubt that they are part of the process of advancing knowledge. The conflicts generated among disciplinary leaders often produce problem-solving energy, if not disciplinary passions. In fact, they can drive entire disciplinary conferences and whole programs of investigation. In this sense, disciplinary controversies are not just "error" or an indication of the absence of knowledge; they can be viewed as a positive part of the discipline, a generator of disciplinary vigor if not purpose.

If this is true, then de-emphasizing the controversial elements of a discipline is de-emphasizing a vital part of the discipline itself. Students may learn accepted aspects of the discipline, but they may not learn, at least directly, the disputed aspects. This de-emphasis may not only produce an incomplete or inaccurate sense of the discipline, but it may also mislead the student to understand the field as more sterile, less emotional, and less "messy" than it truly is. The more rational, factual side is clearly important, perhaps even the more important. The question, however, is: do these more settled and perhaps rational aspects of the discipline have to monopolize courses for beginning students?

Another way to put the question might be: couldn't some portion of the course be devoted to the more controversial, thus allowing the student to engage the field in a more emotional manner? In some sense, the more settled and accepted the information is, the less students can feel they are truly participating in the disciplinary enterprise. After all, this information is already decided; there is no room for involvement in developing and "owning" the information. Students may even assume they will be punished for challenging the disciplinary status quo.

Specific Educational Benefits

Engaging the Discipline. When controversy is placed in the foreground of an educational experience, it gives disciplinary novices (students) permission to participate in and perhaps even form their own positions on some of the issues in the field. After all, some issues have not been addressed; some problems have not been solved. As Brittany put it, she was ready to "do battle" with the alternative position, even though she was quite aware of the arbitrariness of her own positional assignment. She was aware that something was at stake; something was to be decided.

In other words, it is the very *lack* of resolution in a controversy that invites students to make sense of the issues themselves and perhaps even venture their own thoughts. Obviously, students should be encouraged to be humble about these positions, understanding that their perspective is fledgling, but even novice positions can facilitate greater engagement with the materials. In a sense, the controversy, and thus a vital part of the discipline, becomes their own, as the example of Brittany illustrates. She not only "owned" a disciplinary position, she used it as a conceptual bridge to engage other settled and unsettled aspects of the discipline.

Appreciating the Messy. Students can also experience the messiness of disciplines using *Taking Sides*. I use the term "messiness" because conventional texts are notorious for representing the field too neatly and too logically, as if there were no human involvement. If disciplines are more than their settled aspects, there are also unsettled elements, including poorly defined terms and inadequately understood concepts, which also need to be appreciated. This messiness is what led Brittany to "drill down" and do "extra research" in her preparation for her panel discussion. She knew that some of the basic terms and understandings were at play.

Good conventional texts may attempt to include these unsettled aspects, but they typically do so in a deceptively logical fashion, as though the controversy is solely rational. This presentation may not only distort these aspects of the discipline but also deliver merely a secondhand report. By contrast, *Taking Sides* books—in pitting two authors against one another—facilitate an *experience* with actual published authorities, who are struggling with the issues from completely different perspectives. In reading both articles, students cannot help but struggle *with* the authors. They do not need to be *told* that the terms of the debate are problematic; the students *experience* these terms as problematic when they attempt to understand what is at stake in the authors' positions.

Preventing Premature Closure. The *Taking Sides* structure also serves to prevent students from prematurely closing controversies. Premature closure can occur by underestimating the controversy's depth or deciding it without a proper appreciation for the issues involved. *Taking Sides* prevents this prematurity by helping the student to experience how two reasonable and highly educated people can so thoroughly disagree. In other words, premature closure is discouraged because real experts are countering each other, sometimes point by point.

A student would almost have to ignore one side of the controversy, one of the experts, to prematurely close the issue. Brittany, for example, reported that she became "absolutely convinced" of the validity of the first authors' position, only to have the second reading put this position into question! Obviously, if the issue could be closed or settled so easily, presumably the experts or leaders of the discipline would have done so already. Controversies are controversies because they are *deeply* problematic, so it is important for the student to appreciate this, and thus have a more profound understanding of the disciplinary meanings involved.

Rehabilitating the Dialectic. One of the truly unique benefits of the *Taking Sides* experience is its rehabilitation of the age-old educational tradition of the dialectic. Since at least the time of Socrates, educators have understood that a *full* understanding of any disciplinary meaning, explanation, or bit of information requires not only knowing what this meaning or information is but also knowing what it isn't. The dialectic, in this sense, is the educational relation of a concept to its alternative (see Rychlak, 2003). As dialectician Joseph Rychlak (1991) explains, all meanings "reach beyond themselves" and are thus clarified and have implications beyond their synonyms. It may be trivial to note, for example, that one cannot fully comprehend what "up" means without understanding what "down" means. However, this dialectic is not trivial when the meanings are disciplinary, such as when the political science student realizes that justice is incomprehensible without some apprehension of the meaning of injustice.

One of the more fascinating educational moments, when using *Taking Sides* books, occurs when students recognize that they cannot properly understand even one side of the controversy without taking into account another side. Brittany described learning very quickly that she clarified and even became aware of important aspects of her own position only *after* she understood the alternative to her position. This dialectical awareness is also pivotal to truly critical thinking.

Facilitating Critical Thinking. I say "truly" critical thinking because critical thinking has sometimes been confused with rigorous thinking (see Slife et al, 2005). Rigorous thinking is the application of rigorous reasoning or analytical thinking to a particular problem, which is surely an important skill in most any field. Still, it is not truly *critical* thinking until one has an alternative perspective from which to criticize a perspective. Recall that Brittany did not participate in class until she developed a perspective to view other perspectives. In other words, one must have a (critical) perspective "outside of" or alternative to the perspective being critiqued. Otherwise, one is "inside" the perspective being critiqued and cannot "see" it as a whole.

As many recent educational formulations of critical thinking attest, this approach means that critical thinkers should develop at least a dialectic of perspectives (one plus an alternative). That is to say, they should have an awareness of their own perspective *as facilitated by* an understanding of at least one alternative perspective. Without an alternative, students assume either they have no position or their position is the *only* one possible. A point of comparison, on the other hand, prevents the reification of one's perspective and allows students to have a perspective on their perspectives. A clear strength of the *Taking Sides'* juxtaposition of alternative perspectives is that it facilitates this kind of critical thinking.

These five benefits—engaging the discipline, appreciating the messy, preventing premature closure, rehabilitating the dialectic, and facilitating critical thinking—are probably not exclusive to controversy. However, they are, I would contend, a relatively unique *package* of educational advantages that students can gain with the inclusion of a Taking Sides approach in the classroom. Controversy, of course, is rarely helpful on its own; settled information and sound reasoning must buttress and perhaps even ground controversy. Otherwise, it is more heat than light. Even so, an *exclusive* focus on the settled and more cognitive aspects can deprive students of the vitality of a discipline and prevent the ownership of information that is so important to real learning.

Contents In Brief

Contents

The references to all selected articles in *Taking Sides: Clashing Views in Sustainability* can be found on the Web at www.mhhe.com/cls.

Correlation Guide

The *Taking Sides* series presents current issues in a debate-style format designed to stimulate student interest and develop critical-thinking skills. Each issue is thoughtfully framed with an issue summary, an issue introduction, and key end-of-issue instructional and discussion tools. The pro and con essays—selected for their liveliness and substance—represent the arguments of leading scholars and commentators in their fields.

Taking Sides: Clashing Views in Sustainability is an easy-to-use reader that presents issues on important topics such as *global issues, policy, natural resources, energy,* and *business and society.* For more information on *Taking Sides* and other *McGraw-Hill Contemporary Learning Series* titles, visit www.mhhe.com/cls.

This convenient guide matches the issues in **Taking Sides: Sustainability** with the corresponding chapters in two of our best-selling McGraw-Hill Environmental Science textbooks by Cunningham/Cunningham.

TAKING SIDES: Sustainability 2/e	Principles of Environmental Science, 7/e by Cunningham/ Cunningham	Environmental Science: A Global Concern, 12/e by Cunningham/Cunningham
Issue 1: Is Sustainability a Realistic Objective for Society?	**Chapter 1:** Understanding Our Environment **Chapter 2:** Environmental Systems: Matter and Energy of Life	**Chapter 1:** Understanding Our Environment **Chapter 25:** What Then Shall We Do?
Issue 2: Is Sustainability More About Politics Than Science?	**Chapter 16:** Environmental Policy and Sustainability	**Chapter 24:** Environmental Policy, Law, and Planning
Issue 3: Are Western Values, Ethics, and Dominant Paradigms Compatible with Sustainability?	**Chapter 1:** Understanding Our Environment **Chapter 4:** Human Populations **Chapter 15:** Economics and Urbanization	**Chapter 7:** Human Populations **Chapter 22:** Urbanization and Sustainable Cities **Chapter 23:** Ecological Economics **Chapter 24:** Environmental Policy, Law, and Planning
Issue 4: Does Sustainability Mean a Lower Standard of Living?	**Chapter 4:** Human Populations **Chapter 15:** Economics and Urbanization	**Chapter 22:** Urbanization and Sustainable Cities **Chapter 23:** Ecological Economics **Chapter 24:** Environmental Policy, Law, and Planning **Chapter 25:** What Then Shall We Do?
Issue 5: Can India and China Reduce Their Dependence on Coal?	**Chapter 4:** Human Populations **Chapter 15:** Economics and Urbanization **Chapter 16:** Environmental Policy and Sustainability	**Chapter 7:** Human Populations **Chapter 22:** Urbanization and Sustainable Cities **Chapter 23:** Ecological Economics

TAKING SIDES: Sustainability 2/e	Principles of Environmental Science, 7/e by Cunningham/Cunningham	Environmental Science: A Global Concern, 12/e by Cunningham/Cunningham
Issue 6: Is Poverty Responsible for Global Environmental Degradation?	**Chapter 15:** Economics and Urbanization **Chapter 16:** Environmental Policy and Sustainability	**Chapter 22:** Urbanization and Sustainable Cities **Chapter 23:** Ecological Economics
Issue 7: Is Limiting Consumption Rather Than Limiting Population the Key to Sustainability?	**Chapter 4:** Human Populations **Chapter 16:** Environmental Policy and Sustainability	**Chapter 6:** Population Biology **Chapter 7:** Human Populations **Chapter 9:** Food and Hunger **Chapter 24:** Environmental Policy, Law, and Planning
Issue 8: Can Technology Deliver Global Sustainability?	**Chapter 12:** Environmental Geology and Earth Resources	**Chapter 23:** Ecological Economics
Issue 9: Is Monetizing Ecosystem Services Essential for Sustainability?	**Chapter 15:** Economics and Urbanization **Chapter 16:** Environmental Policy and Sustainability	**Chapter 23:** Ecological Economics **Chapter 24:** Environmental Policy, Law, and Planning
Issue 10: Does the Market Work Better Than Government at Transitioning to Sustainability?	**Chapter 16:** Environmental Policy and Sustainability	**Chapter 24:** Environmental Policy, Law, and Planning
Issue 11: Does Sustainable Urban Development Require More Policy Innovation and Planning?	**Chapter 15:** Economics and Urbanization **Chapter 16:** Environmental Policy and Sustainability	**Chapter 22:** Urbanization and Sustainable Cities **Chapter 23:** Ecological Economics **Chapter 24:** Environmental Policy, Law, and Planning
Issue 12: Should Water Be Privatized?	**Chapter 11:** Water: Resources and Pollution	**Chapter 14:** Geology and Earth Resources **Chapter 17:** Water Use and Management
Issue 13: Can Our Marine Resources Be Sustainably Managed?	**Chapter 3:** Evolution, Species Interactions, and Biological Communities **Chapter 11:** Water: Resources and Pollution **Chapter 12:** Environmental Geology and Earth Resources	**Chapter 14:** Geology and Earth Resources **Chapter 17:** Water Use and Management **Chapter 18:** Water Pollution
Issue 14: Is Natural Gas Hydraulic Fracturing Safe?	**Chapter 8:** Environmental Health and Toxicology **Chapter 13:** Energy **Chapter 16:** Environmental Policy and Sustainability	**Chapter 20:** Sustainable Energy **Chapter 24:** Environmental Policy, Law, and Planning
Issue 15: Can Species Preservation Be Successfully Managed?	**Chapter 16:** Environmental Policy and Sustainability	**Chapter 24:** Environmental Policy, Law, and Planning
Issue 16: Can Sustainable Agriculture Feed the World?	**Chapter 7:** Food and Agriculture	**Chapter 9:** Food and Hunger
Issue 17: Can Nuclear Energy Be Green?	**Chapter 13:** Energy	**Chapter 19:** Conventional Energy

(Continued)

TAKING SIDES: Sustainability 2/e	Principles of Environmental Science, 7/e by Cunningham/ Cunningham	Environmental Science: A Global Concern, 12/e by Cunningham/Cunningham
Issue 18: Is Corporate Sustainability More Public Relations Than Real?	**Chapter 15:** Economics and Urbanization **Chapter 16:** Environmental Policy and Sustainability	**Chapter 24:** Environmental Policy, Law, and Planning
Issue 19: Are Social Concerns Taken Seriously in the "Triple Bottom Line" of Sustainability?	**Chapter 15:** Economics and Urbanization	**Chapter 23:** Ecological Economics
Issue 20: Can Cities Be Made Sustainable?	**Chapter 15:** Economics and Urbanization	**Chapter 22:** Urbanization and Sustainable Cities

Topic Guide

This topic guide suggests how the selections in this book relate to the subjects covered in your course. You may want to use the topics listed on these pages to search the Web more easily. On the following pages a number of websites have been gathered specifically for this book. They are arranged to reflect the issues of this *Taking Sides* reader. You can link to these sites by going to www.mhhe.com/cls.

All issues and their articles that relate to each topic are listed below the bold-faced term.

(Continued)

Country Issues

Culture

Developing Countries

Economic Issues

Ecosystems

Energy

Energy Conservation

Environment

16. Can Sustainable Agriculture Feed the World?
20. Can Cities Be Made Sustainable?

Environmental Justice

19. Are Social Concerns Taken Seriously in the "Triple Bottom Line" of Sustainability?
20. Can Cities Be Made Sustainable?

Ethics

3. Are Western Values, Ethics, and Dominant Paradigms Compatible with Sustainability?
8. Can Technology Deliver Global Sustainability?
9. Is Monetizing Ecosystem Services Essential for Sustainability?
19. Are Social Concerns Taken Seriously in the "Triple Bottom Line" of Sustainability?

Food

13. Can Our Marine Resources Be Sustainability Managed?
16. Can Sustainable Agriculture Feed the World?

Forestry

13. Can Our Marine Resources Be Sustainably Managed?
15. Can Species Preservation Be Successfully Managed?
16. Can Sustainable Agriculture Feed the World?

Globalization

3. Are Western Values, Ethics, and Dominant Paradigms Compatible with Sustainability?
7. Is Limiting Consumption Rather Than Limiting Population the Key to Sustainability?
9. Is Monetizing Ecosystem Services Essential for Sustainability?
12. Should Water Be Privatized?
16. Can Sustainable Agriculture Feed the World?
20. Can Cities Be Made Sustainable?

Global Warming/Climate Change

2. Is Sustainability More About Politics Than Science?
5. Can India and China Reduce Their Dependence on Coal?
10. Does the Market Work Better Than Government at Transitioning to Sustainability?
14. Is Natural Gas Hydraulic Fracturing Safe?
15. Can Species Preservation Be Successfully Managed?
17. Can Nuclear Energy Be Green?
20. Can Cities Be Made Sustainable?

Natural Resources

2. Is Sustainability More About Politics Than Science?
6. Is Poverty Responsible for Global Environmental Degradation?
7. Is Limiting Consumption Rather Than Limiting Population the Key to Sustainability?
9. Is Monetizing Ecosystem Services Essential for Sustainability?
12. Should Water Be Privatized?
13. Can Our Marine Resources Be Sustainably Managed?
14. Is Natural Gas Hydraulic Fracturing Safe?
16. Can Sustainable Agriculture Feed the World?
17. Can Nuclear Energy Be Green?

Open Space

11. Does Sustainable Urban Development Require More Policy Innovation and Planning?

Pollution

5. Can India and China Reduce Their Dependence on Coal?
7. Is Limiting Consumption Rather Than Limiting Population the Key to Sustainability?
10. Does the Market Work Better Than Government at Transitioning to Sustainability?
17. Can Nuclear Energy Be Green?

Policy

2. Is Sustainability More About Politics Than Science?
6. Is Poverty Responsible for Global Environmental Degradation?
11. Does Sustainable Urban Development Require More Policy Innovation and Planning?
12. Should Water Be Privatized?
17. Can Nuclear Energy Be Green?
19. Are Social Concerns Taken Seriously in the "Triple Bottom Line" of Sustainability?

(Continued)

Population

Regional Issues

Technology

Toxics

Water

Introduction

Sustainability has emerged as an overreaching concept for the twenty-first century that has major environmental, economic, social, cultural, and ethical implications. A National Science Foundation funded report, "*Toward a Science of Sustainability*" (2009), stated that a major challenge of society will be "to foster a transition to sustainability—toward patterns of development that promote human well-being while conserving the life support systems of the planet." The study of sustainability traverses all disciplines—the liberal arts, natural sciences, physical sciences, and the social sciences. It is derived from the basic term, to sustain or maintain over a long period of time, and in this case refers to the ability of humans to maintain their natural ecosystems while improving human welfare over the long term. It has a number of distinct approaches. First, it can be viewed as a science of analyzing and measuring human impacts on natural ecosystems; second, it is a call to action to change practices that are jeopardizing human survivability; third, it establishes a foundation for both corporate and public decision making; and lastly, it is a way for people to approach their daily tasks in a more sustainable manner. Its most common definition is derived from the *Brundtland Commission Report* of 1987 that called it development "that meets the needs of the present without compromising the ability of future generations to meet their own needs." But Stephen Wheeler notes in *Planning for Sustainability* (2004) a number of other definitions based on specific themes:

- *Carrying capacity of ecosystems*—"Sustainable development means improving the quality of human life while living within the carrying capacity of supporting ecosystems." (World Conservation Union, 1991)
- *Maintain natural capital*—"Sustainability requires at least a constant stock of natural capital, construed as a set of all environmental assets." (David Pearce, 1988)
- *Maintenance and improvement of systems*—"Sustainability . . . implies that the overall level of diversity and overall productivity of components and relations in systems are maintained or enhanced." (Richard Norgaard, 1988)
- *Not making things worse*—"Sustainability is the ability of a system to sustain the livelihood of the people who depend on that system for an indefinite period." (William Rees, 1988)
- *Sustaining human livelihood*—"Sustainability is 'the ability of a system to sustain the livelihood of the people who depend on that system for an indefinite period." (Otto Soemarwoto)
- *Protecting and restoring the environment*—"Sustainability equals conservation plus stewardship plus restoration." (Sim Van der Ryn, 1994)

- *Oppose exponential growth*—"Sustainability is the fundamental root metaphor that can oppose the notion of exponential material." (Ernest Callenbach, 1992)
- *Grabbag approach*—"Sustainable development seeks . . . to respond to five broad requirements: (1) integration of conservation and development, (2) satisfaction of basic human needs, (3) achievement of equity and social justice, (4) provision of social self-determination and cultural diversity, and (5) maintenance of ecological integrity." (International Union for the Conservation of Nature, 1986)

As sustainability has grown with public acceptance, attempts to define it have been challenging. Andrew Basiago (*Sustainable Development*, vol. 3, pp. 109–119, 1995) regards it as "tantamount to a new philosophy, in which principles of futurity, equity, global environmentalism and biodiversity must guide decision making." He states that its meaning is based on different disciplinary settings:

- *In* ***biology,*** *sustainability has come to be associated with the protection of biodiversity. It concerns itself with the need to save natural capital on behalf of future generations.*
- *In* ***economics*** *it is advanced by those who favor accounting for natural resources. It examines how markets, as conventionally conceived, fail to protect the environment.*
- *In* ***sociology*** *it involves the advance of environmental justice in situations where some groups make decisions over the use of natural resources and other groups are affected in their daily lives.*
- *In* ***planning*** *it is the process of urban revitalization where there is a pursuit of a design science that will integrate urbanization and nature preservation.*
- *In* ***environmental ethics*** *it means alternatively preservation, conservation, or "sustainable use" of natural resources. This probes the domain where humans ponder whether they are part of, or apart from, nature, and how this should guide moral choice.*

Perhaps one of the best attempts to define sustainability was undertaken by Kates, Parris and Leiserowitz in the April 2005 edition of *Environment* entitled "*What is Sustainable Development?*" They noted that sustainable development was a concept that is "enmeshed in the aspirations of countless programs, places, and institutions," but difficult to define due to its ambiguity. But, they concluded that a key reason for its successful diffusion is precisely due to its malleability and its "ability to serve as a grand compromise between those who are principally concerned with nature and environment, those who value economic development, and those who are dedicated to improving the human condition." As a result, it has been used to "address very different challenges, ranging from the planning of sustainable cities to sustainable livelihoods, sustainable agriculture to sustainable fishing, and the efforts to develop common corporate standards . . ."

Origins of Sustainable Development

Although the underlying ideas of sustainability can be traced back to Thomas Malthus (1798) and his famous treatise on the impact of exponential population growth on food supply, and other responses of the late eighteenth and early nineteenth century to the ecological effects of the industrial revolution, it was not until the 1970s that the modern concept of sustainability emerged. The 1970s saw a huge increase in world population as a result of better medical technology and increasing affluence. This laid the groundwork for a new perspective on development that was posited by Meadows and other researchers from MIT in the book, *Limits to Growth* (1972). They modeled future population growth, resource consumption, and pollution and found that under existing development scenarios the human system will crash. But, if humans could alter their course, it was still possible to establish a condition of ecological and economic stability that was sustainable. Meanwhile, in the United Kingdom, Goldsmith and other editors of the British journal, *The Ecologist*, published a more rhetorical work in 1972, *Blueprint for Survival*, which called for a"radical change" due to increased population and increased consumption that was disrupting ecosystems, depleting resources, and undermining human survival.

The United Nations and other international bodies carried the concept of sustainability into the mainstream. In the 1972 Stockholm Convention the conflict between environment and development was first acknowledged. It noted: "*We see around us growing evidence of man-made harm in many regions of the earth: dangerous levels of pollution in water, air, earth and living beings; major and undesirable disturbances to the ecological balance of the biosphere; destruction and depletion of irreplaceable resources; and gross deficiencies, harmful to the physical, mental and social health of man . . .*"

But it was only with the convening of the World Commission on Environment and Development by the General Assembly of the United Nations in 1982 and its report, *Our Common Future*, published in 1987 that sustainability became an accepted alternative development model. The prime minister of Norway at that time, Gro Harlem Brundtland, chaired the group which later became known as the "Brundtland Commission" and authored the *Brundtland Report*. They argued:

> *The environment does not exist as a sphere separate from human actions, ambitions, and needs, and attempts to defend it in isolation from human concerns have given the very word "environment" a connotation of naivety in some political circles. The word "development" has also been narrowed by some into a very limited focus, along the lines of "what poor nations should do to become richer," and thus again is automatically dismissed by many in the international arena as being a concern of specialists, of those involved in questions of "development assistance." But the "environment" is what we all do in attempting to improve our lot within that abode. The two are inseparable.*

The Brundtland Commission laid out the most famous definition of sustainable development as development which "meets the needs of the present without compromising the ability of future generations to meet their needs." While emphasizing intergenerational equity, this definition also considers limits imposed by technology and social organization and advocates social equity through effective citizen participation. These ideas were carried forward in the United Nations Conference on Environment and Development in Rio de Janeiro in 1992, the so-called *Earth Summit*, which enunciated *Agenda 21*, a set of principles through which sustainable development could be enacted. And finally, with the *World Summit on Sustainable Development in Johannesburg, South Africa in 2002* sustainable development became entrenched as an international model for alternative development.

Chronology of an Idea—Key Events and Writings in the Development of Sustainability

1798	Thomas Robert Malthus	*An Essay on the Principles of Population*
1854	Henry David Thoreau	*Walden*
1848	John Stuart Mill	*On the Stationary State*
1864	George Perkins Marsh	*Man and Nature*
1848	Aldo Leopold	*A Sand County Almanac*
1962	Rachel Carson	*Silent Spring*
1971	Barry Commoner	*The Closing Circle*
1972	Donna Meadows et al.	*Limits to Growth*
1972	Edward Goldsmith et al.	*Blueprint for Survival*
1972	UN Conference on Environment	Stockholm, Sweden
1973	Herman Daly et al.	*Toward a Steady-State Economy*
1974	World Council of Churches	Conference—First Use of "Sustainability"
1976	Robert L. Stivers	*Sustainable Society: Ethics and Economic Growth*
1980	World Conservation Strategy	International Union for the Conservation of Nature
1987	WCED	"The Brundtland Report"
1989	CERES Founding	Valdez Principles—Business Sustainability
1992	World Summit on SD	Johannesburg, South Africa
1990	ICLEI Founding	Local Governments for Sustainability
1991	UN—Rio de Janeiro	"Earth Summit"
1992	WBCSD	Business Sustainability Council
1996	President's Council on SD	Report, "Sustainable America"
1996	United National Habitat II	"City Summit"
1996	Wackernagel & Rees	*Our Ecological Footprint*
2002	"Melbourne Principles"	Sustainable Cities
2009	National Science Foundation	*Toward a Science of Sustainability*

Conflicting Perspectives on Sustainability

There are five perspectives on sustainability that can be readily identified. These perspectives point to some of the major controversies and differences posed by the concept of sustainability. The first is the natural capital perspective. Natural capital consists of the stock of natural ecosystems that produces a flow of valuable goods and services. It consists of mineral resources, biogeochemical cycles critical for life such as the nitrogen, phosphorous, water, carbon, and oxygen, and ecosystem services that yield a flow of valuable goods and services. The United Nations Millennium Ecosystem Assessment (2004) noted four categories of ecosystem services: *provisioning,* such as the production of food and water; *regulating,* such as the control of climate and disease; *supporting,* such as nutrient cycles and crop pollination; and *cultural,* such as spiritual and recreational benefits. Generally, the conflict within this perspective pertains to the capability to adequately assign monetary values to natural capital since many ecosystem services are basic to human survival. The proponents of natural capital accounting point to the need for assigning value according to classical economic criteria so that depletion and pollution can be measured and thereby effectively managed through economic analysis. The opponents of this perspective look at the intrinsic value of nature and reject the notion that nature can be relegated to commodity status.

A second perspective is the split between soft and hard sustainability. Proponents of soft sustainability tend to view sustainability through a neoliberal focus. They have faith in technology, scientific rationality, and economic growth as the best way to implement sustainability. By measuring and quantifying environmental impacts, developing suitable regulatory apparatus, and using market mechanisms to encourage business toward desired goals, they believe that sustainability as a societal goal can be achieved. Hard sustainability advocates have less faith in traditional capitalist economics, view nature as intrinsic and basic to human survival, and espouse a "deep ecology" environmental ethics. They tend to support the idea of bio-equality of species and that humans cannot and should not be separated and stand apart from nature.

A third perspective is the way that different regions of the world view the concept of sustainability. This divergence is often between the views of rich countries, generally referred to as "developed countries" and poorer countries, referred to as "developing countries." The poorer countries of the world view the attempts of the richer countries to control population growth and material consumption as restrictions on their ability to improve their standard of living. Instead, they view issues of sustainability dealing with social equity, justice, and economy as more weighted than a pure ecological view. They look to the large ecological footprints of richer nations compared with their smaller footprints and wonder why there is not more attention directed at this inequality than the issues that preoccupy richer countries, that is, global population stabilization and the reduction of carbon footprints. There is no doubt that a major challenge to sustainability will be the growing energy and material resource demands of emerging economies such as China, India, and Brazil.

The growing middle classes in these countries will increase the demand for finite resources and enhance the global ecological footprint.

A fourth perspective is the controversy over localism versus globalization. A key perspective in sustainability is the advocacy of sustainable communities. This focus often collides with advocates of globalization and free trade. It holds that sustainability involves production and trade for local needs, and is opposed to large-scale urban development and international trading regimes. It views cities as heterotrophic ecosystems, not being able to produce energy and materials to meet their needs. Thus, they must import energy and materials from national or international bioregions making them unsustainable. Autotrophic ecosystems, on the other hand, can produce energy and materials locally and are highly sustainable. As the cost of transportation increases with higher fuel prices, the sustainable communities' perspective is increasingly attractive. But, issues related to what extent can communities meet their requirements locally and to what extent is the communities' perspective scalable still need to be determined.

A fifth and last perspective is the difference between sustainability as a science and sustainability as a social movement. The scientific approach to sustainability relies on the capacity to build a unifying methodology, a set research agenda, and utilizing science and technology to advance global sustainability. As stated by Clark and Levinin *Toward a Science of Sustainability* (Princeton, 2009), the goal is to "develop core concepts, methods, models, and measurements that . . . (would support all sectors) . . . by advancing fundamental understanding of the science of sustainability." Sustainability as a social movement, defined as people sharing a common ideology, tends to be more inclusive in views, more concrete in application, more subject to immediate action, and generally applicable to a wide range of diverse issues, that is, living simply to animal rights. Yet, both perspectives have a "common ground" of goals: to preserve the basic life support systems of the planet and to foster human development; to develop a set of indicators to benchmark changes in the natural environment and the human condition; and to implement a set of practices that reduces the carbon footprint, minimize waste, and lessens the reliance on finite resources. The social movement approach is led by international organizations such as the United Nations, by nongovernmental organizations such as the International Council for Local Environmental Initiatives, and by business groups like the World Business Council for Sustainable Development. The science of sustainability approach is a major feature of the American Association of the Advancement of Science and university-based programs.

Underlying Principles of Sustainability

Regardless of the various ambiguities of sustainability, there are two types of thinking that are basic to the concept: systems thinking and resilience thinking. A system is a collection of separate parts that are connected to form a coherent whole. They consist of *stocks*, such as food and natural resources which are essential for human system; *sources*, the place where food, natural resources, and energy originate; *flows*, the processes by which materials and energy flow through the system; *sinks*, the places where materials end up, namely the air, earth, or water; and *feedback loops*, the messages that are sent to the system to indicate its present condition. Ecosystems and the biosphere are complex adaptive systems where changes in biotic composition and the relationships among elements have consequences for system-level properties of interest. An example of systems thinking is illustrated by the relationship between the loss of biodiversity and climate change. As Levin (Princeton, 2009) states, "unless we can make the connections between the two we cannot determine what aspects of biodiversity are important for mitigating climate change." Hence, through systems thinking system-wide effects can be understood thus allowing for sustainability scientists to construct early-warning indicators for impending problems. Feedback loops are essential for systems sustainability since they represent signals sent out in by an ecosystem that indicate stresses. If appropriate action is taken, then the ecosystem can adapt and maintain itself. If no action or delayed action is taken, causing the ecosystem to exceed its threshold (its range of tolerance), then irreparable damage occurs.

Resilience thinking is a further elaboration of systems thinking that can extend the understanding of the adaptability of the processes and patterns of natural ecosystems to human systems. As explained by Brian Walkerin *Resilience Thinking* (www.peopleandplacenet, 2008), "resilience is the capacity of a system to undergo change and still retain its basic function and structure." All systems, whether human or natural, undergo adaptive cycles. There cycles are: rapid growth, conservation, release, and reorganization. Systems develop and then conserve themselves until some event causes them to reorganize or adapt. The characteristics of a healthy system are its ability to withstand disturbance or shock. If systems are diverse and redundant, they can resist severe regime change and maintain their health. Both systems thinking and resilience thinking provide an essential foundation for sustainability studies. They demonstrate how human–environmental systems are adaptable and manageable.

Global Issues and Sustainability

Population growth and material and energy consumption are major issues facing global sustainability. Mathis Wackernagel and William Rees (*New Society Publishers*, 1996) advanced the notion of "ecological footprint" as an accounting method to "estimate resource consumption and waste assimilation requirements of a defined human population or economy in terms of a corresponding productive land area." It constitutes a useful tool for measuring the "load" imposed on the environment to sustain consumption and waste disposal. The authors noted that "while 20 percent of the world's population enjoys unprecedented material well-being, at least another 20 percent remain in conditions of absolute poverty." On a global scale, the authors saw the increase in the ecological footprints of rich countries, while viewing overall "earth shares," a measure of ecological demands stated on a per capita basis, as having dwindled. This discrepancy illustrates a major challenge to global sustainability.

Global population growth has always been considered as a major challenge for sustainability, going back to Malthus. The neo-Malthusians have tended to emphasize the relationship between population growth and resource depletion, or the "limits to growth." Western writers such as Paul Ehrlich in *The Population Bomb* (1968) are representative of the view that population control is the primary issue for sustainability. While pointing out the negative impacts of a rapidly populating world, the *Brundtland Report* supported the notion that as societies develop and prosper, fertility levels will decline. This "demographic transition" concept is essentially optimistic on the population issue and relies on the capacity for new technologies to rescue the world. This argument essentially believes that "limits of growth" can be countered through the adoption of new technologies and the "substitution" of finite and scarce materials with ubiquitous and renewable resources.

Levels of consumption, already discussed under the ecological footprint, also reflect a major sustainability issue. The earth's carrying capacity consists of the "load" that population imposes on its resource base. This is often determined by consumption levels. For instance, while human population has grown by a factor of 2.2 between 1960 and 2006, consumption expenditures per person have tripled. As stated in the 2010 *State of the World* (Worldwatch Institute), "humanity now uses the resources and services of 1.3 Earths. Hence, people are using about a third more of Earth's capacity than is available, undermining the resilience of the very ecosystems on which humanity depends. This issue is accentuated with the so-called IPAT model (Environment Impact equals Population Affluence Technology). In this model of the key factors causing environmental degradation or resource depletion, population has always been considered the key variable. But, as Robert Kates and others have stated, the driving force of higher levels of consumption need to be addressed even more than population, as consumption produces an exponential increase in real population. Also, the role of new technologies as a panacea for sustainability has been challenged. According to some resilience thinkers, technology often can produce a level of efficiency that actually encourages resource use and non-sustainable practices. For instance, new deep water drilling technologies can allow us to go down farther for oil and gas extraction. Also, more efficient energy technologies can lead to a "rebound effect" where society uses more of the resource.

Another issue to view is poverty. Poor people tend to rely on natural resources for their livelihoods. It is estimated that 85 percent of the population in the rural areas of developing countries depend on natural resources for daily sustenance. This can lead to environmental depletion and deterioration. Unsustainable farming practices causes soil degradation as the poor seek marginal lands to till, resulting in landslides. Poor people use firewood for cooking, denuding local forests, and often overgraze their land. The *Brundtland Report* stated that "poverty reduces people's capacity to use resources in a sustainable manner," and thereby intensifies pressure on the environment. The idea that economic growth can ultimately reduce environmental degradation, the environmental Kutznet's curve idea, is an accepted notion in developing countries. Emerging economies such as China and India, which in 2010 grew at a rate of 10.5 percent and 9.7 percent respectively, are examples of the stresses that global ecosystems will place on a sustainable planet.

Transitions to Sustainability

There are many transitions to sustainability but the two that are exceedingly important today are the transition to a sustainable urban world and the transition to a sustainable economy. Today, nearly 50 percent of world's population lives in cities. By 2030, this percentage will increase to 60 percent and cities of the developing world are expected to absorb 95 percent of this growth as a result of rural to urban migration, transformation of rural settlements into urban ones, and natural population increase. Although comprising only 3 percent of the earth's land area, cities consume 75 percent of global energy, create 80 percent of global greenhouse gas emissions, and intensely concentrate industry, people, materials, and energy. A capacity for cities to transition to sustainability is one of the great challenges of the twenty-first century. While cities constitute a major hazard, they also possess great opportunities for sustainability. Their compact settlement pattern provides economies of scale that can encourage resource and energy efficiency. With people living closer together, public transit can be encouraged, critical infrastructure such as sewers, roads, and electricity can be minimized, resulting in more efficient land and material use. Also, cities encourage innovation and resource efficiency. Building construction, renewal energy advances, and innovative waste management solutions can be successfully adopted.

It can be argued that cities can never be sustainable since they are heterotrophic, meaning that they are ecosystems that do not capture sufficient energy to meet their needs. Most sustainable ecosystems are autotrophic, ones that capture sufficient energy for their needs. Sustainability for cities centers on the strategies, policies, and programs by which cities can become more "photosynthetic," or closer to autotrophic ecosystems (McDonough & Braugart, *Cradle to Cradle*, 2002). Also, sustainability is largely both a response and a modification to both modernist and post-modernist planning styles. It holds at its core an ecological worldview. Modernist planning relies on a rational, comprehensive view of urban development that emphasizes reliance on the efficiency of technological solutions. The horizontal development of twentieth century cities is often the result of the extensive use of one such technology, the automobile, to provide maximum mobility and metropolitan reach. Post-modernist planning tends to emphasize pluralism and localized cultural traditions. Decision-making models for modernist planning in unitary and for post-modernist planning decentralized. Sustainability planning incorporates some of the characteristics of both these approaches, but holds a distinctly ecological worldview. While recognizing pluralism, sustainability planning is centered on systems thinking or the interconnection of people, values, things, events, and resource use. As a planning mode, it uses communication and education to help evolve public understanding; advocacy planning to achieve shared goals; and incentives and mandates to implement agreed upon strategies.

A second major transition is the transition to a sustainable economy. Two writers have pointed out the current problems with unsustainable economic growth. Herman Daly writes (*Scientific American*, 2005, vol. 293, issue 3): "When the economy's expansion encroaches too much on its surrounding ecosystem, we will begin to sacrifice natural capital (such as fish, minerals, and fossil fuels) that is worth more than the man-made capital (such as roads, factories, and appliance) added by the growth. We will then have what I call uneconomic growth, producing "bads" faster than good—making us poorer, not richer." Paul Hawken writes in *The Ecology of Commerce* (1993), "Quite simply, our business practices are destroying life on earth. Given current corporate practices, not one wild life reserve, wilderness, or indigenous culture will survive the global market economy. We know that every natural system on the planet is disintegrating. The land, water, air, and sea have been functionally transformed from life-supporting systems to repositories for waste. There is no polite way to say that business is destroying the world." Yet, according to Robert Taylor in *Making the Business Case for Corporate Sustainable Development* (2008) there are concrete reasons or drivers for businesses to adopt sustainability. The first driver is governmental compliance. Businesses need to comply with the regulatory apparatus of their host countries. In countries that have strict regulatory procedures and strong enforcement, this constitutes a major consideration. Porter (2005) points out that companies that can respond quicker and more efficiently to governmental regulations gain a significant advantage over their competition and provide better share value to their stockholders. A second driver is the marketplace. Businesses can sustain competitive advantage by responding to the need for environmentally friendly products and services. Also, a company's "branding" as socially responsible can increase its customer base and include its investment base as groups seek to add companies that are sustainable to their investment portfolio. And lastly, businesses are often advised by their stockholders to become more sustainable.

And finally, there are four tools available to corporations to transition to sustainability. The most prevalent tool is the ISO 1400 series, an international standard that establishes a corporate environmental management system (EMS) that integrates environmental responsibility into corporate management procedures. A second tool is environmental cost-accounting (ECA), a financial approach that allows corporations to more precisely measure the environmental costs associated with their products. Another is the design for environment (DFE) corporate sustainability tool that addresses environmental health and safety, preservation, and restoration processes in new project development. And lastly, eco-efficiency (EE), which seeks to create value in products through the use of fewer inputs of materials and energy and closely follows the concepts inherent in industrial ecology.

Transitions to sustainability will not come easily. They will require a scientific understanding of the relationships between human systems and natural systems; the ability of engineers to develop new technologies that are more benign; the political will of our public leaders to develop policies and programs that encourage sustainable practices; and the capacity for all of us to understand what needs to be done and then to do it. There will also be the need for trade-offs, what has been referred to as the tension of the "dual mandate." Natural systems have basic requirements as well as human systems, and often, there is a collision between these systems based upon human priorities. Trade-offs will need to be made, and they need to be made with a scientific understanding of their impact on natural ecosystems. In the field of sustainability, there are no easy solutions but only concerted efforts over a long time period.

Robert W. Taylor
Montclair State University

Internet References . . .

U.S. Department of Environmental Protection—Sustainability

This is the website of the U.S. Department of Environmental Protection, which provides information on sustainability.

http://epa.gov/sustainability

United Nations—Division for Sustainable Development

This website provides information on sustainable development and on the Millennium Development goals.

http://www.un.org/esa

Leading a Green Life

This website is a media outlet publishing blogs, daily newsletters, and regular updates on Twitter and Facebook. It discusses the basics of leading a "green life" and also provides the resources for doing so.

http://www.treehugger.com/

Sustainability—An Open-Access Journal

This is the homepage to *Sustainability*—an open-access journal. This journal publishes articles on many topics related to sustainability.

http://www.mdpi.com/journal/sustainability

Environmental Sustainability Index

An Initiative of the Yale Center for Environmental Law and Policy (YCELP) and the Center for International Earth Science Information Network (CIESIN) of Columbia University, in collaboration with the World Economic Forum and the Joint Research Centre of the European Commission. The site provides access to the 2005 Environmental Sustainability Index and its underlying data and map gallery.

http://sedac.ciesin.columbia.edu/es/esi/

Sustainability Information

A website that introduces the reader to basic concepts, definitions, and approaches to sustainability and sustainable development.

http://www.sustainablemeasures.com/sustainability

U.S. Geological Survey

The website of the U.S. Geological Survey provides a good overview of sustainability issues.

http://www.usgs.gov

Natural Resources

This website provides sustainability facts and statistics on natural resources.

http://isri.org

UNIT 1

Principles and Overview

Sustainability means different things to different people. Although somewhat ambiguous, the concept rests on certain fundamental principles and perspectives. It holds at its core the notion that a healthy life-support system for the planet is essential for human well-being. As a new field of study, it is in the process of developing its own unique set of concepts, methods, and models that can work across all sectors. Presently it relies on certain agreed-upon goals and characteristics. These goals include meeting the needs of the present without compromising the capacity of future generations to meet their needs, improving human development within the constraints of our life-support systems, and maintaining natural and social capital. Specific and measureable characteristics are to reduce, reuse, and recycle our waste stream or replace it with a cradle-to-cradle designed system; move to an energy system that centers on conservation, technological innovation and efficiency, and renewable energy; and seek to reduce both our carbon and our ecological footprints.

Accomplishing these goals will not be easy. Policies that encourage sustainability are subject to extended controversy since they often challenge entrenched economic interests and existing social and cultural constructs. Do governments possess the political will necessary to move to a development pattern that encourages more sustainability? How should this be done: through more government leadership or through the market economy? Is there a basic conflict between the goals of human development and maintaining our natural ecosystems? Understanding the nature of the basic controversies over sustainability and coming to agreement on these contrasting views is essential before sustainability can become operational in society.

- Is Sustainability a Realistic Objective for Society?
- Is Sustainability More About Politics Than Science?
- Are Western Values, Ethics, and Dominant Paradigms Compatible with Sustainability?
- Does Sustainability Mean a Lower Standard of Living?

ISSUE 1

Is Sustainability a Realistic Objective for Society?

YES: Sharon Bloyd-Peshkin, from "Built to Trash: Is 'Heirloom Design' the Cure for Consumption?" *In These Times* (November 2009)

NO: Sharon Begley, from "Green and Clueless," *Newsweek* (August 2010)

Learning Outcomes

After reading this issue, you should be able to:

- Gain an understanding of the challenges facing sustainability.
- Describe the concepts of heirloom design, product life extension, and slow consumption.
- Understand the importance of educating the general public about sustainability.
- Discuss the history of consumerism as a civic duty in the United States.
- Discuss the relationship between heirloom design and sustainability.
- Define functional and fashion obsolescence.

ISSUE SUMMARY

YES: Sharon Bloyd-Peshkin, an associate professor of journalism and freelance writer, believes that sustainability is a realistic objective for society but is achievable only through sweeping changes in our economic system. Enticing producers to market products that have a longer life cycle and are repairable would address much of our overconsumption and help move toward a sustainable society.

NO: Sharon Begley, a journalist for the *Wall Street Journal* and *Newsweek*, believes that people have little idea about how to achieve energy efficiency and lead an eco-friendly lifestyle, and fail to understand how a move to sustainability requires major societal steps.

While sustainability can be a noble goal for society, there is debate about whether or not it is realistically attainable. Supporters of sustainability state that people are inherently moral and ethical, that sustainability is economically sensible in the long term, and that technology and innovation will make sustainability possible with long-term commitment and dedication. Those who believe that sustainability is not realistic for society also have strong scientific and cultural support. Naysayers claim that people are lazy and don't want to sacrifice comfort, that scientists cannot agree on what sustainability means, that sustainability does not allow for equality, and that de-growth is nearly impossible without spiraling into an economic recession.

Kathryn Brown, senior vice president of Corporate Responsibility at Verizon, delivered a passionate address denouncing the disconnected relationship between corporations, citizenship, and sustainability. She stated that a core attribute of responsible citizenship is sustainability and that the interrelationship between citizenship and sustainability is vital to creating responsible corporations. She implored corporate representatives to adopt sustainable practices and reduce complexity in the design, production, and marketing aspects of their operations. She also argued that the long-term ideals and characteristics of sustainable practices are the same principles that lead to long-lasting, thriving, and successful corporations (Brown, 2010).

On the contrary, some writers believe that society only supports sustainability from a theoretical perspective and that people are not willing to make the necessary changes that would make it real. Bryan Welch (2010) believes that we are "speaking the green vernacular" but not making any true strides toward sustainability. He states that our consciousness about switching from incandescent to compact fluorescent light bulbs does not make up for our ballooning population size or that our emerging preference for organic food does not negate the destruction of rainforest to produce cheap soybeans and beef. Welch states that to have any lasting movement toward global sustainability, real incentives need to be offered in exchange for choosing to make real alterations to one's personal life. Another tactic for inspiring sustainability is taxation or punishment for those who do not comply with environmentally friendly regulations. David Oxtoby (2010) of Pomona College expressed similar concerns regarding the basis of Leadership in Energy and Environmental Design (LEED) certifications. Oxtoby uses the example of solar panels versus bike racks. Both award you one point but solar panels cost millions and bike racks are practically free. In other words, the panels and the bike racks have the same LEED reward but the initial costs may be prohibitive for solar panels.

Some believe that regulations, like mandatory emissions reductions, will hinder growth. The European Union recently debated increasing their emission reduction goals by the year 2020. Opponents believe that more stringent, mandatory emissions cuts will burden businesses and individuals during a period of economic downturn. The theory is that the large carbon-emitting corporations, like cement and steel producers, will be charged exorbitant amounts of money for their high emissions. Additionally, the price of electricity for the average consumer would increase. A 30 percent reduction in emissions would cost approximately €81 billion, whereas a 20 percent reduction would only cost €48 billion. Opponents also fear that the burden of paying for the increased price tag would not be distributed equally between different emitters (The Economist, 2010).

Zero-growth advocate, Tim Jackson, believes that planned de-growth is the cure for our sustainability woes. Even though the economy will be unstable for a while, planned de-growth will lead to long-term sustainability. Jackson admits, like many people would, that the economic growth and material progress during the last century have vastly improved health and human welfare. But, unfortunately, the benefits of the additional income decrease as citizens become wealthier. Additionally, our extremely high level of consumerism, claims Jackson, has made us "neurotically competitive and obsessively materialistic." In an ideal world, the poorest countries would continue to grow, while the richest countries would be capped. Jackson believes this can only be achieved through a radical new carbon trading model. This model would link carbon emissions to investment return for all corporations. For example, high emitters would have their profits reduced, while the highest emitters would be shut down—regardless of how profitable they are. To prevent spiraling into a deep economic recession during the period of planned de-growth, people would need to be encouraged to spend their free time on localized, community-based work (Davis, 2010).

Duncan Greene, an experienced policy analyst, does not believe that Jackson's planned de-growth model will achieve its goal of sustainability. Green wonders how we would ever implement de-growth. He suggests that we would need a project similar to the Manhattan Project with a massive amount of investment capital. Green also asks what kind of government would be able to implement this type of program without trampling on democratic rights or political freedom—would violators get a letter in the mail, a hefty fine, or a jail sentence? Additionally, who would decide which countries were poor enough to continue to grow and how would they determine growth (Davis, 2010)?

Another characteristic of sustainable development that would be extremely difficult to achieve, state those who do not believe sustainability is realistic, is transparency. The media and the public cannot have any faith in corporate reporting of sustainability practices without corporate transparency. To truly make a determination about sustainability, corporations would have to be transparent about community impact, employee treatment, ecological footprint and much more than is currently reported (Kossoy, 2010). Sustainability may be achievable but it must be built on transparency.

The most important obstacle that will need to be overcome in order to create a sustainable society is commitment. At the moment, sustainability is a hot media topic—everyone from housewives to politicians to elementary school students to businessmen talk about being environmentally friendly and sustainable. Today, being green is "cool." But the public's attention span is limited, and the news media will eventually move on to other "hot topics." In this scenario, will people continue to strive for sustainability when no one is watching them? Will governments and corporations continue to provide rebates and subsidies for sustainable products? Will investment in innovation and new technologies for sustainability be maintained? Sustainability can never be reached if there is no commitment from one generation to the next (Balta, 2010).

Oxtoby believes that the concept of offsetting your carbon footprint merely supports apathy and laziness. He believes it creates a "false sense of sustainability" and that it is extremely arrogant to continue to gobble up resources while paying someone else to reduce their usage. Sustainability cannot be achieved without commitment and change from everyone. As Wayna Balta, vice president of Corporate Environmental Affairs and Product Safety at IBM, writes, "Point your moral compass in the right direction, but be sure to stay the course. We're in it for the long haul" (2010).

In the YES selection, journalism professor and freelance writer Sharon Bloyd-Peshkin presents a proposal for achieving sustainability through the move to heirloom design. The idea is that sustainability can be achieved partially through stopping the endless supply of factory-to-waste products. Manufacturers design products that have functional and fashion obsolescence. Functional obsolescence is a decision made by manufacturers to create a product that does not last and the consumer will need to purchase a replacement. Fashion obsolescence refers to the new features and aesthetic changes to products that entice consumers to purchase a new one before the original has worn out. Heirloom design calls for products that are durable, upgradable, and repairable, which would reduce the quantity of the energy required for production and distribution.

In the NO selection, journalist Sharon Begley notes that many scholars think that sustainability is unachievable because people just do not care. She explains that this is due to basic psychology. People are not willing to give up their electronic toys, luxury items, etc., or become a nation of pedestrians. The best chance we have is to discover a way to satisfy our materialistic needs using less electricity, fuel, and natural resources. The news media and environmental groups need to better educate the general public on changes they can make in their lives that will help achieve sustainability without sacrificing their standard of living.

Sharon Bloyd-Peshkin

Built to Trash: Is "Heirloom Design" the Cure for Consumption?

As the middle-class daughter of a refugee mother and a Depression-era father, I grew up straddling two worlds. My parents could afford much more than they were willing to buy. Most things that broke could be and were repaired. My German grandmother's aphorisms lingered in the air: "Waste not, want not," "A penny saved is a penny earned," "A stitch in time saves nine."

By the time my own children were born, America was flooded with cheap and cheaply made goods. So while my parents continued working at the sturdy antique desks they inherited from my grandparents and sleeping beneath a hand-crocheted bedspread, my children and their friends became the first and last owners of a seemingly endless supply of plastic toys and particle-board furniture.

I was part of the transitional generation. Building blocks were still made of wood. Comforters were still filled with down. I recall the meticulously machined pencil sharpeners with "made in West Germany" stamped on their sides that lasted until I lost them. Even the cheap items—the ones "made in Japan"—tended to hold up pretty well.

Now nearly everything is produced in China and made to be discarded. According to a 2008 report by the Economic Policy Institute, the United States imported $320 billion in Chinese goods in 2007. In that year alone, this country imported $26.3 billion in apparel and accessories, $108.5 billion in computers and electronic products, and $15.3 billion in furniture and fixtures from China.

The manufacture, distribution, and disposal of an ever-growing mountain of short-lived consumer goods have taken an enormous environmental toll. Annie Leonard's website "The Story of Stuff," which has garnered more than 7 million views in less than two years, has helped spread awareness of that cost far beyond the usual environmentalist circles.

We can't, however, only blame the quantity and quality of Chinese goods for the environmental and other consequences of this transoceanic factory-to-waste stream. For that we can blame the two horsemen of the modern consumer apocalypse: functional obsolescence and fashion obsolescence.

Functional or planned obsolescence is the purposeful decision by designers and manufacturers to ensure things don't last, so that consumers must buy new ones. Fashion obsolescence is the related decision to offer new features and aesthetic changes to entice consumers to discard their old items in favor of updated and supposedly better ones.

Ironically, product obsolescence was once seen as the remedy for what ailed our country. Lizabeth Cohen, chair of the history department at Harvard University and author of *A Consumers' Republic: The Politics of Mass Consumption in Postwar America* (Vintage, 2003), traces the origins of mass consumption to the period immediately before and after World War II, when a demand-driven economy was seen as the key to our nation's recovery and prosperity.

"In the 1940s and '50s, there was a much closer connection between consumer demand and factories and jobs," Cohen says. "That was a completed circle more than it is today. When people were buying things, they were buying things that were made by American workers."

The only way to guarantee continued demand was to ensure that people would keep replacing the things they owned. The literature on planned obsolescence makes frequent reference to statements by industry analysts and strategists of that era. "Our enormously productive economy . . . demands that we make consumption our way of life, that we convert the buying and use of goods into rituals, that we seek our spiritual satisfaction, our ego satisfaction, in consumption," retailing analyst Victor Lebow said in 1948. "We need things consumed, burned up, worn out, replaced and discarded at an ever increasing rate."

This applied to male as well as female consumers, and to styling lines on cars as well as hemlines on skirts. Allied Stores Corporation's chairman B. Earl Puckett, speaking to fashion industry leaders in 1950, said, "Basic utility cannot be the foundation of a prosperous apparel industry. We must accelerate obsolescence." And General Motors' design chief Harley Earl said in 1955, "The creation of a desire on the part of millions of car buyers each year to trade in last year's car on a new one is highly important to the automobile industry."

Business people and politicians weren't the only ones pushing this idea, Cohen says. "Labor really bought into this package. Purchasing power was the answer to how people would be employed and have a better life. Consumers would fuel the powers of factories that would provide jobs that would put money in peoples' pockets."

Since then, Cohen argues, we've conflated our concepts of ourselves as good consumers and as good citizens. The idea of consumption as our country's economic engine continues to this day. Indeed, after the attacks of September 11, 2001, President Bush implored Americans to go shopping. And frugal as I am and as green as I try to be, during the recent economic downturn I've found myself feeling that every major purchase I make is a perverse kind of civic duty. The notion of the citizen-as-consumer clearly runs deep.

But things have changed since the 1940s and '50s. "When people were making goods that lasted [back then], they were benefiting from the explosion of global capitalism and the expanding of markets," Cohen says. "Now that we have this global recession, it's problematic. Where do these companies go if they are going to build goods that last? How do they profit if they don't sell new goods? I don't know the answer to this, but it's a problem that policy makers, economic planners, labor unions—everybody has to think about."

A Radically Obvious Idea

Although the greening of the American consumer has fostered some deceptive greenwashing campaigns seeking to capitalize on our good intentions, it has also made it possible for us to make better ecological and economic choices.

A host of clever websites now enables consumers to calculate their own ecological footprints and offer advice on how to reduce the toll. These include

- **MyFootprint.org**, where you can find out how many Earths would be necessary if everybody on the planet shared your lifestyle;
- **H20Conserve.org**, where you can tally your water footprint;
- **Wattzon.com**, where you can calculate the energy required to sustain your lifestyle.

Some of these calculations become conceptually complex as they try to measure the energy required for the extraction and transportation of raw materials, and the manufacturing, distribution, and ultimate disposal of products. It can all get abstract quite quickly, but there's a far simpler message embedded in all that complexity: Buy stuff that lasts.

Saul Griffith, a 2007 MacArthur Fellow, serial inventor and cofounder of WattzOn, refers to this as "heirloom design"—a term he introduced during a talk at the February 2009 Greener Gadgets conference in New York. The best way to lower the quantity of energy required to manufacture and distribute consumer goods, he argues, is to make those products not only durable, but repairable and upgradable.

Griffith shares this radically obvious idea with Tim Cooper, head of the Centre for Sustainable Consumption at Sheffield Hallam University in Sheffield, England, and editor of the . . . book, *Longer Lasting Solutions* (Gower, June 2010). The Centre, which Cooper founded in 1996, conducts research into consumers' behavior as well as the environmental effects of the choices they make.

Cooper argues for "product life extension"—making things more durable; using them properly; and ensuring they are maintained, repaired, upgraded, and reused. A key obstacle, he says, is the perception (supported by public policies) that higher levels of consumption yield greater happiness. After all, an increase in the GNP is considered healthy for the economy and can only be achieved if consumers increase their spending.

Cooper calls for "slow consumption," the consumer purchasing equivalent of the Slow Food movement (which seeks to build consumer awareness and appreciation of food and its connection to community and the environment). "The issue to address is what kind of economy is going to be sustainable in its wider sense—economically, environmentally and socially," he says. "The current economy is not sustainable. The sheer throughput of energy and materials cannot be continued."

If products were more durable, Cooper argues, some jobs lost due to the decrease in consumption would be offset by the addition of more highly skilled maintenance and repair jobs. And whereas the lost jobs might be overseas, the repair jobs would be local. "We need to look at new business models that move away from manufacturing and selling more and more products," he says. Such models might include "products that last longer but have associated services attached to them, so that the supplier guarantees to maintain, repair and upgrade the products for a certain period."

This might be a hard sell for consumers, however. Cooper cites the results of a survey in which British homeowners were asked what they considered the disadvantages of longer-lasting appliances. Twenty-three percent stated concerns about price, while 30 percent said they feared these products would become "out of date." He found that consumers were often disinclined to have products repaired because of the high cost of labor compared with the low cost of replacement, thanks to the quantity of consumer goods manufactured in countries with low wages and lax environmental regulations.

When I mentioned this conundrum to one of my ecologically conscious friends, she sheepishly admitted she had just discarded her old DVD player because the repair estimate was higher than the cost of a new one. "The present economic system does give an advantage to the current economy," Cooper says, "and for the consumer, replacement is often the cheaper option." That would have to change.

Close Encounters of the Durable Kind

Most of us have had an heirloom design or product life extension epiphany at one point or another.

Years ago, along with an untold number of other caffeine addicts, I succumbed to the then-ubiquitous ads for Gevalia coffee. Buy a couple pounds of beans and get a free drip coffeemaker. That became the first of a steady stream of plastic automatic drip coffee machines of various makes that took up residence on our countertop, none of which lived to see their second birthdays. Each time one broke, my husband and I found that the features available to us had multiplied. We could buy coffeemakers with built-in bean grinders, brew strength controls, and programmable timers. We had been cornered by a combination of functional obsolescence and fashion obsolescence.

Then we discovered what any self-respecting Italian coffee drinker knew all along: A $30 cast aluminum stovetop espresso maker lasts forever. Replace the rubber gasket every couple of years, and you'll stay happily caffeinated for life. We bought a used '70s model on eBay several years ago and have been using it every day ever since.

Not everything older is better, of course. Visit your local thrift store, and you'll be confronted by an exhibit of the unnecessary and the obsolete. Did anybody ever need a bread machine, an ice cream maker, or an Atari game console?

And yet, thrift stores are also the repositories of time-tested items. The garments sold there are Darwinian success stories. They've survived the wrath of washers and dryers and still have significant life left in them. The dishes may be mismatched, but they are dishwasher and hand-washer safe. And I'm convinced that this is where the world's missing teaspoons come to rest. If heirloom design has a line of boutiques, this is it.

Heirloom design has its adherents in the design and manufacturing worlds, too.

Patagonia, based in Ventura, Calif., was founded by avid mountain climbers who began selling clothing to support their barely profitable climbing-hardware business. From the start, the company was grounded in concern for the environment, and was an early adopter of several socially and environmentally responsible corporate policies, from donating a percentage of profits to environmental groups to offering employees on-site daycare.

Patagonia products are designed to last, and when they don't hold up, the company stands behind them. Its "ironclad guarantee" states: "If one of our products does not perform to your satisfaction, return it to the store you bought it from or to Patagonia for a repair, replacement or refund."

Sixteen years ago, my older brother gave me a Patagonia fleece jacket his children had outgrown. He purchased it around 1987 for his oldest daughter, who wore it until she outgrew it and handed it down to her younger sister. When she outgrew it, my daughter wore it, and then my son. At some point, the zipper broke, so I sent it to Patagonia, which repaired it at no cost. The jacket never wore out. That's heirloom design.

The alternative to durability and repair is remanufacturing. After more than two decades in the modular carpet business, Ray Anderson, founder and chair of Interface Inc. of Lagrange, Ga., heard a talk by environmentalist Paul Hawken and was inspired to green up his company. In addition to other ecological efforts on the materials and production sides, in 1995 the company introduced an "evergreen lease" arrangement, essentially turning carpet into a service instead of a product. By taking responsibility for retrieving and remanufacturing carpet no longer wanted by its customers, Interface was able to keep used carpet out of the waste stream and reduce the need for new materials.

Unfortunately, the leasing concept proved complicated and expensive. The company eventually gave up on it but continued to aggressively pursue discarded carpet—both its own and that of other companies—so that the materials could be reclaimed and remanufactured. Andrew King, a visiting fellow in mechanical engineering at the University of Bristol and consultant with the Centre for Remanufacturing and Reuse, notes that remanufacturing is preferable to recycling because it preserves most of a product's embodied energy while bringing it back to its original quality. And it creates jobs.

Here's the kicker: By emphasizing product durability, service, and remanufacturing, both of these companies have earned extra dividends in the form of corporate image and customer loyalty.

In Search of Solutions

The question, then, is what would it take to overcome our dependence on cheap goods? Even though obsolescence is no longer a boon for this country's manufacturers, cheap products are essential for consumers who can barely afford to put food on the table. If a durable coffeemaker costs twice as much as a breakable one but lasts four times as long, it's still less attractive to someone who doesn't have the additional cash up front.

Policy would have to play a key role in reversing this unfortunate check-out counter calculation. Legislation on extended producer responsibility (EPR), requiring manufacturers to account for the full life-cycle of their products from extraction to disposal, could affect consumer culture by making disposable items more expensive and reviving an interest in repair.

Such legislation is complicated, however, by the ongoing pressure to protect industrial production. "Legislation tends to get watered and watered until it gets to be almost a hindrance to these breakthrough changes because in order for Big Business to buy into it, it has to become easy for them," King says. EPR legislation would only be effective if it created a financial incentive for industry to produce more durable goods and for consumers to favor them.

Consumers are certainly influenced by price, but Cooper holds out hope that they also can be persuaded by having more of a connection to the objects they purchase—something referred to as "emotionally durable design." If that sounds too touchy-feely for a coffee machine, consider the difference between a pair of shoes custom-made for you by your local cobbler and an off-the-rack pair from the shoe store. Which would you be more likely to clean, resole, and repair?

Part of the solution might also be having more products available without the burden of ownership. Tool rentals, car sharing, and even Laundromats diminish the number of products that need to be manufactured and place a premium on durability and longevity. (This can even be done informally. We've shared a lawnmower with one of our neighbors for years.) "We should have much more attachment to certain products but for others we should see that they are services," King says.

Cooper warns, however, that rental can backfire in some areas, such as electronics. "The danger of the rental model with technology is that as advances are made, people who rent them might upgrade even more quickly than they would at the moment," he says. But even electronics could have a smaller ecological footprint if they could be updated through the use of modular design instead of being casually discarded. "The trick will be to understand what does and what doesn't change," King says.

King sees consumers playing a large role in putting pressure on industry to make the necessary changes. "The real issue is creating the demand," he says. "I work with a large number of large multinationals. When they assign their designers to the challenge—design this for two lives—they rise to the challenge. They just start to think in a different way."

Ultimately, environmental and economic sustainability won't be possible until we become less dependent on consumer spending, which currently comprises 70 percent of the U.S. economy. We can't just keep churning out, buying and disposing of stuff.

"We can diversify the range of goods that are underpinning our economy and providing us with jobs and some prosperity," Cohen says. "It doesn't just have to be commodities for the individual consumer. That would be the best hope."

Sharon Begley

NO

Green and Clueless

You could practically hear a collective groan from enviros across the world yesterday, when *The New York Times* reported on city apartment dwellers who leave their air conditioning running for days and days when they are not even home: with "utilities included" in their rent, these model citizens don't pay for it, and they want to walk into a nice cool room when they get back from vacation or just a tough, hot slog from the subway. So much for all those *50 Things You Can Do* books, magazine articles, and Web sites, all of which patiently explain that it would be really, really helpful if we didn't run appliances when we're not using them. Apparently, that message—which green groups have been disseminating for at least 20 years—can't hold a candle to people's apathy, ignorance, and selfishness.

But the problem goes beyond the fact that people don't care about, or perhaps understand, the fact that wasting energy and using it inefficiently accounts for a good chunk of the greenhouse-gas emissions that cause global warming. (In one 2009 analysis, scientists led by Thomas Dietz of Michigan State University estimated that household-based steps—as opposed to national policies like cap-and-trade—such as weatherizing homes, upgrading furnaces, switching to higher-mpg cars, changing air filters in a furnace, and not wasting power would cut U.S. carbon emissions by 123 million metric tons per year, which is 20 percent of household direct emissions and 7.4 percent of U.S. emissions.) Despite the millions of words that have been written on how to save energy and use it more efficiently, people basically have no idea what to do.

Scientists led by Shahzeen Attari of the Earth Institute at Columbia University surveyed 505 Americans (recruited through Craigslist), asking them to name the best ways to conserve energy. The most common answers had to do with curtailing use (by turning off lights or driving less, for instance) rather than improving efficiency (installing more efficient lightbulbs and appliances, say). But it is energy efficiency that offers the only possibility for dialing back our voracious consumption of energy and the fossil fuels that generate it. The reason is basic psychology: we are just not going to become a nation of pedestrians, let alone do without all our electronic toys. The only hope is therefore to continue satisfying those materialistic needs but with less electricity and gasoline. Yet as Attari and her colleagues report in a study in the current issue of the *Proceedings of the National Academy of Sciences,* only 12 percent of participants mentioned efficiency improvements as "the most effective way" to conserve energy, while 55 percent mentioned curtailing use. Specifically, 20 percent said turn off lights, but only 3.6 percent said use more efficient bulbs; 15 percent said drive less or use public transit, but only 3 percent said use a more efficient car. No wonder Americans are so resistant to taking personal steps to mitigate climate change: they think it means doing without.

And the ignorance continued. The scientists next asked people to estimate how much energy different appliances used and how much different behaviors saved. More said line-drying clothes saves more than changing the washing-machine settings (the reverse is true). Most people also think trucks and trains that transport goods use about the same energy; in fact, trucks use 10 times more to move one ton of goods one mile. Most people also said that making a glass bottle takes less energy than making an aluminum can (the reverse is true: a glass bottle requires 1.4 times as much energy as the can when virgin materials are used, and 20 times as much when recycled materials are used; making a recycled glass bottle actually takes more energy than making a virgin aluminum can).

The higher the energy used by an appliance, the more wrong people were in their estimates. "In other words," the scientists write, "people's understanding may be worse where the potential for CO_2 reductions is large."

Here's my favorite: participants who said they did lots of environmentally responsible things on the energy front actually had less accurate perceptions of all this—suggesting that while people may think they're doing the planet good, they are not. The notion of making "informed choices" is great, but it kind of requires being, well, informed. What we have instead, it seems, is rampant ignorance. The real problem, Attari told me, is that when people pick the easy things, the low-hanging fruit, they figure they've done their bit for the environment and then don't take steps that could actually make a difference.

Why the ignorance? As usual, the press and green groups bear some of the blame, for promulgating simple feel-good but ultimately almost-useless steps such as turning off your cell-phone chargers. (Yes, it does add up, but a typical cell-phone charger draws one watt of power, so over a day that's 24 watt-hours, or about one 40th of a kilowatt-hour, for a grand total of about 10 cents per day in savings.) The press and enviros have also spread the comforting myth that we can shop our way to saving the planet, a notion I have pilloried before. Whoever's to blame, the consequences are clear: even people who want to conserve energy have barely a clue how to do it, and lots of people don't even want to. No wonder those apartment ACs are running full tilt while nobody's home.

EXPLORING THE ISSUE

Is Sustainability a Realistic Objective for Society?

Critical Thinking and Reflection

1. Is sustainability a concept that is contrary to human nature? Is quality of life built on materialism too entrenched in our DNA to change? Do we have examples of how cultural shifts can modify or change behavior?
2. If society moves to "heirloom design," how would this affect economic growth and development? Would business interests ever consider such a change?
3. What is "planned obsolescence" and what does it have to do with sustainability? William McDonough and Michael Braungart (2002) have introduced the concept of "cradle to cradle" to discuss how products can be re-designed to achieve eco-effectiveness. How does this relate to "planned obsolescence"?
4. What public policies can make sustainability an operational goal for society? Are there any examples of existing policies that have moved society toward sustainability?
5. Do you think that people are apathetic about the environment and sustainability? What steps can be taken to make you live a more sustainable life?

Is There Common Ground?

Sustainability will clearly be an important topic in future centuries and will, hopefully, become a driver for responsible decision making. Most people can agree that sustainability is an important topic—whether they believe technology and innovation will prevent any serious harm to human lifestyle or whether they believe we need to make sweeping changes to our lifestyles immediately, no matter how painful or inconvenient. The difficulty is determining whether or not sustainability is truly a realistic goal for society.

Sustainability is something we should strive for regardless of how realistic or unrealistic it is. Carefully planned changes to our economic systems that include incentives for a reduction in Western consumerism, rewards for innovative use of recycled/reused or sustainably harvested materials, increasing monitoring restrictions on corporate emissions, and subsidies for sustainable business practices will bring society much closer to sustainability.

Certain societal characteristics, like apathy and laziness, commitment-phobia, and corporate transparency, make sustainability virtually impossible.

Perhaps if the appropriate economic and corporate changes for sustainability are implemented, the seemingly insurmountable challenges will seem like minor setbacks and society will be far better off than it is now. Is there any harm in making careful, measured steps toward becoming a sustainable society?

Internet References

Balta, W. (2010) Will you be sustainable when no one is watching? *Huffington Post*. Retrieved on November 5, 2010 from www.huffingtonpost.com/wayne-balta/will-you-be-sustainable-w_b_754788.html

An article posted on the Internet that examines the recent popularity of the environmental movement. Balta wonders whether or not people will continue to strive toward sustainability when environmental issues are no longer a popular new item.

David Suzuki Foundation. (2004, Feb 5) Environmental sustainability possible within a generation. *EarthEasy*. Retrieved on November 4, 2010 from http://eartheasy.com/ article_sustainability_possible.htm

An article posted on the Internet that ponders the possibility of achieving sustainability within a generation. The article postulates that if governments work with industry and public policy groups to address major issues like energy efficiency, Canada can reach economic and environmental sustainability within one generation.

Koch, W. (2010, Jun 1) Getting to the heart of 'simple living'. *USA Today*. Retrieved on November 2, 2010 from www.usatoday.com/LIFE/ usaedition/2010-06-02-urbanska02_ST_U.htm

An article posted on the Internet that examines tactics for leading an eco-friendly lifestyle. Simplicity guru, Wanda Urbanska, discusses her ways of disconnecting and leading a low-stress life.

The Ecologist. (2010, Sept 10) 'Priveleged' opposition holding back wind farm development. *The Ecologist*. Retrieved on November 1, 2010 from www.theecologist.org/News/news_round_ up/592516/privileged_opposition_holding_back_wind_farm_development.html

An Internet article that discusses the likelihood of local opposition to wind farms. The article states that the communities likely to oppose are characterized by higher life expectancy, higher percentage of voting in national elections, and a lower crime rate.

The Ecologist. (2010, Aug 10) Human response to climate change is making matters worse. Retrieved on January 9, 2011 from www.theecologist.org/News/news_round_up/568432/human_response_to_climate_change_is_making_matters_worse.html

An article posted on the Internet that postulates that humans may actually be increasing their carbon footprint while attempting to become more sustainable. For example, biofuels may cause more harm to the environment than good.

ISSUE 2

Is Sustainability More About Politics Than Science?

YES: Bill McKibben, from "Hot Mess: Why Are Conservatives So Radical About the Climate?" *The New Republic* (October 2010)

NO: Huub Spiertz, from "Food Production, Crops, and Sustainability: Restoring Confidence in Science and Technology," *Current Opinion in Environmental Sustainability* (November 2010)

Learning Outcomes

After reading this issue, you should be able to:

- Find out whether sustainability is more about politics than science.
- Know the role played by politics in sustainability.
- Know the role played by science in sustainability.
- Find out whether sustainability can be achieved in the current political and scientific scenario.
- Explore what requires to be done to turn the scientific work into sustainable solutions to the sustainability issues facing the world today.

ISSUE SUMMARY

YES: Noted environmental writer Bill McKibben discusses how money and vested political interests undermine efforts toward sustainability and how this is reflected in politics.

NO: Huub Spiertz, a professor of crop ecology and past-president of the International Crop Science Congress, elaborates on how applicable agro-technologies and bio-technologies can address global food and population issues and offers an example of how science provides a more sustainable world.

Sustainability means different things to different people. While it can be a difficult idea to define, it does possess certain agreed-upon characteristics. First, it is generally related to ecological system integrity, or the relationship of people to the natural environment. This, usually, is related to issues of human natural resource use, the physical impact of humans on the environment, and the capacity of people to adapt to complex planetary processes through scientific knowledge and technological innovation. Science is particularly useful in measuring and understanding the complexities of the human–environment relationship. Another characteristic of sustainability is that it has become a banner under which groups interested in environmental preservation and social justice have coalesced into a substantial social movement with political ramifications. Also, but in a different direction, business leaders have sought to discredit science or to exert political influence on decision making that would minimize or contradict scientific findings. An example of this would be the climate change debate. This issue deals with the debate over which approach, the scientific or the political, takes precedence in sustainability.

Tim O'Riodan in "Environmental Science, Sustainability, and Politics," (2004) discusses how sustainability issues, that is, climate change, environmental degradation and pollution, food insecurity, epidemics, population growth, political unrest, and wars based on resource competition, are all associated with political decision making. Sustainability today, he states, is in the hands of "formal governments at global, multinational, national and sub-national levels, and an increasing range of quasi-formal governing arrangements ranging from devolved institutions to non-departmental arms-length agencies, to private sector in various partnership guises, to a mass of semi-organized community-based organizations." He draws the connection between the lack of sustainability and political and economic insecurity.

Political support for sustainability is often displayed in public policies. The principles behind "smart growth," the planning of more efficient land use and the more rigid management of urban development, have led to legislation in many U.S. states to control urban sprawl through land-use controls and zoning regulations. Also, a social movement to ensure that poor and powerless communities not be the exclusive repository for hazardous waste and environmental and public health risk forged the notion of "environmental justice." This movement was based on the recognition that sustainability dealt with more than environmental and ecological concerns, but also included issues of social justice and equity. As a result, national and state mandates were enacted to ensure that the location of hazardous and undesirable uses was to be equitably shared by people in all communities, and not just in communities of color and poor people.

Perhaps the most politicized issue in sustainability is the climate change debate. Seth Schulman in "Smoke, Mirrors & Hot Air," presents a case study of how Exxon-Mobil, in an effort to deceive the public about the reality of global warming, "has underwritten the most successful disinformation campaign since the tobacco industry misled the public about the scientific evidence linking smoking to lung cancer and heart disease" (2007). He points out that ExxonMobil is a powerful player in the world's gas and oil business and that it was ranked sixth in the world in carbon dioxide in 2005. As a result, "ExxonMobil has played the world's most active corporate role in underwriting efforts to thwart and undermine climate change regulation." Another example of the politicalization of climate change is when the chief government scientist in the United Kingdom delivered a warning to the American Association for the Advancement of Science that climate change constitutes a bigger global threat than terrorism. In this case, the prime minister's office requested him to not discuss this issue publically because it could impact negatively on the UK's relationship with the United States, which, at the time, viewed terrorism as its prime concern (O'Riordan, 2004).

Another reason why sustainability is politicized is the issue of money and vested interests. The lack of support of certain oil interests in supporting climate change legislation as discussed above is a case in point. Yasarata et al. (2010) found that support for sustainable tourism in Cyprus was correlated to the political demands of the local leadership. It was supported as long as it did not conflict with traditional tourist businesses. Also, the issue of money and protection of vested interests was noted in a study done by Zulu (2009). He studied the decision and policy-making process in community-based forest management in Malawi and found that policies that conflicted with the interest of local elites were not supported even though they were unsustainable over the long term. And finally, O'Riordan (2004) noted that politics is required to advance the subject of sustainability. He suggested that scientists and geographers need to embrace the politicization of sustainability if it is going to succeed as a major social and political movement.

The notion of sustainability as a science is well supported. William C. Clark and Simon A. Levin in "Toward a Science of Sustainability," state that "Fostering a transition toward sustainability—toward patterns of development that promote human well-being while conserving the life support systems of the planet—is one of the central challenges of the twenty-first century" (2009). They go on to say, "Building a science of sustainability . . . requires a truly multi-disciplinary approach that integrates practical experience with knowledge and know-how drawn from across the natural and social sciences, medicine and engineering, and mathematics and computation." The rise of sustainability science points to the belief that sound research and the use of scientific methodologies can lead to better policies and that sustainability and science cannot be separated.

Sound science has already made a substantial contribution to the advancement of sustainability by generating data, leadership, and new technologies. Larsen et al. (2007) have looked at the role of science in the Large-scale Biosphere-Atmosphere experiment in Amazonia (LBA), which was intended as a guide for environmental decision making. He praises the project but notes that although it has produced many scientific papers, it has failed in its goal of transitioning the data to real sustainability initiatives, especially at the local level. Rametsteiner et al. (2011) discuss the challenges of developing scientific indicators for sustainability, and notes that scientists and politicians need to engage in greater collaboration. In a paper linking carbon sequestration science with local sustainability, Yin et al. (2007) discuss the role of science in forest resource management and local sustainability enhancement in China. He supports the role of science in developing sustainability approaches in forestry through the use of appropriate technology for monitoring, modeling, and measuring carbon sequestration.

While there is general agreement that scientists have much to contribute to sustainability, it is widely recognized that they must communicate better with the public and enlist the support of social and political leaders. For science to contribute more to sustainability, most authors agree on the need to create forums for discussion of scientific material related to sustainability and to involve more people in the scientific process. And finally, there is a consensus that both politics and science are "bedfellows" and that both have an important role to play in sustainability.

In the YES selection, that sustainability is more about politics than science, noted environmental writer Bill McKibben discusses how money and vested political interests undermine efforts toward sustainability and how this is reflected in politics. Published a few weeks before the 2010 midterm elections, it is designed to garner support for climate change legislation. It discusses how some politicians choose to discredit the science of climate change and believe that it is a hoax perpetrated on the American people. He also discusses the linkages between vested economic interests and politics, and how some politicians twist or deny scientific truths so as to pursue their own parochial interests. He speaks directly to the issue of how denial of climate change and the lack of political commitment to deal with it will lead to future problems.

In the NO selection, that sustainability is more about science, Professor of crop technology Huub Spiertz addresses the need for a scientific approach to sustainability that takes into consideration dwindling natural resources like arable land. He points out that in order to sustain a growing world population, scientists need to harness new breakthroughs in agro-technologies and biotechnologies. He discusses how food security could be pursued by combining new insights in genetics, systems functioning, climate change, and multiple stresses to guide development of new cultivars. Although science and technology could foster food security, there are the social challenges due to the people refusing to use these new technologies. In the agricultural sector, he notes that a large number of people have refused to buy the products of genetically modified agriculture due to safety concerns, or for religious and ethical reasons. He insists that science needs to restore confidence in modern agro-technologies and biotechnologies.

YES 

Bill McKibben

Hot Mess: Why Are Conservatives So Radical About the Climate?

One interesting fact heading into the mid-term elections: Almost none of the GOP Senate candidates seem to believe in the idea that humans are heating the planet. A few hedge their bets—John McCain says he's no longer sure if global warming is "man-made or natural." (In 2004, he told me: "The race is on. Are we going to have significant climate change and all its consequences, or are we going to try to do something early on?") Most are more plainspoken. Marco Rubio, for instance, attacks his opponent Charlie Crist as "a believer in man-made global warming," explaining, "I don't think there's the scientific evidence to justify it. The climate is always changing." The most likely cause of that change, according to Ron Johnson, who is leading the Senate race in Wisconsin: "It's far more likely that it's just sunspot activity."

The political implications are clear. Climate legislation didn't pass the current Congress, and it won't have a prayer in the next one. If the Republicans take the Senate, James Inhofe has said that the Environment and Public Works Committee will "stop wasting all of our time on all that silly stuff, all the hearings on global warming." And in the House, Representative Darrell Issa says that he would turn his Oversight and Government Reform Committee over to the eleventeenth investigation of Climategate, the British e-mail scandal. But, for the moment, it's less the legislative fallout that interests me than what this denial of climate change says about modern conservatism. On what is quite possibly the single biggest issue the planet has faced, American conservatism has reached a near-unanimous position, and that position is: pay no attention to all those scientists.

The few exceptions prove the rule. Ronald Bailey, the science writer at *Reason,* converted a few years ago to belief in global warming and called for a carbon tax. His fellow libertarians weren't impressed: Fred Smith, the head of the Competitive Enterprise Institute, suggested that Bailey had been "worn down by his years on the lecture circuit." Jim Manzi, a software exec and contributing editor at *National Review,* wrote a piece asking conservatives to stop denying the science. Even though he's also downplayed the risks of warming, it was enough to earn a brushback pitch from Rush Limbaugh: "Wrong! More carbon dioxide in the atmosphere is not likely to significantly contribute to the greenhouse effect. It's just all part of the hoax." For the most part, even

Manzi and Bailey's own colleagues pay them no mind: *National Review* maintains a *Planet Gore* blog devoted to—well, three guesses.

In any event, the occasional magazine column has had no impact at all. Only 10 percent of Republicans think that global warming is very serious, according to recent data. Conservative opinion has been steadily hardening—for decades Republicans were part of the coalition on almost every environmental issue, but now it's positively weird to think that as late as 2004, McCain thought it would make sense for a GOP presidential candidate to position himself as a fighter for climate legislation. And all of that is troubling. Because we're going to be dealing with climate change for a very long time, and if one of the great schools of political thought in this country has checked out completely, that process is going to be even harder. I don't have any expectation that conservatives will mute their tune between now and November—but it is worth thinking in some depth about what lies beneath this newly overwhelming sentiment.

One crude answer is money. The fossil-fuel industry has deep wells of it—no business in history has been as profitable as finding, refining, and combusting coal, oil, and gas. Six of the ten largest companies on earth are in the fossil-fuel business. Those companies have spent some small part of their wealth in recent years to underwrite climate change denialism: Jane Mayer's excellent *New Yorker* piece on the Koch brothers is just the latest and best of a string of such exposés dating back to Ross Gelbspan's 1997 book *The Heat Is On*. But while oil and coal contributions track remarkably close to political alignment for many senators, they are not the only explanation. Money only exerts political influence if it can be connected to some ideological stance—even Inhofe won't stand up and say, "I think global warming is a hoax because my campaign treasurer told me to." In fact, some conservatives have begun to question endless fossil-fuel subsidies—since we've known how to burn coal for hundreds of years, it's not clear why the industry needs government help.

Another easy answer would be: Conservatives possess some new information about climate science. That would sure be nice—but sadly, it's wrong. It's the same tiny bunch of skeptics being quoted by right-wing blogs. None are doing new research that casts the slightest doubt on the scientific consensus that's been forming for two decades, a set of conclusions that grows more robust with every issue of *Science* and *Nature* and each new temperature record. The best of the contrarian partisans is Marc Morano, whose *Climate Depot* is an environmental *Drudge Report:* updates on Al Gore's vacation homes, links to an op-ed from some right-wing British tabloid, news that a Colorado ski resort is opening earlier than planned because of a snowstorm. Morano and his colleagues deserve their chortles—they're winning, and doing it with skill and brio—but not because the science is shifting.

No, something else is causing people to fly into a rage about climate. Read the comments on one of the representative websites: Global warming is a "fraud" or a "plot." Scientists are liars out to line their pockets with government grants. Environmentalism is nothing but a money-spinning "scam." These people aren't reading the science and thinking, *I have some questions about this*. They're convinced of a massive conspiracy.

The odd and troubling thing about this stance is not just that it prevents action. It's also profoundly unconservative. If there was ever a radical project, monkeying with the climate would surely qualify. Had the Soviet Union built secret factories to pour carbon dioxide into the atmosphere and threatened to raise the sea level and subvert the Grain Belt, the prevailing conservative response would have been: Bomb them. Bomb them back to the Holocene—to the 10,000-year period of climatic stability now unraveling, the period that underwrote the rise of human civilization that conservatism has taken as its duty to protect. Conservatism has always stressed stability and continuity; since Burke, the watchwords have been tradition, authority, heritage. The globally averaged temperature of the planet has been 57 degrees, give or take, for most of human history; we know that works, that it allows the world we have enjoyed. Now, the finest minds, using the finest equipment, tell us that it's headed toward 61 or 62 or 63 degrees unless we rapidly leave fossil fuel behind, and that, in the words of NASA scientists, this new world won't be "similar to that on which civilization developed and to which life on earth is adapted." Conservatives should be leading the desperate fight to preserve the earth we were born on.

Trying to figure out what is happening in the conservative movement requires more than just recording inanity. John Raese, who is leading the Senate race in West Virginia and who says proudly that he was "a tea partier before the Tea Party existed," recently declared that "one volcano puts out more carbon dioxide than everything that man puts out." This doesn't tell you anything about volcanoes. (Actually, humans emit about one hundred times more carbon than volcanoes.) It tells you that, buried beneath the nonsense, there's a powerful structure of argument, one that needs to be taken seriously.

Part of the conservative creed has always been that markets, left to themselves, accomplish most tasks more efficiently than government regulation. That's true, of course, just as it's true that markets don't do everything you want. (That's why we have cheap deregulated airlines and yet retain the Federal Aviation Administration.) But conservatives have grown more insistent on the deification of markets in recent years; Rand Paul is ever less an outlier. If markets do damage, that's okay—it's creative destruction à la Schumpeter.

But even if you accept that process absolutely within the economic sphere (and very few of us do, which is why Rand Paul just might lose), it doesn't follow that it works outside of it. Destruction of the planet's fundamental physical systems isn't creative—it's just destruction. If Microsoft disappears, innovators will take its place. If Arctic ice disappears, no young John Galt is going to remake it in his garage. The essential question is: Is the environment a subset of the economy, or is it the other way around? Or, more combatively, you really think you can out-argue physics? Hayek's good, but atmospheric chemistry is a tough opponent.

If conservatives acknowledged the crisis, they could make a powerful contribution to the solution. One option for tackling global warming is for the government to regulate just about everything. Or, we could limit government's role to simply imposing a price on fossil fuel that reflects the damage it does. This wouldn't even need to be a traditional tax: One proposal gaining ground is to take every dollar produced by such a levy and rebate it to each citizen, using government as a kind of pass-through. You'd get the signal from your electric bill to start insulating, and the numbers on the gas pump would urge you in the direction of a hybrid car, but most people would come out ahead. It's a plan designed with real deference to a conservative understanding of human nature.

Instead of that kind of debate, though, most of the movement has decided to describe any regulation of carbon as eco-fascism. (Recently, for instance, a British green group released a purportedly tongue-in-cheek video in which environmentalists blow up people who don't believe in global warming. According to some right-wing websites, the video wasn't merely a vile political stunt but a preview of impending government policy.) As Christine O'Donnell put it in her attack on the nanny state, "You're not the boss of me." But here's the thing: Carbon dioxide mixes easily and freely in the atmosphere. If the climate change you caused followed you around like Pigpen's cloud, then no problem. But it doesn't—your Navigator drowns Bangladeshis. Given the magnitude of the changes now underway, and the way they will foreclose individual choices unto the generations, it's possible to argue that this is the greatest attack on freedom we've ever witnessed.

In response, there's a kind of right-wing nationalism that demands we take no action until China, India, and the rest have played their part. But that doesn't even make mathematical sense—China's per capita emissions are one-quarter of ours. If leadership in the world means anything, then that imposes certain burdens on us. But it feels like resentment is becoming the leitmotif of conservatism, in a way that makes it ever more cramped and ever less noble. In this worldview, environmentalists are seen as scolds or even traitors. A recent poll asked right-wing bloggers to name the worst people in American history. President Obama came in second. The victor? Jimmy Carter, ten spots ahead of John Wilkes Booth (Al Gore was thirteenth, tied with Al Sharpton, Noam Chomsky, Jane Fonda, and Harry Reid). If Jimmy Carter was the worst guy the country ever produced, we're doing pretty well—but surely it was his nagging reminders that there were limits to our national power that account for his ranking. *New York Times* columnist Ross Douthat wrote an embarrassed piece earlier this fall about the failure of conservatives to take climate change seriously—it was the '70s, "a great decade for apocalyptic enthusiasms," that turned many of them off, he concluded. That's not much of an argument—it's like saying "conservatives mostly got it wrong on civil rights, so let's never listen to them again about liberty and freedom."

I hold out no particular hope that anyone in the conservative movement will listen to me. But I do hold out some hope that, over time, the conservative intellectual tradition will come to grips with climate change.

There are already parts of the generally conservative body politic that have begun to get the message. Our religious institutions, for instance, which send hundreds of thousands of Americans overseas every year to do relief and mission work. Those good-hearted people are starting to understand that development can't happen in a world wracked by increasing climatic chaos. The leaders of many evangelical seminaries and some of the country's largest churches have signed a letter calling for action on global warming with "moral passion and concrete action" and explaining that "Christians must care about climate change because we are called to love our neighbors."

The military has also started to pay close attention. Earlier this year, the Quadrennial Defense Review noted that climate change could "act as an accelerant of instability or conflict." The authors were looking ahead to destabilizing events similar to this year's flood on the Indus—the sharpest blow that Pakistan has suffered in years (which is saying something) and precisely the kind of catastrophe every climate model predicts will become more common in a warmer, wetter world.

What missionaries and militaries have in common is that they have to deal with reality. In fact, that was always the trump card of conservatism: It refused to indulge in sentimentality and idealism, insisting on seeing the world as it was. But, at the moment, it's the right that is indulging in illusion, insisting, fists balled up and face turning red, that the reports from scientists simply can't be true.

I understand why those reports are bad news. Dealing with climate change really will be the most difficult thing we've ever done. (Too many progressives are clinging to their own illusion that we can simply rip the internal combustion engine out of our economy, toss in a windmill, and carry on as before.) The only thing harder than dealing with it will be not dealing with it and inheriting a world radically changed.

Conservatives in much of the rest of the world have figured this out. The new Tory government in England is doing at least as much as its Labour predecessors; in Germany, Angela Merkel is presiding over one of the greatest renewable-energy buildouts ever. And eventually, I imagine, American conservatism, too, will come around and make its vital contributions to the task of figuring out what needs to be done to protect the civilization we should cherish. But we don't have until eventually. We have, the scientists say, a very short time to make very big changes. So let's hope the fever passes quickly.

In the meantime, many of us are rolling up our sleeves and getting down to work. On October 10, in thousands of communities around the country, we're holding a Global Work Party to put up solar panels and dig community gardens and lay out bike paths. We don't think we can stop climate change this way—that will take action to reset the price of carbon. But we do think we can show the way. Not with a Tea Party, but with a work party. Which, in a different era, would have appealed to conservatives above all.

Huub Spiertz

Food Production, Crops, and Sustainability: Restoring Confidence in Science and Technology

By 2050, the global food requirement will increase significantly, driven by a population increase to more than nine billion and by a richer diet. There is a need for agricultural and food systems that are not only more productive, but also sustainable. Currently, progress is hampered by a lack of understanding how to close the yield and sustainability gap. The consequence is stagnation in implementing policies and regulations that meet future needs. The challenge of meeting global food security in a sustainable way requires a knowledge-intensive approach and the use of advanced technologies. The confidence in modern agrotechnologies and biotechnologies should be restored by sound science, transparency and regulatory institutions.

Introduction

Agriculture currently appropriates a substantial portion of the Earth's natural resources. Land used for crop production, pasture and livestock grazing systems amounts to 38% of total land area [1]. Population growth and expanding demand for agricultural commodities constantly increase the pressure on scarce land and natural resources [2]. The drop in cropland in relation to population is very evident. Until the middle of the 20th century, available cropland was in the order of 0.45 ha per person; by 1997 it had been reduced by almost a factor 2 resulting in 0.25 and the projection for 2050 is 0.15 ha per person [3]. The question is if we can produce enough food, feed, fibre and fuel to meet the needs of a 50% larger global population in 2050 in a sustainable manner. Cassman *et al.* [4••] concluded that '*avoiding expansion of cultivation into natural ecosystems, increased N-use efficiency, and improved soil quality are pivotal components of a sustainable agriculture that meets human needs and protects natural resources.*' More resources are required for meeting the demands of the growing human population. Recent studies [5••] indicate that significant systems improvements and efficiency gains in agriculture are needed worldwide in the next decades, to be able to feed the increasing global population and at the same time to circumvent large-scale degradation of natural ecosystems and deterioration of ecosystem services through agricultural activities. The sustainability framework, comprising the balance

Spiertz, Huub. From *Current Opinion in Environmental Sustainability,* December 2010, pp. 439–443.

between short-term and long-term objectives with respect to profitability, ecological health and social-ethical acceptance gives guidance to research directions and policy measures. The conceptual framework of a sustainable gap was presented by Fischer *et al.* [6•]. They suggest a hierarchy of considerations with the biophysical limits of the Earth setting ultimate boundaries. The question is, if this concept with the 'economies' embedded in 'human societies' does fit for major food production systems with free trade as drivers at a global scale.

Transitions in agriculture are a response to external and/or internal 'events' that provide the incentive for structural change [7••]. Possible events or *'driving forces'* for transitions in agriculture include gradual and sudden processes, like population pressure, changes in natural conditions (climate, diseases, and flooding), changes in markets and market prices, innovations and applications of new technology. Transitions in agriculture involve large-scale structural changes, which have a distinct impact on society. The difficulties in understanding the causes and effects of changes in agriculture arise from the diversity and complexity of agriculture, and the multitude of factors that affect agriculture [7••]. Demands by society, economy and environment determine the direction of change in agriculture. Decision making requires intensive mutual interaction and discussion to identify the challenges, trade-offs, discrepancies, and possibilities for synergy. A more effective strategy for the transition towards sustainable agriculture is setting suitable goals with clear targets and indicators to measure progress, when the gap between (socio-)economic and ecological targets is too big. To meet the challenges of a global food security in a sustainable way requires the intensification of knowledge-intensive approaches and the use of modern agrotechnologies and biotechnologies.

In this paper the following topics are addressed: first, transitions in food production systems; second, to what extent are emerging technologies and sustainable agriculture compatible? third, prospects to integrate technology and sustainability.

Causes and consequences of a lack of trust in science and technology as well as the prospects for restoring trust are presented.

Transitions in Food Production Systems

The focus on food security and the establishment of a free trade world market triggered a strong intensification and specialization of agricultural production and technological innovations contributed to a rise in production and labor productivity. Intensification and specialization of agriculture in most industrialized countries took place from the 1950s onwards to raise the profits. Environmental side-effects were neglected in the beginning. Nitrogen fertilizers comprise almost 60% of the global reactive N load attributable to human activities. This resource use has a major impact on the functioning of the ecosystems and human well-being. From 1985 onwards, a series of environmental policies and measures have been implemented in EU-countries, especially constraining the use of nitrogen (N) and phosphorus (P), insecticides, fungicides, heavy metals, and the use of land near nature conservation areas [8].

A major point of concern for many intensively managed agricultural systems with high external inputs is the low resource-use efficiency on the plot and field level, especially for water and nitrogen [9,10•]. These efficiencies are even lower when nutrient flows of the whole food chain are taken into account. Ma *et al.* reported that average N use efficiencies in China for crop production, animal production and the whole food chain amounted to 26, 11 and 9%, respectively [11•]. A high input combined with a low efficiency ultimately results in environmental problems such as degradation of resources (reduced stocks of fresh water and phosphorus), eutrophication and emissions of greenhouse gases [12,13•].

Cultivated land should also provide ecosystem services to society, such as biodiversity, water conservation, wildlife and mitigating climate change [14,15•]. A framework for the assessment of ecosystem goods and services was presented by Posthumus *et al.* [16•]; they explored six alternative floodplain management scenarios and found that there are both synergies and conflicts between ecosystem services. For example, there is a conflict between agricultural production and environmental outcomes [17•] such as water quality, GHG, habitat and species. Johnson [18] analysed the policies of two competing visions on food and agricultural sustainability; one to promote organic and local food and another to continue the productionist hegemony, emphasizing biotechnology and technological panaceas. He contends that the political will to promulgate radical agricultural policies that break the productionist hegemony are lacking. The dichotomy between 'intensive agriculture' and 'eco-agriculture' was denied by others. Brussaard *et al.* [19•] suggest that biodiversity loss undermines the provision of ecosystem services on which agriculture itself depends on. However, this hypothesis is not underpinned by a quantitative analysis of intensive and eco-based farming systems. In their paper they present a figure based on the work of Swift *et al.* [20••] that nicely shows the trade-offs between agricultural production and biodiversity. I agree with their conclusions that there are prospects for synergy (symbiosis, facilitation, etc.), but these should be present not only in low-productive systems but also in systems that are able to produce high yields per unit of land. Hard evidence for synergy in high-yielding systems was not shown. In the recent past, increased productivity of the major cereal crops has been derived from genetic improvement by conventional breeding and greater use of external inputs (fossil energy, fertilizers, feed, pesticides, and irrigation water). Now, the focus should be on enabling technologies and transitions in cropping systems to increase overall resource-use efficiencies under biotic and abiotic stress conditions [10•], including climate change, in order to prevent environmental degradation.

Are Emerging Technologies and Sustainable Agriculture Compatible?

A sense of urgency to apply emerging technologies in solving some of the most severe problems (drought, salinity, diseases, pests, weeds, nutrient acquisition, etc.) in food production is lacking. The issue of genetically modified crops—GM crops—has been highly controversial since the introduction of the recombinant DNA technology in the 1970s [21]. It was shown that the debate on genetically modified organisms (GMOs) extended in terms of actors involved and concerns reflected. A GMO is *'an organism in which the genetic material has been altered in a way that does not occur naturally by mating and/or natural recombination'* (EU Directive 2001/18). The concerns on developing and commercializing GMOs have been worldwide [22••], but are most prominent in the European Union and many African countries. European countries have been very defensive to adopt GMO crops based on the precautionary principle. In a reconstruction to analyze the complexity of concerns and actors involved Devos *et al.* [21] distinguished four phases:

a. how recombinant-DNA technology evolved in a dynamically changing context from laboratory science to societal concerns on ethics, ecological risks and socio-economic conflicts;
b. the impact of these concerns for the growing involvement of actors in the debate;
c. the change and dynamics in the content of the concerns;
d. how scientific objectives became intertwined with extra-scientific objectives.

Depending on underlying values and ideals the appreciation of new technologies will differ between stakeholders. Scientists and end-users (e.g. modern farmers) tend to be more positive about the benefits of GM-crops. However, the growing awareness of educated citizens about food safety, sustainability and equity has led to an increasing distrust in multinationals, but even in private funded research by scientific institutions [22••,23,24]. By scaling up scientific and technological developments into commercial activities, techno-scientific developments are entering society more directly, exposing it at large to potential risks and benefits. Science and technology have become public goods.

Recently, the academic institutions take part in the public discussions by publishing reports on the prospects and challenges of plant genomics in the 21st-century [25]. The essential claim is that *'plants form the basis of the food web that sustains all other forms of life.'* They focus on the following themes: improving food crops (food quality and plant-pathogen arms race), biofuels and bioenergy goals, environmental stewardship (saving on water and fertilizer use, biocontrol of pests, and plant defense) and biomedical (drug discovery and immune systems). The plea is to maximize the potential of the plant genome sciences in contributing significantly to human health, energy security and environmental stewardship. Strikingly, food production is not listed as one of the major challenges; however, climate change and the world food crisis bring a *'sense of urgency'* in the debate on meeting the demands of a growing global and wealthier population.

In Europe the rejection of GM crops is more categorically than it is the case for the application of biotechnology in producing medicines for human health. The opposition is complex and rooted in ethical-religious, environmental and social-economic objections [24]. It seems that this opposition is not generic for all modern technology (e.g. IT and nanotechnology) because the latter technologies do not affect the integrity and composition of food directly.

Prospects to Integrate Technology and Sustainability

The compatibility of modern biotechnology and sustainability was recently addressed by Ervin *et al.* [22••]. They presented a comprehensive analysis based on a sustainability framework that includes the full spectrum of environmental, economic and social impacts. A review on each impact revealed that '*crop biotechnology cannot fully be assessed with respect to fostering a more sustainable agriculture due to key gaps in evidence, especially for socio-economic distributive effects.*' First generation GM crops generally showed progress in reducing agriculture's environmental footprint and improving farmers profits; however, these crops fall short of the technology's capacity to develop a more sustainable agriculture. The latter was based on the presupposition that all stakeholders should be engaged and salient equity issues should be addressed. For realization of the potential of biotechnology, fundamental changes are required in the way public and private research and technology development and commercialization are structured [26]. More public/private partnerships in advanced research using enabling technologies are needed during the pre-competitive phase. Furthermore, transparency in objectives and methodologies should be realized through an open dialogue with all stakeholders. A good example is the concerted action taken by CGIAR research institutes (CIMMYT and ICARDA), advanced research institutes and national institutes in Ethiopia and Kenya to deal with the outbreak of the race Ug99 of the stem rust *Puccinia graminis tritici* causing severe epidemics in wheat. A major threat to wheat production not only regionally but also globally, because of the susceptibility of the existing plant material and the rapid spread by air of spores over large distances in North Africa, Middle East and West-South Asia [27]. Within 10 years new seed material could be released as a result of rigorous screening in labs and the field on resistance for the race. Combining genetics, molecular assisted selection and modern breeding made it possible to control a disease that potentially could destroy half of the global wheat production.

A plea for radically rethinking agriculture for the 21st century was presented by advocating systems that close the loop of nutrient flows from microorganisms and plants to animals and back [28]. By making better use of sunlight and seawater it would become possible to decrease the land, fossil energy, and fresh water demands of agriculture, while at the same time ameliorating the pollution currently associated with agricultural chemicals and animal waste. A combination of scenario development and back-casting will be required to identify ways where science and technology can contribute effectively. The study of solar-powered drip irrigation of vegetables in the rural food-insecure Sudano-Sahel region of West Africa [29•] nicely shows the potential of modern technologies to augment both household income and nutritional intake in a cost-effective manner compared to conventional technologies.

The prospects to integrate enabling technologies and sustainability while securing the needs for food, feed, fibre and fuel should be explored at various scales: molecular, cell, plant, field, agroecosystem and landscape. Integration of crop modeling into genetic and genomic research facilitates '*breeding by design,*' because the impact of changing traits on crop performance can be explored for various scenarios of environmental conditions and climate change. The use of robust crop models to understand Genotype × Environment × Management (G × E × M) quantitatively did get more attention recently [30••]. Assessments of the relationship between crop productivity and climate change rely upon a combination of modeling and measurement [31•]. It was argued that the generation of knowledge for adaptation should be based on reliable quantification of uncertainty, combining diverse modeling approaches and observations and judicious calibration of models. This approach is not just an improvement of the methodology, but it also contributes to more transparency.

On the level of the crop and field a science-based understanding of the dynamics of phenology, plant physiological processes and soil conditions is required to implement precision agriculture. Adapting inputs (water, fertilizers, pesticides, etc.) site-specifically allows a better use of resources in crop production, while preventing emissions to the environment [32]. A dedicated approach with modern technologies (sensors, IT, machinery, etc.) and knowledge-intensive decision support systems (DSS) can enhance resource-use efficiencies, and enhance the quantity and quality of agricultural produce [33]. There are no easy generic solutions that fit to all agroecosystems as was nicely shown for conservation agriculture in Africa [34•]. Technologies that integrate bio-physical and ecological processes into the framework of sustainable food production by an efficient use of natural resources (land, climate, and water) and minimizing the use of non-renewable inputs (especially fossil energy and phosphorus) should get strong support by making use of knowledge transfer and modern communication means engaging all actors in the food chain [35•,36,37•].

Conclusions

The concerns for the impact of agriculture on the environment are valid. However, it is not realistic to benchmark the environmental load of agroecosystems with those of nature areas. Specific threshold values are needed that meet the standards of food safety and environmental health.

Technology, especially plant breeding and crop management, and government policies have contributed to counter-balance the explosive growth in food demand during the last four decades [38•]. On average, food availability per capita improved despite the doubling of the global population in the recent past. To meet the huge future demands during the next four decades, it will be necessary to make use of the best science and technology to raise crop productivity per unit of land on average with 2% per year and resource-use efficiencies of water and nutrients by a factor 2 [39].

New insights in genetics, systems functioning, climate change and multiple stresses can guide the development of improved cultivars and highly productive farming practices to close the yield gap [40••]. So far, a combination of advanced plant breeding, systems innovations, development of best practices and legislation turned out to be effective in developing more environment-friendly agricultural systems that are profitable, ecologically safe and socially acceptable.

All references for articles included in *Taking Sides: Clashing Views in Sustainability* can be found on the Web at www.mhhe.com/cls.

EXPLORING THE ISSUE

Is Sustainability More About Politics Than Science?

Critical Thinking and Reflection

1. Is there enough technology to tackle our sustainability issues?
2. What role does politics play in sustainability?
3. What role does science play in sustainability?
4. Is sustainability achievable in the current political and scientific atmosphere?
5. How can science be translated into sustainable solutions for our current sustainability issues?

Is There Common Ground?

While most writers take either the yes or the no position, a number of other writers maintain that there is a "common ground" within the issue. These writers believed that both science and politics are important to sustainability and cannot be separated. They view science and politics as "bedfellows" that are intertwined. In fact they are so interconnected that one could say that there exists both the "politics of environmental science and sustainability" and the "science of environmental and sustainability politics." The most effective way to bridge the gap between science and politics is for scientists and politicians to work together to achieve agreed-upon sustainable goals for society. Scientists have been reticent in the past to become politically active as there has been a perception that activist scientists relinquish their objectivity. The scientific method trains scientists to be cautious in their assertions and to never state discoveries with 100 percent certitude. Politicians have a different perspective. They know that their publics like clear and definitive answers to questions, even when those questions are very complex. Hence, politicians generally stay clear of the subtleties and nuances of scientific discovery. For the two to bridge the gap requires new methods of collaboration. Questions to ask are: How can scientists successfully link with politicians? How can politicians successfully link with scientists? Who plays the dominant role in sustainability, is it the scientists or the politicians? How can scientific research be translated into practical solutions in sustainability?

Internet References

www.newint.org/books/no-nonsense-guides/nn_green_politics_foreword_contents_introduction.pdf

This is a link to a forward by Derek Wall of England on his book titled *The no nonsense guide to green politics*. He touches the key issues discussed in the book including global green politics and green philosophy.

http://sspp.proquest.com/archives/vol1iss1/0410-011.cohen.html

This site is home to a good journal paper that discusses how politics impacts on sustainability. One good example is when President Bill Clinton formed the president's council on sustainable development (PCSD), which President Bush disbanded as soon as he took power. This is a good example to show that politics has a big role to play in sustainability.

www.sustainability.org/

The site explains what sustainability is and how it can be achieved

www.mdpi.com/journal/sustainability

This is home to the sustainability open access journal. This journal publishes articles on many topics related to sustainability.

www.epa.gov/sustainability/

This site hosts Environment Protection Agency's (EPA's) sustainability information. Different aspects of sustainability are discussed.

ISSUE 3

Are Western Values, Ethics, and Dominant Paradigms Compatible with Sustainability?

YES: Jo Kwong, from "Globalization's Effects on the Environment—Boon or Bane?," *Lindenwood Economic Policy Lecture Series* (July 2004)

NO: Erik Assadourian, from "The Rise and Fall of Consumer Cultures," *2010 State of the World—Transforming Cultures from Consumerism to Sustainability* (The Worldwatch Institute, 2010)

Learning Outcomes

After reading this issue, you should be able to:

- Acquire knowledge of the basic elements of the dominant social paradigm (DSP).
- Know how the DSP impacts decisions for sustainability.
- Understand the relationship between globalization and sustainability.
- Explore how economic development can improve environmental quality.
- Understand how culture influences sustainability.

ISSUE SUMMARY

YES: Jo Kwong, vice president of institute relations at the Atlas Economic Research Foundation in Fairfax, believes that globalization is a basic part of the solution of the global problems that plague the developing world. Greater movement of goods, services, people, and ideas can lead to economic prosperity, improved environmental protection, and a host of other social benefits.

NO: Erik Assadourian, a senior researcher at the Worldwatch Institute and the project director of *2010 State of the World,* believes that Western culture is the origin of consumer culture and the consumption trend and, therefore, leads to a global culture of excess and is emerging as the biggest threat to the planet. Higher levels of consumption can affect the environment and, in the long run, limit economic activity. As a matter of fact, higher levels of consumption require larger inputs of energy and material to produce and therefore generates a high volume of waste products. It also increases the extraction and exploitation of natural resources.

In the West, the Enlightenment initiated a set of beliefs that placed an emphasis on rationality and science and the thought that society could be improved through hard work and the application of technology and moral development. Before the 1950s, the prevailing viewpoint in the West was that humans could and should modify nature to their advantage, and that the environment was virtually limitless and resilient. Nature was to be studied, catalogued, tamed, and exploited. However, technological optimism, a key driver of Western development, began to falter by the 1960s as awareness of environmental problems grew and the need to manage the impacts of humans on the environment was realized. In 1965, Adlai Stevenson, the then U.S. ambassador to the UN, stated that the world is a fragile, unique, closed environment in which both rich and poor countries were multiplying beyond their life-support capacity. No one was in control and both were vulnerable to each other and increasingly to natural hazards. Through the Brundtland Commission Report of 1983 and the Rio Earth Summit in 1992, the concept of sustainable development was posited with two main goals: to reduce environmental degradation and natural resource deterioration by learning to live within the carrying capacity of supporting ecosystems and to reduce global poverty through economic and social development. One of the problems in the early definition of "sustainable development" was the inherent conflict between the two words. Can development, as traditionally interpreted within a Western context, meaning economic growth and consumption, be sustainable, realizing the rapid growth of world population?

Culture constitutes the values, beliefs, customs, norms, and institutions that guide how humans both perceive and fashion reality. One of the important components of Western culture has been the dominant social paradigm (DSP). Zachary Smith defines the DSP in *The Environmental Policy Paradox* (2004) as those clusters of beliefs, values, and ideals that influence our thinking about society, government, and individual responsibility. The most important elements of the DSP are free-market economics, faith in science and technology, the growth orientation common in Western democracies, and a sense of separation from the natural environment. It is within the context of the DSP that we ask this question: Are Western values and belief systems compatible with the goals of sustainability? On the surface it seems that the DSP clashes with important characteristics of sustainability. The primacy of humans in nature, a key element of the DSP, seems to give humans the moral justification to conquer or use the environment to foster the development of the species, with little concern for ecological balance. Yet, an argument can also be made that human intelligence can be harnessed to live within the constraints of the Earth's ecosystem without destroying it, which is the main objective of sustainability.

Western dominant paradigms reflect the idea that technology is the method by which nature is tamed and controlled for human objectives and the tools of development are contained in such concepts as globalization, property rights, privatization, democracy, and the rule of law. A critical concern is that many developing countries are busy trying to catch up with wealthy countries and it raises the question whether the achievements of these wealthy countries can be emulated. Take the Western growth concept of globalization for instance. Some writers believe that globalization is a potential solution to the poverty that plagues countries in the developing world. Based on the economic concept of comparative advantage, the idea is that free trade will enable poor countries to harness their advantages, that is, low-cost social and natural capital, to spur economic growth. A technology transfer from wealthy countries to poorer countries will provide an economic benefit to these countries.

Is the DSP compatible with sustainability? Kilbourne et al. (2002) have suggested that as one's belief in the DSP increases, concern for the environment decreases. But, on further investigation, as concern for the environment increases, perception of necessary changes and willingness to change to achieve environmental balance also increase. Birdyshaw et al. (2007) state that privatization reduces the exploitation of natural resources and improves environmental quality. Udo and Jansson (2009) have suggested that nations that are struggling to survive are less inclined to adopt environmental sustainability than advanced nations. Also, property rights, a key element of the DSP, can also be a tool for sustainability. Rodgers et al. (2009) state that property rights lead to environmental stewardship.

Writers who believe that the DSP is not compatible with sustainability note that consumption has grown dramatically over the past five decades. As a result, more fossil fuels, minerals, and metals have been mined from the earth; more trees have been cut down; and the exploitation of these resources to maintain ever higher levels of consumption has put increasing pressure on Earth's systems. This process has dramatically disrupted the ecological systems on which humanity and countless other species depend. Orecchia Carlo (2010) in "Consumerism and Environment: Does Consumption Behavior Affect Environmental Quality?" states that the increase in per capita income affects per capita consumption. Higher levels of consumption can affect the environment and, in the long run, limit economic activities. Therefore, he does not support the environmental Kuznets curve, which states that an inverted-U relationship exists between environmental degradation and economic growth. In this theory, environmental quality improves with economic development. However, journalist Suzanne Goldenberg's summary "Cult of Greed Is Now a Global Environmental Threat" in the Worldwatch Institute's *2010 State of the World* clearly asserts that the Western cult of consumption and greed poses a global environmental threat. Also, Hamilton et al. (2010) have argued that consumerism and individualism may lead to new ecological consequences.

Aggarwal et al. (2005) examine the various mechanisms through which globalization can lead to loss in ecosystem resilience and thus increase the vulnerability of poor people. Rees et al. (2006) opine that globalization and trade exacerbate the situation by shuffling resources around and short-circuiting the negative feedback that would otherwise result from local resource degradation. This allows population and material growth within each individual trading region to exceed local biophysical limits. This, in turn, accelerates the depletion of natural capital everywhere and ensures that all trade-dependent regions hit global limits simultaneously.

In the article for the YES position, that Western values are compatible with sustainability, researcher and writer Jo Kwong provides an overview of views both supporting and opposing globalization. Globalization is one of the key drivers of Western-based modernization. She points out that it is not globalization's impact on the global environment that is the issue, but rather the lack of key institutions in developing countries, that is, rule of law, property rights, and free and open markets. She explores the validity of the environmental Kuznets curve, which holds that countries can improve environmental quality through economic growth. She quotes sources that estimate that for each 1 percent increase in per capita income in a nation, pollution falls by 1 percent and, as free trade expands, each 1 percent increase in per capita income tends to drive pollution concentrations down by 1.25–1.5 percent. This is done through clean technology.

The article for the NO position, that Western values are not compatible with sustainability, is presented by Erik Assadourian, Senior Fellow at the Washington-based Worldwatch Institute. He states that consumption has grown dramatically over the past five decades. This has produced a disruption in ecological systems. The world's richest 500 million people are currently responsible for 50 percent of the world's carbon dioxide emissions, while the poorest 3 billion are responsible for just 6 percent. High-income countries accounted for 78 percent of consumption expenditures but have just 16 percent of world population. If everyone lived like Americans, the Earth could sustain only 1.4 billion people. He addresses issues relating values to consumer practices that are unsustainable, such as the large increase in bottled water, fast foods, disposable paper products, and so forth. He expresses some optimism for the future as he sees a shift to new eco-friendly habits and the adoption of sustainable practices.

Jo Kwong

Globalization's Effects on the Environment—Boon or Bane?

In recent years, globalization has become a remarkably polarizing issue. In particular, discussions about globalization and its environmental impacts generate ferocious debate among policy analysts, environmental activists, economists and other opinion leaders. Is globalization a solution to serious economic and social problems of the world? Or is it a profit-motivated process that leads to oppression and exploitation of the world's less fortunate?

This paper examines alternative perspectives about globalization and the environment. It offers an explanation for the conflicting visions that are frequently expressed and suggests elements of an institutional framework that can align the benefits of globalization with the objective of enhanced environmental protection.

What Is "Globalization" and Why Are Some So Concerned About Its Impacts?

Globalization, free of the emotional rhetoric, is simply about removing barriers so goods, services, people, and ideas, can freely move from place to place. At its most rudimentary level, globalization describes a process whereby people can make their own decisions about who their trading partners are and what opportunities they wish to pursue.[1]

While this may seem fairly innocuous, globalization certainly raises many concerns. In developed nations, some people worry about globalization's impacts on culture, traditional ways of living, and indigenous control in less developed parts of the word. They wonder, "What's to stop profit-motivated companies from developing some of the pristine environments and fragile natural resources found in the developing world?" These critics of open trade fear that residents of developing nations will be the losers in more ways than one—stripped of their land's natural resources and hopelessly in debt to exploitative developed countries. This group takes a rather paternalistic view of the problems facing the world's poor.

Kwong, Jo. From *Lindenwood Economic Policy Lecture Series* 5, July 2004, pp. 1–15, address by Kwong, Jo at Lindenwood University, St. Charles, MO, cohosted by the Institute for Study of Economics and the Environment (ISEE) and the School of Business and Entrepreneurship.

Others—free marketers—believe that the developed world can produce positive benefits by exporting knowledge and technology to the developing world. By avoiding mistakes made in the developed world, it is argued, developing countries can advance in manners that sidestep some of the errors that occurred in others' development processes. Third-world poverty is cited as an important reason to foster greater economic growth in the developing world. To proponents of globalization, trade is seen as a way to lift the third world from poverty and enable local people to help themselves.

Moreover, there are divided views within the developing world. Some argue against so-called "eco-imperialism." "Why are others dictating whether or not we can develop our own resources? Who are these environmental activists that say billions of people in China shouldn't have cars because this will greatly accelerate global warming?" they ask. But others question, "Who are these corporations that come in and buy huge tracts of land in third-world interiors and develop large-scale forestry or oil developments, seemingly without concern about the impact on the local environment?"

In many ways, these alternative perspectives can be viewed as a "conflict of visions," to steal a phrase from Thomas Sowell.[2] Some people simply view the world fundamentally differently. In the globalization context, for example, one view values the protection of indigenous ways of life, even if that means living with greater poverty and fewer individual choices. Others believe economic efficiency is key—getting the most from our resources to provide the greatest amount of financial wealth and opportunity. Most likely, however, most people fall somewhere in between.

This discussion will offer an additional factor other than a "conflict of visions" that can help us understand the broad disparities in perspectives and understandings about the question, "Is globalization good for the environment?" In particular, it raises the possibility that perhaps we are not asking the right questions to address the set of concerns at hand.

Is Globalization Good for the Environment?

In the 1990s, a number of economists sought to empirically answer the question of whether globalization helps or harms the environment.[3] Some of the most often-cited findings are those from economists Gene Grossman and Alan Krueger. Grossman and Krueger investigated the relationship between the scale of economic activity and environmental quality for a broad set of environmental indicators. They found that environmental degradation and income have an inverted U-shaped relationship, with pollution increasing with income at low levels of income and decreasing with income at high levels of income. The turning point at which economic growth and pollution emissions switch from a positive to a negative relationship depends on the particular emissions and air quality measure tracked. For NOx, SOx and biological oxygen demand (BOD), the turning point appears to be around $5,000 per capita gross domestic product (GDP). This observation supports the view that countries can grow out of pollution problems with wealth.[4]

Figure 1

Environmental Kuznets Curve

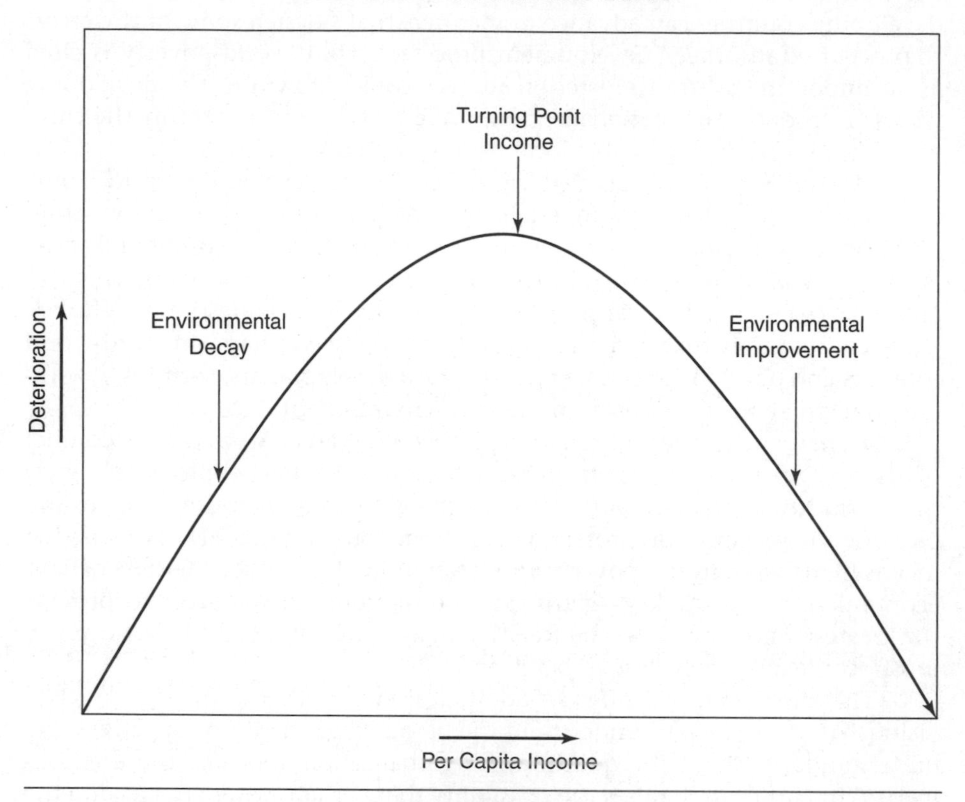

Source: "The Environmental Kuznets Curve A Primer," by Bruce Yandle, Maya Vijayaraghavan, and Madhusudan Bhattarai, Political Economy Research Center, PERC Research Study 02-1, May 2002.

These findings were followed by further studies that examined this "Environmental Kuznets Curve,"[5] as this inverted U-shaped curve was labeled. (See Figure 1.) The research generated a new set of policy implications that supported the idea that trade can be good for the environment.[6] If economic growth is good for the environment, policies that stimulate growth (trade liberalization, economic restructuring, and free markets) should also be good for the environment.

The most basic description of how this inverted curve can occur is to think about the types of activities that countries experience as they develop. At the most rudimentary level, people are burning cow dung and other readily available materials for heat and cooking sources. No controls are in place; the pollutants are released directly into the air. As economic activity increases, the economy reaches a point at which people begin making investments—catalytic converters, furnaces, etc.—pollution levels are reduced, and hence the inverted curve results.

In "Poverty, Wealth and Waste," Barun Mitra compares patterns of waste distribution in India to those of the developed world.[7] He addresses the myth that poor countries have lower levels of pollution:

> The painstaking efforts to recycle materials do not mean that a poor country like India is pollution-free. Indeed, the low quantity of waste generated in an economy with little capital and technological backwardness keeps the waste industry from graduating above small-scale local initiatives. And higher pollution occurs because there isn't the technology to capture highly dispersed waste such as sulfur dioxide from smokestacks or heavy metals that flow into wastewater.

A number of possible explanations for this observed relationship between pollution and income were advanced:

- As local economies grow and develop, they will inevitably change the way they use resources, creating different types of impacts upon the environment. A simple example is the pollution trade-offs involved from our transition in transportation modes from horses to cars. Horses generated plenty of pollution in terms of manure, carcass disposal, etc. Cars, of course, generate an entirely different brand of pollution concerns. In other words, some environmental degradation along a country's development path is inevitable, especially during the take-off process of industrialization.[8]
- Growth is associated with an increasing share of services and high-technology production, both of which tend to be more environment-friendly than production processes in earlier stages of industrialization.[9]
- Knowledge and technology from the developed world can help ease this transition and lessen its duration, moving countries more quickly to the levels at which pollution will be decreasing. Free trade can promote a quicker diffusion of environment-friendly technologies and lead to a more efficient allocation of resources.
- The prosperity generated from economic activity will lead to more investments and higher standards of living that enable still greater investments in cleaner and newer technologies and processes. When a certain level of per capita income is reached, economic growth helps to undo the damage done in earlier years.[10] As free trade expands, each 1 percent increase in per capita income tends to drive pollution concentrations down by 1.25 to 1.5 percent because of the movement to cleaner techniques of production.[11]
- As individuals become richer they are willing to spend more on non-material goods, such as a cleaner environment.[12] This point is made by Indur M. Goklany, in his description of earlier stages of development:

> Society [initially] places a much higher priority on acquiring basic public health and other services such as sewage treatment, water supply, and electricity than on environmental quality, which initially worsens. But as the original priorities are met, environmental problems become higher priorities. More resources are devoted to solving those problems. Environmental degradation is arrested and then reversed.[13]

These findings and explanations, unsurprisingly, generated an outpouring of negative response from environmental activists and anti-globalization proponents. "How can these economists be serious?" they, in effect, asked. "Do they really think it is wise to advocate policies that predictably increase pollution? Are we supposed to believe pollution will eventually decrease if we continue with the polluting activities? How absolutely ludicrous!"

Typical responses to the "growth is good" thesis include

- Globalization will result in a "race to the bottom" as polluting companies relocate to countries with lax environmental standards.
- Trading with countries that do not have suitable environmental laws will lower environmental standards for all countries.
- Multinationals will exploit pristine environments in the developing world, reaping the resources for short-term growth, and then pulling out to repeat the process elsewhere—growth ruins the environment.
- Free trade provides a license to pollute—it is bad for the environment. Stronger environmental regulations at national and international levels are needed.

The Sierra Club[14] summarized the widespread critiques to the Grossman and Krueger studies, drawing from research studies produced by the World Wide Fund for Nature and others.[15] It argued that the findings were sufficiently over-generalized to dispense with any notion that they justify complacency about trade and the environment, pointing out that

- The empirical estimates of where "turning points" occur for different pollutants vary so widely as to cast doubt on the validity of any one set of results. For instance, where Grossman and Krueger found turning points for certain air pollutants at less than $5,000 per capita, others found turning points above $8,000 per capita.
- For some air pollutants, Grossman and Krueger found that emissions levels don't follow an inverted U-curve, but follow an S-curve that starts to rise again as incomes rise. For instance, they found that sulphur dioxide emissions start to rise when income increases above $14,000 per capita. The implication is that efficiency gains from improved technology at medium levels of per capita income are eventually overwhelmed by the growing size of the economy.
- Since most of the world's population earns per capita incomes well below estimated turning points, global air pollution levels will continue to rise for nearly another century. By that time, emissions of some pollutants will be anywhere from two to four times higher than current levels.
- Even for the limited number of pollutants that Grossman and Krueger study, they only demonstrate a correlation between changing per capita income and changing levels of environmental quality. They do not demonstrate a causal connection. The positive relationships they describe could actually be caused by non-economic factors, such as the adoption of environmental legislation.

Both camps seem to have reasonable grounds for their views. Clearly there is a conflict of visions that is rooted in very different value systems. Can these two opposing perspectives be reconciled sufficiently to reach some type of consensus?

The Wealth of Nations and the Environment

As noted earlier, many studies have re-examined the Environmental Kuznets Curve since the publication of the Grossman and Krueger analysis in 1991, each attempting to prove or disprove the relationship between economic growth and environmental quality, or to isolate variables that may explain the observed relationships. In that same year, a fascinating monograph was published in London, called *The Wealth of Nations and the Environment.* Author Mikhail Bernstam set out to analyze the contention that economic growth negatively impacts the environment by examining how institutional structure impacts this relationship.

Bernstam examined and contrasted the impact of economic growth upon the environment in both capitalist and socialist countries. Interestingly, he found that the environmental Kuznets curve does in fact exist, but it does not apply to countries across the board. The Kuznets curve, he found, applies to market economies, but not to socialist ones. The difference, according to Bernstam, has its roots in the different structures of incentives and property rights of these two economic systems.

Under market economies with secure property rights and open trade, the pursuit of profits leads to the husbanding of resources. These capitalist economies use fewer resources to produce the equivalent level of output and hence do less damage to the environment. In contrast, in socialist countries, the managers of state enterprises operate under incentives that encourage them to maximize inputs, with little regard towards economic waste or damage to the environment.[16]

More recently, a 2001 study by economists Werner Antweiler, Brian R. Copeland, and M. Scott Taylor asked, "Is Free Trade Good for the Environment?"[17] They analyzed data on sulfur dioxide over the period 1971 to 1996, a time when trade barriers were coming down and international trade was expanding. They found that countries that opened up to trade generated faster economic growth. Although economic growth produced more pollution, the greater wealth and higher incomes also generated a demand for a cleaner environment.

To separate these effects, the Antweiler model looked at the negative environmental consequences of increases in economic activity (the scale effect), the positive environmental consequences of increases in income that lead to cleaner production methods (the technique effect), and the impact of trade-induced changes in the composition of output upon pollution concentrations (the composition effect). When the scale, technique, and composition effects estimates were combined, the Antweiler et al. model yielded the conclusion that free trade is good for the environment. For example, when analyzing sulfur dioxide, the authors estimate that for each 1 percent increase in per capita income in a nation, pollution falls by 1 percent.

The critical explanatory factor is that wealthier countries value environmental amenities more highly and enhance their production by employing environmentally friendly technologies.[18] However, like Bernstam, these authors specified that it is important to distinguish between communist and non-communist countries. Communist countries provided the exception to their rule about globalization's positive impacts upon the environment.

The studies, which consider the impact of institutional structures, make an important contribution to our understanding of the "economics vs. environment" debates. They suggest we consider other factors in our analysis of the effects of globalization. It is true that we often do find examples of disastrous environmental conditions, particularly when we look at socialist countries. But it is misleading to attribute the disasters to globalization. Instead we need to examine the institutional arrangements in a particular country to see what role they play in economic development and environmental protection.

Positive Globalization

As described earlier, at its most rudimentary level, globalization simply embodies a process of free and open trade, whereby people can make decisions about who their trading partners are and what opportunities they will choose to pursue.

But the cautions of the environmental activists are worthy of consideration. Free trade, in and of itself, will not guarantee positive outcomes. We also need guiding rules that essentially create the terms for fair and civil interaction.

In *Property Rights: A Practical Guide to Freedom & Prosperity,*[19] Terry Anderson and Laura Huggins describe the importance of institutional rules. They use the example of children playing together and inventing games. In essence, the children work together to form rules that are fair. When they cannot agree on rules, chaos typically results and their play breaks down. The same is true for civil society. Institutional rules, in the form of constitutions, common law, and so on, provide the structure for human activity.

The critical role of institutions in shaping human behavior gained international attention in 1993 when Douglass C. North received the Nobel Prize in economics. North's groundbreaking research in economic history integrated economics, sociology, statistics, and history to explain the role that institutions play in economic growth.

For several decades, North looked at the question, "Why do some countries become rich, while others remain poor?" In seeking answers to this query, he came to understand that institutions establish the formal and informal sets of rules that govern the behavior of human beings in a society. His research showed that, depending on their structure and enforcement, institutional arrangements can either foster or restrain economic development.[20]

For the past nine years, the *Index of Economic Freedom,* jointly published by the Wall Street Journal and the Heritage Foundation (Washington, DC), has provided fascinating empirical evidence of the relationship of various institutions to economic prosperity. The study analyzes and ranks the economic freedom of 161 countries according to 10 institutional factors (trade policy, property rights, regulation, and black market activities, for example) in an effort to trace the path to economic prosperity.

The key finding of the research, supported year after year, is that countries with the most economic freedom enjoy higher rates of long-term economic growth and prosperity than those with less economic freedom. But, more relevant to this discussion, is the finding that economic freedom, which enables people to choose who and where their trading partners are, ultimately leads to more efficient resource use.

In another comparative index, *Economic Freedom of the World 2002,* published by the Fraser Institute in conjunction with public policy institutes around the world, Nobel laureate Milton Friedman describes the importance of private property and the rule of law as a basis for economic freedom. He spells out the three ingredients key to establishing economic freedom as follows:

> First of all, and most important, the rule of law, which extends to the protection of property. Second, widespread private ownership of the means of production. Third, freedom to enter or to leave industries, freedom of competition, freedom of trade. Those are essentially the basic requirements.[21]

These same factors also provide a framework for positive environmental development.

In the 1980s, a team of economists affiliated with the Property and Environmental Research Center (PERC) in Bozeman, Montana, began developing a new paradigm for environmental policy. Their model, which eventually was coined "Free Market Environmentalism," described how incentives are the key to environmental stewardship. Not surprisingly, people who face little or no consequences for environmentally destructive actions face no incentive to protect the environment. Alternatively, people who are rewarded for good stewardship are much more likely to invest in environmental protection. The key, according to economists John Baden, Richard Stroup and Terry Anderson, are the very same three elements that Milton Friedman mentioned for economic prosperity: free and open markets, clearly established property rights, and rule of law.[22]

Free and Open Markets

One of the most important benefits produced by a market economy is information, conveyed in the form of prices. Prices of natural and environmental resources provide clear signals about their availability. As a resource becomes more scarce, its price increases. And of course, the reverse is also true—when a resource becomes more abundant, the price decreases.

Many people fear that the profit motive leads to the depletion or degradation of environmental resources. As counterintuitive as it may sound, the profit motive actually works to the benefit of the environment.

Businesses face incentives to carefully consider the prices of the various natural resources that they use in their production processes. If a particular resource is in short supply, its price will be higher than others that are more readily available. It makes little sense for a producer to over utilize, or "waste," a high-priced resource.

High prices also encourage the search for, and development of, appropriate substitutes or alternatives. As companies search for ways to reduce costs, they naturally tend toward utilizing lower-priced, more abundant resources. Thus, the pursuit of profits is actually a driving force to conserve resources. In essence, under free market systems, entrepreneurs compete in developing low cost, efficient means to solve contemporary resource problems.

Property Rights

Clearly-established property rights generate another incentive for environmental stewardship. It makes no sense for private landowners, for example, to exploit and destroy their own property. Ownership creates a long-term perspective that leads to preserving and protecting property.

Careless destruction, however, does make sense for those who are only loosely held accountable for their actions. Politicians, bureaucrats, or others, who may be shortterm managers, face the incentive to maximize immediate returns, even if this means long-term environmental damage. Even managers with longer tenures realize they can simply turn to the federal government for more funds to address the problems that shortsighted decision making may have created.[23]

Rule of Law

In many ways, the "rule of law" is the glue that holds market transactions and property rights together. Freedom to exchange is meaningless if individuals do not have secure rights to property, including the fruits of their labors. Failure of a country's legal system to provide for the security of property rights, enforcement of contracts, and the mutually agreeable settlement of disputes will undermine the operation of a market-exchange system. If individuals and businesses lack confidence that contracts will be enforced and the returns from their productive activity protected, their incentive to engage in innovative activities will be eroded.[24]

With these elements in place, the economists' explanations prevail—globalization will enable local cultures to pick and choose the development and environmental paths that they wish to traverse. But without these institutional arrangements, the likelihood of negative consequences increases.

In countries that lack property rights and rule of law and that promote barriers to trade, an institutional structure develops that fosters destruction of the environment.[25] For example, in Liberia, former President Charles Taylor rapidly sold off many of the nation's natural resources in order to fund his dictatorship. In the lawless structure of that country, Taylor was able to exploit the environment and his people. In a country that has clear property rights and rule of law, such corrupt options are closed off. Neither can corporations force a village, or a state, or a country to destroy its natural resources against the will of the people.

We see this illustrated in an ongoing controversy in Peru.[26] In the 1990s, when then-bankrupt Peru opened its statist economy to foreign investment, the nation drew almost $10 billion in mining capital. That sector now accounts for half of Peru's $8 billion in exports, and Peru has become one of the world's largest gold producers. Yet, the opening economy does not necessarily mean that multi nationals can run rough shod over the locals. It all depends on the institutional arrangements that are in place.

In the small town of Tambogrande, Peru, a Canadian mining company holds the rights to tap into $1 billion worth of cooper and zinc beneath the town. To do so, however, requires demolishing many local homes. In a referendum held in 2002, the town residents voted to turn down the mining company's offer to build new homes in a different location. If the country's laws hold firm to the property rights of the villagers, the mining company will not be allowed to develop the copper mine without local consent. But, if the rule of law and respect of property rights are not upheld, then the foreign firm can force its will on the indigenous people.

Property rights provide a powerful incentive for people to carefully assess their options—in this case, whether the loss of their existing houses and the village is compensated for by the new homes they would be receiving. The nature of the property rights institutions indeed affects the range of outcomes. If the local government owned the rights to the housing, rather than individuals, we would expect an entirely different outcome. Local politicians likely would gain by acquiescing to the mining firm's proposal because the villagers, not the politicians, would incur the costs.

Unfortunately, in many developing countries, corruption and back door deal making, enabled by weak rule of law and property rights, proliferates. The result is that a few leaders come out ahead and the locals get short changed. Local protests are reportedly stalling at least 10 mining-investment projects in Peru that are worth $1.4 billion—and for good reason. The noted Peruvian economist, Hernando de Soto, author of the best-seller, *The Mystery of Capital,* comments that although the mines in some towns pay double the prevailing minimum wage, they do not compensate for "the loss of their sense of environmental and economic sovereignty." Consequently, the National Society of Mining, Petroleum & Energy is urging the government to adopt reforms that immediately give at least 20 percent of the royalties to on-site communities instead of sending all these funds to Lima. Manhattan Minerals, one of the companies interested in Tambogrande, thinks local communities should receive an even bigger cut, making these towns, in effect, feel more like shareholders. In other words, they need to give the locals an interest—or property right—in the operations.

In the southern Andean town of Lircay, Huanavelica—Peru's poorest state—residents are concerned that the mine will threaten adjacent agricultural lands. To show their anger, they have resorted to street demonstrations and setting fire to government installations. Their actions seem less extreme in light of previous experiences. For decades, state-owned mining created many environmental problems that residents are rightly worried about. This cultural legacy is a key factor for private mining companies as they hammer out new relationships and try to move forward.

Fortunately, positive examples are evolving. The La Oroya copper smelter in the central Andes region was purchased by the Doe Run company—based in St. Louis. The Peruvian government gave the company 10 years to clean up the environmental mess that the government created. Doe Run has reportedly spent $40 million so far, including money for a program to reduce high blood-lead levels in area children.

Peru needs to continue to open its doors to foreign investment—or what some would call globalization—to lift its people out of poverty. It must establish institutions—rule of law, property rights and open markets—that create a safe investment climate and allow corporations to prosper. Simultaneously, companies need to conduct business in a way that will benefit the local residents as well.

As another example of how incentives—and disincentives—can impact the environment, consider the case of India's automobile industry. Disincentives generated by the government's regulatory policies contributed to a stagnant, non-innovative industry which caused harm to the Indian economy and environment for decades.

Although Indian automobile manufacturers began producing cars in the 1930s, there was very little development and growth in that industry for over 50 years. Auto manufacturing was heavily regulated, licensed and protected. In addition, consumers faced high taxes and duties on imported automobiles and on gasoline. The upshot was that very little competition developed in India's automobile industry—autos with low fuel efficiency and high air emissions became the norm.

In recent years, however, the automobile sector has been slowly liberalizing, allowing some major multi-national corporations to set up shop in India. As a result of the increased competition and relaxed barriers, more efficient and less-polluting automobiles are becoming available to Indian consumers. A free trade regime, from the outset, would have increased access to vehicles for consumers, lowered the cost of transportation, enabled the best technologies to be locally available, and improved air quality.[27]

In other words, incentives matter. And the structure of institutions plays a key role in the nature of incentives that are in effect.

Institutional Reform for Positive Globalization

Economists have raised interesting empirical questions by developing the Environmental Kuznets Curve, but, as the World Wide Fund for Nature (WWF) study and others suggest, there is no one curve that fits all pollutants for all places and times. Economist Bruce Yandle of Clemson University describes it this way, "There are families of relationships, and in many cases the inverted-U Environmental Kuznets Curve is the best way to approximate the link between environmental change and income growth."[28]

Additionally, environmental activists are right in pointing out globalization's potentially negative impacts upon the environment. Income growth alone is insufficient to reduce environmental harms and may even increase these harms if a core set of institutional features are not in place.

As the Antweiler model indicates, economic growth creates the conditions for environmental protection by raising the demand for improved environmental quality and by providing the resources needed for protection. Whether environmental quality improvements materialize or not, or when, or how they develop, depends critically on government policies, social institutions, and the strength of markets. Better policies, such as removing distorting subsidies, introducing more secure property rights over resources, and using market-like mechanisms to connect the costs of pollution to prices paid for pollution-producing goods will lower peak environmental harm (flatten the underlying Environmental Kuznets Curve). These improved policies may also bring about an earlier environmental transition.[29]

While it may seem to be an overwhelming challenge to accomplish the institutional reforms described above, the good news is that it is happening in some very unlikely parts of the world. Consider, for example, exciting changes that have recently been occurring in Rwanda, Africa.

Lawrence Reed, the founder and president of the Mackinac Center for Public Policy in Midland, Michigan, recently toured eastern Africa, home to the remaining wild mountain gorillas left in the world. Here, approximately 670 gorillas live on a string of lush, rain-forested volcanoes along the Rwanda border with Uganda and the Congo.

To Reed's surprise, native-owned and locally staffed companies conduct all gorilla safaris. Part of the fee goes to the government for salaries for national park employees and for programs that protect gorilla habitat. (These programs also are substantially supplemented by the efforts of private, non-profits that get support from around the world.) Two Rwandan entrepreneurs started the firm, Primate Safaris, three years ago. With six employees, they provide everything a gorilla safari enthusiast could hope for—a competent guide with a four-wheel drive vehicle, good meals and comfortable accommodations.

In fact, Reed's experience with Private Safaris was only the tip of the iceberg. Rwanda, he learned, is engaged in the continent's most ambitious privatization campaign. After experiencing the kind of stifling, socialist rule that consigned virtually all of Africa to grinding poverty for decades, this nation is now embracing the private sector with deliberate policy and enormous enthusiasm.

Imagine Reed's surprise when, shortly after landing in Rwanda, he came across a sign at the airport outside the capital of Kigali which reads, "Privatization: A Loss? No Way." Further down the road, another sign says, "Privatization fights laziness, privatization fights poverty, privatization fights smuggling, and privatization fights unemployment."

Several of the country's privatization efforts have had direct positive impacts on the environment. For example, in 1999, Shell Oil bought a portion of the assets of Petrorwanda (the bankrupt state oil company) and completely renovated 14 of the defunct firm's decrepit and environmentally hazardous gasoline stations.[30]

An interesting development in Uganda suggests signs of similar institutional reforms. An English language, African-based band named "Afrigo" released a song entitled, "Today for Tomorrow," which celebrates the benefits of privatization. Here is a sample of the lyrics:[31]

> Privatization, the surer route to economic emancipation/ Yeah, businessmen run businesses/government govern the nation/ You and I didn't create the situation/ Let's unite/check the economy/ a better future for our children.

Apparently, citizens of Rwanda and Uganda are embracing private property rights and other economic and political changes to better their lives and those of the next generation. Environmental protection surely will fare better in this setting than in the failed socialist systems being replaced.

Conclusion

Is globalization good for the environment? Viewing globalization as the destroyer or savior of the environment misses the point. The problem is not globalization *per se*. A lack of key institutions, rule of law, property rights, free and open markets, is the real villain in the tale. These institutions hold people accountable for their actions, and at the same time, reward them for positive behavior. They create conditions in which market competition rewards innovation and efficiency, and in which economic development and increased wealth can fuel improved environmental quality.

Globalization—free trade and multinational investments—can advance these institutional changes, leading to enhanced social and political stability. Concerns that multinational corporations might be engaged in a "race to the environmental bottom" seem unlikely in these circumstances. To the contrary, where these institutions are in place, the result can be a "race to the top," as jurisdictions compete to improve the quality of life for their constituents.[32]

Globalization can be a means to accelerate learning about the importance of market institutions to economic growth. Environmental protection can be one of many important benefits resulting from such a transition.

Getting back to my earlier comment about "a conflict of visions," I certainly hold a contrasting view from opponents of globalization. Critics believe globalization underlies many of the problems that plague the developing world. On the other hand, I see globalization as a basic part of the solution to these problems. Greater movement of goods, services, people and ideas can lead to economic prosperity, improved environmental protection, and a host of other social benefits.

All references for articles included in *Taking Sides: Clashing Views in Sustainability* can be found on the Web at www.mhhe.com/cls.

Erik Assadourian

The Rise and Fall of Consumer Cultures

In the 2009 documentary *The Age of Stupid,* a fictional historian who is possibly the last man on Earth looks at archival film footage from 2008 and contemplates the last years in which humanity could have saved itself from global ecological collapse. As he reflects on the lives of several individuals—an Indian businessman building a new low-cost airline, a British community group concerned about climate change but fighting a new wind turbine development in the area, a Nigerian student striving to live the American dream, and an American oilman who sees no contradiction between his work and his love of the outdoors—the historian wonders, "Why didn't we save ourselves when we had the chance?" Were we just being stupid? Or was it that "on some level we weren't sure that we were worth saving?" The answer has little to do with humans being stupid or self-destructive but everything to do with culture.[1]

Human beings are embedded in cultural systems, are shaped and constrained by their cultures, and for the most part act only within the cultural realities of their lives. The cultural norms, symbols, values, and traditions a person grows up with become "natural." Thus, asking people who live in consumer cultures to curb consumption is akin to asking them to stop breathing—they can do it for a moment, but then, gasping, they will inhale again. Driving cars, flying in planes, having large homes, using air conditioning . . . these are not decadent choices but simply natural parts of life—at least according to the cultural norms present in a growing number of consumer cultures in the world. Yet while they seem natural to people who are part of those cultural realities, these patterns are neither sustainable nor innate manifestations of human nature. They have developed over several centuries and today are actively being reinforced and spread to millions of people in developing countries.

Preventing the collapse of human civilization requires nothing less than a wholesale transformation of dominant cultural patterns. This transformation would reject consumerism—the cultural orientation that leads people to find meaning, contentment, and acceptance through what they consume—as taboo and establish in its place a new cultural framework centered on sustainability. In the process, a revamped understanding of "natural" would emerge:

it would mean individual and societal choices that cause minimal ecological damage or, better yet, that restore Earth's ecological systems to health. Such a shift—something more fundamental than the adoption of new technologies or government policies, which are often regarded as the key drivers of a shift to sustainable societies—would radically reshape the way people understand and act in the world.

Transforming cultures is of course no small task. It will require decades of effort in which cultural pioneers—those who can step out of their cultural realities enough to critically examine them—work tirelessly to redirect key culture-shaping institutions: education, business, government, and the media, as well as social movements and long-standing human traditions. Harnessing these drivers of cultural change will be critical if humanity is to survive and thrive for centuries and millennia to come and prove that we are, indeed, "worth saving."

The Unsustainability of Current Consumption Patterns

In 2006, people around the world spent $30.5 trillion on goods and services (in 2008 dollars). These expenditures included basic necessities like food and shelter, but as discretionary incomes rose, people spent more on consumer goods—from richer foods and larger homes to televisions, cars, computers, and air travel. In 2008 alone, people around the world purchased 68 million vehicles, 85 million refrigerators, 297 million computers, and 1.2 billion mobile (cell) phones.[2]

Consumption has grown dramatically over the past five decades, up 28 percent from the $23.9 trillion spent in 1996 and up sixfold from the $4.9 trillion spent in 1960 (in 2008 dollars). Some of this increase comes from the growth in population, but human numbers only grew by a factor of 2.2 between 1960 and 2006. Thus consumption expenditures per person still almost tripled.[3]

As consumption has risen, more fossil fuels, minerals, and metals have been mined from the earth, more trees have been cut down, and more land has been plowed to grow food (often to feed livestock as people at higher income levels started to eat more meat). Between 1950 and 2005, for example, metals production grew sixfold, oil consumption eightfold, and natural gas consumption 14-fold. In total, 60 billion tons of resources are now extracted annually—about 50 percent more than just 30 years ago. Today, the average European uses 43 kilograms of resources daily, and the average American uses 88 kilograms. All in all, the world extracts the equivalent of 112 Empire State Buildings from the earth every single day.[4]

The exploitation of these resources to maintain ever higher levels of consumption has put increasing pressure on Earth's systems and in the process has dramatically disrupted the ecological systems on which humanity and countless other species depend.

The Ecological Footprint Indicator, which compares humanity's ecological impact with the amount of productive land and sea area available to supply key ecosystem services, shows that humanity now uses the resources and services of 1.3 Earths. In other words, people are using about a third more of Earth's capacity than is available, undermining the resilience of the very ecosystems on which humanity depends.[5] . . .

The shifts in one particular ecosystem service—climate regulation—are especially disturbing. After remaining at stable levels for the past 1,000 years at about 280 parts per million, atmospheric concentrations of carbon dioxide (CO_2) are now at 385 parts per million, driven by a growing human population consuming ever more fossil fuels, eating more meat, and converting more land to agriculture and urban areas. The Intergovernmental Panel on Climate Change found that climate change due to human activities is causing major disruptions in Earth's systems. If greenhouse gas emissions are not curbed, disastrous changes will occur in the next century.[6] . . .

The adoption of sustainable technologies should enable basic levels of consumption to remain ecologically viable. From Earth's perspective, however, the American or even the European way of life is simply not viable. A recent analysis found that in order to produce enough energy over the next 25 years to replace most of what is supplied by fossil fuels, the world would need to build 200 square meters of solar photovoltaic panels every second plus 100 square meters of solar thermal every second plus 24 3-megawatt wind turbines every hour nonstop for the next 25 years. All of this would take tremendous energy and materials—ironically frontloading carbon emissions just when they most need to be reduced—and expand humanity's total ecological impact significantly in the short term.[7]

Add to this the fact that population is projected to grow by another 2.3 billion by 2050 and even with effective strategies to curb growth will probably still grow by at least another 1.1 billion before peaking. Thus it becomes clear that while shifting technologies and stabilizing population will be essential in creating sustainable societies, neither will succeed without considerable changes in consumption patterns, including reducing and even eliminating the use of certain goods, such as cars and airplanes, that have become important parts of life today for many. Habits that are firmly set—from where people live to what they eat—will all need to be altered and in many cases simplified or minimized. These, however, are not changes that people will want to make, as their current patterns are comfortable and feel "natural," in large part because of sustained and methodical efforts to make them feel just that way.[8]

In considering how societies can be put on paths toward a sustainable future, it is important to recognize that human behaviors that are so central to modern cultural identities and economic systems are not choices that are fully in consumers' control. They are systematically reinforced by an increasingly dominant cultural paradigm: consumerism. . . .

Institutional Roots of Consumerism

As long ago as the late 1600s, societal shifts in Europe began to lay the groundwork for the emergence of consumerism. Expanding populations and a fixed base of land, combined with a weakening of traditional sources of authority such as the church and community social structures, meant that a young person's customary path of social advancement—inheriting the family plot or apprenticing in a father's trade—could no longer be taken for granted. People sought new avenues for identity and self-fulfillment, and the acquisition and use of goods became popular substitutes.[9]

Meanwhile, entrepreneurs were quick to capitalize on these shifts to stimulate purchase of their new wares, using new types of advertising, endorsements by prominent people, creation of shop displays, "loss-leaders" (selling a popular item at a loss as a way to pull customers into a store), creative financing options, even consumer research and the stoking of new fads. For example, one eighteenth-century British pottery manufacturer, Josiah Wedgwood, had salespeople drum up excitement for new pottery designs, creating demand for newer lines of products even from customers who already had a perfectly good, but now seemingly outdated, set of pottery.[10]

Still, traditional social mores blocked the rapid advance of a consumerist mindset. Peasants with extra income traditionally would increase landholdings or support community works rather than buy new fashions or home furnishings—two of the earliest consumer goods. Workers whose increased productivity resulted in greater pay tended to favor more leisure time rather than the wealth that a full day at increased pay might have brought them.[11]

But over time the emerging consumerist orientation was internalized by a growing share of the populace—with the continued help of merchants and traders—redefining what was understood as natural. The universe of "basic necessities" grew, so that by the French Revolution, Parisian workers were demanding candles, coffee, soap, and sugar as "goods of prime necessity" even though all but the candles had been luxury items less than 100 years earlier.[12]

By the early 1900s, a consumerist orientation had become increasingly embedded in many of the dominant societal institutions of many cultures—from businesses and governments to the media and education. And in the latter half of the century, new innovations like television, sophisticated advertising techniques, transnational corporations, franchises, and the Internet helped institutions to spread consumerism across the planet.

Arguably, the strongest driver of this cultural shift has been business interests. On a diverse set of fronts, businesses found ways to coax more consumption out of people. Credit was liberalized, for instance, with installment payments, and the credit card was promoted heavily in the United States, which led to an almost 11-fold increase in consumer credit between 1945 and 1960. Products were designed to have short lives or to go out of style quickly (strategies called, respectively, physical and psychological obsolescence). And workers were encouraged to take pay raises rather than more time off, increasing their disposable incomes.[13]

Perhaps the biggest business tool for stoking consumption is marketing. Global advertising expenditures hit $643 billion in 2008, and in countries like China and India they are growing at 10 percent or more per year. In the United States, the average "consumer" sees or hears hundreds of advertisements every day and from an early age learns to associate products with positive imagery and messages. Clearly, if advertising were not effective, businesses would not spend 1 percent of the gross world product to sell their wares, as they do. And they are right: studies have demonstrated that advertising indeed encourages certain behaviors and that children, who have difficulty distinguishing between advertising and content, are particularly susceptible. As one U.S. National Academy of Sciences panel found, "food and beverage marketing influences the preferences and purchase requests of children, influences consumption at least in the short term, is a likely contributor to less healthful diets, and may contribute to negative diet-related health outcomes and risks among children and youth."[14]

In addition to direct advertising, product placement—intentionally showing products in television programs or movies so that they are positively associated with characters—is a growing practice. Companies spent $3.5 billion placing their products strategically in 2004 in the United States, four times the amount spent 15 years earlier. And, like advertising, product placements influence choices. Research has found, for example, a causal relationship between cigarette smoking in the movies and the initiation of this behavior in young adults in a "dose-response" manner, meaning that the more that teenagers are exposed to cigarette smoking in the movies, the more likely they are to start smoking.[15]

Other clever marketing efforts are also increasingly common tools. In "word of mouth" marketing, people who are acting as unpaid "brand agents" push products on unsuspecting friends or acquaintances. In 2008, U.S. businesses spent $1.5 billion on this kind of marketing, a number expected to grow to $1.9 billion by 2010. One company, BzzAgent, currently has 600,000 of these brand agents volunteering in its network; they help to spread the good word about new products—from the latest fragrance or fashion accessory to the newest juice beverage or coffee drink—by talking about them to their friends, completing surveys, rating Web sites, writing blogs, and so on. In Tokyo, Sample Lab Ltd. recently brought this idea to a new level with a "marketing café" specifically created to expose consumers to samples of new products. Companies now even harness anthropologists to figure out what drives consumers' choices, as Disney did in 2009 in order to better target male teens, one of their weaker customer bases.[16]

Any of these marketing strategies, taken alone, stimulates interest in a single good or service. Together these diverse initiatives stimulate an overall culture of consumerism. As economist and marketing analyst Victor Lebow explained in the *Journal of Retailing* over 50 years ago, "A specific advertising and promotional campaign, for a particular product at a particular time, has no automatic guarantee of success, yet it may contribute to the general pressure by which wants are stimulated and maintained. Thus its very failure may serve to fertilize this soil, as does so much else that seems to go down the drain." Industries, even as they pursue limited agendas of expanding sales for their products, play a significant role in stimulating consumerism. And whether intentionally or not, they transform cultural norms in the process.[17]

The media are a second major societal institution that plays a driving role in stimulating consumerism, and not just as a vehicle for marketing. The media are a powerful tool for transmitting cultural symbols, norms, customs, myths, and stories. As Duane Elgin, author and media activist, explains: "To control a society, you don't need to control its courts, you don't need to control its armies, all you need to do is control its stories. And it's television and Madison Avenue that is telling us most of the stories most of the time to most of the people."[18]

Between television, movies, and increasingly the Internet, the media are a dominant form of leisure time activity. In 2006, some 83 percent of the world's population had access to television and 21 percent had access to the Internet. In countries that belong to the Organisation for Economic Cooperation and Development, 95 percent of households have at least one television, and people watch about three to four hours a day on average. Add to this the two to three hours spent online each day, plus radio broadcasts, newspapers, magazines, and the 8 billion movie tickets sold in 2006 worldwide, and it becomes clear that media exposure consumes anywhere from a third to half of people's waking day in large parts of the world.[19]

During those hours, much of media output reinforces consumer norms and promotes materialistic aspirations, whether directly by extolling the high-consumption lives of celebrities and the wealthy or more subtly through stories that reinforce the belief that happiness comes from being better off financially, from buying the newest consumer gadget or fashion accessory, and so on. There is clear evidence that media exposure has an impact on norms, values, and preferences. Social modeling studies have found connections between such exposure and violence, smoking, reproductive norms, and various unhealthy behaviors. One study found that for every additional hour of television people watched each week, they spent an additional $208 a year on stuff (even though they had less time in a day to spend it).[20]

Government is another institution that often reinforces the consumerist orientation. Promoting consumer behavior happens in myriad ways—perhaps most famously in 2001 when U.S. President George W. Bush, U.K. Prime Minister Tony Blair, and several other western leaders encouraged their citizens to go out and shop after the terrorist attacks of September 11th. But it also happens more systemically. Subsidies for particular industries—especially in the transportation and energy sectors, where cheap oil or electricity has ripple effects throughout the economy—also work to stoke consumption. And to the extent that manufacturers are not required to internalize the environmental and social costs of production—when pollution of air or water is unregulated, for example—the cost of goods is artificially low, stimulating their use. Between these subsidies and externalities, total support of polluting business interests was pegged at $1.9 trillion in 2001.[21]

Some of these government actions are driven by "regulatory capture," when special interests wield undue influence over regulators. In 2008, that influence could be observed in the United States through the $3.9 billion spent on campaign donations by business interests (71 percent of total contributions) and the $2.8 billion spent by business interests to lobby policymakers (86 percent of total lobbying dollars).[22]

A clear example of official stimulation of consumption came in the 1940s when governments started to actively promote consumption as a vehicle for development. For example, the United States, which came out of World War II relatively unscathed, had mobilized a massive war-time economy—one that was poised to recede now that the war was over. Intentionally stimulating high levels of consumption was seen as a good solution to address this (especially with the memory of the Great Depression still raw). As Victor Lebow explained in 1955, "our enormously productive economy demands that we make consumption our way of life, that we convert the buying and use of goods into rituals, that we seek our spiritual satisfactions, our ego satisfactions, in consumption."[23]

Today, this same attitude toward consumption has spread far beyond the United States and is the leading policy of many of the world's governments. As the global economic recession accelerated in 2009, wealthy countries did not see this as an opportunity to shift to a sustainable "no-growth" economy—essential if they are to rein in carbon emissions, which is also on the global agenda—but instead primed national economies with $2.8 trillion of new government stimulus packages, only a small percentage of which focused on green initiatives.[24] . . .

Cultivating Cultures of Sustainability

Considering the social and ecological costs that come with consumerism, it makes sense to intentionally shift to a cultural paradigm where the norms, symbols, values, and traditions encourage just enough consumption to satisfy human well-being while directing more human energy toward practices that help to restore planetary well-being.

In a 2006 interview, Catholic priest and ecological philosopher Thomas Berry noted that "we might summarize our present human situation by the simple statement: In the 20th century, the glory of the human has become the desolation of the Earth. And now, the desolation of the Earth is becoming the destiny of the human. From here on, the primary judgment of all human institutions, professions, and programs and activities will be determined by the extent to which they inhibit, ignore, or foster a mutually enhancing human-Earth relationship." Berry made it clear that a tremendous shift is necessary in society's institutions, in its very cultures, if humans are to thrive as a species long into the future. Institutions will have to be fundamentally oriented on sustainability.[25]

How can this be done? In an analysis on places to intervene in a system, environmental scientist and systems analyst Donella Meadows explained that the most effective leverage point for changing a system is to change the paradigm of the system—that is to say, the shared ideas or basic assumptions around which the system functions. In the case of the consumerism paradigm, the assumptions that need to change include that more stuff makes people happier, that perpetual growth is good, that humans are separate from nature, and that nature is a stock of resources to be exploited for human purposes.[26]

Although paradigms are difficult to change and societies will resist efforts to do so, the result of such a change can be a dramatic transformation of the system. Yes, altering a system's rules (with legislation, for instance) or its flow rates (with taxes or subsidies) can change a system too, but not as fundamentally. These will typically produce only incremental changes. Today more systemic change is needed.[27]

Cultural systems vary widely, as noted earlier, and so too would sustainable cultures. Some may use norms, taboos, rituals, and other social tools to reinforce sustainable life choices; others may lean more on institutions, laws, and technologies. But regardless of which tools are used, and the specific result, there would be common themes across sustainable cultures. Just as a consumerism paradigm encourages people to define their well-being through their consumption patterns, a sustainability paradigm would work to find an alternative set of aspirations and reinforce this through cultural institutions and drivers.

Ecological restoration would be a leading theme. It should become "natural" to find value and meaning in life through how much a person helps restore the planet rather than how much that individual earns, how large a home is, or how many gadgets someone has.

Equity would also be a strong theme. As it is the richest who have some of the largest ecological impacts, and the very poorest who often by necessity are forced into unsustainable behaviors like deforestation in a search for fuelwood, more equitable distribution of resources within society could help to curb some of the worst ecological impacts. Recent research also shows that societies that are more equitable have less violence, better health, higher literacy levels, lower incarceration rates, less obesity, and lower levels of teen pregnancy—all substantial bonus dividends that would come with cultivating this value.[28]

More concretely, the role of consumption and the acceptability of different types of consumption could be altered culturally as well. Again, while the exact vision of this will vary across cultural systems, three simple goals should hold true universally.

First, consumption that actively undermines well-being needs to be actively discouraged. The examples in this category are many: consuming excessive processed and junk foods, tobacco use, disposable goods, and giant houses that lead to sprawl and car dependency and to such social ills as obesity, social isolation, long commutes, and increased resource use. Through strategies such as government regulation of choices available to consumers, social pressures, education, and social marketing, certain behaviors and consumption choices can be made taboo. At the same time, creating easy access to healthier alternatives is important—such as offering affordable, easily accessible fruits and vegetables to replace unhealthy foods.[29]

Second, it will be important to replace the private consumption of goods with public consumption, the consumption of services, or even minimal or no consumption when possible. By increasing support of public parks, libraries, transit systems, and community gardens, much of the unsustainable consumption choices today could be replaced by sustainable alternatives—from borrowing books and traveling by bus instead of by car to growing food in shared gardens and spending time in parks.

The clearest example of this is transportation. Reorganizing infrastructure to support walkable neighborhoods and public transit could lead to a dramatic reduction in road transportation—which pollutes locally, contributes about 17 percent to total greenhouse gas emissions, and leads to 1.3 million deaths from accidents each year. The centrality of cars is a cultural norm, not a natural fact—cultivated over decades by car interests. But this can once again be redirected, extracting cars from cities, as Masdar in Abu Dhabi, Curitiba in Brazil, Perth in Australia, and Hasselt in Belgium have already started to demonstrate. For example, the Hasselt city council, facing rapid growth in car usage and budget shortfalls, decided in the mid-1990s to bolster the city's public transit system and make it free for all residents instead of building another expensive ring road. In the 10 years since then, bus ridership has jumped 10-fold, while traffic has lessened and city revenues have increased from an enlivened city center.[30]

Third, goods that do remain necessary should be designed to last a long time and be "cradle to cradle"—that is, products need to eliminate waste, use renewable resources, and be completely recyclable at the end of their useful lives. As Charles Moore, who has followed the routes of plastic waste through oceans, explains, "Only we humans make waste that nature can't digest," a practice that will have to stop. The cultivation of both psychological and physical obsolescence will need to be discouraged so that, for example, a computer will stay functional, upgradable, and fashionable for a decade rather than a year. Rather than gaining praise from friends for owning the newest phone or camera, having an "old faithful" that has lasted a dozen years will be celebrated.[31]

Having a vision of what values, norms, and behaviors should be seen as natural will be essential in guiding the reorientation of cultures toward sustainability. Of course, this cultural transformation will not be easy. Shifting cultural systems is a long process measured in decades, not years. Even consumerism, with sophisticated technological advances and many devoted resources, took centuries to become dominant. The shift to a culture of sustainability will depend on powerful networks of cultural pioneers who initiate, champion, and drive forward this new, urgently needed paradigm.[32]

EXPLORING THE ISSUE

Are Western Values, Ethics, and Dominant Paradigms Compatible with Sustainability?

Critical Thinking and Reflection

1. Is energy from fossil fuels a necessary evil for emerging economies?
2. Will China overtake the United States as the number one economy of the world?
3. Will China become a dust bowl?
4. Is renewable energy a feasible option for emerging economies?
5. How will China effect resource consumption in the world?

Is There Common Ground?

Westernization, modernization, mand globalization are all terms that embody the values of scientific advancement, the sanctity of private property, free trade, and economic growth. These values have produced a global lifestyle for developed countries based on material affluence and convenience that developing countries such as India and China seek to emulate. Is it possible for a world of nine billion people by 2050 to add another two billion automobiles using existing fossil-fuel based technology and not produce environmental disaster? Can and should the developing countries follow the same development model of developed countries? Does the global community really have any tools or resources that can stop this convergence with all of its environmental consequences? Perhaps a common ground can be reached based on the ethic of sustainable "global harmony"—an ethic that encourages both developed and developing economies to consider their global environmental impacts when developing their future development goals. This will need to be done without significantly curtailing the economic development objectives of developing countries that seek to bring millions of people out of poverty. This new global ethic of sustainable "global harmony" will need to harness the applied sustainability tools of a environmentally benign technology and incentive program that rewards intelligent sustainable development.

Internet References

Aggarwal, R.M., 2005. "Globalization, local ecosystems, and the rural poor." *World Development* Vol. 34, No. 8, (2006) pp. 1405–1418.

Birdyshaw, E., and Ellis, C., 2007. "Privatizing an open-access resource and environmental degradation." *Ecological Economics* 61 (2007), pp. 469–477.

Kilbourne, W. E., Beckmann, S. C., Thelen, E. , 2002. "The role of dominant social paradigm in environmental attitudes a multinational examination." *Journal of Business Research* 55 (2002), pp. 193–204.

Orecchiay, C., 2010. "'Consumerism' and environment: does consumption behavior affect environmental quality?" CEIS Working Paper No. 261, November 1, 2007.

Rees, W.E., 2006. "Globalization, trade and migration: Undermining sustainability." *Ecological Economics* 59 (2006), pp. 220–225.

Rodgers, C., 2009. "Property rights, land use and the rural environment: A case for reform." *Land Use Policy* 26S (2009), pp. S134–S141.

Udo, V. E., and Jansson, M.P., 2009. "Bridging the gaps for global sustainable development: A quantitative analysis." *Journal of Environmental Management* 90 (2009), pp. 3700–3707.

This is the website of the International Monetary Fund. It provides good and credible information about economic scenarios of the world in general and different countries of the world in particular.

http://www.imf.org

The website of ScienceDirect provides the latest journal articles about science. The latest journal articles about energy, both alternative and renewable, can be searched at this website.

http://www.sciencedirect.com

The website of environmentcomplete provides the latest journal articles about fields related to the environment. Journal articles about sustainability can be searched at this website.

http://www.environmentcomplete.com

ISSUE 4

Does Sustainability Mean a Lower Standard of Living?

YES: Will Wilkinson, from "In Pursuit of Happiness Research: Is It Reliable? What Does It Imply for Policy?" *Policy Analysis* (April 11, 2007)

NO: Saamah Abdallah et al., from "Unhappy Planet Index 2.0: Why Good Lives Don't Have to Cost the Earth" (2009), http://happyplanetindex.org

Learning Outcomes

After reading this issue, you should be able to:

- Understand the difference between one's standard of living and quality of life.
- Know the problems with measuring happiness.
- Understand the political issues with the Easterlin Paradox.
- Understand the need to start using happiness indexes over just the GDP.

ISSUE SUMMARY

YES: Will Wilkinson, a policy analyst at the Cato Institute, staunchly supports the economist's perspective that happiness and standard of living are related to economic growth.

NO: British psychologists Saamah Abdallah and Sam Thompson, writing for the New Economics Foundation who developed the Happy Planet Index, argue that we need to get away from focusing on GDP and instead measure a successful society by supporting life satisfaction that doesn't cost the earth.

A common perception is that only through a reduction in the use of natural resources and energy, in essence, a self-imposed no-growth focus, can societies reduce their ecological footprint sufficiently enough to become sustainable. Does this mean that sustainability requires a lower standard of living? The answer would be yes if we define standard of living solely by measuring the gross domestic product (GDP), the primary scale by which the Western world judges growth and development. The answer becomes no if you include quality-of-life, nonmonetary indicators—life expectancy, access to nutritious food, safe water supply, and availability of medical care. Generally, sustainability deals with people's well-being and not just with how much they can produce, but well-being is very subjective and difficult to measure. The GDP is, in contrast, very easy to measure and, according to some researchers, very closely related to well-being.

In 1974, Richard Easterlin found that, within a country, a citizen's happiness increased as their incomes increased. But when comparing countries, happiness increased only until their subsistence was taken care of, and did not increase substantially beyond that, regardless of income. He found that as people accumulate more, a phenomenon called the "hedonic treadmill" or "affluenza" overcomes them, where one quickly develops a take-for-granted attitude toward money and desires more, instead of being grateful for what they have gotten. We get greedy and insatiable. Those afflicted with affluenza "are inclined to be less satisfied with their lives in general than those who focus their energies elsewhere"(Easterlin, 1991).

Since the Easterlin Paradox, countries such as Bhutan, Australia, China, Thailand, and the United Kingdom have developed happiness indexes. Some studies have been affecting policy decisions. A study in England suggests taxing "conspicuous consumption" to make more people happier by reducing envy. Bhutan has decided 60 percent of the country should remain forested. Happiness indexes are showing up in many departments of the United States as well—U.S. Bureau of Labor Statistics, U.S. Department of Energy Benefits Analysis, and at Yale University.

The indicators used to define well-being have also been refined. Leiserowitz et al. have found that happiness is affected by trends in living standards, educational opportunity, availability of well-paying jobs, working conditions, health and survival of self, children, family, and others (Leiserowitz et al., 2006). Spirituality has proven to be a major factor in an individual's happiness. A recent Gallup survey of more than 550,000 people found religious Americans to have the highest rates of well-being (Banks, 2010). Martha Nussbaum, a Western philosopher, developed the capabilities approach, which is based on basic "substantive freedoms" rather than emotions like happiness and resources like income and commodities. Nussbaum suggests all democracies should support people's ability to live life fully, with bodily health, bodily integrity; ability to use their senses, imagination and thought; emotions; practical reason; ability to affiliate with other humans with self-respect, and with animals, plants, and the world of nature; ability to play; and ability to have control over one's own environment both politically and materially. Her list has inspired the UN's Human Development Index (HDI) and is very influential in developing policy.

The results, however, have political implications, which is a problem with the science of happiness, because it relies on surveys and not empirical data. More research is needed in measuring happiness more precisely. In her article, "The Joy of Economics" (2007), Rana Foroohar suggests that "happiness research could be used to advance authoritarian aims, and . . . that happiness studies should be used to inform only individual choices, not public policy." Ruut Veenhoven's studies generally found that the social democratic countries where incomes are the most equal were the happiest. Researchers have concluded that working less and playing more will make you happier. But hard working America is 17th on the happiness scale, and France, with long vacations is 39th, disproving that hypothesis. Also, if one aspires to achievement, one tends to be happier whether successful or not. Hence, Foroohar suggests that GDP can be seen as a very good indicator of happiness.

Ruut Veenhoven, an eminent happiness scientist, in his paper "Wealth and Happiness Revisited" (2003), tests how income affects happiness. The survey also found that increases in national income do increase national happiness, though not by much. There is more happiness gained in the short term than in the long term. Also, there is more happiness gained per dollar in the countries with lower GDPs than with higher GDPs.

Stevenson and Wolfers find that money does buy happiness. Countries get happier as their GDP rises, and happiness in wealthier countries is more resilient. They point out that weather, culture, and other influences may explain discrepancies. In relation to living sustainably, Stevenson and Wolfers note that utility can increase as consumption decreases (Stevenson, 2008, p. 89). Another interesting point is that happiness is not passed genetically. Through evolution, social stature has meant more for reproduction than happiness. When we ask our genes if we should take an unpleasant, but high paying job—the job wins (Stevenson, 2008, p. 90).

The "2010 Environmental Performance Index" (2010) measures and compares a number of indicators of environmental health for human habitation and ecosystem vitality in 163 countries. The purpose of this report is for policymakers to analyze empirical information for the use of policy formation. The categories and their indicators reflect core environmental issues and trends useful in policymaking in national, regional, and international levels. The authors find that, generally, the countries with a higher GDP do have healthier environments for people. The issue is that all but two of these wealthier countries also have low ecosystem vitality, which is more closely related to sustainability. In fact, environmental health and ecosystem vitality scores actually have a negative correlation. The ecosystem vitality scores are very evenly distributed between wealthy and poor countries. The higher vitality scores are the result of good environmental management and/or lack of development.

"Sustainability Values, Attitudes, and Behaviors: A Review of Multinational and Global Trends" by Anthony A. Leiserowitz, Robert W. Kates, and Thomas M. Parris looks at what values, attitudes, and behaviors are necessary to transition into a sustainable global society without a revolution. They found the most important value to sustain is the life support systems of the environment. Too much attention on economic development has overrun environmental protection and social development. Our behavior, however, does not always carry out our values because our infrastructure, built for a growth economy, presents obstacles, that is, laws, regulations, built on environment.

In their paper, "Personal and Planetary Well-Being" (2008), the authors Jeffrey Jacob, Emily Jovic, and Merlin B. Brinkerhoff conclude that spiritually inclined people can improve their personal well-being and be more apt to be concerned with the planet's well-being through practicing mindful meditation, which is similar to Buddhist meditation practices. Although quality of life is difficult to measure, the practice of meditation has spread around the world, because people find it beneficial.

The article for the YES position, that sustainability can mean a lower standard of living, is presented by policy analyst Will Wilkinson who defends the use of GDP as the barometer for determination of well-being. He suggests that there is no other single, more easily used method of tracking well-being than the GDP. He notes that recent studies found a significant correlation between the GDP and other indicators including health, longevity, education, and volunteering. The GDP also had a significant negative correlation with income inequality, poverty, and child mortality.

The article for the NO position, that sustainability does not mean a lower standard of living, is presented by British psychologists Abdallah and Thompson. They suggest that instead of concentrating on GDP as a measure of well-being societies need to emphasize goals of life satisfaction and preserving the natural resources upon which human systems depend. In their study, they found that once one's subsistence needs are met, that income makes little difference in happiness/life expectancy. Generally, a country's ecological footprint is closely related to their income, but generally happiness increases slowly as income increases beyond a person's subsistence needs. Living sustainably may very well mean a lower income, but people are then given the chance to turn their attention away from money and draw happiness from their social, spiritual, and environmental lives.

Will Wilkinson

In Pursuit of Happiness Research: Is It Reliable? What Does It Imply for Policy?

Introduction: Is the United States a Failure?

"There is a paradox at the heart of our lives," writes Richard Layard, head of the London School of Economics Center for Economic Performance and member of the British House of Lords. "As Western societies have got richer," Layard tells us, "their people have become no happier."[1] Psychologist of happiness David Myers opens his book, *The American Paradox,* on a Dickensian note: "It is the worst of times, and the best of times." We owe the "worst of times," according to Myers, to "radical individualism" and "libertarianism," both civil and economic.[2] Journalist Gregg Easterbrook puts it this way: "We live in a favored age and do not feel favored."[3] His bestselling book, *The Progress Paradox,* set out to explain "why capitalism and liberal democracy, both of which justify themselves on the grounds that they produce the greatest happiness for the greatest number, leave so much dissatisfaction in their wake."[4]

All of those works and many more tap into a rapidly growing body of research on the correlates of human happiness. Starting roughly with University of Southern California economist Richard Easterlin's watershed 1973 paper showing that average happiness levels reported by Americans had not risen for decades despite a doubling in average incomes, economists, sociologists, and psychologists have been busy canvassing the world, handing out "life satisfaction" surveys and customized "experience sampling" Palm Pilots and then running the data through computers with cutting-edge statistical software to tease out the determinants of a satisfying life.[5]

How important is wealth to happiness? How important is marriage? Parenthood? Job satisfaction? Leisure time? Health? The rate of unemployment? The rate of economic growth? Democratic institutions? Social safety nets? The happiness researchers even have their own journal, *The Journal of Happiness Studies,* where all of this, and more, is analyzed at length.

Layard is sufficiently confident in the quality of happiness research to bless it as a "new science." It is claimed that we now *know,* at long last, what really makes people happy. Geoffrey Miller, a psychologist at the University of New Mexico, writes: "In the last ten years, psychology has finally started to deliver the goods—hard facts about what causes human happiness."[6] Scholars like Layard have not hesitated to base dramatic policy recommendations on our alleged newfound facts. Layard argues, for example, that a government that cares about the pursuit of happiness will levy higher taxes on income, impose strict controls on advertising and marketing, mandate generous periods of paid parental leave, and implement "radical" mandatory public school courses covering aspects of life generally left to parents aiming "to produce a happier generation of adults than the current generation,"and much else besides.[7] . . .

An article on happiness research in the *New York Times* reports that George Loewenstein, a leader in "behavioral economics" at Carnegie Mellon University, "doesn't see how anybody could study happiness and not find himself leaning left politically."[8]

Perhaps the most compelling left-leaning arguments based on happiness research are those, such as Robert Frank's in his book *Luxury Fever,* which deemphasize the importance of *absolute* material wealth to happiness and stress instead the importance of *relative* position in the distribution of income and social status. Whereas happiness research has shown a flat trend in happiness *over time,* it also shows that *at any time* wealthier people are more likely to say they are happy. However, so the argument goes, if we *all* run harder to pull ahead in the race for the benefits of higher relative standing, those ahead will just run harder too. In the end, the frantic pace will have left us all harried and exhausted, and average happiness will have remained unchanged.

"Every time [some people] raise their relative income (which they like)," Layard writes,"they lower the relative income of other people (which those people dislike). This is an 'external disbenefit' imposed on others, a form of physical pollution."[9] Layard's proposed solution is a tax on "the polluting activity" or, as economists call it, the "negative externality." The polluting activity here is nothing less than your and my working hard to make more money. But, if it is *relative* standing that matters, the increase in total wealth will not increase happiness on average. There will always be a top half and bottom half. A tax that reduces the monetary benefits of labor and so encourages everyone to ease up in unison will slow the pace of life and reduce incomes. This, the argument goes, will do no harm to happiness, but the time and energy freed to pursue the pleasures of family, friends, and leisure will do a world of good. . . .

The aim of this paper is to demonstrate that happiness research poses no threat to U.S. ideals as they have been historically interpreted and are embodied (albeit imperfectly) in our present socioeconomic system. Happiness research is seriously hampered by confusion and disagreement about the definition of its subject as well as the limitations inherent in current measurement techniques. Happiness research in its present state cannot be relied on as an authoritative source for empirical information about happiness, which, in any case, is not a simple empirical phenomenon but a cultural and historical moving target. Furthermore, happiness is not the only element of human well-being or of a valuable life. At the very least, believing that it is has no standing as a scientific proposition, and there is no *liberal* moral justification for holding up happiness as the sole standard for evaluating policy in a contentiously pluralistic society. Yet the problems with the political uses of happiness research run deeper than methodology. Even if we grant that the findings of happiness research do shed some light on the state of human well-being, few of the main alleged implications for public policy actually follow from a fair reading of the evidence. In a nutshell: even if we put aside charges of questionable science and bad moral philosophy, the United States still comes off as a glowing success in terms of happiness. If any nation deserves an "A" on the "American test," the United States does. . . .

Is There Something Wrong with the United States?

Happiness research presently falls short as good science and fails to get off the ground as an adequate ethical standard for evaluating public policy. These conclusions admittedly follow from a fairly complex train of reasoning starting from a number of contestable assumptions that are impossible to vindicate fully in such a short space. Maybe you're not entirely convinced. That's okay.

Even if you don't buy the foregoing analysis of the complex methodological and philosophical problems that dog happiness research; even if you remain convinced that recent survey-based scholarship really does give us highly reliable and useful information on the determinants of happiness and well-being; even if you think happiness so conceived really is the primary target at which policymakers ought to aim, it remains possible to accept survey-based happiness research at face value and show that the U.S.-style socioeconomic model is not only a recipe for immense riches, which no one disputes, but a winning recipe for happiness.

A casual glance at the comparative international happiness data is enough to make us wary of the claim that there is some special problem with the United States relative to other nations in terms of happiness.

The United States is evidently among the world's happiest nations, on par with Canada, Australia, New Zealand, Ireland, Norway, Sweden, Switzerland, and a few surprises, like El Salvador and Nigeria. Notably, the largest European social democracies, Germany and France, fall one or two ranks below the United States.

In various rankings using different surveys, the United States consistently ranks from the mid-teens to the mid-twenties out of more than 200 countries—in or around the 90th percentile—in terms of average self-reported happiness. A 2005 Harris poll using the Eurobarometer questions found the United States to be happier than every European country, other than Denmark. Or consider Table 1, which lists the top 50 countries in self-reported happiness according to the World Values Survey. The United States ranks higher than Sweden, Norway, Belgium, Finland, Germany, and France. So much for Benjamin Radcliff's claim that "life satisfaction should increase as we move from less to more social democratic welfare states."[10] If we take the data at face value, the obvious conclusion is that the United States is among the happiest places in the world.

Table 1

Subjective Well-Being Rankings of 50 Countries

1.	*Puerto Rico*	*4.67*	26.	**France**	**2.61**
2.	*Mexico*	*4.32*	27.	*Argentina*	*2.61*
3.	**Denmark**	**4.24**	28.	Vietnam	2.59
4.	**Ireland**	**4.16**	29.	*Chile*	*2.53*
5.	**Iceland**	**4.15**	30.	Taiwan	2.25
6.	**Switzerland**	**4.00**	31.	*Domin.Rep.*	*2.25*
7.	**N. Ireland**	**3.97**	32.	*Brazil*	*2.23*
8.	**Netherlands**	**3.86**	33.	**Spain**	**2.13**
9.	**Canada**	**3.76**	34.	**Israel**	**2.08**
10.	**Austria**	**3.69**	35.	**Italy**	**2.06**
11.	*El Salvador*	*3.67*	36.	**E. Germany**	**2.02**
12.	*Venezuela*	*3.58*	37.	Slovenia	2.02
13.	**Luxembourg**	**3.52**	38.	*Uruguay*	*2.02*
14.	**United States**	**3.47**	39.	*Portugal*	*1.99*
15.	**Australia**	**3.46**	40.	**Japan**	**1.96**
16.	**New Zealand**	**3.39**	41.	Czech Rep	1.94
17.	**Sweden**	**3.36**	42.	South Africa	1.86
18.	Nigeria	3.32	43.	Croatia	1.55
19.	**Norway**	**3.25**	44.	Greece	1.45
20.	**Belgium**	**3.23**	45.	*Peru*	*1.32*
21.	**Finland**	**3.23**	46.	China	1.20
22.	**Saudi Arabia**	**3.01**	47.	South Korea	1.12
23.	**Singapore**	**3.00**	48.	Iran	0.93
24.	**Britain**	**2.92**	49.	Poland	0.84
25.	**W. Germany**	**2.67**	50.	Turkey	0.84

Source: Based on Ronald Inglehart, "Subjective Well-Being Rankings of 82 Societies," World Values Survey. http://www.worldvaluessurvey.org/Upload/5_wellbeingrankings.doc Latin American countries, which score higher than predicted given the quality of their economic and political institutions, are in italics. The wealthy OECD countries are in bold. Ex-communist countries are underlined. East and West Germany are scored separately to reflect the effects of their different institutional histories.

As noted in my introduction, psychologist Geoffrey Miller believes that "the utilitarian argument for the rich giving more of their money to the poor is now scientifically irrefutable."[11] Would Americans be happier with a larger and more generous welfare state? If the argument for downward redistribution is "irrefutable," then the evidence for it ought to shine forth in the data. According to Dutch sociologist Ruut Veenhoven, chief of the World Database of Happiness and founder and editor of the *Journal of Happiness Studies,* there is barely a flicker of a finding for a welfare-happiness connection:

> Contrary to expectation there appears to be no link between the size of the welfare state and the level of wellbeing within it. In countries with generous social security schemes people are not healthier or happier than in equally affluent countries where the state is less open-handed. Increases or reductions in social security expenditure are not related to a rise or fall in the level of health and happiness either.[12]

Another Dutch happiness researcher, Piet Ouweneel of Erasmus University, Netherlands, conjectured that at least the *unemployed* would have higher average well-being, according to a number of indicators, in nations that spent a larger percentage of GDP on welfare. But greater welfare spending had no statistically significant effect—*even on the happiness of the unemployed.* While larger welfare states generally do achieve lower levels of income inequality through redistribution. "This apparent income redistribution does not have any significant effect on the subjective well-being of the unemployed. The general picture is that they are neither happier nor healthier in welfare states."[13] Ouweneel concludes that "in first world nations there is no consistent pattern of social security levels having a positive effect on well-being indicators."[14]

By contrast, Notre Dame political scientist Benjamin Radcliff does find a small statistically significant positive effect of generous welfare spending on average happiness, and Harvard economist Rafael Di Tella finds a small boost from generous unemployment benefits. Ouweneel, however, criticizes both studies for comparing a very small set of countries. Further, he notes both were able to achieve statistical significance only by treating successive years in these countries as independent data points—a methodological faux pas.[15]

The most conservative inference to draw from the existing happiness literature is that if the redistributive openhandedness of the state has any effect on happiness at all, it is a surpassingly small one. When slightly different econometric techniques using slightly different datasets generate weak correlations in opposite directions, the correct lesson to draw is that the variable barely matters at all. If Americans are less happy on average than citizens of some other nations—and we are happier than all but a handful—the scope and generosity of the welfare state has little or nothing to do with it.

If relatively lavish welfare spending fails to increase happiness, other forms of government spending might nevertheless succeed. One reason big government could have a positive effect on happiness, aside from progressive redistribution, is that a higher rate of government spending as a percentage of GDP might indicate better provision of the kinds of public goods that unaided market institutions are often thought to be incapable of providing.[16] But economists Christian Bjornskov, Axel Dreher, and Justina Fischer find that "life satisfaction *decreases* with higher government spending."[17] Intriguingly, they also find that the "negative impact of the government is stronger in countries with a leftwing median voter"—which is to say, in places where voters most *want* big government.

Advocates of progressive taxation and income redistribution through welfare transfer payments often attempt to justify those policies not only in terms of the increased well-being of the least well-off, but in terms of the overall importance of reducing income inequality generally. In theory, high relative income and social status are important to happiness, and we are therefore aggravated by the conspicuous display of goods we cannot afford. If true, it would make sense for a leveling of incomes to have some positive effect on happiness. But, again, empirical evidence is hard to come by.

Alberto Alesina, Rafael Di Tella, and Robert MacCulloch have found that inequality in the United States has *no* effect on the self-reported happiness of the poor. "Probably the most striking result of all is the complete lack of any effect of inequality on the happiness of the American poor and the American left," they report. There is a very small statistically significant negative effect in the United States, but it is driven almost entirely by the effect of inequality on *the rich.* This is so striking in part because it is basically the reverse of the situation in Europe, where there is a much larger negative effect of inequality, due mostly to the dislike of inequality by the poor and left-wing voters.[18]

This contrast between Europe and the United States raises a crucial point about the interpretation of the effect of macroeconomic variables on happiness—namely, their effect is culturally and ideologically mediated. The authors argue that the different effects of inequality in Europe and the United States are due largely to different prevailing attitudes about income mobility. Whether or not their beliefs are realistic, Americans—even the poor and the left-wingers—have a strong faith in the possibility of upward mobility, at least compared to Europeans. Under those conditions, the fabulously rich demonstrate to other Americans just how astronomically high it is possible to rise and stand out as figures of admiration and emulation. (This is perhaps why rich-bashing populism is a perpetual electoral failure in U.S. politics.) The American rich also believe strongly in mobility, but where there are wide income disparities, they see all too clearly how far they could fall. Of course, it is possible to criticize Americans for having unrealistic beliefs about mobility, for they *are* generally unrealistic, especially among the poor. But if inequality has no negative effect on happiness independent of attitudes toward mobility, it seems that a happiness-promoting policymaker would want to encourage, not discourage, the American conviction in mobility.

By most measures, income inequality has been rising in the United States. However, inequality in *happiness* has declined. Rising income inequality, then, does not imply a widening gap in satisfaction with life. On the contrary, Americans are becoming more equal in happiness even as the income gap widens. Sociologist Jan Ott finds that rising average levels of happiness go together with decreasing levels of happiness *inequality* because the "level and equality of happiness depend eventually on the same institutional conditions." And the institutions of wealth creation are among the most important: "Wealth contributes to higher levels of happiness and creates ample possibilities to reduce inequality in happiness," Ott writes.[19]

The creation of wealth depends on a complex system of underlying economic, legal, and cultural institutions. Other things equal, nations that ensure their citizens' greater economic freedom are also wealthier. The United States is the most visible embodiment of the ideals of economic freedom on the world stage. However, emphasis on the importance of distinctively *economic* freedom in the happiness literature is relatively new, which is perhaps one reason the ideals of relatively unhampered markets and open exchange—ideals strongly associated with the United States on the world stage—have yet to get adequate emphasis in popular accounts of happiness research.

Ott finds that economic freedom as measured by both the Heritage Foundation and the Fraser Institute correlates strongly with high and highly equal levels of happiness—as strongly as almost any variable.[20] Veenhoven finds that economic freedom correlates more strongly with happy-life-years (HLY) than any variable other than wealth (as measured by purchasing power per capita) and degree of social tolerance (i.e., acceptance of pluralism).[21] And in the largest study on economic freedom and happiness yet conducted, economists Tomi Ovaska and Ryo Takashima find that economic freedom is the variable *most* highly correlated with self-reported happiness. According to Ovaska and Takashima:

> Compared to the GDP per capita measure, the index of economic freedom—personal choice, freedom to compete and the security of privately owned property as its core components—turned out to be about four times as important, as measured by elasticities. This indicates that the newly found interest of economics and of policymakers in measures of institutional quality is well placed. Based on the regression results, economic freedom holds some promise in serving as one of the policy tools that could be potentially used to increase the SWB of a nation's population.[22]

According to Ovaska and Takashima, "The results suggest that people unmistakably care about the degree to which the society where they live provides them opportunities and the freedom to undertake new projects, and make choices based on one's personal preferences."[23]

According to the Heritage Foundation's 2007 *Index of Economic Freedom,* the United States ranked fourth, behind Hong Kong, Singapore, and Australia. And according to the Fraser Institute-Cato Institute 2006 *Economic Freedom of the World Report,* the United States was in a three-way tie for third with Australia and Switzerland, behind Hong Kong and Singapore. The evidence is extremely strong that the outstanding level of economic freedom in the United States has a strong effect on its high showing in the international happiness comparisons. Any policy package aiming to improve American happiness should have initiatives to improve economic freedom at the forefront.

A fair look at a number of the most recent, and most sophisticated, happiness studies suggests that the United States, far from being a problem country, exemplifies many of the institutional virtues that strongly predict high levels of national self-reported happiness. Although the United States is routinely criticized by Europeans and its own political left for a stingy welfare state and high levels of income inequality, there is *almost no evidence* in the happiness literature to support the idea that Americans would be better off with either lower levels of income inequality or a policy of more generous welfare transfers. The high levels of economic growth and economic freedom in the United States are effectively increasing the average American level of happiness while decreasing inequality in life satisfaction between its citizens.

If we take the current international comparisons at face value, one very clear picture emerges: advanced, liberal-democratic market economies are the happiest places on Earth. However, if we descend from such rarefied heights of generality, the picture goes blurry. The data are too coarse to distinguish among packages of specific policies—to tell us, for example, that we would be happier with mandatory paid maternity leave, or with greater restrictions on the content of advertising to children. As with inequality, we will often find that the effect of policy on happiness is mediated by culturally specific beliefs and attitudes. So, beyond a general recommendation to increase economic freedom and eliminate policies that hinder economic growth . . . , there is almost no *specific* guidance here for a policymaker. The picture that does emerge from the data is most emphatically *not* a picture of American misery, nor does it even hint at a problem with America's conduciveness to happiness relative to the European social democracies. . . .

Getting Rich, Getting Happy

The relative position hypothesis also helps drive the animus toward economic growth. If we're grinding away to get ahead, and everybody gets richer, but money doesn't make us happier, and no one on average gets ahead anyhow, then what's the point of everybody getting richer? What's the point of a high rate of GDP growth? We could grind away a lot less, be a bit less rich, but also a bit happier. We should relax more instead: build model airplanes, spend time with the kids, adopt a highway, or whatever—and policy should help make this easier. Money isn't everything, and there's something wrong with a government that doesn't seem to understand that.

"GDP is a hopeless measure of welfare," Layard concludes. "For since the [Second World] War that measure has shot up by leaps and bounds, while the happiness of the population has stagnated."[24] Elsewhere he writes, "We desperately need to replace GDP, however adjusted, by more subtle measures of national wellbeing."[25] This also is the kind of thinking that led Andrew Oswald to write, "Economists' faith in the value of growth is diminishing. That is a good thing and will slowly make its way down into the minds of tomorrow's politicians."

It has, in fact, made its way to the minds of *today's* politicians. As British prime ministerial hopeful David Cameron announced last spring, "It's time we admitted that there's more to life than money, and it's time we focused not just on GDP, but on GWB—General Wellbeing." In Cameron's plea one can hear echoes of Robert Kennedy's famous attack on national income accounts as a measure of human well-being. Stumping for president just months before his tragic murder, Kennedy lamented that a measure like GDP

> does not allow for the health of our children, the quality of their education, or the joy of their play. It does not include the beauty of our poetry or the strength of our marriages; the intelligence of our public debate or the integrity of our public officials. It measures neither our wit nor our courage;neither our wisdom nor our learning; neither our compassion nor our devotion to our country; it measures everything, in short, except that which makes life worthwhile. And it tells us everything about America except why we are proud that we are Americans.[26]

Kennedy was right that national income statistics don't tell us much about "our wit" or"the joy of our children's play." Nevertheless, if we're looking for a single socioeconomic variable that tracks with most objective indicators of well-being, GDP per capita is hard to beat. Even if it does not measure everything that makes life worthwhile (because nothing does), it most definitely relates positively to measures of a lot of good things, including happiness. But before looking at the effects of money on happiness, I will examine how important high average individual wealth, as measured by GDP per capita, can be to non-subjective indicators of well-being.

A large recent study by OECD economists Romina Boarini, Asa Johansson, and Marco Mira d'Ecole focused on the relationship between GDP per capita and alternative measures of well-being in the OECD nations. The authors found significant positive correlations of GDP per capita with self-sufficiency, average years of schooling, life expectancy at birth, healthy life expectancy at birth, mortality risks, and volunteering. Further, GDP per capita was significantly negatively correlated with income inequality, relative poverty, child poverty, and child mortality.[27] As economists Vito Tanzi and Hamid R. Davoodi show, GDP per capita is also significantly positively correlated with lower levels of corruption—so GDP may have something to say about the "integrity of our public officials" after all.[28]

Given the serious charge that high-growth market societies erode "social capital" and fray the social fabric, it is important to note that even if GDP per capita is not significantly positively associated with most indicators of social cohesion other than rates of volunteerism and a decrease in crime, neither does it appear to accompany symptoms of social breakdown. According to the authors, "indicators of crime victimization, prisoners and suicides—as well as of divorces, drug use and road accidents—are not significantly correlated with GDP per capita," either positively or negatively.[29]

In his recent book, *The Moral Consequences of Economic Growth,* Harvard economist Benjamin Friedman emphasizes that in addition to the litany of its astonishing humanitarian benefits, economic growth is also a powerful force for the encouragement of broadly liberal social and political aims. "The value of a rising standard of living lies not just in the concrete improvements it brings to how individuals live," Friedman writes, "but in how it shapes the social, political, and ultimately moral character of a people. Economic growth—meaning a rising standard of living for the clear majority of citizens—more often than not fosters greater opportunity, tolerance of diversity, social mobility, commitment to fairness, and dedication to democracy."[30]

And as Tyler Cowen has detailed at length, wealthier societies produce more paintings, poems, films, songs, operas, and sculptures. They build more museums, support more symphonies, and patronize more artists than do less wealthy societies. Economists can't tell a skeptic about economic growth whether the poetry is *beautiful,* but at least there is more of it, and most of us, economists or not, recognize that much of it is in fact beautiful.[31]

So economic growth makes us healthier, better educated, and more public spirited; fosters social toleration; increases the integrity of our public institutions; and produces a surfeit of art and culture. But does economic growth make us *happier*?

It is impossible to review the happiness literature without constantly tripping over the fact that GDP per capita, or some other proxy for average wealth, dominates almost all variables in terms of the strength of correlation with a society's average happiness. As we have already seen, at any time and place, individuals with higher relative income are more likely to say they are "very happy." But, as we are constantly reminded, the idea that average happiness has not increased with average income is a bedrock finding of happiness research. It is also false.

On average, wealthier nations have happier people. . . . Moreover, the most recent statistical work on the relationship between wealth and happiness, using larger sets of data and more sophisticated techniques of analysis, show unequivocally that we are getting happier as we get richer.

In a recent debate with Richard Easterlin in the journal *Social Indicators Research,* Michael Hagerty and Ruut Veenhoven have argued that increasing wealth *is* making us happier. Much of the debate centers on small esoteric points of statistical methodology and how to rhetorically frame the conclusions. However, Veenhoven and Hagerty's methods do appear to be a marked improvement over most past happiness studies, and their well-argued interpretation of their findings goes mostly unchallenged by Easterlin, which augurs ill for the anti-growth crowd. They argue that the data are inconsistent with the predictions of strong relative position theories and that although adaptation does reduce the rate of increase in happiness, it does not wash out all absolute gains in happiness from increasing wealth.[32] There are non-relative and non-evaporating gains from wealth. They conclude:

> Happiness is apparently not a zero-sum game and can be raised by growth in national income. This has been a central but until recently untested belief of economists and public policy analysts. Not too long ago unhappiness was deemed the normal human condition. Since expulsion from Paradise, humans could only hope for happiness in the after-life. Promises of greater happiness in earthly existence were dismissed as overly simplified utopianism. The current research on happiness allows empirical tests of this, and has shown that entire nations can become happier with economic growth and its covariates.[33] . . .

Despite the apparently overwhelming evidence that wealthy, high-growth societies are the happiest places in the world, and only getting happier, there is no lack of hand-wringing about the spiritual emptiness of "materialism" in liberal market societies, and some of the hand-wringing is motivated by putatively scientific findings. In his 2004 book *The High Price of Materialism,* Knox College psychologist Tim Kasser presents his research with Richard Ryan showing that "extrinsically motivated" people who care predominantly about material acquisition are more likely to find themselves unhappy and dissatisfied with life than are "intrinsically motivated" people devoted to personally meaningful work and relationships. Kasser's research on the negative effects of "materialistic" value orientation on happiness seems sound and conforms to common sense.

However, Kasser barely takes a breath before taking an awesome leap in logic from the micro to the macro level of diagnosis. Because extrinsically motivated individuals with predominantly materialistic values are more likely to be unhappy, market societies, which create unparalleled opportunities for material accumulation and consumption and which motivate the production of goods and services others value extrinsically with profits and paychecks, must be unhappy.[34] But this is a simple non sequitur. Kasser boldly equivocates on the meaning of the word "materialistic," implying that consumer demand for material consumption requires a widespread "materialistic" attitude in his special theoretical sense. However, capitalist consumer societies—and markets in general—don't require materialistic monomania in order to operate. They require only that people want things, for good reasons or bad, and that they are willing to trade what they have produced to get them. Showing that there is a problem with materialistic monomania says nothing about capitalist societies, nor does it imply that denizens of capitalism are more likely to be materialistic than others.

Recent studies by Stephanie M. Bryant, Dan Stone, and Benson Weir have developed a new theoretical construct called Financial Self-Efficacy, which they define as "the belief that one can competently manage one's finances."[35] The authors find that individuals high in FSE are more likely to treat money as an *instrument* for the achievement of other, nonmaterialistic aims. People high in FSE tend to have higher levels of debt but more intrinsic motivation for carrying it (e.g., a student loan, a family home, a trip to a foreign country, etc.), and they are more likely to have high levels of life satisfaction. The upshot is clear: the aspiration to make and spend money is neither good nor bad. What matters is our *attitude* toward our financial goals, and their content. If we want money simply for its own sake, to impress friends, or to buy gadgets as palliatives for boredom and ennui—Kasser's "materialism"—money won't do us good. But if we regard money as a mere *tool* with which to achieve more meaningful ends, more money will help us do more of what we find meaningful.

University of Michigan political scientist Ronald Inglehart's work shows that nations with a rising level of per-capita GDP tend to shift culturally from "materialist" values, "which emphasize economic and physical security," to "post-materialist" values, "which emphasize self-expression and quality of life."[36] According to Inglehart, the cultural shift includes a significant time lag, because "to a large extent, one's basic values reflect the conditions that prevailed during one's pre-adult years."[37] Inglehart finds that there has been a *large* shift from materialist to post-materialist values in wealthy Western liberal market democracies.

> For example, in the earliest U.S. survey, materialists outnumbered postmaterialists by 24 percentage points; in West Germany, they outnumbered postmaterialists by 34 points. During the three decades following 1970, a major shift occurred: by the 1999–2001 surveys, postmaterialists had become more numerous than materialists in all nine countries.[38]

This, of course does not mean that younger generations spend all their time shopping for "fair trade" coffee and performing sun salutations (though there is surely more of that). But the shift away from economic scarcity increases the emphasis on self-definitional and self-expressive consumption. "The rise of postmaterialism does not mean that materialistic issues vanish," Inglehart and Christian Welzel write:

> The publics in postindustrial societies have developed more sophisticated forms of consumerism, materialism, and hedonism. . . . New forms of consumption no longer function primarily to indicate people's economic class. Increasingly, they are means of individual self-expression.[39]

Inglehart is not using "materialist" in precisely the same way as Kasser, but the rough idea—an emphasis on material acquisition as opposed to meaning—is the same. If you want fewer materialists, the way to go is to make *more material* readily available to people, at which point they'll stop worrying about it so much and start worrying instead about things like happiness and the meaning of life.

Many people seem to think that a government's emphasis on measurements like GDP indicate a kind of collective affirmation of materialist goals, encouraging a narrowly materialist attitude at war with more exalted values. But this is simply a mistake. The very *function* of money is to serve as a neutral medium of exchange. It is a shape-shifting embodiment of almost *any* value. The same $100 can be spent on a prostitute or donated to an HIV/AIDS clinic. The relative value neutrality of money is precisely why the measurement of per-capita wealth is well suited to pluralistic liberal societies; it doesn't beg many questions about competing conceptions of the good life. Money can't be converted into *anything* that someone might value, but it is of the nature of money to be *convertible* into a phenomenally broad range of values. Societies with high levels of average income and wealth are societies in which people have more resources at their disposal to achieve their aims, no matter what those aims might be, which is why it should be no surprise that, other things equal, people with more money are more satisfied. By measuring GDP, household wealth, and the like, government is *not* affirming one set of values over others. It is, in fact, embodying an ideal of liberal neutrality by measuring something that is valuable in varying degrees to all of us.

Conclusion

The United States is not failing the Founders' test. The happiness-based evidence points unambiguously to the conclusion that those of us lucky enough to live in the United States . . . are succeeding fairly well in the pursuit of happiness. Whether or not our Founders would recognize—or even *like*—their country, Americans are indeed living up to the promise of our founding.

So, we are left with a puzzle. If we're so happy, then why are we so ready to be persuaded by claims that we are suffering from a world-historical spiritual malaise, despite all the evidence to the contrary?

In his bestselling 2004 book *The Paradox of Choice,* Schwartz argues that capitalist consumer culture gets us down by offering *too* many choices. It's not just that the onslaught of new brands of toothpaste, breakfast cereal, chocolate bars, and books about happiness taxes our frail deliberative capacities, but when our set of options explodes, each new choice requires *not* choosing so many other things.[40] The perceived cost of making any choice and sticking with it seems higher and higher the more alternatives there are to forgo. On this score, Schwartz points us to Robert Lane's claim in *The Loss of Happiness in Market Democracies:*

> There are too many life choices . . . without concern for the resulting overload; and the lack of constraint by custom, [and] demands for self-actualization, that is, demands to discover or create rather than accept a given identity . . . all adds to the stress.[41]

To be sure, it is a hassle to have to discover or create our identities instead of being "given" one—or having one forced upon us. But this is, in essence, what it means to be postmaterialist in Inglehart's sense. Instead of slipping into pre-assigned, traditional social roles, we are able to sit atop mountains of wealth and survey the vast horizon of possibility, with a heretofore unthinkable independence from custom, wondering what kind of person we would like to be. And then we become agoraphobic.

Our problem is that there are both too many *and* too few choices. There are particular goods that would specially benefit and satisfy each of us, but which don't exist. Yet it is hard to identify the specially fitting goods that already do exist in the panoply of choice. If we weren't so diverse, we wouldn't require so much diversity. One kind of shoe, one kind of bread, one kind of antacid would be universally satisfactory. But we *are* diverse, and, for the first time in history, we are liberated from ancient demands of conformity, because, for the first time in history, we now come into the world at a sufficiently safe distance from scarcity to permit us to express and experiment with our singular natures. In fine post-materialist fashion, we demand that our consumption express our self-conceptions-in-progress, and so we *need* diversity. But we also don't know exactly who we want to be before we get to the store. So we can easily feel lost in the consumer cornucopia, as though we are sorting through a landfill for a diamond etched with just our name.

As John Maynard Keynes wrote in his startlingly prescient essay "Economic Possibilities for Our Grandchildren," there may be a sense in which we have already solved (we lucky few in the advanced liberal democracies, that is) the economic problem of scarcity. But *then* what?

> Thus for the first time since his creation man will be faced with his real, his permanent problem, how to use his freedom from pressing economic cares, how to occupy the leisure, which science and compound interest will have won for him, to live wisely and agreeably and well.[42]

And this, our permanent problem, we have yet to solve, and it weighs on us. Our culture has not yet caught up to the new, happier world of science and compound interest, and we do not yet see how our inherited visions of the good life fit into it. So it seems plausible to most of us that something is wrong, even if so much is right. If happiness research is going to be good for anything, it is not going to be for guiding well-meaning technocrats who seek to make us happier by pulling this policy lever or pushing that policy button. Rather it is going to be good for providing insight in how "to live wisely and agreeably and well." This is insight we all badly need, and it is not the government's to give.

All references for articles included in *Taking Sides: Clashing Views in Sustainability* can be found on the Web at www.mhhe.com/cls.

Saamah Abdallah et al.

Unhappy Planet Index 2.0: Why Good Lives Don't Have to Cost the Earth

Introduction

In the final year of the first decade of the third millennium, humanity stands at a crossroads. Depending on the choices we make now, future generations will either look back at our time with anger or with gratitude. Currently, we are set on the former course. Should we continue our reckless over-consumption of resources and destruction of the environment, driven by an insatiable appetite for economic growth, our descendants will face a world of scarcity, uncertainty and conflict.

However, over the last few years the first signs have emerged that we may be able to find a different path, one which future generations will look back on with gratitude and relief. That path, should we take it, will not only ensure we halt catastrophic environmental damage, but will also support good lives for all. A path where our understanding of progress and prosperity takes account of the needs of humans, and the needs of the planet. In short, it will lead us towards better, more meaningful lives that do not cost the Earth.

The first Happy Planet Index (HPI) was launched by **nef** (the new economics foundation) in July 2006 to help steer us along this path.[2] It presented a completely new indicator to guide societies, one that measures the ecological efficiency with which happy and healthy lives are supported. Even then, its message resonated with hundreds of thousands of people around the world—the report was soon downloaded and read in over 185 countries worldwide.[3] Now, in 2009, with the world facing the triple crises of economic turmoil, impending peak oil and continually bleaker predictions of the impacts of climate change, the message of the HPI is more timely than ever before. We *need* to strive for good lives that do not cost the Earth and we need indicators that can help get us there.

HPI 2.0 takes advantage of new and improved data for 143 countries around the world, to determine which countries are closest to achieving sustainable well-being. It also looks back over time to see how we've been faring over the last 45 years—and looks forward to see where we need to get to.

It reveals that most countries are woefully far from where they need to be. Indeed the largest countries of the world appear to be moving in the wrong direction; as with the first HPI report, the graffiti on the front cover is therefore still appropriate. However, there are exceptions—countries that appear to be supporting good lives for their citizens whilst living close to their fair share of the world's resources. Based on the data at the national level, and at the individual level, it appears that good lives that do not cost the Earth really are possible. So, alongside this report, we are launching a charter (www.happyplanetindex.org) calling for governments, organisations and individuals around the world to work towards making this possibility a reality.

The End of the End of History

2008 marked the end of an era. As the world's major financial institutions collapsed around us, the economic leaders of the time pronounced *mea culpa*. In October, the former chair of the US Federal Bank Alan Greenspan admitted to the US Congress that he had found a 'flaw' in our guiding economic ideology.[4] In March 2009, the UK Prime Minister Gordon Brown admitted that he should have taken steps to control the UK financial markets during his time as Chancellor.[5] At the World Economic Forum in Davos in 2009, confident self-satisfaction had been replaced with a far greater degree of humility and uncertainty. Many believe that the economic crisis spells the death of neoliberalism. Some go even further. Professor Anthony Giddens, often regarded as the architect of Tony Blair's *Third Way,* has declared it the 'end of the end of history.'[6] For him, the crisis highlights that we need to 'think seriously about the nature of economic growth.' Perhaps even more surprising are the words of Thomas Friedman, long-time advocate of growth and globalisation:

> Let's today step out of the normal boundaries of analysis of our economic crisis and ask a radical question: What if the crisis of 2008 represents something much more fundamental than a deep recession? What if it's telling us that the whole growth model we created over the last 50 years is simply unsustainable economically and ecologically and that 2008 was when we hit the wall—when Mother Nature and the market both said: 'No more.'[7]

For those versed in ecological economics—a discipline which recognises the dependence of our economic systems on the Earth's resources—it is tempting to adopt a smug 'I told you so' attitude. As far back as 1972, the Club of Rome's *Limits to growth,* highlighted the impossibility of an ever-growing economy on a finite planet.[8,9] **nef**'s 2003 *Real World Economic Outlook* predicted 'collapse in the credit system of the rich world, led by the United States, leading to soaring personal and corporate bankruptcies.'[10] It was obvious that our economic system was doomed to another cycle of bust. The added element of approaching the Earth's resource limits threatens to make this cycle the worst for over 100 years.

One should not forget that it is people with average incomes whose quality of life will be hit most. In early 2009, the International Labour Organization (ILO) estimated that 18 million people worldwide could be made unemployed as a result of the crisis, whilst 200 million more people in developing countries are expected to be driven into extreme poverty.[11] The sad truth is that our current economic system relies on continuous growth—when this comes to a halt, it is those who are already deprived who bear the brunt.[12,13]

Given the huge attention the crisis has attracted, it is easy to forget that the world was far from a perfect place before the credit crunch hit. Despite 60 years of constant economic growth, in 2005, more than half of the world's population(56.6 per cent) lived on less than the equivalent of $2.50-a-day.[14] The benefits of growth have been wildly disproportionate. For every $100 worth of growth, only $0.60 contributes to reducing poverty for the more than one billion people living below $1-a-day.[15] Worldwide, one in thirteen children dies before the age of five. For people living in twenty-two of the poorest countries, this rate is over one in seven.[16]

Even in rich countries, our system has not been a constant tale of success. Inequality has been rising in OECD countries over the last 20 years—before the recession kicked in, disparities in income in the UK were the highest since records began in the 1960s.[17] Real median incomes have actually remained stagnant in many countries, including the USA. People do not report being any happier or more satisfied with life than they did 20 or even 40 years ago.[18,19] Commentators on both the left and right talk of a 'social recession.'[20,21] In the UK, child poverty still remains a shameful reality, and the Government has abandoned its ambitions to halve child poverty by 2010. Our model of progress has failed to deliver even what it claims to deliver best: money in people's pockets.

And where it does worst, the current model has done very badly indeed. The UN Millennium Ecosystem Assessment found 60 per cent of the world's ecosystems to be degraded. Concentrations of CO_2 in the atmosphere stood at 387 parts per million (ppm) in 2008. This is the highest they have been for the last 650,000 years. With the annual rate of CO_2 emissions actually *increasing* in recent years, it is no wonder that the Intergovernmental Panel on Climate Change (IPCC) predicts that the 'most likely' global increase in temperature, in a 'business as usual' scenario, would be 4°C above 1990 levels—double the 2°C target that climate scientists and indeed the EU have strived to meet to avoid positive feedback loops leading to the climate spiralling out of control. Indeed, many scientists, including NASA's top climatologist Jim Hansen, now feel that only by returning to a level of 350 ppm can we prevent this happening.[22] In other words, to preserve the climatic conditions which human civilisation has enjoyed since it began, not only do we need to stop emitting fast, we also need to physically remove CO_2 from the atmosphere.

A Crisis Is a Terrible Thing to Waste[23]

And yet, as Hazel Henderson, one of the leading figures of the Club of Rome has recently highlighted, with crisis comes opportunity. A remarkable transformation has occurred over the past five or so years, whereby concerns over resource depletion and fear of climate change are no longer the domain of fringe environmentalists, but rather the norm in many developed countries. Tangible impacts from climate change on both development in poorer countries,[24,25] and the economy of the developed world are fast becoming recognised.[26] Where electoral systems allow, green parties are gaining ground; for example, in elections in April 2009, the Green party in Iceland entered the ruling coalition with over 21 per cent of the vote—the highest percentage that any Green party has won in national elections to date. In the same month, US Secretary of State Hilary Clinton publicly accepted the USA's substantial contribution to climate change.[27] Earlier in the year, US Director of National Intelligence Dennis Blair stated that climate change is a 'top threat to . . . national security.'[28] In a recent Eurobarometer survey, EU citizens rated climate change as *the* most serious problem currently facing the world as a whole, above poverty and international terrorism. Perhaps not surprising when the evidence suggests climate change will exacerbate both these problems: 62 per cent of those surveyed ranked climate change amongst their top two global concerns. The debate is no longer about whether climate change is an issue, but about how best to deal with it.

And yet, for all this acknowledgement of the problem, we are still moving in precisely the wrong direction. Global CO_2 emissions are rising year-on-year, and our ecological debt, as measured by the ecological footprint, continues to grow.[29] What's going wrong? In the words of the recent film on climate change starring Pete Postlethwaite: are we really that stupid?[30] Are we to collectively assume the role in global history that was played by the person who chopped down the last tree in Easter Island, as described so eloquently in Jared Diamond's *Collapse*?[31]

Unfortunately, our all-too-human fear of change is currently trumping our scientifically endorsed fear of global warming. A wealth of evidence suggests that we could reduce our resource consumption whilst maintaining or even improving our quality of life, but this cannot overcome the paralysis caused by our desire to maintain exactly the way of life to which we have become accustomed. Less consumption, less growth and fewer emissions is hardly a rabble-rousing mantra for change. Rendered impotent by fear, we need a *positive* vision of what progress could look like. Martin Luther King may well have had nightmares in his life, but it was for his dream that he will be remembered.[32]

The HPI plays a part in illuminating the path towards that dream. By stripping the economy down to what it really should be about—providing long and happy lives for all today, without infringing on the chances of future generations to do the same—it goes to the heart of what we should be measuring. If the second half of the twentieth century was about the pursuit of economic growth and material goods, the twenty-first century should be defined by the pursuit of good lives that do not cost the Earth. The former was measured more or less adequately by GDP. To achieve the latter we need the HPI.

Defining Our Goals—the HPI

Human Goals

How does one measure well-being in terms of happy and healthy lives? The health aspect is (relatively) straightforward—the best-known headline indicator being life expectancy at birth. The 'happy' part has been debated since the time of Aristotle. In recent years, the debate has moved from philosophy to the realm of science, with a growing body of research identifying what it means to be happy, what drives it and how to measure it. For us, being 'happy' is more than just having a smile on your face—we use the term *subjective well-being* to capture its complexity. Aside from feeling 'good,' it also incorporates a sense of individual vitality, opportunities to undertake meaningful, engaging activities which confer feelings of competence and autonomy, and the possession of a stock of inner resources that helps one cope when things go wrong. Well-being is also about feelings of relatedness to other people—both in terms of close relationships with friends and family, and belonging to a wider community.[33]

Encapsulating all of these aspects of well-being precisely requires detailed measurement, and **nef** has called for governments to collect thorough and regular National Accounts of Well-being to do so.[34] However, extensive data has already been collected in surveys worldwide and over the last forty-five years on one fundamental aspect of well-being—life satisfaction.

Life satisfaction is typically measured with the following question:

> All things considered, how satisfed are you with your life as a whole these days?[35]

Responses are made on numerical scales, typically from 0 to 10, where 0 is dissatisfied and 10 is satisfied. Years of research have demonstrated that, despite its apparent simplicity, the question produces meaningful results. Individuals' responses correlate with the size and strength of their social networks, relationship status, level of education, presence of disability, as well as with their material conditions, such as income and employment.[36,37,38] The averages for countries tend to be higher where people within that country enjoy higher levels of social capital, better climate, richer natural resources, higher life expectancy, better standards of living, and more voice within government.[39,40]

Furthermore, responses to this question correlate well with other attempts to assess well-being. People who say they are satisfied with their life tend also to make other positive assessments, such as reporting more frequent good moods, are described by their loved ones as being satisfied, are observed to smile more often, and are less likely to commit suicide later on in life.[41,42,43] Importantly, reported life satisfaction also correlates with all the complex aspects of well-being described earlier, such as feeling autonomous and being resilient.[44]

In 2008, two years after the HPI was launched, the UK Department for Environment, Food and Rural Affairs (Defra), built subjective well-being measures including life satisfaction into its set of sustainable development indicators providing official acknowledgement that they may be useful in assessing progress towards human goals.[45]

The Dutch sociologist Ruut Veenhoven has developed an approach to combining life satisfaction with life expectancy in a term we call 'happy life years' (HLY)—which can be seen as happiness-adjusted life expectancy.[46,47] Doing so ensures both the subjective and objective elements of well-being are captured. It recognises that a satisfying life is not ideal if it is very short, but also that a long life is not ideal if it is miserable.

Respecting Ecological Limits

The last few paragraphs have focused on what we want societies to enable us to achieve—their human goals. There was no mention of the means with which they do this, or of the inputs required. Yet consideration of these issues is essential, given that how we ensure our well-being now will affect whether others around the world can also secure their own well-being, and whether *any* of us can do so in the future. This is the 'sustainable' aspect of sustainable well-being. No moral framework would accept high well-being if it was at the expense of others living today and/or future generations. Such considerations are particularly relevant where limited resources are required to support well-being. And the most finite limited resources that we currently rely on are natural ones.[48]

Jared Diamond's *Collapse* takes its reader through a potted history of societies that overtook their ecological limits, and collapsed as a result. The most poignant example is that of Easter Island, famed for its sombre giant stone statues—*moai*. It is not certain, but the moai appear to have been built as part of status competition between the various tribes on the island, with bigger moai demonstrating greater power. The early seventeenth century was likely the pinnacle of Easter Island culture, the time the biggest moai were being built, an echo of the skyscrapers going up across the world from Canary Wharf to Kuala Lumpur. However, moai construction consumed a lot of resources, particularly wood for transport and energy. By 1650, the last tree had been felled. By the time Europeans arrived on the island's shores in 1722, the numbers of Easter Islanders had fallen dramatically, and they had been reduced to petty wars and cannibalism.

Easter Island reminds us of the danger of only measuring what we consider to be human goals to the exclusion of factors affecting sustainability. In the first half of the seventeenth century, given the archaeological evidence, quality of life in Easter Island may well have been at its highest ever. If Easter Islanders had been measuring well-being, they may well have been seeing ever growing life expectancy and reported life satisfaction. Despite that, or rather because of it, disaster was just around the corner. It appears no one in Easter Island was measuring their environmental impact.

Their society was particularly vulnerable, being separated by over 2000 km from the next inhabited island. Such dramatic collapses are rare in a world where societies are and have always been interconnected. Resources are traded, people migrate, and empires are conquered. But the Earth itself is also an island. The nearest other island, inhabited or otherwise, is 40,000,000 km away. If there is one lesson we must learn above all others, it is to not let the Earth go the way of Easter Island.

In such a complex world, it is not a simple matter to measure our impact on the planet. How can one compare the impact of using a gallon of oil with a gallon of water, or a tonne of potatoes with a tonne of potassium? The best available approach is currently the ecological footprint, developed by ecologists Mathis Wackernagel and William Rees, and championed by a range of organisations including the Global Footprint Network and WWF.[49] The EU statistical agency

Eurostat is considering incorporating the ecological footprint into its sustainable development indicator set,[50] whilst the Welsh Assembly Government has already adopted it as one of five headline indicators of sustainability.

The ecological footprint of an individual is a measure of the amount of land required to provide for all their resource requirements plus the amount of vegetated land required to sequester (absorb) all their CO_2 emissions and the CO_2 emissions embodied in the products they consume. This figure is expressed in units of 'global hectares.' The advantage of this approach is that it is possible to estimate the total amount of productive hectares available on the planet. Dividing this by the world's total population, we can calculate a global per capita figure on the basis that everyone is entitled to the same amount of the planet's natural resources. Using the latest footprint methodology—and it should be noted that this is a developing methodology—the figure is 2.1 global hectares.[51] This implies that a person using up to 2.1 global hectares is, in these terms at least, using their fair share of the world's resources—one-planet living.

In 2005, the per capita footprint for the rich OECD nations was 6.0 global hectares. The implication: we are living as if we had almost three planets' worth of resources.

Such large footprints are in part possible by relying on poorer countries to provide us with raw materials—they represent the ecological debt owed by rich countries to poor ones.[52] This raises the stark reality that it is pointless for poorer countries to aspire to becoming 'more like the West'—it is simply impossible for everyone on the planet to live as Westerners do today. We would indeed need three planets to do so. We still only have one. For this reason, the ecological footprint is also useful for understanding social justice. Improving living standards in poorer countries can only be achieved in parallel with declining resource consumption in richer ones.

The average per capita footprint worldwide also highlights a serious problem. At 2.3 global hectares it is just above the world's sustainable capacity, and has been since the mid-1980s. This ecological overshoot in part represents the unsustainable emission of CO_2 into the atmosphere at a rate faster than the planet can re-absorb it.

Society as a System

If well-being is our goal, and the planet itself defines our resource limits, we should not lose sight of all that happens in the middle.

[Let's portray] human society as a system with inputs, means and ends. The means are vitally important—a field of grass is not converted into happy, healthy, meaningful lives without complex systems of agriculture, trade, culture, education and much more. They are, however, only 'instrumentally' important: they are important because they play a role in helping us achieve our ends.[53] Debates about what makes the best economy, education or governance systems should ultimately be decided in terms of which supports the provision of the highest, fairest and most sustainable well-being.[54] As such, all means should be considered as *strategies* to achieving our ends, as pathways to sustainable well-being. The success of these strategies *does* need to be assessed and measured. But that is not the role of the HPI. In simple terms, the HPI measures what goes in and what comes out—not what happens in the middle. In doing so, of course, it provides us clues as to what we need to do in the middle to achieve society's objectives.

It is worth reminding ourselves what we have been doing over the last 60-odd years. We have focused on a few strategies, specifically technology, healthcare, employment, and above all the economy—defined very narrowly in terms of GDP—so as to provide for well-being. We have tended to pay less attention to some other strategies such as values, leisure time and social capital. Worse still, we have fully measured neither our inputs nor our ends. As a result, we have not been able to assess whether gains with respect to some strategies may have caused losses with respect to others. We have not been able to determine whether we have achieved progress in real terms, which ultimately comes down to people's experiences of their lives.

HPI in Equation Form

In essence, the HPI is an efficiency measure: the degree to which long and happy lives (life satisfaction and life expectancy are multiplied together to calculate happy life years) are achieved per unit of environmental impact.

$$\text{Happy Planet Index} \sim \frac{\text{Happy Life Years}}{\text{Ecological Footprint}}$$

. . . [E]ach of these components will be considered in a little more depth. . . . Certain statistical adjustments are required to ensure that no single component dominates the indicator and to produce an easy-to-interpret figure ranging from 0 to 100.[55]

The Rise and Fall of a Foolish Myth

> Every society clings to a myth by which it lives. Ours is the myth of economic growth.[56]

The myth of economic growth as progress has held sway for over half a century. But now, stimulated by the ongoing economic crisis and impending environmental and resource crises, alternative visions of progress, such as that represented by the HPI, are gaining popularity. They are still not the dominant view, but the tipping point may not be far off. In this chapter, we sketch out how we got to become obsessed with 'more'—in terms of economic growth and GDP as the indicator of it, the damage it has caused, and how alternate visions are now gaining ground.

Our Obsession with 'More'

Writing during an economic crisis, it may seem inopportune to question the centrality of economic growth. Now more than ever, governments around the world are desperate to restart growth by any means possible. GDP is even more omnipresent in public discourse than usual. In the UK alone it was referred to in 3590 articles in national newspapers in the eight months following the escalation of the crisis in September 2008—more than double the frequency for the same time period in previous years.[57]

And yet we should not lose sight of the fact that economic growth is just one strategy to achieve well-being and, in terms of natural resources, a demonstrably ineffcient one. Rather than pursuing growth at all costs, even if detrimental to well-being or sustainability, leaders should be striving to foster well-being and pursue sustainability, even if detrimental to growth. The horse and the cart need to be returned to their rightful places. As the UK's Sustainable Development Commission, a public body that directly advises the Prime Minister's office on sustainable development issues, eloquently points out in its report *Prosperity Without Growth?*:

> . . . the state has become caught up in a belief that growth should trump all other policy goals. But this narrow pursuit of growth represents a horrible distortion of the common good[58]

Things have not always been like this. For most of the history of humanity, economic growth was a minor phenomenon: a side-effect, where it existed, of the pursuit of other goals.[59] It only attained its quasi-mystical role when GDP was placed atop the podium of indicators with the development of the United Nations System of National Accounts, in 1947. At that time, focusing on productivity growth made sense. Much of the world needed to be rebuilt following the war, and that required growing economies.[60] Furthermore, economic growth helped avoid distributional debates. The rising voice of the working classes demanded more of the material cake. The only way elites could respond to that voice without having to give up anything themselves was by growing the cake.

Some time since then, economic growth *per se* became less pressing a need for developed countries.[61,62] Europe, Japan and other regions ravaged by the war had been rebuilt, and living standards had been raised. Increasing growth has not seemed to reduce inequalities any further, and in some cases may have been contributing to their worsening. But systems carry their own momentum, and even the wealthiest countries still pursue economic growth as if they were still struggling to recover from the war. The European Union's focal strategy—the Lisbon Agreement—is pivoted on growth. From 2005, OECD publishes an annual report entitled *Going for growth* which attempts to untangle how member states can quicken the pace of their GDP growth.

The notion of GDP growth almost seems to have a halo around it. It has reached the status of motherhood and apple pie. Even as early as 1967, the economist E.J. Mishan noted:

> Among the faithful . . . any doubt that, say, a four per cent growth rate . . . is better for the nation than a three per cent growth rate is near-heresy; is tantamount to a doubt that four is greater than three.[63,64]

Of course, that is not to say that economic growth can be simply 'switched off' without consequence—our economic system will require substantial change for a steady state economy to succeed.[65] Also for many developing countries, economic growth may indeed be required. The problem is that the dominant economic paradigm of the last 30 years has not always been successful in delivering the benefits of growth to where it is needed—by far the greatest share goes where it isn't needed, i.e. to those who are already wealthy.[66]

The Consequences of Myopia

Biologists talk about physical growth as a process which has an optimum level beyond which further growth is not beneficial, and can indeed turn malignant. Economic growth can be subjected to the same analysis. Aside from the obvious environmental impacts which we have already discussed, there is gathering evidence that an obsession with growth may have led us to ignore other aspects of life critical to our well-being.

To maintain growth, Western capitalist economies have a structural need to sustain demand for consumption.[67,68,69] But this feature of the system sets it at odds with a widely noted fact about human nature—that once our basic material needs are comfortably met, more consumption tends to make little difference to our well-being. This is not just folk wisdom, although it is certainly the case that throughout history, and across all cultures and religions, people have cautioned against an excessive focus on wealth and material possessions. Research suggests that in most reasonably developed countries, material circumstances such as wealth and possessions play only a small role in determining levels of happiness—some psychologists estimate that they explain only around 10 per cent of variation in happiness at the aggregate level.[70] Beyond a certain level of income, increasing wealth makes little difference.[71] Much more significant are factors relating to individual differences in outlook and to the kinds of activities that people engage in: socialising, participating in cultural life, having meaningful and challenging work and so on.

But the requirement to maintain consumption growth at all costs has led to a situation in which, for decades, we have been presented with a poisonous combination of messages. First, we are constantly bombarded with messages from advertisers and marketers, all pushing the idea that buying this or that new product will make us happier. Added to this, in many countries we have been offered staggeringly easy access to credit with which to keep up our level of consumption. Quite apart from the environmental impacts, this has served us very poorly in a number of ways.

For one thing, levels of debt have soared in recent years; in 2007 and 2008, for the first time on record, UK personal debt exceeded total GDP.[72] As recent research from the Institute of Psychiatry in London shows, debt is a large contributing factor to a person's chances of developing clinically significant anxiety and depression, largely irrespective of their income.[73] It is not hard to imagine why this might be. The stress of working just to keep up repayments is exhausting, the fear of defaulting constant and gnawing, and that's without having to deal with the feelings of despair and inadequacy for having failed.

But there is also a more subtle and no less damaging aspect to all this focus on personal consumption. People who are strongly motivated by the idea of getting rich and famous are what psychologists refer to as *materialistic.* Using an engaging metaphor, psychologist and author Oliver James describes them as having caught the 'affluenza' virus.[74] The scientific evidence for the negative impacts of materialism is overwhelming; they range from poorer personal relationships through fewer good moods and lower self-esteem, to increased prevalence of psychological symptoms.[75] In short, people whose main aspiration is to be wealthy are inclined to be less satisfied with their lives in general than those who focus their energies elsewhere.[76] What is worrying, but perhaps unsurprising, is the extent to which materialism is on the rise. . . .

However, it is not just individuals who are harmed by this myopia. Various scholars have argued that the 'social recession' that burdens modern capitalist societies can be attributed to a shift towards individualism.[77] A striking statement of this thesis can be found in no less a journal than the unrepentantly free-market *The Economist.* Attempting to explain why well-being does not keep rising in line with consumption, it suggests that 'there are factors associated with modernisation that, in part, offset its positive impact.'[78] Specifically, it argues that alongside consumption growth

> [a] concomitant breakdown of traditional institutions is manifested in the decline of religiosity and of trade unions; a marked rise in various social pathologies (crime, and drug and alcohol addiction); a decline in political participation and of trust in public authority; and the erosion of the institutions of family and marriage.

Two things are significant about the cultural changes highlighted here. The first is that they involve factors known to determine well-being—in particular, feelings of social and community relatedness and trust. The second is the suggestion that these changes have occurred *as a result of* the modernisation process. In other words, the pursuit of consumption has systematically undermined not only the environmental conditions on which future well-being depends, but also certain social conditions (e.g., family, friendship, community, trust) that are critically important for well-being *now.*

Gaining Ground

The need for a new vision of progress is being felt in many places. A UK poll found 81 per cent of people supported the idea that the Government's prime objective should be the 'greatest happiness' rather than the 'greatest wealth.'[79] An international survey found that three-quarters of respondents believed health, social, and environmental indicators were just as important as economic ones and should be used to measure national progress.[80]

The French Commission on the Measurement of Economic Performance and Social Progress was set up in January 2008 by President Nicholas Sarkozy to respond to these opinions, by reassessing GDP's role. It includes amongst its number three Nobel Prize winners—Joseph Stiglitz, Amartya Sen and Daniel Kahneman.

It comes at a good time. Even during the peak of the economic boom in 2006, 61 per cent of people in 20 European nations felt that 'for most people in their country, life was getting worse.'[81] Recognising this brewing dissatisfaction, the Commission notes:

> There is a huge distance between standard measures of important socio-economic variables like growth, inflation, inequalities etc. . . . and widespread perceptions. . . . Our statistical apparatus, which may have served us well in a not too distant past, is in need of serious revisions.[82]

In a fundamental sense, it should not be surprising that economic growth does not epitomise all that people want from life. Paraphrasing Michaela Moser of the European Anti-Poverty Network, 'no one wakes up dreaming they lived in the country with the highest economic growth.' In a cross-cultural study across 26 countries, health and happiness were consistently rated above affluence as most important to people.[83,84] In that sense, there is nothing radical about defining progress in terms of the HPI—once sustainability is considered, it is more or less how people define what they want from life anyway.

But the public's implicit recognition of the importance of well-being is only half the story. Politicians and governments still labour under the illusion that economic growth defines success. There are signs that this could change. Well-being was first given legislative muscle in 2000, when the UK Local Government Act gave local authorities the power to promote social, economic and environmental well-being. However, it is only in the last three years, after the publication of the first HPI report, that the first glimmers of its growing international recognition are emerging.

In 2007, the Conservative Party published a *Blueprint for a green economy,* highlighting quality of life, and not economic growth, as a priority. Conservative Leader David Cameron has explicitly mentioned the HPI as a better measure of progress.[85] The Austrian *Lebensministerium* is calling for the strategy that will replace the Lisbon Treaty from 2010 to focus on quality of life, rather than growth.[86] Most exciting is the UK Sustainable Development Commission's ground-breaking new report *Prosperity without growth?* The report advocates a new vision of prosperity around the themes of sustainability and well-being, and questions the relevance of economic growth to these goals. It may have been a long time coming, but credit is certainly due to the UK government for allowing debate on what has, until now, been an entirely taboo subject.

Emerging amongst these new visions are a burgeoning number of initiatives aimed at developing indicators to measure progress towards them. In June 2007, OECD hosted an international conference in Istanbul on *Measuring the progress of societies,* leading to the Istanbul Declaration, signed by many inter-governmental organisations including the UN and the EU. The OECD is continuing to engage with experts, national governments, and statistical agencies to try to support the development of new initiatives for measuring progress. Their draft proposal for a framework, based on work by Robert Prescott-Allen,[87] has at its heart human well-being. The OECD is organising another conference in October 2009 in South Korea, entitled *Charting progress, building visions, improving life.*

The European Commission also has its own process, entitled *Beyond GDP,* which was launched with a conference in Brussels in November 2007, where the President of the European Commission, José Manuel Barroso called for 'a breakthrough that adapts GDP, or complements it with indicators that are better suited to our needs today, and the challenges we face today.' In the same year, Eurostat, the European statistical agency commissioned a consortium of experts, including **nef**, to consider the feasibility of a well-being indicator for Europe. As has already been mentioned in Chapter 2, the UK has already incorporated pilot well-being measures into its assessment of sustainable development.

Most recently, in March 2009, the UK's *All Party Parliamentary Group on Wellbeing Economics* had its first meeting. The group's aims are to promote the enhancement of well-being as an important government goal, encourage the adoption of well-being indicators as complementary measures of progress to GDP, and promote policies designed to enhance well-being.

This list of initiatives is by no means exhaustive. In the UK, the Office of National Statistics has followed Defra's lead and started exploring how well-being could be measured. Other projects setting out to develop better measures of well-being and progress include WellBeBe in Belgium, QUARS in Italy, the Canadian Index of Well-Being, Measures of Australia's Progress, and of course the Bhutanese measure of Gross National Happiness.[88]

This growing momentum makes it quite clear that GDP's days as our sole indicator of progress are numbered. The stage is set for the HPI.

All references for articles included in *Taking Sides: Clashing Views in Sustainability* can be found on the Web at www.mhhe.com/cls.

EXPLORING THE ISSUE

Does Sustainability Mean a Lower Standard of Living?

Critical Thinking and Reflection

1. Answer for yourself these questions posed by Colin Beavan writer of "No Impact Man": "To what extent do we—as individuals and as a culture—prioritize what really makes life worth living? How much time do we not spend with our kids or friends, for example, because we're trying to get rich so that we can later . . . have the leisure time to spend with our kids and our friends? How much time—and resources—do we spend on big houses or better cars when really we just want to watch the sunrise?" (Abdallah, 2009, p. 41).
2. Developing countries have the opportunity to choose the Western path of high consumption, high environmental impact, and degradation of social capital, or they can safeguard the natural and social capital they presently enjoy, and make a conscience effort to develop sustainably. Should there be a cutoff point to affluence because it wastes more of our resources than it benefits human well-being?
3. How bad would it be for individuals in the Western world if their GDP dropped? How far a drop produces what response? How long-lasting is the resulting trauma? If everyone in the higher income range had a lowering of their standard of living, would that affect the amount and/or breadth of the trauma? What would be their quality of life changes?
4. Which person lives a more rewarding life—a South American farmer or a North American wealthy enough to be able to spend his time playing computer games?

Is There Common Ground?

The common ground in the topic of "sustainability in standard of living" lies in the realm of the Porter hypothesis (PH). Porter hypothesized that well-designed environmental regulation can enhance profitability by making corporations innovative with their solutions. An important point is that it is not a hypothesis that is always applicable. The regulations written need to be "well designed" and the benefit does not always offset the expense in research and development. Through these and other qualifications, the PH has a great deal of relevance to our developing business world. It opens the door for corporations, most interested in the bottom line, to get ahead by trying to be environmentally friendly. It can work and has worked. And politically it "refutes the idea that environmental protection is always detrimental to economic growth."

Internet References

Abdallah, S., Thompson, S., Michaelson, J., Marks, N., Steuer, N. (2009). Happy Planet Index 2.0: Why Good Lives Don't Have to Cost the Earth. New Economics Foundation. Retrieved from http://www.happyplanetindex.org/public-data/files/happy-planet-index-2-0.pdf.

Akst, D. (November 23, 2008). A talk with Betsy Stevenson and Justin Wolfers. *Boston Globe*. Retrieved from http://www.boston.com/bostonglobe/ideas/articles/2008/11/23/a_talk_with_betsey_stevenson_and_justin_wolfers/?page=full.

Ambec, S., Cohen, M.A., Elgie, S., Lanoie, P. (2010). The Porter Hypothesis at 20: Can Environmental Regulation Enhance Innovation and Competitiveness? *Cirano*. Retrieved from http://www.cirano.qc.ca/pdf/publication/2010s-29.pdf.

Banks, A.M. (November 4, 2010). Study: Spirituality may help to boost feelings of happiness. *Houston Belief*. Retrieved from: http://www.chron.com/disp/story.mpl/life/religion/7279963.html.

Datta, L. (2009). FisheriesMarineGPIReport. Hawaii-Pacific Evaluation Association. Retrieved from http://h-pea.org/2009/HPEAConference2009_Datta.pdf.

Easterlin, R.A. (1974). "Does Economic Growth Improve the Human Lot?" In: Paul A. David and Melvin W. Reder, eds., *Nations and Households in Economic Growth: Essays in Honor of Moses Abramovitz*. New York: Academic Press.

Emerson, J., Esty, D.C., Levy, M.A., Kim, C.H., Mara, V., de Sherbinin, A., and Srebotnjak, T. (2010). 2010 Environmental Performance Index. New Haven: Yale Center for Environmental Law and Policy.

Foroohar, R. (April 5, 2007). The Joy of Economics. *Newsweek*. Retrieved from http://www.newsweek.com/2007/04/04/the-joy-of-economics.html.

Hagerty, M., Veenhoven, R. (2003). Wealth and Happiness Revisited: Growing Wealth of Nations Does Go with Greater Happiness. *Social Indicators Research,* vol. 64, pp. 1–27.

Jacob, J., Jovic, E., Brinkerhoff, M. (2008). Personal and Planetary Well Being: Mindfulness Meditation, Pro-environmental Behavior and Personal Quality of Life in a Survey from the Social Justice and Ecological Sustainability Movement. Retrieved from http://www.springerlink.com/content/93138gm25k228612/.

Leiserowitz A.A., Kates R.W., and Parris T.M. (2006). Sustainability Values, Attitudes, and Behaviors: A Review of Multinational and Global Trends. *Annual Review of Environmental Resources* vol. 31, pp. 413–444. doi: 10.1146 annurev.energy.31.102505.133552.

Leiserowitz, A., Maibach E., Roser-Renouf, C. (2009). Climate change in the American mind: Americans' climate change beliefs, attitudes, policy preferences, and actions. Yale Project on Climate Change.

Retrieved from http://www.climatechangecommunication.org/images/files/Climate_Change_in_the_American_Mind.pdf.

Nussbaum, Martha & Amarty Sen, *The Quality of Life* (Oxford: Clarendon Press 1993)

Pannozzo, L., Coleman, R. (2008). The 2008 Nova Scotia Genuine Progress Index. Retrieved from http://www.gpiatlantic.org.

Stevenson, B, Wolfers, J. 2008. Economic Growth and Subjective Well-Being: Reassessing the Easterlin Paradox. Retrieved from http://bpp.wharton.upenn.edu/jwolfers/Papers/EasterlinParadox.pdf.

Wilkinson, W. (April 11, 2007). In Pursuit of Happiness Research: Is It Reliable? What Does It Imply for Policy? *Policy Analysis,* vol. 590. Retrieved from http://www.cato.org/pubs/pas/pa590.pdf.

Internet References . . .

International Monetary Fund

This is the website of the International Monetary Fund. It provides a source of credible information about economic growth in different countries of the world.

http://www.imf.org

World Bank

The website of the World Bank provides authentic and valuable information about different projects initiated by the World Bank around the world.

http://www.worldbank.org

Scientific and Global Issues

This website discusses scientific and social issues affecting the earth.

http://www.globalissues.org/

Center for Biological Diversity

The Center for Biological Diversity links human welfare to nature. It explains how mankind exists with mutual symbiosis with the plants and animals of the biota, so that diversity is maintained.

http://www.biologicaldiversity.org/campaigns/overpopulation/index.html

Steady State Economy

This website, The Center for Advancement of the Steady State Economy (CASSE). CASSE discusses the relationship between economic growth and environment.

http://steadystate.org/

Population Issues

This website discusses the impact of overpopulation and overconsumption.

http://www.overpopulation.org/

Global Ecological Footprint Network

Global Footprint Network is an international group that focuses on advancing sustainability through use of the ecological footprint analysis and measurement.

http://www.footprintnetwork.org

World Wildlife Fund

The website of the world's leading conservation organization.

http://www.worldwildlife.org

Culture and Sustainability

This is a website about the struggle for indigenous land, language, and culture.

http://www.culturalsurvival.org

UNIT 2

Global Issues

Issues related to sustainability are often global in nature. They require both a global perspective and a global decision-making capacity. There is good reason why the United Nations has been a strong force in advancing sustainability. It is one of the few institutions that possess the necessary reach to discuss and perhaps solve problems across sovereign borders. Since the planet constitutes a single ecosystem, the decisions and development patterns of one country influence the social and environmental conditions of another. The most obvious global issues are related to poverty, climate change, population, the disparity of global wealth, and contrasting patterns of world consumption. Although there is a global consensus on maintaining ecosystems, reducing pollution, and enhancing social capital, there are a wide variety of strategies to achieve these goals. For instance, what type of trade-offs are necessary to reduce dependence on oil and coal as major sources of energy? Is it realistic for some countries to make more sacrifices than others to move toward global sustainability?

It is important to first understand how the global ecosystem functions. What are the environmental thresholds and how resilient are global ecosystems? Debates over these issues will guide the decisions of countries in the future.

- Can India and China Reduce Their Dependence on Coal?
- Is Poverty Responsible for Global Environmental Degradation?
- Is Limiting Consumption Rather Than Limiting Population the Key to Sustainability?
- Can Technology Deliver Global Sustainability?

ISSUE 5

Can India and China Reduce Their Dependence on Coal?

YES: M. Asif and T. Muneer, from "Energy Supply, Its Demand and Security Issues for Developed and Emerging Economies," *Renewable and Sustainable Energy Reviews* (2007)

NO: Yun Zhou, from "Why Is China Going Nuclear?" *Energy Policy* (March 2010)

Learning Outcomes

After reading this issue, you should be able to:

- Comprehend the challenges facing the emerging economies for achieving sustainability.
- Understand the relationship between booming economic growth and the energy crisis of China and India.
- Understand fossil fuel consumption for energy security as a necessary evil.
- Comprehend the challenges in making renewable energy a feasible option on a large scale.
- Understand the increasing consumption of depleting natural resources by China and India.
- Fathom the ever widening gap between shrinking agricultural fields and increasing urbanization.

ISSUE SUMMARY

YES: Professors M. Asif and T. Muneer of the School of Engineering, Napier University, Edinburgh, UK, indicate that emerging economies like China and India are moving toward renewable energies and will need to continue to do so if they want to stem the environmental degradation due to global warming and climate change.

NO: Yun Zhou, a Nuclear Security fellow at the Belfer Center's Project on Managing the Atom and International Security Program at John F. Kennedy School of Government, Harvard University, sees a continuation of the use of coal in China with its environmental consequences due to its increased demand for cheap energy. He sees nuclear as the only alternative to coal.

Rapidly developing countries are increasing their demand for global natural resources. The emerging economies of China and India possess the largest populations in the world with growing middle classes. China is the leading consumer of grain, meat, coal, and steel in the world and is second only to the United States in oil consumption (Brown, 2006). China's economic growth was 10.5 percent in 2010 and is projected to grow at 9.6 percent for 2011. India's growth rate was 9.7 percent in 2010 and is projected to grow at 8.4 percent for 2011. China surpassed Japan as the second largest economy of the world in the third quarter of 2010 and is on track to become the world's largest economy by 2025. China and other emerging economies have emulated the Western world in utilizing fossil fuels for energy production, especially coal. This has made China the second largest producer as well as second largest consumer of energy in the world. This concentration on fossil fuel combustion energy means that China has become the largest carbon emitter in the world, surpassing the United States. The emerging economies of China and India, as well as the other BRIC's, Brazil and Russia, are adding to the global environmental stresses.

The most urgent task that confronts global sustainability is the reduction of carbon and the need to shift from fossil fuel dependence to renewable energy resources. World population will pass seven billion people in 2011 and most of this growth will occur in developing countries. The pressure that this growing population places on fossil fuel use and agricultural resources is alarming. The ability of developed economies to shift to renewable energy is a challenge, much less the emerging economies that are just beginning to experience significant economic growth. China is surging ahead and writers such as Li et al. (2010) emphasize the need to conserve energy. He discusses the concept of circular economy, which has three main principles, that is, to reduce, reuse, and recycle. According to him, efficiency of process industries needs to be increased as well as the total recycling of wastes.

With the rise of automobile use, China has become a net importer of oil (Hu et al., 2010). Hu advocates better automobile fuel millage and discusses how traffic congestion is adding to fuel waste and pollution and needs to be addressed. He emphasizes that public transport should be increased in addition to the exploration of alternative fuels for road transport. Varun and Singal (2007) note that India also depends on fossil fuels for their energy needs. They point out that India is rich in renewable resources and should utilize solar, biomass, wind, and small hydropower energy for sustainable growth. They believe that the major limitation to the production of renewable energy in India is high installation costs. These costs could be overcome through advances in technology.

There are other practices that emerging economies could develop that would be sustainable. There is a huge potential in utilizing waste biomass in

India. Rao et al. (2010) observe that renewable energy from biomass is efficient and cheap as compared with other renewable sources of energy. Wastes can be utilized from various sources, that is, agricultural waste, crop residue, waste water sludge, industrial waste. Moreover, raw materials are locally available, so there is less price fluctuation as there is in imported materials and resources such as oil.

Yet, it is difficult for emerging economies like China, which possesses large reserves of coal, to move from this cheap energy source without economic and perhaps social ramifications. In China, irrespective of all the detrimental effects of coal-based energy on the environment, coal is hard to substitute as an energy source because of its low price. The technologies to make coal energy a clean energy are still in their infancy. Natural gas–based plants cannot compete economically with coal and can only survive with governmental assistance. The price of electricity is more or less set and based on the lower costs of coal, so natural gas is at a severe economic disadvantage. Even though China is making a commitment to renewable energy, sources of energy like wind and solar are not feasible in large-scale applications due to lack of sufficient grid infrastructure to transport electricity from their sources in central China to the large-consumption cities along the coast.

There seems to be a consensus that it will be difficult for China to move away from coal. Chen and Xu (2010) emphasize that energy from coal, although harmful to the environment, cannot be easily substituted in China in the near future. They conclude that coal will remain a major contributor to the energy needs of the rapidly growing economy. They discuss different technologies like Integrated Gasification Combined Cycle (IGCC) and Carbon Capture and Storage (CCS), among others that can be incorporated into the coal energy to produce clean and environmentally safe energy. According to these researchers, China should fund more research and development of these technologies for large-scale feasibility.

China's neighbor India, also developing at a fast pace, is also meeting its energy demands through fossil fuels, which is an unsustainable option to Bhattacharyya (2010). India has to look at alternative sources of energy that are renewable. There should be conservation of energy and utilization of locally available energies, better governance, and management in order to reach a sustainable solution to the energy crisis. His solution is for India to limit bureaucratic obstruction and unnecessary political interference in new energy projects to attract foreign investors in the energy sector.

Zhang (2010) says that as China overtook the United States as the world's largest carbon emitter in 2007, there is increasing global pressure on China to cut carbon emissions. So far, China has not committed internationally post-Kyoto to major carbon emission reductions. He believes that any arm-twisting tactics may prove counterproductive. He recommends that the United States take the lead and work together with China to lower carbon emissions.

Elizabeth Economy in 2006 wrote that China's rapid economic growth in the last two decades has led to environmental degradation. China has launched various programs to deal with the environment, that is, Green Gross Domestic Product (GDP), the National Environmental Model City Program, and tradable permits for sulfur dioxide emissions. But, the way to calculate a Green GDP is not quite clear. Presently, Chinese leadership favors economic development over sustainable development, and there is a lack of transparency in some of the tradable permit programs designed to increase environmental quality.

The YES selection, that China and India can reduce their dependence on coal and transition to sustainability, is the work of Professors Asif and Muneer of the School of Engineering, Napier University, Edinburgh, UK. They indicate that emerging economies like China and India are moving toward renewable energies and will need to continue to do so if they want to stem the environmental degradation due to global warming and climate change. China has a huge potential in renewable energy resources like wind, solar, hydropower, tidal, and biomass energy, which can meet its energy requirement. The growth of renewable energy in China is the fastest in world and will continue to grow. India, with a wide gap between its energy production and consumption, can ill afford to depend on fossil fuels and needs to fully utilize its potential in renewable energy. India, like China, has significant potential in renewable energy, particularly solar, wind, biomass, and small hydropower.

The NO selection, that emerging economies such as China and India will not move away from cheap coal, is the work of Yun Zhou, a Nuclear Security fellow at the Belfer Center's Project on Managing the Atom and International Security Program at John F. Kennedy School of Government, Harvard University. Zhou sees a continuation of the use of coal in China with its environmental consequences due to its increased demand for large amounts of low-cost energy. He views nuclear energy as the only alternative to coal. Coal will remain the major source for electricity generation in the near future in spite of its considerable environmental constraints. Liquefied natural gas, solar and wind energy are either too costly or are not feasible in large-scale usage, and hydropower has been stagnant due to public sentiment against large-scale dam construction after the Three Gorges Dam project.

M. Asif and T. Muneer

Energy Supply, Its Demand and Security Issues for Developed and Emerging Economies

Introduction

Energy drives human life and is extremely crucial for continued human development. Throughout the course of history, with the evolution of civilizations, the human demand for energy has continuously risen. The global demand for energy is rapidly increasing with increasing human population, urbanization, and modernization. The growth in global energy demand is projected to rise sharply over the coming years. The world heavily relies on fossil fuels to meet its energy requirements—fossil fuels such as oil, gas and coal are providing almost 80% of the global energy demands. On the other hand presently renewable energy and nuclear power are, respectively, only contributing 13.5% and 6.5% of the total energy needs. The enormous amount of energy being consumed across the world is having adverse implications on the ecosystem of the planet. Fossil fuels, the main source of energy, are inflicting enormous impacts on the environment. Climatic changes driven by human activities, in particular the production of Greenhouse Gas emissions (GHG), directly impact the environment. According to the World Health Organization (WHO) as many as 160,000 people die each year from the side-effects of climate change and the numbers could almost double by 2020. These side effects range from malaria to malnutrition and diarrhea that follow in the wake of floods, droughts and warmer temperatures [1].

Presently employed energy systems will be unable to cope with future energy requirements—fossil fuel reserves are depleting, and predominantly the developed countries employ nuclear power. Fossil fuel and nuclear energy production and consumption are closely linked to environmental degradation that threatens human health and quality of life, and affects ecological balance and biological diversity. It is therefore clear that if the rapidly increasing global energy needs are to be met without irreparable environmental damage, there will have to be a worldwide drive to exploit energy systems that should not endanger the life of current and future generations and should not exceed the carrying capacity of ecosystems. Renewable energy sources that use indigenous

Asif, M. and Muneer, T. From *Renewable and Sustainable Energy Reviews,* September 2007, pp. 1389–1391, 1393–1394, 1396–1406, 1408, 1412–1413.

resources have the potential to provide energy services with almost nil emissions of both air pollutants and greenhouse gases.

Renewable energy sources such as solar energy, wind power, biomass and geothermal energy are abundant, inexhaustible and widely available. These resources have the capacity to meet the present and future energy demands of the world. . . . The development and use of renewable energy sources can enhance diversity in energy supply markets, contribute to securing long-term sustainable energy supplies, help reduce local and global environmental impacts and provide commercially attractive options to meet specific energy service needs, particularly in developing countries and rural areas, creating new employment opportunities. The cost of energy generated from these renewable resources is significantly coming down while the cost of fossil fuel produced energy is in an increasing mode. Over the last two decades solar and wind energy systems have experienced rapid growth. This is being supported by several factors such as declining capital cost; declining cost of electricity generated and continued improvement in performance characteristics of these systems. The fossil fuel and renewable energy prices, and social and environmental costs of each, are heading in opposite directions, and the economic and policy mechanisms needed to support the widespread dissemination and sustainable markets for renewable energy systems are rapidly evolving. . . .

Human Civilization and Energy Use

Energy is one of the most basic of human needs. The accomplishments of civilization have largely been achieved through the increasingly efficient and extensive harnessing of various forms of energy to extend human capabilities and ingenuity. Providing adequate and affordable energy is essential for eradicating poverty, improving human welfare, and raising living standards worldwide.

With the exception of humans, every organism's total energy demand is its supply of energy in the form of food derived directly or indirectly from the sun's energy. For humans the energy requirements are not just for heating, cooling, transport and manufacture of goods but also those related to agriculture. However, the ingenuity of humans is such that throughout history there has been an exponential increase in the carrying capacity of land, particularly during the past few decades. . . . What is apparent [is] that in the long-term future availability of energy will increasingly become linked with agriculture production.

The estimates of world population produced by the UN suggest that six and a half billion people were living in 2005 and that the next 45 years will see a further increase of 40% to 9.1 billion.

Other estimates suggest that the world population is expected to double by the middle of this century [2]. Most of the population increase will take place in developing countries. Three years ago the United Nations Population Division had estimated the 2050 population at 8.1 billion. This rapid increase will take place mostly in the developing world. With the growing world population and people's innate aspirations for improved life, a central and collective global issue in the new century will be sustaining economic growth within the constraints of our planet's limited natural resources while at the same time preserving our environment—the so-called global sustainability issue. The population growth and the inevitable need to expand economic output would place enormous demands on our stock of natural and environmental resources. To frame it in a better perspective, global economic output doubles approximately every 30 years, accompanied by an increased demand on our natural resources and a greater impact on the environment. But it is not simply population growth that makes the future compounded. Urbanization is occurring even faster, as impoverished people seek opportunity by migrating to already crowded cities. These factors will place demands on increased energy output as well, as per capita energy consumption will rise with economic growth. For instance, assuming that sometime before the end of the 21st century the global average per capita energy consumption reaches half that of the US today (which is approximately the same as Eastern Europe in today's terms) and former Soviet Union today, the global annual energy consumption will be five times the present value.

Since, the utilization of energy is essential to the survival of human civilization, the challenge for the 21st century is to develop methods of generating and using energy that meet the needs of growing global population while protecting the planet.

Transition in Energy Use

Energy, being a crucial feature of human life, has evolved to match with contemporary human development and requirements. It has been estimated that the global population in 1800 was approximately 1 billion, an uncertain estimate given that the first population census had just been introduced around that time in Sweden and England. Estimates of past energy use based on historic statistics and current energy use in rural areas of developing countries suggest that energy use per capita typically did not exceed some 20 GJ as a global average. Over 200 years later, the global population has risen by a factor of 6 while the per capita energy consumption is estimated to have risen by a factor of 20 [3]. A 20-fold increase, far in excess of world population growth, constitutes the first major energy transition, a transition from penury to abundance. This transition is far from complete and is characterized by persistent spatial and temporal heterogeneity. This transition in energy quantities is also closely linked to corresponding energy transition in terms of energy structure as well as in terms of energy quality. . . . Given the past record of developed countries in their profligate use of energy, developing nations tend to mimic their energy consumption pattern to match those of the developed nations. . . .

One of the most significant transitions in global energy systems is that of decarbonization, an increase in energy quality. Considering the case of fossil fuels, the dominating energy resource over the course of human history, each successive transition from one source to another—from wood to coal, from coal to oil—has entailed a shift to fuels that were not only harnessed and transported more economically, but also had a lower carbon content and higher hydrogen content. It is also evident that at each step greater energy density is being achieved. The third wave of decarbonization is now at its threshold, with natural gas use growing fastest, in terms of use, among the fossil fuels. The fourth wave, the production and use of pure hydrogen, is certainly on the horizon. Its major drivers are technological advances, renewed concern about the security and price of oil and gasoline, and growing pressure to address local air pollution and climate change. . . .

Renewable Energy

Renewable energy as the name implies is the energy obtained from natural sources such as wind power, solar energy, hydropower, biomass energy and geothermal energy. Renewable energy sources have also been important for humans since the beginning of civilization; Biomass, for example, has been used for heating, cooking and steam production; wind has been used for moving ships; both hydropower and wind have been used for powering mills to grind grains. Renewable energy sources that use indigenous resources have the potential to provide energy services with zero or almost zero emissions of both air pollutants and greenhouse gases. Renewable energy resources are abundant in nature. They are presently meeting almost 13.5% of the global primary energy demands and are acknowledged as a vital and plentiful source of energy that can indeed meet the entire world's energy demand.

Renewable energy sources have enormous potential and can meet many times the present world energy demand. They can enhance diversity in energy supply markets, secure long-term sustainable energy supplies, and reduce local and global atmospheric emissions. They can also provide commercially attractive options to meet specific needs for energy services (particularly in developing countries and rural areas), create new employment opportunities, and offer possibilities for local manufacturing of equipment.

Renewable sector is now growing faster than the growth in the overall energy market. Approximately US$22 billion was invested in renewable energy worldwide in 2003. Annual investment in renewable energy has grown almost fourfold from $6 billion in 1995, while cumulative investment since 1995 is of the order of $110 billion. The 2003 investment shares in the renewables sector was roughly 38% for wind power, 24% for solar PV, and 21% for solar thermal hot water. Small hydropower, biomass power generation, and geothermal power and heat made up the remaining 17%. Total renewable power capacity stood at roughly 140 GW as of 2003, excluding large hydro. This represents slightly less than 4% of the world's total electric power capacity. About 40% of total renewable power capacity is installed in developing countries. Worldwide, wind capacity increased by 26% and grid-connected solar PV capacity increased by an incredible 50% (365 MW) in 2003. These growth rates far outpace those for traditional electric power, currently 1–3% in most countries, except China, where traditional power capacity is growing at rates of 7–9% [4]. Some long-term scenarios postulate a rapidly increasing share of renewable technologies made up of solar, wind, geothermal, modern biomass, as well as the more traditional hydro. Under such a scenario, renewables could reach up to 50% of the total share by the mid-21st century with appropriate policies and new technology developments. . . . We may observe that the latter two forms of renewable energy are complementary and thus, while nations close to the equator may be blessed with solar power, wind energy has good prospects for nations closer to the poles. An appropriate combination of the above two sources would obviously be a feasible solution for any given nation.

Growing Energy Demand

The demand for the provision of energy is increasing worldwide and will continue to rise due to rapidly rising human population and modernisation trends across the world. The International Energy Outlook projects strong growth for worldwide energy demand up to 2025 [5]. Total world consumption of marketed energy is expected to expand by 57% over the 2002–2025 time period. In the *IEO2005* mid-term outlook, the emerging economies account for nearly two-thirds of the increase in world energy use, surpassing energy use in the mature market economies for the first time in 2020. In 2025, energy demand in the emerging economies is expected to exceed that of the mature market economies by 9%. Much of the growth in energy demand among the emerging economies is expected to occur in emerging Asia, which includes China and India; demand in this region is projected to more than double over the forecast period. Primary energy consumption in the emerging economies as a whole is projected to grow at an average annual rate of 3.2% up to 2025. In contrast, in the mature market economies—where energy consumption patterns are well established—energy use is expected to grow at a much slower average rate of 1.1% per year over the same period. In the transitional economies of Eastern Europe and the former Soviet Union, growth in energy demand is projected to average 1.6% per year.

Energy-Related Challenges

The present energy situation, led by fossil fuels, has four major concerns: depletion of fossil fuel reserves, global warming, energy security concerns and rising energy cost.

Fossil Fuels Depletion

World ultimate conventional oil reserves are estimated at 2000 billion barrels. This is the amount of production that would have been produced when production eventually ceases. The global daily consumption of oil equals 71.7 million barrels. Different countries are at different stages of their reserve depletion curves. Some, such as the United States, are past their midpoint and are in terminal decline, where as others are close to midpoint such as UK and Norway. However, the five major Gulf producers—Saudi Arabia, Iraq, Iran, Kuwait and United Arab Emirates—are at an early stage of depletion and can exert a swing role, making up the difference between world demand and what others can supply. The expert consensus is that the world's midpoint of reserve depletion will be reached when 1000 billion barrels of oil have been produced—that is to say, half the ultimate reserves of 2000 billion barrels. It is estimated that around

Table 1

Various Projections of Global Ultimate Conventional Oil Reserves and Peak Year (Billions Barrels)

Author	Affiliation	Year	Estimated Ultimate Reserves	Peak Year
Hubert	Shell	1969	2100	2000
Bookout	Shell	1989	2000	2010
Mackenzie	Researcher	1996	2600	2007–2019
Appleby	BP	1996		2010
Invanhoe	Consultant	1996		2010
Edwards	University of Colorado	1997	2836	2020
Campbell	Consultant	1997	1800–2000	2010
Bernaby	ENI	1998		2005
Schollenberger	Amoco	1988		2015–2035
IEA	OECD	1998	2800	2010–2020
EIA	DOE	1998	4700	2030
Laherrere	Consultant	1999	2700	2010
USGS	International Department	2000	3270	
Salameh	Consultant	2000	2000	2004–2005
Deffeyes	Princeton University	2001	1800–2100	2004

1000 billion barrels have already been consumed and 1000 billion barrels of proven oil reserves are left in the world. It was reported in the year 2003, that reserve to production ratio of fossil fuels for North America, Europe and Eurasia, and Asia Pacific were 10, 57 and 40 years, respectively [6]. Research conducted at the University of Uppsala in Sweden claims that oil supplies will peak soon after 2010, and gas supplies not long afterwards, making the price of petrol and other fuels rocket with potentially disastrous economic consequences unless people have moved to alternatives to fossil fuels [7]. Similarly, a growing number of opinions among energy experts suggest that global conventional oil production will probably peak sometime during this decade, between 2004 and 2010 as shown in Table 1. Declining oil production will cause a global energy gap, which will have to be filled by unconventional and renewable energy sources [8].

Global Warming

There is an intimate relationship between energy and environment. The production and use of all energy sources results in undesirable environmental effects, which vary based on the health of the existing ecosystem, the size and health of the human population, energy production and consumption technology, and chemical properties of the energy source or conversion device. A shorthand equation for the environmental impacts of energy production and use has been provided by Solomon, as following [9]:

$$I = PAT,$$

Figure 1

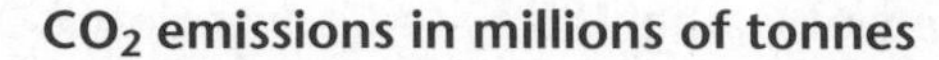

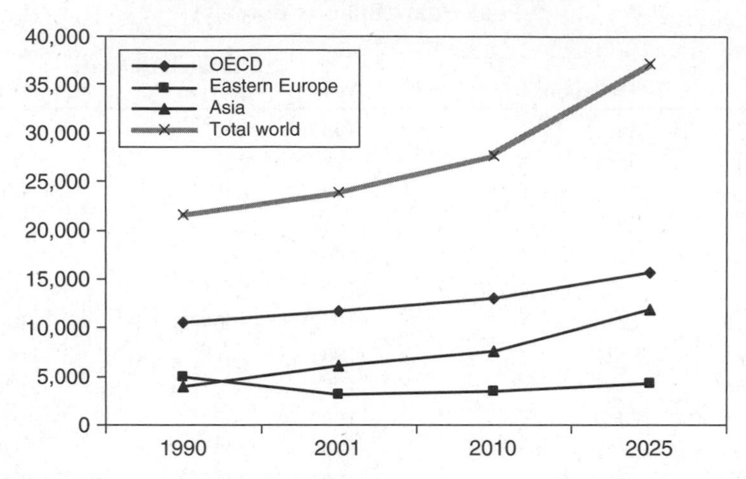

where *I* is the environmental impact; *P* the size of human population; A the affluence of the population (e.g., per capita income and/or energy use); and *T* the Technology (e.g., energy efficiency, emission rate of air and water pollution).

The balance of evidence suggests that there is a discernible human influence on global climate. Figures on carbon dioxide (CO_2) illustrate how the waste deluge has grown. Global CO_2 production climbed to 23.9 billion tons in 2001 from 21.5 billion tons in 1990. In business as usual scenario CO_2 level is projected to rise to as much as 37.1 billion tonnes by the year 2025 as indicated in Figure 1. Global mean temperature is forecast to rise by between 1°C and 4.5°C by 2100, with best estimates somewhere between 2 and 3 degrees [2]. All projections produce rates of warming that are greater than those experienced in the last 10,000 years. Sea level is projected to rise by about 50 cms by 2100 (with a range of 15–95 cms), due to thermal expansion of the oceans and melting glaciers and ice sheets. Temperature and sea level changes will not be globally uniform. Land areas, particularly at high latitudes, will warm faster than the oceans, with a more vigorous hydrological cycle potentially affecting the rate and scale of various extreme events such as drought, flood and rainfall. Impacts on natural and semi-natural ecosystems, agriculture, water resources, human infrastructure and human health are subject to many uncertainties, but all will be subject to stresses which will exacerbate stresses from other sources such as land degradation, pollution, population growth and migration, and rising per capita exploitation of natural resources.

Climate change is responsible for huge economical consequences. Between the 1960s and the 1990s, the number of significant natural catastrophes such as floods and storms rose ninefold, and the associated economic losses rose by a factor of nine. Figures indicate that the economical losses as a direct result of natural catastrophes over 5 years between 1954 and 1959 were

US$35 billion while between 1995 and 1999 these losses were around US$340 billion [10]. Natural catastrophes associated to global warming killed over 190,000 people in 2004, twice as many as in 2003, with an economic cost of US$145 billions. The August 2005 Hurricane Katrina is being held responsible for more than 1000 human lives. Hurricane Katrina caused at least $125 billion in economic damage and could cost the insurance industry up to $60 billion in claims. That is significantly higher than the previous record-setting storm, Hurricane Andrew in 1992, which caused nearly $21 billion in insured losses in today's dollars [11]. Some economists believe, the physical and psychological damage caused by Katrina is likely to reverberate across the global economy in ways that will curb growth well into [the future]. Katrina shut down large portions of oil and gas production in the Gulf of Mexico at a time when worldwide energy output was already stretched thin. While the storm's impact was most acute in the United States, it also sent fuel costs higher around the globe, squeezing consumers in Europe and Asia [12].

Energy Security

The economies of all countries, and particularly of the developed countries, are dependent on secure supplies of energy. Energy security means consistent availability of sufficient energy in various forms at affordable prices. These conditions must prevail over the long-term if energy is to contribute to sustainable development. Attention to energy security is critical because of the uneven distribution of the fossil fuel resources on which most countries currently rely. The energy supply could become more vulnerable over the near term due to the growing global reliance on imported oil. Of the trillion barrels of the proven oil reserves currently estimated, 6% are in North America, 9% in Central and Latin America, 2% in Europe, 4% in Asia Pacific, 7% in Africa, 6% in the Former Soviet Union. Presently 66% of global oil reserves are distributed amongst Middle Eastern countries: Saudi Arabia (25%), Iraq (11%), Iran (8%), UAE (9%), Kuwait (9%), and Libya (2%) [13]. The oil and gas reserves in non-Middle East countries are being depleted more rapidly than those of Middle East producers. If production continues at present rate, many of the largest, non-Middle Eastern, producers in 2002, such as Russia, Mexico, US, Norway, China and Brazil will cease to be relevant players in the oil market in less than two decades. At that point, the Middle East will be the only major reservoir of abundant crude oil—within 20 years or so about four-fifths of oil reserves could be in the hands of the Middle Eastern countries as shown in Figure 2. Many of these leading oil producing countries are politically unstable.

The Middle Eastern region as a whole has quite a volatile geopolitical situation as it has seen a number of conflicts over the past few decades. The oil factor cannot be ruled out in some of the major conflicts in the area. There are serious reservations regarding security of oil; production and supply channels of some of the Middle Eastern countries like Iraq, that is the second largest oil-producing country in the world, are regarded as the legitimate targets of radical elements because of various internal and external conflicts.

Figure 2

BP statistical review of world energy: world reserves for 2002 and 2020

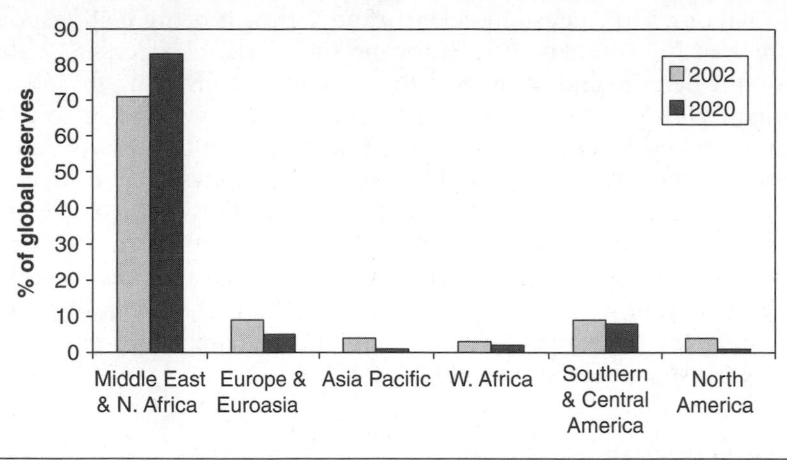

Getting oil from the well to the refinery and from there to the service station involves a complex transportation and storage system. Millions of barrels of oil are transported every day in tankers, pipelines and trucks. This transportation system has always been a possible weakness of the oil industry, but it has become even more so in the present volatile geopolitical situation, especially in the Middle East region. The threats of global terrorism have made the equation more complex. Tankers and pipelines are quite vulnerable targets. There are approximately 4,000 tankers employed, and each of them can be attacked in the high seas and more seriously while passing through narrow straits in hazardous areas. Pipelines, through which about 40% of the world's oil flows, are no less vulnerable and due to their length, they are very difficult to protect. This makes pipelines potential targets for terrorists.

The above analysis indicates that the present situation is not sustainable, and that it cannot guarantee secure supplies of energy. The psychological effects are felt already, as demonstrated by significant fluctuation of oil prices over the recent years, due to relatively minor events in the Middle East. A more serious event, such as the sinking of an oil tanker in one of the busy shipping lanes or disruption of a major pipeline, would have a much more catastrophic effect on the oil prices and, hence, the world economy.

Rising Oil Prices

The rise in oil price has made headlines across the globe throughout the year 2005. Increasing demand especially from countries like China and India, geopolitical features across the world especially in the Middle Eastern region and weather related supply shocks have fuelled the continual rise in crude oil prices. The cost of gulf oil, which was US$17 per barrel at the end of 1999, had

Figure 3

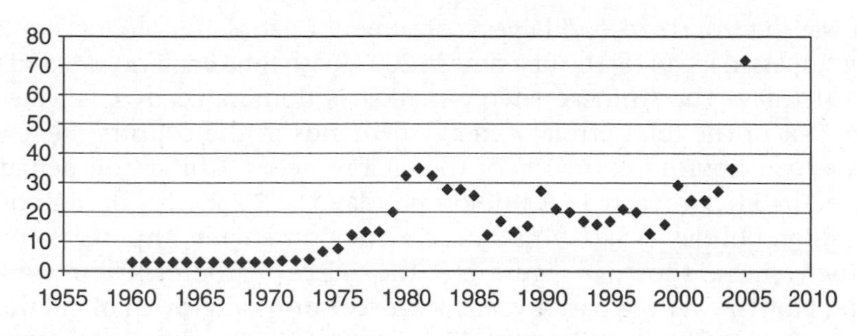

reached US$35 per barrel by the end of 2004. By the middle of 2005 the price stood well over US$60 per barrel. Oil prices have especially skyrocketed after July 2005 for various reasons.

The crude oil prices are extremely sensitive to a number of geopolitical factors. Political unrest, military conflicts or extreme weather events, all have played their role in causing a rapid rise in global oil prices. Having a look at Figure 3 it is evident that several such issues like Yom Kippur War (1973), Iranian Revolution (1979), Iran/Iraq War (1980), First gulf War (1991), unrest in Venezuela (2002) and Second Gulf War (2003) have all contributed to a rapid increment in crude oil price. The oil prices were noticed to make an immediate jump from US$57 to more than US$65 at the news of the death of King Fahad—the late king of Saudi Arabia. Natural catastrophes have started emerging as another crucial element in the price game—the US hurricane Katrina caused prices to rise to an all time record high, almost US$72/barrels shown in Figure 3.

The global demand for oil is continuously increasing especially after the growth trend of some of the emerging economies of the world such as China and India that started appearing towards late 1990s. The production capacity on the other hand is facing the reverse trend. It has been observed that the loss of production capacity of various oil rich countries in the world such as Iraq and Venezuela combined with increased production to meet growing international demand led to the erosion of excess oil production capacity. In mid 2002, there was over 6 million barrels per day of excess production capacity, but by mid 2003 the excess was below 2 million. During much of 2004 and 2005 the spare capacity to produce oil has been less than one million barrels per day. A million barrels per day is not enough spare capacity to cover an interruption of supply from almost any OPEC producer [14]. In a world that consumes well over 70 million barrels per day of petroleum products that adds a significant risk premium to crude oil price.

The Crucial Energy Economies

China

China was the world's second largest consumer of petroleum products in 2004, having surpassed Japan for the first time in 2003, with total demand of 6.5 million barrels per day. The Chinese energy market is dominated by coal, meeting almost 58% of the total primary energy demands in the country. Renewable energy is contributing up to 7% of the energy needs. China's oil demand is projected by EIA to reach 14.2 million bbl/day by 2025, with net imports of 10.9 million bbl/day [15]. China's dependency on energy imports is consistently increasing as shown in Figure 4. As the source of around 40% of world oil demand growth over the past 4 years, with year-on-year growth of 1.0 million bbl/day in 2004, Chinese oil demand is a key factor in world oil markets.

There is enormous potential for renewable energy in China. Renewable energy is seen as crucial and there is enormous international interest in China's potential as a huge market for wind power and other renewable energy technologies. China has similarly huge potential for solar, wave, tidal and biomass power and with energy efficiency could meet all its needs solely from clean energy. By 2004, China has been equipped with 110 GW installed hydropower capacity, 760 MW installed wind power capacity in 43 interlinked plants, and about 60 MW solar photovoltaic batteries. Solar energy water heaters cover 65 million square meters, accounting for more than 40% of the world's total. Moreover, China has built over 11 million household biomass pools in rural areas and around 2,000-odd large and medium-sized biomass projects, with annual fuel output reaching 5.5 billion cubic meters [16].

Renewable energy is developing rapidly in China, with an annual growth rate of more than 25%, the highest in the world. The growth of the wind

Figure 4

China's oil production and consumption (1980–2005), thousand barrels per day

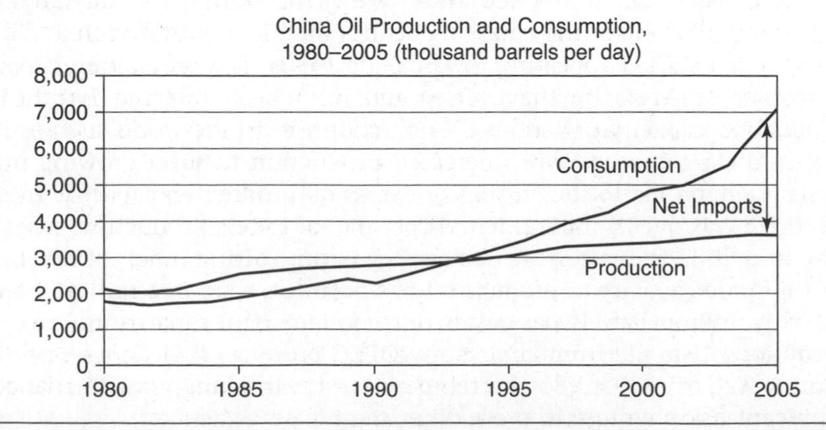

Table 2

Expected Growth of Renewables in China Under the New Law

Sector	Current Capacity (End 2004) (MW)	Expected Capacity in 2020 (MW)
Wind power	560	20,000
Solar energy	50	1000
Biomass	2000	20,000
Hydropower	7000–8000	31,000

energy in China in 2004 was 35%. In February 2005, China has approved its renewable energy promotion law. According to the plan by 2015, China will annually develop new and renewable energy resources amounting to 2% of the country's total energy consumption [17]. The renewable energy targets, set under the new law, to be achieved by the year 2020 have been highlighted in Table 2.

India

India accounted for 3.5% of world primary consumption and 12% of total primary energy consumed in the Asia-Pacific region in 2002 [18]. The country is the world's sixth largest energy consumer and indeed a net energy importer. India is a vast country with a diverse mix of resources. The energy consumption is attributed to a number of fuels with varying use patterns. Coal dominates the energy mix in India, meeting about 60% of energy requirements in the country. Nearly 30% of India's energy needs are met by oil, and more than 60% of that oil is imported. Natural gas has experienced the fastest rate of increase of any fuel in India's primary energy supply. Presently, the natural gas contribution is met entirely by domestic production. However, the gap between demand and supply is set to widen unless major gas discoveries are made. It is expected that by 2010 almost three-quarters of India's oil and gas needs will be met by imports [19].

For India to tackle the economic and environmental challenge of its energy demand growth it is important to have a good understanding of how these and other factors shape energy use in various sectors of the economy. Detailed and coherent information is needed in order to judge the potential for energy efficiency improvements or to measure the progress of already implemented policies. India's rapidly growing economy will drive energy demand growth at a projected annual rate of 4.6% through to the year 2010 [20]. It has been reported that India's electric power demand is likely to increase threefold by the year 2051. Indications are that the electric power demand is expected to grow at around 10% per annum in next 15 years requiring about 10,000 MW of capacity addition every year over this period [21]. Other studies indicate that by 2020, India's demand for commercial energy is expected to increase by a factor of 2.5 [22]. Underpinning this trend will be the ongoing growth in population, urbanization, industrial production and transport demand.

Energy shortages in India have been increasing in the past few years, from 5.9% in 1998/99 to 7.8% in 2000/01. The peak energy demand during 2001–02 touched 86 TWh, of which only 75 TWh could be met. India will continue to experience this energy supply shortfall for at least another 15 years. This gap has been exacerbated since 1985, when the country became a net importer of coal. The growing gap between the demand and supply of energy, and environmental externalities associated with energy use are the key issues today. The energy demand, GDP and population are predicted to increase significantly within the next two decades as shown in Figure 5. Almost half of India's trade deficit is due to petroleum imports, the cost of which also limits capital that could be invested in the economy [20]. High economic development requires rapid growth in the energy sector. This implies substantial increases in electric power generation and transmission capacities, and exploitation of new avenues of energy supply.

Nuclear energy contributes 2% to India's total power generation. Indications are that the share would be even less in 2010 due to many unsolved problems—high cost, radioactive wastes and decommissioning costs [23]. In 2002, India's oil consumption rose to 97.7 million tonnes, more than 2.6 times the domestic production [18]. Estimates indicate that oil imports will meet 75% of total oil consumption requirements and coal imports will meet 22%

Figure 5

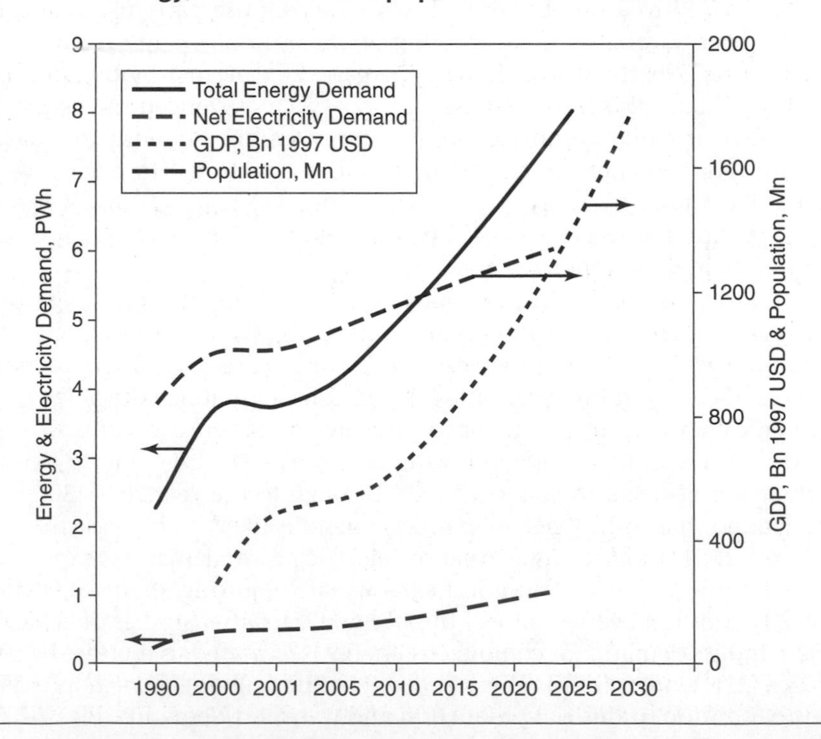

Table 3

Renewable Energy Potential and Achievements in India

Energy Sources	Potential	Achievement as on 31.12.2002	India's Position in the World
Biogas plants	12 million	3.37 million	Second
Improved chulhas	120 million	33.9 million	Second
Wind	45,000 MW	1702 MW	Fifth
Small hydro	15,000 MW	1463 MW	Tenth
Biomass Power/Cogeneration	19,000 MW	468 MW	Fourth
Biomass gasifiers		53 MW	First
Solar PV	20 MW/sq. km	107 MW_p	Third
Waste-to-energy	2,500 MW	25 MW_e	
Solar water heating	140 million sq.m Collector area	0.68 million sq.m Collector area	

Subscript p stands for peak and e for electrical.

of total coal consumption requirements in 2006 [23]. India can ill afford to overly depend on fossil fuels considering the high cost of imported petroleum products. India's fossil fuel resources are limited compared to global reserves.

There is a large potential of renewable energy resource available in India, an estimated aggregate of over 100,000 MW, which needs to be harnessed in a planned and strategic manner to mitigate the gap between demand and supply. In the present scenario renewable energy is contributing to about 3.5% of the total installed electric capacity of about 3700 MW. It is planned to increase this to 10% of the total power generation capacity by the year 2012. Today India has the largest decentralized solar energy programme, the second largest biogas and improved solar stove programmes, and the fifth largest wind power programme in the world. Table 3 describes India's renewable energy potential and achievements [24]. However, only a fraction of the aggregate potential in renewables, and particularly solar energy, has been utilized so far. The achieved wind power capacity is less than 4% of the potential from this source. Similarly, the respective installed biomass power and small hydropower capacities are about 2.4% and 9.8% of their potential. . . .

Discussion and Conclusions

The global energy supply is facing an array of severe challenges in terms of long-term sustainability, fossil fuel reserve exhaustion, global warming and other energy related environmental concerns, geopolitical and military conflicts surrounding oil rich countries, secure supply of energy and fuel price increase. Renewable energy sources are capable of meeting the present and future energy demands with ease without inflicting any considerable damage to the global ecosystem. The renewable energy sector, currently meeting 13.5% of the global energy demand, is now growing faster than the growth in the overall energy market. Some long-term scenarios postulate a rapidly increasing share of renewable technologies. Under these scenarios, renewables could reach up to 50% of the total share by mid-21st century with appropriate policies and new technology developments. A large number of renewable-based activities are being undertaken on national and international platforms around the world. . . .

The five candidate countries chosen for this study are amongst the top seven energy-consuming nations in the world. They alone are responsible for consuming 49% of the world's total energy. The United States of America and the United Kingdom, respectively, are the first and fourth largest economies in the world. China and India jointly constitute almost one third of the world's entire population and have fast growing economies. Russia is amongst the leading economies of the world and is an oil rich country—in 2004 Russia was the second largest oil producing nation in the world after Saudi Arabia with a capacity of 8.8 million barrels/day. Four out of the five countries studied in the present work, China, India, UK and USA, are all net importers of energy and heavily depend on fossil fuel to meet their energy requirement as shown in Fig 12. Their local fossil fuel reserves are close to exhaustion. For example, China, India, UK and USA have local oil reserves that will only last 9, 6, 7 and 4 years, respectively.

The UK has pledged to increase its present capacity of renewable energy production, 3% in 2004, to 10% and 20%, respectively, by the year 2010 and 2020. China is seeing renewable energy as playing an active role in its future energy scenario. China has renewable energy growing at an exemplary pace in the world—in 2004 renewable energy in China grew by 25% against 7–9% growth in electricity demand. While in the same year, wind energy in China saw a growth of 35%. China is also leading the global solar thermal market as it has already installed solar collectors over 65 million square meters, accounting for more than 40% of the world's total collector area. India, having immense potential for renewable energy, has also taken steps to utilize the immense potential for renewable energy sources. The goals of Indian energy planning include the promotion of decentralised energy technologies based on renewable resources in the medium term, and the promotion of energy supply systems based on renewable sources of energy in the long term. . . .

It is therefore concluded that with increasing concern regarding climate change, depleting fossil fuel reserves and human quest for development of cleaner forms of energy and ingenuity, a switch over to renewable energy sources is quite achievable.

All references for articles included in *Taking Sides: Clashing Views in Sustainability* can be found on the Web at www.mhhe.com/cls.

Yun Zhou

Why Is China Going Nuclear?

Introduction

In the past several years China has made an extraordinary commitment to nuclear energy development. It currently has 11 nuclear power units in commercial operation, a small stake compared to the 104 operational reactors in the United States and the 59 operational reactors in France, but its nuclear energy output is expected to grow substantially in the coming decades.

As of 2004, China's nuclear power plants had a capacity of 7 GWe and produced 50.4 TWh, accounting for 2.3 percent of the nation's electricity generation (National Bureau of Statistics of China, 2004). A slate of subsequent policy initiatives proposed building on this total. In 2006, China's State Council approved the National Development and Reform Commission (NDRC)'s "Medium and Long-Term Nuclear Power Development Plan (2005–2020)," which outlined plans to increase the nation's nuclear capacity to about 40 GWe by 2020, raising to 4 percent nuclear's share of the national electricity generation capacity. A 2007 State Council Information Office White Paper, "China's Energy Conditions and Policies," further enshrined nuclear energy as an indispensable energy option (State Council Information Office, 2007). Recent reports suggest that the country's installed nuclear power capacity might even exceed 60 GWe by 2020 due to faster than expected construction (China Daily, 2008).

While a 4–6 percent share of national generation capacity would be relatively small, the absolute quantity is remarkably large. To implement the plan, China will have to construct at least three 1 GWe nuclear power units each year for the next 16 years. This is a tremendous growth rate, particularly in contrast to growth in Western countries, many of which have pledged to phase out nuclear power or are waiting for the "nuclear renaissance" to begin.

Such rapid nuclear expansion will affect China financially, environmentally, politically, and even socially. Yet the circumstances under which China has developed these nuclear energy policies are not well understood. Why did China transform its nuclear energy policy so quickly? Is nuclear energy necessary to meet China's huge and growing energy demands? How will the Chinese nuclear expansion unfold? To address these questions requires a review of China's energy profile and challenges, especially coal uses in China, which has dominated China's energy mix for decades and will continue to dominate through 2030. A comparative study between coal and other energy sources cannot be neglected in any of China's energy policy studies.

Can Coal Suffice to Meet China's Growing Energy Needs?

An Increasing Demand for Coal

Since its economic reforms in 1978, China's gross domestic product (GDP) has grown by about 10 percent per year (Bergsten et al., 2006). This growth has quadrupled China's total energy consumption (National Bureau of Statistics of China, 1978–2009).

Figure 1 shows the total electricity generation and contributions from major generation sources from 1990 to 2009. In 2009, China had a total installed electricity generation capacity of 874 GW and generated 3650 TWh of electricity (National Bureau of Statistics of China, 2009). The rapid growth in electricity demand spurred significant investment in new power stations. Since 2004, the total installed capacity has increased at an average rate of 90 GWe per year.

Although China's rapid growth in electricity capacity makes it the second largest country for installed capacity and electricity generation in the world, China still suffers from severe power shortages. This is most apparent in the summer, when China's coastal regions have to rotate daily power blackouts in industrial and residential areas to ease electricity load. China is aiming to quadruple its 2000 GDP by 2020, which would be equivalent to a 7.2 percent annual GDP growth rate. Projections for China's 2020 electricity demand range from 2254 to 5200 TWh depending on differing assumptions about the relationship between electricity demand growth and GDP growth, and about the electricity elasticity of GDP (Li et al., 2004; Hu et al., 2005; EIA, 2006).

Figure 1

The total electricity generation and components in China from 1990 to 2009

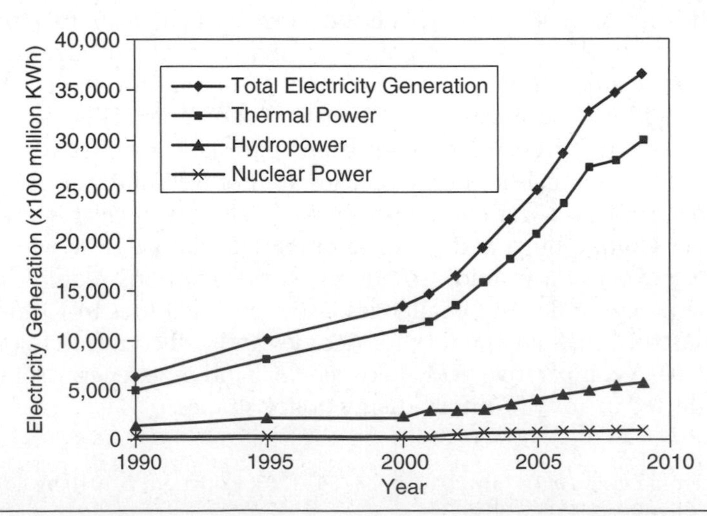

While electricity demand projections are relatively uncertain, coal-fired generation will definitely play a fundamental role in China's energy mix and electricity generation. Coal, as China's primary energy resource, has supported most of its energy growth to date. Coal-fired generation accounted for 80 percent of China's electricity generation in 2007 (National Bureau of Statistics of China, 2007). Another way of characterizing China's dependence on coal is to note that electricity generation is responsible for 68 percent of the increase in coal consumption during the past 15 years; coal-fired generation has met on average 80 percent of increases in electricity demand during the same period (National Bureau of Statistics of China, 1990–2007). China's reliance on coal is only expected to increase as a consequence of rapid economic growth. Although China aims to reduce the percentage of coal use in its total primary energy consumption from 69.5 to 40 percent by 2050, coal-fired power generation will remain dominant.

On one hand ramping up coal generation is a sensible way to address energy demands. Compared with other energy generation options, coal-fired power generation in China has lower investment costs, shorter construction periods, and lower electricity production costs. The pollution penalties for utilities are low, and coal-fired plants often choose to pay the fee rather than invest in cleaner generation technologies. On the other hand, China's reliance on coal has slowed the diversification of China's energy sector and has caused a range of problems, from heavy air pollution to clogged transportation routes to unsafe coal mines.

China pursues comprehensive energy conservation and efficient energy use to lower its energy demands. China's 11th Five-Year energy development plan aims to limit China's total primary energy consumption to 2.7 billion tons of standard coal by 2010. However, in 2008, China's coal production had already reached 2.72 billion tons of coal, and total energy consumption reached 2.91 billion tons of standard coal in 2008 (see Figure 2, National Bureau of Statistics of China, 2003–2009). China's Medium and Long-Term Energy Conversation Plan (2004) warned that when national coal consumption approaches 3 billion tons of standard coal, society and the environment would be pushed to a critical point, posing tremendous costs and pressures on energy infrastructure construction, water resources, and transportation capability. Existing environmental problems also would be aggravated to an intolerable level. If China's energy consumption continues to grow at the current pace, continued heavy reliance on coal would result in energy security challenges.

It is not clear that China would expand its domestic coal production to meet demand, even if all the attendant problems associated with coal could be solved. As part of its long-term energy policy, China has said it would like to meet 90 percent of its energy demand with domestic resources and generating capacity. Simply importing coal from abroad to meet its growing energy demand is thus not an option.

Coal Transportation and Price

In January 2008, the worst winter snowstorm in five decades hit central, eastern and southern China. China's coal-dominated energy infrastructure

Figure 2

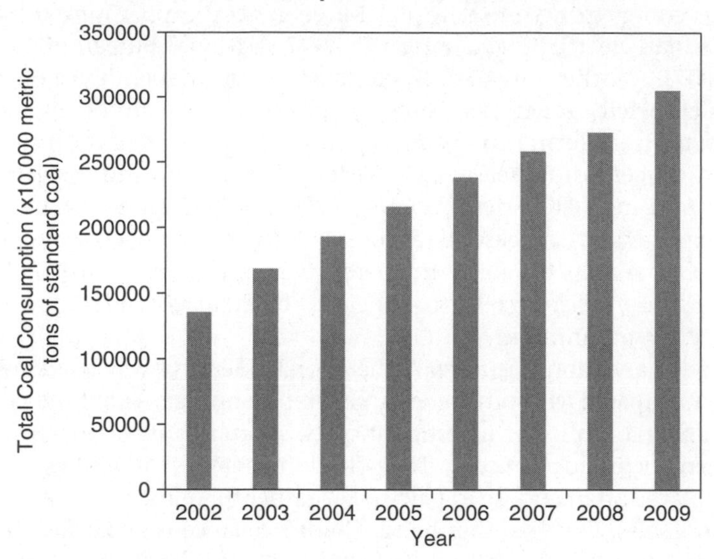

exacerbated the disastrous consequences of the extreme weather. Snow caused bad road conditions that prevented coal from being transported from the inland regions to the major population centers on the coast. The cold weather dramatically increased electricity demand throughout the country. Coal-fired plants in several provinces, such as Zhejiang Province, suffered a sharp decline in coal reserves. At the shortage's most severe point, coal stockpiles were sufficient to generate only three days of electricity. Some regions had to cut the power supply to their industrial areas to ensure local residents would survive.

Even in good weather, coal transportation is a very serious challenge to China's power industries. Approximately 80 percent of China's coal resources are located in mountainous regions far away from industrial centers and highly populated coastal regions. . . . Shaanxi, Shanxi, and Inner Mongolia, the three largest coal-producing provinces, contribute more than 50 percent of China's total coal output, while coastal regions such as Shanghai, Zhejiang, Guangdong, and Fujian account for a majority of coal consumption. As of 2007, the rail network had the capacity to transport more than 1.17 billion metric tons of coal per year, which amounts to more than 47 percent of the total railway transportation capacity (Xinhua News Agency, 2008a, b). Yet this capacity has proved insufficient to deal with rising coal output and is one of the major factors leading to rising coal prices in recent years (National Bureau of Statistics of China, 2003–2008). From 2000 to 2008, the rate of coal freight carried by national rail grew only 8.8 percent annually, much lower than the 13.7 percent coal production growth rate over the same period (National Bureau of Statistics of China, 2000-2008). The shortage of railway networks, their inferior foundations, and slow renovation schedules suggest that railway transport capacity will remain nearly saturated along all the main coal transport routes for the foreseeable future, except along the Daqin Railway, the largest coal haul railway in China and the railway with the largest annual freight volume in the world (Liu, 2007). The Daqin line uses an electrified double track system with advanced heavy-haul transportation technologies that allowed it to increase its annual load to 300 million tons of freight in 2007. The line is capable of maximally supporting the transport of 400 million tons of freight annually (Xinhua News Agency, 2009). According to the Ministry of Railways, demand for the rail transport of coal is expected to be 1700–1800 Mtpa by 2010 and 2000-2200 Mtpa by 2020, which is larger than rail capacity is expected to grow in the next decade. Therefore, bottlenecks are expected to persist, particularly in Shaanxi and the central southern part of Shanxi as coal production becomes further concentrated in these provinces and Inner Mongolia (IEA, 2009).

Until recently, the Chinese government controlled domestic coal prices as a way to guarantee enough cheap resources to support energy use and economic development. The Chinese government shifted course in 1993 and adopted a "dual track" system. At the end of every year, the government announces its "guided" utility coal price, which reflects some increase based on market dynamics but is still below the commercial coal price. Table 1 gives the average coal prices from key state-owned mines under the Dual Track system (Pan, 2004–2008; Pan and Zhang, 2003; Pan and Wang, 2009). In December 2005, the NDRC went a step further and abolished its use of a guiding price for utility coal. However, electricity prices remained fixed while coal prices soared, hurting utility companies. The government was forced to temporarily intervene, for example, by capping price increases at 8 percent in 2005 and capping them again in 2008. As demonstrated in Table 1, the gap between coal

Table 1

Average Commercial Prices of Coal vs. Prices of Utility Coal During 2002–2008

Average commercial price of coal (Yuan/ton)							
Year	2002	2003	2004	2005	2006	2007	2008
	168	174	207	270	302	331	463
Increase rate		6.0% ↑	15.7% ↑	30.4% ↑	11.9% ↑	9.6 ↑	39.9 ↑
Average price of coal for electricity (Yuan/ton)							
Year	2002	2003	2004	2005	2006	2007	2008
	137	139	162	212	216	246	313
Increase rate		1.5% ↑	11.7% ↑	30.9% ↑	2% ↑	13.9% ↑	27.2% ↑
Difference between two prices (Yuan/ton)							
Year	2002	2003	2004	2005	2006	2007	2008
	31	33	45	58	86	100	150
Increase rate		6.5% ↑	36.4% ↑	28.9% ↑	48.3% ↑	16.3% ↑	50% ↑

prices for electricity generation and coal prices for other users increased and reached 32.4 percent in 2008, which showed the government's continued ability to mediate potential conflicts between coal producers and coal-fired plants. Nevertheless, the government has emphasized that it still plans on liberalizing coal pricing, ending government intervention, and encouraging long-term contracts between utility companies and mines. As part of coal industry reforms, the Chinese government also raised the tax rate for coal resources from 1 to 3 percent, and the state council proposed requiring resource-based enterprises, such as the coal industry, to set up a reserve fund system for sustainable development. In addition, the government collects an environmental recovery deposit and contributes to a production safety fund.

Coal Safety and Environmental Impacts

The central government is simultaneously attempting to reform safety regulations for coal mining. Due to insufficient safety measures and management, nearly 6000 Chinese coal miners died on the job in 2005. Such accidents are so commonplace that only larger accidents that kill hundreds hit the news (Oster, 2006). Small coal mines are especially dangerous. Because they currently account for about one-third of China's coal production but two-thirds of the deaths in the coal industry, the State Council proposed shutting down more than half of them and bringing their total number below 10,000 before 2010 (NDRC, 2008a, b). Though the campaign might increase safety, it is also likely to reduce total coal output capability by 250 million metric tons and it may take time for large state-owned mines to lift their output to compensate the shortfalls (NDRC, 2008a, b).

While coal output, transportation capacity, and mine safety can be improved gradually, the environmental impact of massive coal use is irreversible and devastating. Coal use is responsible for as much as 90 percent of China's sulfur dioxide emissions and 50 percent of its particulate emissions (Economy, 2007). Coal-generated air pollution has contributed to a recent, sharp rise in the number of people suffering from respiratory illness caused by particulates. Sulfur dioxide discharged from plants has increased acid rain that has reportedly damaged soil quality across one-third of China's landmass; and in 2005, acid rain fell in more than half of the 696 cities and counties under air-quality monitoring (Reuters News, 2006).

The environmental effects of China's heavy coal use are not China's problem alone. Acid rain is already a regional problem in China, South Korea, and southwest Japan (Streets et al., 1997). The government estimates that the costs of environmental damage from coal mining, including wasted resources, environmental pollution, ecological destruction, and surface subsidence, total about RMB 30 billion per year (IEA, 2009). In addition, carbon dioxide emissions are impacting global climate. Studies show that China overtook the United States as the largest emitter of carbon dioxide in 2009 (Auffhammer and Carson, 2008). According to the Intergovernmental Panel on Climate Change (IPCC), to avoid severe and irreversible consequences to the climate, nations need to stabilize the atmospheric concentrations of carbon dioxide at 550 ppmv (double the pre-industrial level) or even 450 ppmv by 2050. To reach that goal, reductions in annual carbon dioxide (CO_2) emissions need to begin by 2020. China has ratified the Kyoto Protocol so it should be trying to reduce carbon emissions even if the treaty does not mandate reductions for developing countries. In addition, China set as a goal to cut CO_2 emissions intensity by 40–45 percent below 2005 levels by 2020 as part of the agreement to emerge from the 2009 United Nations Climate Change Conference. So far, though, China has failed to lower the environmental costs of its economic growth (SEPA, 2006). Without stronger and more effective measures, China will also fail to achieve the environmental goals of its 11th Five-Year Plan, which is to reduce by 10 percent the emission of major pollutants by 2010.

The sum of these challenges—the rapidly increasing demand for coal, transportation constraints, coal cost increases, and environmental costs—are driving China to diversify its energy resources and to pursue comprehensive energy conservation and efficient energy use.

China's Energy Alternatives

To implement its energy self-sufficiency principle, relieve environmental and social pressures, and promote sustainable development, China needs to develop a range of domestic power generation alternatives. These alternatives have to be economically feasible, environmental friendly, publicly acceptable, and capable of being implemented on a large scale. Of the possible development directions, this analysis examines natural gas, hydropower (and other renewable energy options such as solar and wind), clean coal technologies, and nuclear power.

Natural Gas

Natural gas accounted for about 3 percent of total Chinese energy consumption in 2005. The U.S. Energy Information Administration predicts that use of natural gas in China will double by 2010 (EIA, 2005). As part of this growth, China has built several large-scale liquefied natural gas (LNG) power plants in its southern and eastern regions. Compared with coal-fired power plants, LNG plants have a number of advantages. They emit only 42 percent of the carbon dioxide, 21 percent of the nitrogen oxides, and relatively little of the sulfur dioxide that a coal-fired power plant of comparable size does. LNG plants require less human, land, and water resources to construct and operate, which reduces construction investment. Capital costs of LNG plants are one-third lower than those required by same size coal-fired plants. Additionally, the energy efficiency of advanced LNG plants can surpass 55–58 percent, a much higher rate than a coal-fired power plant, and these plants can be deployed on a large scale (Wu et al., 2002).

LNG plants also have several financial limitations. Fuel costs account for 60 percent of the total generating costs of LNG plants. As such, the cost and profitability of LNG power generation is highly sensitive to gas prices. In recent years, natural gas prices have only increased, while the price of coal for electricity generation remained cheap. An example of this imbalance can be found in Zhejiang Province, where the price of natural gas has been more than three times the price of coal, and where the price of electricity from natural gas generation has been twice as expensive as from coal generation. Without strong subsidy support from the Chinese government, LNG plants will have a difficult time competing with coal-fired plants (Wu et al., 2002; Yang et al., 2007; Zhu, 2007).

Insufficient domestic sources of natural gas are another major constraint to its economic viability. Chinese deposits of natural gas are distributed mainly within western provinces. China built a 4200 km-long pipeline called "west gas transport east" to bring natural gas to eastern regions, such as Shanghai and Zhejiang, yet this infrastructure would not be able to supply the large-scale LNG plants located along the pipeline were natural gas's contribution to electricity generation to grow.

The lack of domestic supply has forced LNG plants to purchase natural gas from the international market at much higher prices. Without supply stability, many LNG plants have had to shut down after several months of operations. Additionally, natural gas suppliers and consumers typically sign long-term contracts with a fixed price. Under this kind of agreement, suppliers are unable to guarantee the provision of enough gas to meet periods of peak demand, weakening the source's competitiveness (NDRC, 2006). All of these characteristics suggest that natural gas will not play a key role in reducing China's reliance on coal and helping it mitigate emissions during the coming decades.

Renewable Energy

Renewable energy accounts for only a small fraction of China's total primary energy supply. The Chinese government plans to increase the share of domestic renewable energy consumption to 10 percent of total energy consumption by 2010, according to the 11th Five-Year Plan, and it aims for a 30 percent share by 2050 (China Daily, 2007). Due to grid constraints and an insufficient long-distance transmission infrastructure, large-scale implementation of renewable energy will be a challenge, and its role in industrial and highly populated regions will remain limited for the foreseeable future.

One bright spot for China has been wind power. Total installed wind capacity reached 12.2 GW in 2008, exceeding the planned capacity outlined in the 11th Five-Year Plan (NDRC, 2008a, b). Although this rapid growth makes China the fourth largest wind market in the world, only 60 percent of this capacity is connected to grids (Forbes, 2009). Most wind resources are located in the north and the west of China, in areas such as Xinjiang, Inner Mongolia, and Gansu. This poses a huge challenge as the government will need to construct a grid and transmit the electricity generated by wind to the coastal areas where demand is high. In addition, although the recent Renewable Energy Law sets up a preferential price for electricity generated by wind power and priority access for wind power to connect to grids, it does not guarantee all electricity generated by renewable resources the priority access to connect to electricity grids. Grid companies do not have financial motivation to accommodate wind power as a non-baseload power source. Extensive policy support and law enforcement are needed to help wind power compete with other energy options. Thus, despite China's excellent wind resources, wind power's total percent contribution to electricity generation will probably remain low, particularly in the eastern coastal provinces.

Although China's solar power potential is enormous, the Chinese solar power industry is in its infancy. Grid-connected solar photovoltaic capacity in China is still marginal, just a few megawatts in 2006 (Martinot and Li, 2007). A large plant tied to the Chinese grid, such as the 1000 MW plant installed in Germany in 2006, is still at least a decade away, as the cost of solar photovoltaic technology declines further and as conventional power costs rise (Martinot and Li, 2007).

Of the renewable sources available, only hydropower presently makes a significant contribution, accounting for about 16.4 percent of China's total electricity generation in 2008. China has employed hydropower for thousands of years and is currently the world's third largest consumer and producer, after Canada and Brazil. But concerns about the environmental and social impact of implementing further hydroelectric projects might ultimately limit their growth.

China has built small- and medium-sized dams without great environmental and social impact, but its large-scale dam projects, such as the Three Gorges Dam, have been controversial. The Three Gorges project was completed as planned in 2008 and is expected to produce 84.7 billion kilowatt hours of electricity, about one-tenth of China's projected electricity consumption. Critics of the dam allege that it could damage the local environment, culture, and historical resources by, for example, changing the course of the Chang Jiang River, affecting water quality and local climate, and inducing biodiversity loss. Between 1.2 and 1.9 million people have already been forced to leave their homes and resettle elsewhere as a consequence of the dam's construction (Schreurs, 2007).

Strong public opposition has slowed down several other large-scale hydroelectric projects, including the Nu River project, and has made future development uncertain. Another potential challenge for hydropower is that most plants are located in western regions. As discussed previously, transmitting energy and electricity from western regions to the east has the potential to aggravate pressures on the environment and transportation infrastructure.

Advanced-Coal and Decarbonization Technologies

In recent years, the Chinese government has improved its advanced-coal technological capabilities, including clean coal power technology, pollution-control technology, coal gasification technology, coal liquefaction technology and coal gasification-based co-production technology and deployed certain technologies, such as pollution-control technologies. For example, the total installed capacity of flue gas desulfurization technologies increased from 53 GW in 2005 to 270 GW in 2007, accounting for more than 50 percent of total installed thermal power capacity (Wang et al., 2008). However, the development and deployment of coal gasification, coal liquefaction and coal gasification-based co-production technologies are still very limited due to weak technological innovation capabilities. In addition, due to insufficient regulation enforcement and lax emission standards in China, the deployment of pollution-control technologies did not necessarily control the increase of pollutant emissions. For example, even though a lot of flue gas desulfurization units have been installed in coal-fired plants, it is not clear that the equipment is always in operation (Zhao and Gallagher, 2007). Enforcing regulations and standards, raising regulatory standards, and improving monitoring measures remain huge challenges. Since coal will continue to dominate China's energy mix for decades, decarbonization technologies cannot be ignored as a part of the solution. Innovative decarbonization technologies are well understood but have yet to be demonstrated together at commercial scale. The cost of capturing, transporting, and disposing of carbon dioxide is still high, and the environmental impacts are largely unknown.

Decarbonization technology is still far away from the deployment stage. Liu and Gallagher (2009) briefly described three phases for the development and deployment of carbon capture and storage (CCS) technology in China. By 2020, pilot-scale demonstration projects should start up, and early commercial deployment might be possible. Between 2020 and 2030, CCS could be a commercialized technology for an emerging low-carbon economy. Beyond 2030, the adoption of CCS could become standard practice for all large stationary fossil fuel installations.

The Nuclear Energy Option

The development of additional nuclear energy capacity in China promises to overcome many of the barriers that confront the energy sources discussed above. Though China's reliance on nuclear energy has been limited to date, it has built an extensive industrial base of nuclear and technical capabilities that is poised to support substantial growth.

China built its first heavy water research reactor and cyclotron in 1958 and connected its first indigenously designed, constructed, and managed pressurized water reactor to its electricity grid in 1991. Since then, nuclear power growth has been slow. In 2004, China's nuclear power plants produced only 50.4 TWh of electricity, accounting for 2.3 percent of national generation. In comparison, South Korea's and Japan's nuclear power sectors account for 40 and 30 percent, respectively, of total electricity generation.

In contrast to other potential energy sources, nuclear reactors are a fully developed and low-carbon emission electricity generating option that has the potential for large-scale expansion. Despite the large cost of nuclear power plants, China's booming economy has helped to ensure enough capital investment for planned projects. The ongoing global finance crisis has affected China, but it did not decrease Chinese investment in nuclear energy development. Instead, the government has increased the amount of financial aid and guaranteed loans for the nuclear industry. China has not participated in any international nuclear liability regime, but China set up its nuclear insurance pool in 1999, which is a community comprising 15 major non-life insurance companies and four reinsurers.

Nuclear power could be cost competitive with other forms of electricity generation, except where utilities have direct access to low-cost fossil fuels, such as coal and natural gas. Yes, the cost of building nuclear power reactors is relatively high, but the operating costs are relatively low. Additionally, nuclear fuel costs are a minor portion of total generating costs, while they make up 40 and 60 percent of costs for coal-fired and LNG plants, respectively. This insulates the price of electricity generated from nuclear reactors to fuel price escalation. Standardized designs, shorter construction times, and high capacity factors have also lowered reactor construction costs to the point that even without environmental subsidies, nuclear reactors can be competitive with other power options over the their operating lifetimes (WNA, 2008). . . .

Nuclear energy has several other practical advantages for China. Nuclear fuel, predominantly comprised of uranium, has the advantage of being a highly concentrated source of energy that requires less transport capacity. For example, a 1 GWe pressurized water reactor refills only one-third of its fuel assemblies per year, and the total quantity of fuel is much less than the amount of coal or oil needed to generate an equivalent amount of energy. Additionally, extreme weather or seasons affect nuclear power plant operations very little.

Adequate and affordable uranium resources form the foundation of China's proposed nuclear expansion. China's estimated uranium resources, about 100,000 tons (CAEA, 2007), should enable it to satisfy uranium demands for the next decade.[1] To meet its long-term needs, it will need to strengthen its domestic uranium exploration and mining capacity. The China National Nuclear Group, the only state-owned nuclear corporation with uranium exploration and mining capabilities, recently announced that it had verified a large uranium ore deposit in the Inner Mongolia Autonomous Region. And it claims that the amount of newly proven uranium found each year in China outpaces the country's growing demand. Assuming it will be able to mine and process these deposits at a reasonable pace, the group expects to be able to fuel Chinese nuclear power development in the long run (Xinhua News Agency, 2008a, b). In addition the IAEA's *Uranium 2007: Resources, Production and Demand,* also known as the "Red Book," shows that an estimated 5.5 million tons of global uranium resources exist, 130 times the global production of uranium estimated for 2007 (IAEA, 2008). Unconventional uranium sources, such as those in phosphate rocks and in seawater, are available to explore when cheap uranium sources become scarce and uranium prices increase. In addition to using natural uranium resources, a number of Chinese researchers are looking at the reprocessing of spent nuclear fuel and advanced nuclear technologies, such as fast breeder reactor designs, as a way to significantly extend existing uranium supplies. Uranium resources are thought to be geographically more evenly distributed than any other energy resource, though a relatively few countries—including Australia, Canada, and countries in Central Asia, hold the largest shares of the most economical, high-grade uranium ores. Given this distribution of uranium resources, the risk of supply disruption is minimal as compared to oil and natural gas reserves, which are concentrated in the Middle East (EIA, 2009). In addition, countries can maintain stockpiles of nuclear fuel with relative ease, given that uranium fuel storage requires far less space than for fossil fuels. Lastly, nuclear fuel costs are only about 5 percent of total generating costs, while fuel costs for coal-fired and natural gas-fired plants make up 40 percent and 60 percent of costs, respectively (NEA, 2008). All of these arguments suggest that the availability of nuclear fuel should not constrain future nuclear expansions.

Building enough new nuclear power plants and operating them long enough to make a significant contribution to China's growing energy needs will require greater acceptance from the Chinese public. Public opposition has been a major impediment to nuclear development in the West, but as a consequence of China's relatively centralized government and government-driven economy, the Chinese public is typically informed of nuclear decisions only after they are made. Chinese authorities are taking steps to increase public involvement in nuclear energy decisions. The State Environmental Protection Administration, China's top environmental body, recently initiated limited public involvement in the nation's environmental impact assessment process. Local governments are now required to release environmental impact assessment reports and allow public feedback during a public comment period before starting construction of large-scale projects. This system has so far been ineffective and inefficient. Gradually, public participation on nuclear projects should improve because of improved regulatory transparency.

In general, the Chinese public seems willing to accept and embrace nuclear technologies and the role they will play in the country's continued development. Further nuclear development, for instance, is likely to provide thousands of jobs in local communities, which has set off a scramble among local governments eager to have nuclear power plants built in their regions. In contrast to Japan, for example, where local officials have fought to keep nuclear facilities out of their regions, local Chinese officials believe that nuclear power can positively impact the local economy, increase the local tax base, and resolve electricity shortfalls and have aggressively initiated cooperation with nuclear investment corporations, such as the China National Nuclear Group. . . .

Meeting Rising Demand

For a country like China, there is no single approach to accomplish the goals of reducing environmental issues and meeting growing energy demands at the same time. It is hard for one particular technology alone to ease all issues on a timescale or size scale. Of all the energy technologies discussed, nuclear has the clearest potential to contribute to increased Chinese demand and climate change mitigation strategies. Nuclear should not be left out of China's energy mix. . . .

All references for articles included in *Taking Sides: Clashing Views in Sustainability* can be found on the Web at www.mhhe.com/cls.

Note

1. The annual mass of fuel, in metric tons of heavy metal (MTHM), which must be loaded into one PWR reactor is obtained as

$$M = \frac{P \times CF \times 365}{\eta th \times B}$$

where, M: mass of fuel loaded per year (NTHM/year); P: installed electric capacity (GWe); CF: capacity factor; ηth: thermal efficiency (GWe/GWth), and B: discharge burnup (GWd/MTHM). In this paper, the installed capacity of 60 GWe is assumed in 2020; the capacity factor is 85 percent; the thermal efficiency is 33 percent; and the discharge burnup is 50 GWd/MTHM. From the calculation, China needs 7090 tons uranium fuel from 2006 to 2020 considering initial fuel load for new installed capacities.

The mass of natural uranium required for fuel production can be obtained by considering the enrichment process. The required enrichment level for a given burnup can be calculated given the amount of feed material (natural uranium 0.711%) by

$$F = p \times \left(\frac{X_p - X_w}{X_f - X_w}\right)$$

where, X_f = weight fraction of U-235 in the natural uranium; here, X_f = 0.711%; X_p = weight fraction of U-235 in the enriched uranium fuel; here, X_p = 4.5% for the time period of 2006–2020; X_w = weight fraction of U-235 in the waste stream; here, X_w = 0.3%; F = the amount of natural uranium; P = the amount of product enriched.

From the calculation, China needs 72,328 tons natural uranium from 2006 to 2020.

EXPLORING THE ISSUE

Can India and China Reduce Their Dependence on Coal?

Critical Thinking and Reflection

1. Is energy from fossil fuel a necessary evil for emerging economies?
2. Will China overtake the United States as the number one economy of the world?
3. Will China become a dust bowl?
4. Is renewable energy a feasible option for emerging economies?
5. Is China a huge consumer of resources in the world?

Is There Common Ground?

Sustainability practices for emerging economies like China and India are feasible. The alternative sources of energy and renewable energy can become major sources of energy for the emerging economies provided there is more research and development in these energy sectors. The sustainability practices could not be feasible for the emerging economies as they have to depend upon energy sources derived from fossil fuels only which have the capability to satisfy the ever growing energy demands. The midway approach could be a mix of conventional (fossil fuels) and unconventional (alternative and renewable) energy sources to fuel the growing economies of China and India. There could be a mix of public and private transport that would lead to less environmental pollution than solely private transport. There could be conservation practices of resources without getting too stingy.

Internet References

International Monetary Fund

www.imf.org

It is the website of International Monetary Fund. It provides credible information about economic scenarios around the world.

World Bank Group

www.worldbank.org

It is the website of the World Bank. It provides authentic and valuable information about different projects initiated by the World

Bank around the world especially in developing countries. It also gives information about economic scenarios of the world.

ScienceDirect

www.sciencedirect.com

The website of Science Direct provides the latest journal articles about science. Latest journal articles about energy, alternative and renewable energy could be searched at this website.

Environment Complete

http://www.ebscohost.com/academic/environment-complete

The website of Environment Complete provides the latest journal articles about fields related to the environment. Journal articles about sustainability can be searched at this website.

Web Of Knowledge

http://wokinfo.com/

The website of ISI Web of Knowledge provides the latest journal articles about science. They provide articles about energy, alternative and renewable energy, could be searched at this website.

ISSUE 6

Is Poverty Responsible for Global Environmental Degradation?

YES: J.B. (Hans) Opschoor, from "Environment and Poverty: Perspectives, Propositions, Policies," *ISS Working Paper* 437 (Institute of Social Studies, Netherlands, 2007)

NO: John Ambler, from *Attacking Poverty While Improving the Environment: Towards Win-Win Policy Options* (East Asia Program Development, Social Science Research Council, New York, 2004)

Learning Outcomes

After reading this issue, you should be able to:

- Know how poverty can cause environmental degradation.
- Discuss how sustainable development relates to both ecological preservation and the need for development.
- Understand how the poverty-environment nexus relates to the environment-development system.
- Discuss the nature of the relationship between environmental protection and economic development.

ISSUE SUMMARY

YES: Professor J.B. (Hans) Opschoor of the Dutch Institute of Social Studies views the relationship between environmental quality and poverty within the wider context of the environmental-development system. He sees poverty as both an agent of environmental degradation and a cause of deepened poverty.

NO: Researcher John Ambler, Director of East Asia Program Development, Social Science Research Council, dispels various myths on poverty and environmental degradation and points to how specific policies can produce a "win-win" situation.

Environmental degradation is the loss in environmental quality from pollutants and other activities and processes such as improper land use and natural disasters. The foundation upon which environmental degradation rests is related to the dynamic interplay between socioeconomic, institutional, and technological activities. But, poverty remains a root cause of several environmental problems. The poor are often forced into unsustainable practices by farming on marginal lands; clearing tracts of rainforest for its timber value; and putting at risk fragile ecosystems for their basic survival. These destructive practices often lead to soil degradation, erosion, and landslides. It is argued that environmental conservation is a luxury that the poor cannot afford because their livelihood is at stake, and this produces an unsolved vicious circle of poverty and environmental degradation (Durraiappah, 1998). Marginalization of the poor is one of the factors that leads to socioeconomic or environmental change in both developed and developing countries. The adverse impact of population growth and economic marginalization produces a negative downward spiral for the poor. This results in unusual stress on natural resources and land use and often forces the relocation of the poor to less productive or marginal areas. This causes a radical alteration of the environment.

The role of poverty in fostering environmental degradation has been recognized by the United Nations. This linkage was not made to "blame the victim," the poor, but to show the connection between poverty and environmental degradation. According to this argument, if the livelihoods of the poor can be uplifted, then environmental quality will improve. This relates to the environmental Kuznets Curve where economic growth eventually produces better environmental quality.

The causes of environmental degradation, particularly in developing countries, are many and varied. Globalization can increase the wealth of developing countries but it can also produce environmental impacts. For instance, much highly polluting manufacturing has moved to emerging economies like China and India, but their products are consumed in the West, especially the United States. Hence, this produces a shift of environmental degradation from the West to poorer developing countries. Also, the West often uses developing countries as a source for waste disposal. It is well noted that electronic waste from the United States has been shipped overseas for "recycling," which adds to the environmental health problems in receiving countries.

Other factors cause environmental degradation in developing countries. The move to manufacturing and the increased demand for energy has placed a heavy emphasis on the use of fossil fuels, particularly coal. This causes an increase in atmospheric pollution. Coal-fueled thermal plants produce large amounts of sulfur byproduct, which combines with water to form acid rain. Menz et al. (2004) discussed how acid rain was a serious environmental problem in both Europe and the United States until strict environmental standards on power plants reduced the emission of sulfur. Also, a successful emission trading system was established in the United States to deal with this problem in the 1990 Clean Air Act. Before this policy reform, acid rain had a serious effect on aquatic ecosystems with the acidification of inland water bodies and groundwater. In the United States, acid rain gained attention with the reporting of acidic lakes in the Adirondack Mountain region in New York (Schofield, 1976). Today, acid rain is a global issue. The heavy reliance of China on coal-burning thermal plants is pushing sulfur across the Pacific to the western United States.

Urbanization is also having an impact on the environment in developing countries. Today, nearly 50 percent of the world's population lives in cities. By 2030, this percentage will increase to 60 percent and cities of the developing world are expected to absorb 95 percent of this growth as a result of rural-to-urban migration, transformation of rural settlements into urban ones, and natural population increase. Although comprising only 3 percent of the earth's land area, cities consume 75 percent of global energy, create 80 percent of global greenhouse gas emissions, and intensely concentrate industry, people, materials, and energy (Taylor, 2011). The ability to successfully plan for this urbanization will be one of the great challenges of the twenty-first century.

Effective public policy can increase environmental quality in developing countries. Robert Taylor in "The Key Role of Civil Society in Managing Environmental Sustainability in Asia" believes that the many environmental compliance systems in developing countries are modeled on Western systems and they do not work well in poorer countries. He states that in poorer countries the demands for economic growth outweigh concerns for environmental protection at the various levels of practical decision making. As a result, emerging market economies have two options. They can seek to strengthen their regulatory enforcement regimes so that environmental laws and standards, already in place in most emerging market countries, can actually produce a stabilized or reduced amount of environmental degradation. Or these same countries, especially ones that are so positioned with the necessary determinants, a robust democratic process, open print media, available and affordable communications technology, and active public participation in governance, can utilize a second option, a civil society environmental management regime. Taylor recommends that poorer countries include more civil society collaborations in their environmental management regimes.

Mukadasi Buyinza from the Faculty of Forestry and Nature Conservation at the University of Makerere in Uganda discusses how the problems of poverty and environmental degradation are closely related in developing countries (2010). He notes that developing countries have a high dependency on primary economic activities such as agriculture, mining, and forestry. This places the natural environment at more risk because it is the primary source of wealth. With population increase there has been more pressure on agricultural production in Uganda, which has led to the utilization of marginal lands. Cultivation of steep slopes has produced erosion and soil degradation. Poor farmers are forced into unsustainable farming practices on marginal land, while rich farmers are able to cultivate the most fertile land. He shows the linkages between land degradation and soil conservation on the poverty level of rural households in Uganda.

Today we see a number of schemes designed to elevate people out of poverty. Natural resources are often used by the poor in an unsustainable manner because it is their only source of income, that is, farming on marginal lands, illegal logging, etc. Through creative policy, natural resources can be used both as a source of income and as a means to protect the environment. This can be done through policies such as the Payment for Environmental Services (PES) scheme. The aim behind PES is that people should be given incentives to provide environmental services, and those who use the service should pay the costs. Therefore, this will generate "net environmental income." PES would directly confront some of the reasons why environmental degradation occurs in developing countries, that is, lack of secured resource rights and lack of incentives to protect the environment.

In the YES selection, that poverty is responsible for global environmental degradation, Professor Hans Opschoor of the Dutch Institute of Social Studies views the relationship between environmental quality and poverty within the wider context of the environmental-development system. He examines the two popular views, first that poverty is a vector or agent of environmental degradation and the second that poverty causes a deepened poverty or the "poor as victim" perspective. He concludes that the two, added together in a dynamic setting, yield the vicious downward spiral that the Brundland Report stated needed to be addressed in its sustainable development agenda.

For the NO selection, that global environmental degradation is more attributable to poor governance, researcher John Rambler of the Social Science Research Council in New York states that too much emphasis has been placed on the relationship between poverty and the environment. Instead he believes that improved governance can break the "vicious cycle" of the poverty-environment nexus. He challenges two major assumptions relating to poverty–environmental interactions. First, he attacks the assumption that poverty necessarily leads to environmental degradation. And second, he discards the notion that population growth necessarily leads to environmental degradation. He contends that specific policies can improve the lives of the poor while protecting the environment. These policies include: giving the poor control over key environmental resources; encouraging the development of democratic governance systems at all governmental levels, from local up to national; and providing a way to protect the current assets of the poor through insurance systems that use cash payments as incentives for environmental services.

YES 

J.B. (Hans) Opschoor

Environment and Poverty: Perspectives, Propositions, Policies

Introduction

The theme of 'environment—inequality' appears on several agendas, but its prime context is the discourse on *sustainable development.* The widely embraced Brundtland definition of sustainable development recognizes the issues of equity in two broad perspectives: 1) that of intragenerational equity to which poverty as it is currently manifest is related and 2) that of intergenerational (or broader: intertemporal) equity. Fairness or responsibility in relation to nonhuman species and other forms of biodiversity would be a third equity-related aspect. In what follows that third aspect will, as an issue in itself, be left aside. Of the two Brundtland-perspectives, this paper focuses on the first one. The paper will address both inequality and poverty, but it highlights poverty-related features.

The concern with environmental issues since the 1960s emerged mainly out of the problems experienced by the industrially advanced countries in consequence of their economic growth. Developing countries were interested first and foremost in environmental problems related to: '. . . poverty and (the) very lack of development . . .' of their societies: problems to be overcome by development and growth (as witnessed by the 1971 Founex Report, UNEP 1981). Yet it was recognized there that economic growth in developing countries would also give rise to further environmental pressures to be taken into account by those countries. The WCED-report (1987) saw poverty as both a major cause and an effect of environmental degradation and described many parts of the world as being caught in a vicious downwards spiral: poor people forced to overuse natural resources to survive from day to day—this impoverishment of their environment further impoverishes them, making their survival more difficult and uncertain. . . .

Poverty and the Poor

Poverty can be defined as a social condition of chronic insecurity resulting from a malfunctioning of economic, ecological, cultural and social systems, causing groups of people to lose the capacity to adapt and survive and to live beyond minimal levels of satisfaction of their needs and aspirations. In recent

From *ISS Working Paper,* no. 437 (Institute of Social Studies, January 2007), pp. 5, 6–10, 32, 33–35.

World Bank shorthand, it is 'pronounced deprivation in well-being' (World Bank 2001:15). Dimensions of poverty are many and the complex realities of poverty vary between regions, countries, communities and individuals. Some aspects of it are: illiteracy, ill health, inequality, environmental degradation, gender bias. Within this concept of poverty as deprivation one might articulate the environmental side of this by highlighting 'ecologic poverty': a lack of ecologically healthy natural resources needed for human survival and development (Agarwal and Narain 2000).

When it comes to a more precise understanding, one would have to differentiate, as there are inequalities in vulnerability within societies, having to do with dependency on the resource base, access to options for livelihood generation, gender, etc. One main differentiation is that between the rural and the urban poor, but even that is too abstract for many purposes. Within the rural poor UNEP (Tognelli et al. 1995) distinguished according to access to land (landless population, smallholders; etc.), mode of production (pastoralists, farmers, etc), ethic dimensions (indigenous people, etc.), gender (including female/male headed households), and recognized special categories such as displaced people. An even more fundamental distinction to be made in the light of the concern with sustainable development is that between the present poor and the future poor, or: the issues of intergenerational equity. . . .

Conventional analyses looked at poverty in terms of income and/or consumption. A 'money metric approach' was (and is) used considering someone poor if her/his consumption or income level is below some minimum level (the 'poverty line') necessary to meet basic needs (e.g. $1 or 2 per day). On this basis, measures have been proposed such as: (i) the headcount measure of poverty (the number of people below the PL) and (ii) the poverty gap index (the distance of poor people from the poverty line). Inequality was measured by e.g. the Gini coefficient.

Clearly, not all aspects are captured that way (e.g. public goods, assets and access to assets and certainly immaterial dimensions). In fact, some argue (Saith 2005: 3) that the dominant money metric methodology 'renders a good deal of poverty invisible' and 'distorts the understanding of poverty,' by suppressing a number of social and economic realities, by being exclusionary of a wide range of deficits which are held to be significant by those that experience them (ibid: 16). This, as Saith argues, is remarkable, as the world seems intent on halving poverty by 2015, defined in terms of poverty-line based metrics.

When it comes to measuring poverty, the World Bank nowadays tries to first measure 'income poverty,' and then to introduce indicators of education and health—beyond which it wishes to turn to vulnerability and voicelessness (WB 2001: 16). On 'income poverty' it uses particular measures of incomes, consumption and prices (WB 2001: 17) to arrive at poverty lines in the $1–2 range using a PPP-approach to arrive at internationally comparable metrics. On discussing health and education the World Bank largely notes how inadequate existing measures are (e.g. infant mortality rates and gross primary education enrolment rates) but hardly deals with ways to enhance such measures and integrate them into a wider indicator. The situation is even worse on vulnerability: the Bank reports a 'growing consensus that it is neither feasible nor desirable to capture vulnerability in a single indicator'—but seems to leave the matter at that. And then again it concentrates on income inequality. Thus, widening the number of aspects and hence the set of indicators to measure poverty has been tried but not successfully (World Bank 2001).

Apart from the approach to widen the set of indicators in order to better capture the scope of the concept of poverty, there have been attempts to incorporate *within* the traditional metrics more of these elements, by monetizing them. One proposal in this direction has been to capture the monetary value of environmental services to communities (and especially the poorer sections of societies) in 'environmental income.' . . .

Still, poverty is better defined in terms of vulnerability and deprivation. Vulnerability is related to power/control. Poor people are particularly vulnerable to adverse events outside their control as well as to ill treatment by governmental and societal institutions (i.e. organizations) due to their being excluded from voice and power (e.g. World Bank 2001: 15). Voicelessness and powerlessness are key aspects of poverty, often exacerbated by barriers related to ethnicity, gender, age or occupation. Lack of security is identical to vulnerability taken as defenselessness and exposure to risks or shocks.

Causes of poverty can be found at every level in society, from the individual to the global. At the international level, trade regimes and policies on international finance play a role and, underlying that, the international economic system in which these regimes and policies are embedded; at the national level monetary and fiscal regimes and policies, the roles of the various actors (state, private sector, civil society) further determine the socio-economic environment of people; at the meso-level markets, sectors and socio-economic groups are relevant in shaping the life worlds of communities and individuals. At the micro level poverty is caused by factors such as: (1) low levels of resource availability, (2) unproductive technologies, (3) poor access to markets, (4) vulnerability to shock in revenues, exchange rates, etc (due to disasters, political processes and market instability). Poverty cannot generally be explained by any single element at any one of these levels. Yet, for good reasons it has been suggested that approaches to alleviate and eradicate poverty are to start at the international level with trade issues, debt and the international migration of labor and capital, before policies at the lower levels might become sufficiently ('sustainably,' if one wishes) effective (Pyatt 1999).

The Environment: Provider Functions and Infrastructural Functions of Nature

The biosphere is an envelope for a range of biotic and abiotic processes operating in and between ecosystems. It provides human beings (as well as other species) with economically *directly* relevant resources and with sinks to absorb the waste humans dispose of as a result of their use of natural resources. Besides these flows of goods and services that have an immediate or 'direct' relevance to human activity and well-being, nature (re-)presents to human society a set of *indirectly* relevant assets: systems and processes that enable the (re-)sources and sinks alluded to above, to work over time (e.g. to regenerate resource stocks, to absorb and recycle wastes). In the jargon of ecologists: the biosphere can be seen as the base of a series of so-called *'life-support systems.'* Life-support systems are the ecological processes that shape climate, clean air and water, regulate water flow, recycle essential elements, create and regenerate soil, and keep the planet fit for life (IUCN etc. 1991: 27; see also Siebert 1982). In other words, they contain ecological processes sustaining the productivity, adaptability and capacity for renewal of lands, water and/or the biosphere as a whole. Thus, they provide a *niche* for species, and for *Homo sapiens* in particular; in more pragmatic jargon, they demarcate an *environmental (utilization) space* (or *ecospace*—Opschoor 1995).

More specifically, the natural environment offers a range of services to society. They fall into two categories (Perrings and Opschoor 1994):

1. Infrastructural functions that concern the capacity of an ecosystem to develop and maintain itself, notably (i) the regulation functions (De Groot 1992) related to the capacity of ecosystems to regulate balances between ecological processes such as regulation of the climate system and (ii) carrier functions, through which ecosystems provide space and a suitable substrate or medium for human activities;
2. '*Provider* functions': *production functions and information functions* (De Groot 1992) related to the provision of goods, services and information directly relevant in societal processes.

Fulfillment of the infrastructural functions enables a system to operate its provider functions.

The environmental utilization space is to a significant degree constrained, at any point in time, in terms of the resource flows it can provide sustainably, due to the capacities of the underlying infrastructural (regulatory and carrier) functions. Maintaining and enhancing provider functions for the benefit of present and future human beings has become a key concern in national, and international environmental policy. Such policy would need to manage the (global regional, national) essential life support systems so that they are able to continue to function.

Development

Traditionally, development has been equated with: deliberately stimulated economic growth, economic growth often being defined as non-negative changes in per capita income or gross domestic product. Obviously, however, development is more than economic development and economic development is more than economic growth. It has been thought for decades that it was not too unreasonable to assume that the three somehow ran parallel: that if economic growth occurred, this would imply development in a broad sense, or at least: entail enhanced *possibilities* for development. The assumptions on parallelism between growth and development are no longer valid in a finite biosphere, but apart from that there have been other doubts. Concerns over income distributions have led to the broadening of factors taken into account. Because economic development and development as such might not run parallel, human and social aspects were added to the notion, such as: health, literacy, empowerment. 'Human development' is defined by UNDP as a process of enlarging the set of options that people have to improve their livelihoods and determine their futures, operationally focused on productivity, equity, sustainability and empowerment (UNDP 1995: 12).

In the process of economic development natural resources are used to produce physical capital and consumption goods. An implicit use is made of the infrastructural components in human life support systems. Developing countries generally, but especially as their income levels are lower, have a higher dependency on primary economic activities (mining, fishing, forestry, agriculture) and hence on natural resources. These resources include: forests (livelihoods, wood, etc), land/soils (degradation aspect), water systems (habitats of fish, water for irrigation and drinking). This holds especially for the (rural) poor. Moreover, the quality of the various environmental compartments (air, water, soils) is increasingly threatened by pollution, with impacts on living conditions beyond levels that are considered tolerable by developing countries' populations and administrations. Hence, both the resource aspect and the environmental quality concern have entered into decision making and politics.

The concept of *sustainable development* is a synthesis of concerns over ecological sustainability of natural resource use and of considerations of the need for development and economic growth to meet the other needs and aspirations of societies—especially the poor—now and in future. . . .

Conclusions

. . . Poverty has been proposed to be a vector ('can agent') of environmental degradation where others highlighted degradation as the cause of deepened poverty ('the poor as victims'). The two added together in a dynamic setting yield the vicious downward spiral, referred to in the Brundtland Report that put the need for *sustainable* development on the agenda. The concept of *sustainable development* is a synthesis of concerns over the ecological sustainability of natural resource use and of considerations of the need for development and economic growth, in order to meet societal needs and aspirations now and in future. The World Summit on Sustainable Development was told to prioritize the nexus and to address both focal points simultaneously, as there will be no 'ecology' without 'equity' and no 'equity' without 'ecology' (Sachs 2000). Meanwhile, development 'on the ground' of the planet has taken a direction and a form referred to as globalization, or: 'neoliberal' globalization. Apart from the short term benefits in terms of higher levels of trade, division of labour, and—especially—average income and reductions in absolute poverty 'income poverty'), there are some embedded negatives: globalization in its purely neoliberal forms tends to, *inter alia,* increase inequalities (i.e. gives rise to mounting *relative* poverty), to crowd out care and to increase ecological unsustainability. . . .

. . . Poverty is not a number—it is even more than a 'pronounced deprivation': it is a *social relationship* (Reed 2002: 177) of competition among individuals, social groups and the state in a pursuit of wealth and power. In this perspective, poverty is the result of an inability of people to gain access to life-supporting assets (productive, environmental, cultural) while others are capable of securing the conditions for stable, productive lives. Hence, overcoming poverty may require the changing of social relations as well as changing these capabilities, asset endowments, etc.

Poverty also is a *structural phenomenon,* and a basic requirement for dealing with the structural causes of poverty is to design and implement an appropriate development strategy (Kay 2005), with the associated change at the macro and international level in terms of regimes governing production, consumption, trade, division of labour, and access to the environmental space.

On the environmental side, awareness has been growing as to the complexities underlying the provision of environmental goods and services, and natural resources in the form of infrastructural 'ecological functionings' and 'life support systems.' Economic development affects ecosystem balance and, in turn, is affected by the state of the ecosystem. In addition, critical impact thresholds, and vulnerability to environmental change impacts, are directly connected to social and economic conditions. Poverty can be both a result and a cause of environmental degradation. Infrastructural functions of ecosystems typically are to be preserved for future generations and by definition exceed what economic structures such as markets could manage on their own.

Despite our lack of knowledge at the factual level, the above newer insights have helped in searching for more sophisticated approaches towards learning about environment-poverty relationships and towards addressing problems with the nexus. There now seems—at the idealistic level—more interest in a deeper concept of livelihood, a wider notion of capabilities and a better knowledge of institutions. Also, the interest in variability around treads, and in risks, is overtaking the use of the linear generalizations of not so long ago. . . .

Emerging perspectives include:

1. institutional change as poverty/degradation mitigation
2. sustainable livelihoods strategies (assets and assets management enhancement)
3. the capabilities approach to human and ecological functionings
4. rights-based sustainable development (environmental rights, resource tenure, etc.)
5. resources as assets of the poor and a source of environmental income. . . .

It can be observed that:

All these approaches recognize the need to connect poverty reduction with sustainability in resource use. Still, some emphasize poverty more and others the environment—the need for more, or more sophisticated balancing remains.

They also differ in the keys suggested to open the doors to equitable and sustainable development: property (and related) rights livelihoods/capabilities, or resource base management and marketing.

These perspectives very often tend to take a 'bottom up'—approach starting from local/regional level analyses of manifestations of poverty-environment linkage.

Most (all but perhaps the capabilities approach) seem to stay close to commodification/market oriented views.

When it comes to agency and agents, post Johannesburg mainstream thinking on sustainable development (and poverty-environment within that) reflects the then embraced ideas of *partnerships* between the various categories of agents, where a more critical alternative might emphasize power asymmetries, differentiated access, etc.

One element of a broader strategy might be: enabling the poor to gain control over the resources and to capture a larger proportion of the potential revenues of them through the development of systems of marketing and payment for environmental services (PES). This commodification/market oriented view has been seen to bring in benefits as well as costs, to be an opportunity with several risks attached. Although the effectiveness of PES can be enhanced by auxiliary institutional changes (e.g. resource rights) and additional interventions (e.g. off-farm employment), it will remain vulnerable to all the asymmetries mentioned just above, as it opens up decision making on resource use to powerful market forces. That is why the international and macro-level processes need a place in the analyses as much as in the recipes. While the perspectives discussed above seem to still be very much on the drawing boards of the policy advisers, real development paths being pursued remain very close to a somewhat policy-couched and slightly more context-sensitive, market based, growth oriented, variant to the liberalization/industrialization/trade scenarios that are typical of the current stage in globalization.

All references for articles included in *Taking Sides: Clashing Views in Sustainability* can be found on the Web at www.mhhe.com/cls.

John Ambler

Attacking Poverty While Improving the Environment: Towards Win-Win Policy Options

Background

The poor in rural and urban areas rely on the environment for their livelihood and survival strategies and are affected by the way others around them use environmental resources. At the same time, because most natural resources are exhaustible or degradable, improved management techniques for environmental resources are needed for populations that continue to grow in both numbers and consumption habits.

Concern with the state of the environment often seems to pit environmentalists and policy-makers against the poor. Unfortunately, the link between poverty and environment is uncritically characterized as a "vicious circle" or a "downward spiral." Population growth and inadequate resources are presumed to lead to the migration of the poor to ever more fragile lands or more hazardous living sites, forcing them to overuse environmental resources. In turn, the degradation of these resources further impoverishes them. This does sometimes happen, but as an overarching model it is highly simplistic and often leads to policies that either reduce poverty at the expense of the environment, or protect the environment at the expense of the poor.

Well-planned actions can break this cycle. Based on experience from around the world, "win-win" options exist that can build better institutions and partnerships with poor people creating more robust livelihoods and healthier environments. These options simultaneously pursue two goals: reduced poverty and better social equity, and enhanced environmental protection. Improved governance is an important vehicle for achieving these goals.

Improving Assets for Win-Win Outcomes

While income and consumption are important, many policy options for addressing poverty-environment interactions focus on improving the asset base of the poor. Assets include natural capital (forests, water, land, fish, energy resources, and minerals), social capital (relationships of trust and reciprocity, groups, networks, customary law), human capital (skills, knowledge, beliefs, attitudes, labor ability, and good health), physical capital (basic infrastructure), and financial capital (monetary resources). With improved access to and control over different types of assets, the poor are better able to meet their basic needs and to create more flexible livelihood options.

The concept of poverty is complex and the subject of much debate. In terms of how the government views it, the poor are usually defined as those falling below some standard and, therefore, form a target group for particular policies. Be that as it may, it is important to keep in mind that the location of poverty also influences policy options. For example, although rural non-farm income is rising in importance, the rural poor are usually very dependent upon natural resources. Attacking rural poverty thus relies heavily on improving poor people's ability to derive sustenance and income from these natural resources. In urban areas, the health of the poor is particularly affected by a degraded environment, one characterized by substandard housing, inadequate or polluted water, lack of sanitation systems, and outdoor and indoor air pollution. Ill health leads to a host of problems, including a decreased ability to work. Improving the urban environment positively impacts the health of the poor. Therefore, such improvements may well be a prerequisite for other poverty reduction measures.

Poverty is also a set of relationships. The poor compete with each other and with the non-poor for control over assets. Poverty can also differ within households. Women and children, especially girls, often have the least access to productive assets and are usually the most affected by pollution. Efforts to reduce poverty must also recognize this competition for resources and the differential impact of environmental degradation among and within households.

Revising the Understanding of Poverty-Environment Interactions

New experience with different resource management regimes involving the poor is challenging two major, entrenched assumptions about how the poor relate to the environment.

1. *Poverty does not necessarily lead to environmental degradation.* The linkages between poverty and the environment are complex and require locally specific analysis to be understood—there is no simple causal link. In many areas, the non-poor, commercial companies, and state agencies actually cause the majority of environmental damage through land-clearing, agro-chemical use, water appropriation and pollution. Sometimes privileged groups force the poor onto marginal lands, where, unable to afford conservation and regeneration measures, their land-use practices further damage an already degraded environment. But there are also many examples in which very poor people take care of the environment and invest in improving it. Thus, poverty can sometimes be associated with environmental degradation, but there is not necessarily a direct causal relationship.

2. *Population growth does not necessarily lead to environmental degradation.* Initially, degradation may occur as a population increases. But what happens

next is context-specific. Where people are too poor to invest now, or too poor to wait for the fruits of their investment, further degradation can occur. In other cases, as the cost of land relative to labor increases beyond a certain point, farmers can change their methods of managing plants and animals or make investment in the land to offset initial declines in productivity from more intensive use. In urban areas, some empirical evidence shows that good governance can maintain the environment even as cities grow. Thus, halting population growth or removing people from densely settled areas might improve neither productivity nor resource quality.

Resource Characteristics and Community-Based Action

Developing effective community-based institutions for the collective management of resources is a key factor in determining the success of efforts designed to aid the poor. However, the characteristics of the resource have important implications for building local management institutions. In general, institutional costs tend to be high for managing resources that are mobile (such as water, wild game, and fish), dispersed or vast, slow-growing or fragile, difficult to patrol or monitor, difficult to observe or measure, technically complex, accessed by many people or heterogeneous groups, or highly skewed in their distribution among stakeholders. Conversely, the institutional or organizational cost of collective action tends to be lower when the resource-to-user ratio is high, a common cause creates cohesion, the group is relatively homogeneous and isolated from disruptive external pressures, and access rights to the resource are secure. Well-conceived policy and strategic investment from the State can help reduce high organizational costs. But in all cases, the benefits of organizing must outweigh the costs of maintaining the organization. Efforts that fail to address the institutional issues consistently under-perform, no matter who is the manager.

Implementing Better Policy

Implementing pro-poor and pro-environment policy requires both conceptual and operational shifts, as well as better, specific policies. Some of these shifts are mentioned here.

Conceptual Shifts

Building partnerships with local communities is a new endeavor for many government agencies. It requires not only good will, but also a commitment to experimentation, fine-tuning solutions, and institution building. Some conceptual shifts include:

- empowering the poor to identify their problems and seek their own solutions—not assuming the poor are the problem;
- engaging poor people as partners, not as beneficiaries, and using people-centered frameworks for planning and implementation;

- creating incentives for the poor as well as for private-sector entrepreneurs to mobilize resources for poverty eradication, and move away from simply exhorting poor people to mobilize their resources or providing all the resources from the State; and
- seeing the value of giving the poor real rights and ownership of assets, not just a sense of ownership.

If governments take steps to make these shifts, experience has shown that poor people will be more inclined to view the government as a viable partner in development.

Operational Shifts

Changes in the nature of the environment-poverty nexus have as much to do with how different interests are negotiated and expressed as they do with policy failures. Counteracting the influence of power strongholds can only be achieved through governance reform, as expressed through improving accountability, transparency and representation. Policy and institutional fora must be open and inclusive, and support should be channeled to local institutions that effectively represent the needs and aspirations of marginalized groups. For decentralization to truly reflect the needs of the poor, however, it will be necessary to invest in local-level capacity; otherwise decentralization may place power in the hands of the local elite.

In parallel with partnership approaches, governments need to shift toward more pluralistic approaches to decision making. Such approaches can accommodate different interests and potential conflict, and do not require total consensus before being able to move matters forward. Pluralism has encouraged the development of innovative tools such as resource management contracts and codes of conduct. Such concepts raise issues such as how to achieve needed checks and balances and accountability, especially in the absence of absolute standards or single clear-cut solutions. When equity is also part of the desired solution, they also require weighting attention to the needs of the poor when not all interests can be accommodated equally.

Moving to decentralized planning facilitates participation and maximizes resource mobilization. In turn, this ensures that services will be more relevant to the needs of communities and households. Decentralization implies local plans of action, places accountability and responsibility at appropriate levels, and allows quick action following monitoring and problem solving. However, it cannot be presumed that decentralization is automatically beneficial for all groups. Communities cannot be seen as homogenous and non-hierarchical. Marginalized groups such as poor families and minority ethnic groups may be excluded even in decentralized processes. An understanding of norms, values, attitudes, rules and regulations underlying decentralized decision-making at the community level is necessary to ensure that vulnerable groups, such as women, children or the aged, are not further marginalized.

Experience has shown that decisions made by communities and households result in more sustainable solutions. Developing community-based decision-making and transparent dialogue cannot occur without political will at the highest level. Governments need to help catalyze the formation of people's organizations through enhanced rights and building on local organizational forms. Mechanisms must be developed to ensure feedback of learning from local to national policy levels. The poor also need to be able to draw on networks and links with state, market or civil society actors who will help them to access, defend, and capitalize their assets.

Finally, effective implementation of poverty eradication programmes requires employing a gender analysis lens that takes specific note of the relationships between men and women, including division of responsibilities, labor, as well as access to and control over resources and decision-making. There needs to be a concern in the inherent constraints to the achievement of gender-equality goals in the institutions, structures and processes within each sector. Important information on the responsibilities of women and men gained at household and community levels needs to be fed back to develop macro-level policies, strategies and institutions. The strategy to increase women's involvement also needs to go beyond an analysis of their contribution to the sector relative to menus, and to consider possibilities beyond the existing division of responsibilities.

Specific Policy Recommendations

Policy-makers can be constrained by certain global forces. For example, structural adjustment programmes may limit the ability to provide subsidies for the poor. Macroeconomic reform can help a country become more competitive, but investing in the fastest-growing sectors can draw resources away from long-term investment in the resources of the poor. Changing global markets can exert downward pressures on living standards. In some countries, these have obliged many poor people to increase their pressure on natural resources just to survive. Countries may be tempted to overexploit natural resources to handle balance-of-payment problems. And sometimes new and distant markets can encourage the depletion of local resources to the detriment of the poor.

But much can be done through sound national policy. Policies that support both enhanced livelihoods for the poor and more sustainable management of environmental resources—the win-win scenario—depend on the nature of the resource and the groups involved. Policies that are win-win for the environment and the poor can also mean the non-poor must forfeit some resources or perquisites, and this requires political will. In general, however, there are several categories of win-win policy options that can be pursued. Many promising experiments are still young and have yet to face second generation challenges. Still, enough research has been done and enough experience has been accumulated to suggest a few general principles and some concrete ways forward.

The strategic interventions that are appropriate for any country or region depend on the nature of the environment, the local characteristics of poverty, and the possibilities for developing and strengthening state and civil-society institutions. Not every policy option presented here is feasible or appropriate for all conditions. Most countries have diverse resource bases and face different conditions in different parts of the country and in different sectors. Poverty and environment have many faces, and diverse conditions call for varied and flexible policies.

Even where appropriate policies can be identified, financial or human resources constraints may limit the practical possibilities. The options presented here are intended to provide a menu of alternatives that can be used as a basis for adapting policies to local conditions and capabilities. Many of these options can be experimented with at a local level and scaled up to larger levels as experience is gained. However, almost all of these options require training, adaptation and transfer of new technologies, as well as the strengthening of institutions to be successful.

It is also important to note that many of the options presented here depend on reallocating investments toward the poor and not necessarily on raising investment levels overall. Although increased financing from the State is required in some cases, much can be accomplished immediately with existing human and financial resources. Reallocating existing financial resources to viable projects that benefit the poor is one way. The political process of reallocating funding toward projects that benefit the poor and the environment can be made easier if "pro-environment/pro-poor" guidelines are used for public investments. This includes properly valuing long-term environmental benefits ("greening the internal rate of return") and weighting investment criteria to recognize the fact that a particular monetary return on investment for the poor is more valuable for increasing net well-being than the same return on investment is for the non-poor. Allocating resources for the poor is also not a matter of giving handouts. Investing in the poor can be done in a way that helps them mobilize their own resources, thereby creating not only ownership by the poor, but much larger net gains for the society as a whole.

Access to Assets

Policy Option: Protecting the Current Asset Base of the Poor

In many cases, the poor already have the right to manage key environmental resources. But they are not able to protect their rights. Large-scale farmers may take excess irrigation water; trawlers may scour the fishing grounds of coastal villages; companies may obtain concessions to tribal forests; and municipal funds meant for improvements in poor areas may be siphoned off for other projects. Industrial pollution of rural resources by both large and small enterprises is a major cause of environmental degradation and rural poverty in some areas. The State often does not have the resources to monitor in a timely and effective manner the remote and dispersed resources that the poor depend upon. Corruption, incompetence and indifference can also deprive the poor of the rights they already theoretically enjoy.

The key to protecting the poor from these abuses is encouraging the development of democratic governance systems from local levels upward. The State needs to support representation by institutions that are accountable to the people, so that monitoring of action and enforcement of rights can take place at all levels. Citizen oversight boards, community-level review processes for State-initiated development plans, and ombudsman systems for dispute resolution are examples of such mechanisms. It is also important to strengthen the judicial system as an impartial and independent institution, and to foster the emergence of institutions of civil society that can mediate between different actors.

Another way of protecting the current assets of the poor is through insurance systems. Cash payments, in-kind provisions, or public-works employment for the poor during periods of drought, major crop failure, or natural disaster can provide subsistence needs while reducing the need to over-exploit natural resources. A corollary to this would be to establish formal arrangements for (limited) access to basic environmental resources for temporary migrants and refugees (e.g., from drought, disasters, or war), and to limit local over-exploitation and conflict. Regular programmes in which the poor are put to work improving the environment can be expanded during crisis times to provide a measure of insurance for those most affected.

Policy Option: Expanding the Asset Base of the Poor

A highly unequal distribution of assets often depresses subsequent rates of growth. Reducing income inequality tends to increase aggregate growth and further reduce poverty indirectly. Improving the access of the poor to natural resources and the productivity of those resources not only addresses directly the equity issue, but also provides new markets for other goods and services, thereby stimulating the economy as a whole.

Policy should focus on environmental entitlements. These include the broad set of social structures and networks that allow poor people access to a healthy environment and resources for sustainable livelihoods. This often involves turning resources over to the poor as individuals or to organizations composed of the poor.

Where land distribution is highly skewed, land reform can be one option. Because agricultural productivity and investments in agriculture per unit-area are usually negatively correlated with the size of holdings, in some cases, conducting land reform can both increase production and encourage environmental improvement. Another possibility is regulatory reform, including provisions of rental, lease or harvest (gleaning) agreements for both private and public lands. These can include longer-term rental contracts, explicit agreements about the distribution of benefits from resource improvements, or the granting of formal tenure rights to individuals or groups currently squatting on public lands or in urban areas, so that they can legitimately seek technical assistance, credit and other services and have incentives for conservation.

Historically, governments have had a difficult time with centrally managing dispersed resources in which local people also have a stake. Turnover of the resource to local groups can be one solution. Granting rights to groups involves establishing or strengthening local people's institutions. Where only State or private ownership is allowed by law, legal change may be necessary to accommodate multi-user tenure. Any program to assign rights to resources should be checked for overt or implicit barriers to women and the poor obtaining rights, in both design and implementation. For example, modifications in local property rights that accommodate common property management have been key elements in African "success stories" for land reclamation, forest management, local fisheries management, small-scale irrigation, resource protection, range management, and wetland cultivation.

Awarding resources to poor people sometimes can lead to environmental degradation, especially if their tenurial rights to the resource are not secure. But in numerous other cases, it has been shown that improving poor people's control over the environment provides a powerful incentive for them to protect it. But it is important to understand the local context and to promote equitable local organization. Resources that have been devolved to local levels for community management can be differentially captured by local elites, unless the State plays appropriate monitoring and enforcement roles.

Successful examples of turning over water rights to landless people or to local groups of farmers have been recorded, especially in Asia. In a number of cases, governments have turned over the rights to the forests to local groups, as in Nepal, which led to greater environmental benefits and more income for poor people. These efforts were most successful when the new arrangements built upon the management systems communities were already practicing. Ownership of resources can be fostered through the availability of locally based finance and credit schemes or through selective investments in the resource base at the time of turnover.

In urban areas, one of the key resources that can be awarded to the poor is the right to occupy the land they live on. Many urban poor live in illegal squatter villages where they are systematically denied access to municipal services. In such cases, poor people can benefit from more secure tenure to the land on which they build their houses. This not only permits public utilities to be extended and upgraded in their area, but also encourages the poor to make investments in their housing and surroundings.

The question can arise as to whether people have the technical knowledge to manage a resource well. When people move into new settings or when conditions change, then a period of learning and adaptation is needed. The poor may not always be immediately aware of the effects of gradual and sometimes imperceptible degradation. In general, however, evidence shows that the poor have an enormous store of indigenous technical knowledge and a body of customary law that provides a social platform for collective action. Scientists and extension workers have discovered that many indigenous technologies and management practices are suitable for dissemination or as the basis for improvements. These technologies and local organizational forms have already been ground-tested. Extra effort is needed on the part of government agencies to understand and appreciate this important body of local knowledge. Customary law is not always equitable, but it does constitute an important starting place for negotiating better rights for the poor. Customary rights are location-specific and highly dependent on negotiated solutions. Unless care is taken, poor groups and women can lose out as a result of policies and processes to formalize these relationships into forms recognizable by the State.

Asset Improvement

Policy Option: Co-managing Resources with the Poor

When a resource has multiple stakeholders with conflicting objectives and differential power, the government may wish to work out co-management arrangements. The government seeks to strengthen local organizations, but also to provide technical assistance and to mediate the overlapping and conflicting claims on the resource. This approach is often favored by governments that wish to continue to exercise a regulatory role (important where there are environmental externalities associated with the use of the resource), and to retain control over components of the resource of direct value to the State. Typically, successful co-management partnerships give local people specific benefits in return for involvement in decision making and various duties related to the protection of the resource.

Co-management responses may be more successful than complete turnover when the capacity to manage at the community level has become eroded or broken. The high transaction costs associated with organizing fragmented communities to take on responsibilities within co-management systems require intensive and sustained involvement by external bodies. It is especially important to ensure that the poorest users of the resource are not excluded under any new arrangements.

The challenge in co-management efforts is to reconcile poor peoples' needs and environmental enhancement goals, and to channel the returns to those who bear the costs. Positive examples of forest co-management systems can be found in the Joint Forest Management program in India, and in the game park management efforts for the Campfire program in Zimbabwe. Co-management programmes in forestry have been most successful in villages bordering extensive tracts of degraded forests, where the forest-to-household ratio is relatively high, where ethnically homogeneous communities possess forestry knowledge, and when benefits accrue from minor forest products at a relatively early stage. Joint management has also been successful in the mangrove areas when the protection against flooding and erosion from improved management directly and immediately benefits local people.

For some large-scale or technically complex systems, the State may have some comparative advantage in management. Water supply and sanitation systems, large irrigation systems, and power-generation facilities, for example, all require specialized personnel. But while in many cases these personnel work for the State, better mechanisms of accountability can be set up so that they are more responsible to the people they serve. Efforts toward co-management can include setting up citizen oversight boards, linking salary adjustments to user-evaluated performance, or actually putting the technical staff in the employ of the user group.

Policy Option: Co-investing with the Poor

In other cases, transfer of ownership or management authority is not the issue. Instead, the State co-invests with the poor on the resources they already own. Co-investment with local communities or farmer organizations may be used to mobilize longer-term investments, such as soil conservation or improvement, irrigation and drainage infrastructure, grazing land rehabilitation, land-leveling, or micro-watershed re-vegetation. Notable examples of this can be seen in government programmes to assist hill irrigation systems in Nepal and Bhutan, the *irigasi desa* programmes in Indonesia, and various tank rehabilitation and micro-watershed improvement programmes in India. In some projects, the villagers themselves contribute up to 90 percent of the value of the investments. The key to the success of these systems is the fact that farmers retain control of the authority and responsibility to manage their resources, and that investment from the State catalyzes the mobilization of additional resources from the farmers themselves.

In both rural and urban areas, improving access to better water supply, sanitation and energy services is critical for reducing the health effects associated with indoor-cooking smoke and poor hygiene, and for reducing the illiteracy frequently found in houses with poor lighting. But the poor often face high initial costs in the form of connection fees, the cost of LPG canisters, or other one-time costs. The government can help improve the poor's access to municipal services and modern energy technologies by subsidizing initial costs or by developing innovative financing mechanisms. Expanding the market for advanced technologies such as household lighting systems using photovoltaic technology or efficient biomass or LPG stoves for the poor will also help reduce unit costs. Subsidies may be used in the earliest phase of the program to generate interest and wide participation in unfamiliar technologies. But subsidies beyond management costs are neither necessary nor desirable. The poor can usually pay the monthly costs of electricity, gas and water. Indeed, they usually already pay more for poorer quality services. But the primary obstacle can be the high initial costs associated with better-quality services.

Infrastructure and Technology Development

Policy Option: Supporting Infrastructure Development for the Poor

People living in poor rural areas can benefit from state-financed or subsidized improvements such as rural electrification, feeder roads, irrigation development and long-term investments. It is important, however, to develop or work with local organizations to collaborate with the government on planning these investments, so that the cost of subsequent routine maintenance is largely carried by the users.

In urban areas, poor people suffer greatly from air pollution caused by inefficient transportation systems. Since transportation is one of the fastest-growing sectors of energy use in the developing world, and mobility is linked to access to jobs, the planning of efficient land-use patterns and transport corridors in urban areas will have significant long-term implications for both energy and poverty. Clean fuels and efficient public transportation systems can reduce pollution in urban areas, improving health dramatically. Zoning reform that allows the poor to live closer to the areas in which they work also can reduce pollution and the cost of transportation to the poor. Improved telecommunication systems have had the proven effect of cutting down on trips meant mainly to seek information.

Many accidental injuries, fires and health problems are linked to poor quality, overcrowded housing. Governments can improve the condition of housing for the poor, both through direct investments in construction and allied infrastructure, as well as through indirect means like credit provision and improving tenure rights on the land.

Strengthening the capacity of city and municipal governments to address the lack of sewers, drains, piped water supplies, garbage collection and health services is also a pre-condition for building the national institutional capacity to address air and water pollution, protect natural resources and reduce greenhouse gas emissions. Cities such as Ilo in Peru and Manizales in Colombia have developed local Agenda 21 plans that have brought major benefits for low-income people. Through more democratic local government, Porto Alegre in Brazil has provided nearly all its inhabitants with piped water, regular garbage collection and reasonable sanitation facilities. In these examples, innovative city government has been the driving force. In other cases NGOs have been able to play strong stimulating and intermediary roles in urban improvement efforts.

Policy Option: Developing Technologies That Benefit the Poor

Developing technologies and resource management systems that raise overall productivity and protect or improve the environment requires a conscious reallocation of research funds away from the most favored environments and toward the resources upon which the poor depend most—fragile and rainfed lands, livestock development, agro-forestry systems, and subsistence crops. Technologies need to be tailored for use on specific soil types and climates, requiring a heavy commitment to on-farm adaptive research. New institutional strategies are needed to reduce the cost of this research, by linking extension efforts and people's organizations.

In the energy sector, most countries have a regulatory environment that does not promote the adoption of innovative technologies and approaches. Instead, conventional energy supply options are favored. Policies must be designed to improve the access to energy services for poor people, with incentives offered to private power-developers to make use of the best-suited technology options. In some cases, where grid access is convenient and cost-effective, flat-rate yet low-cost billing can overcome costly metering. Similarly, many problems of theft can be avoided by encouraging local, self-governing institutions to manage distribution of energy services, for instance, through bulk sales to cooperatives.

To improve the health of the poor and reduce stress on the environment, innovation in sanitation and water treatment that does not use chemicals is required. Technical innovation (and changes in regulations) could better-enable human excreta to be recycled for both urban and rural agriculture. Ecological sanitation would prevent disease, conserve and protect water sources, and recover and recycle nutrients in a non-polluting way.

Employment and Compensation for the Poor

Policy Option: Employing the Poor

Some macro-environmental improvements, like watershed protection or nature reserves, are public goods whose benefits accrue only partially to poor local people. Many of these activities are labor-intensive, and offer an opportunity for public and private-sector organizations to provide paid employment to the poor. Longer-term livelihood opportunities for the poor may be integrated into plans for environmental management, such as hiring poor or landless peoples as guards in community and national parks, forests, and biodiversity reserves; as people who can establish and protect wildlife corridors in agricultural regions; or as personnel to monitor local water quality.

Employment of the poor for large-scale resource improvements may be financed through municipal governments for such projects as the protection of water resources, or through temporary public works programmes intended for relief or employment generation. Thus, such employment programmes directly improve the environment in addition to providing direct income for the poor.

Direct employment projects appear most likely to be successful where there are well-established supervisory organizations, reliable funding arrangements, and where the people hired—and who will be using the resources over the long term—are involved in the process of the design and selection of interventions.

Policy Option: Compensating the Poor

In some cases, the rural poor may have few economic incentives to manage their natural resources more sustainably, but other groups have an abiding economic or environmental stake in maintaining or improving the resource. Here, it may be possible for governments or other institutions to develop mechanisms for them to be compensated for the costs incurred in changing their management or use of resources.

Examples of compensating the poor include systems to pay local farmers to control agricultural burning so as to achieve national or international carbon emission or air quality targets, or various tradable-rights systems. Municipal water companies may be able to reduce the cost of acquiring water by paying farmers to use water-conserving practices. Municipal authorities may also invest in watershed improvement to reduce erosion into reservoirs supplying urban areas.

Market and Planning Reform

Policy Option: Intervening to Overcome the Deficiencies of the Market

Market forces may lead to an efficient allocation of resources when maximizing short-term return is the goal. However, markets are not always environmentally friendly, and not always supportive of poor people. In many cases, markets barely reach poor and isolated communities. In other cases, integration of poor areas into national or international economies, or the popularization of products that were formerly consumed only locally, can create demand that outstrips sustainable supply. Resources that had been used only for local consumption can be suddenly over-exploited as markets increase, as happened in the case of the shrimp industry in Southeast Asia.

Trade for industrial or niche export markets often exposes rural households to high levels of risk. This is particularly true where the trade has encouraged people to move away from more diversified and less risky agriculture-based livelihoods. When the government promotes a certain market, it must also avoid playing the middleman. It does this by obliging farmers to sell to government marketing bodies or to traders to whom concessions have been granted—or to compete in the same market from state-owned enterprises, many of whom receive various forms of subsidies.

Market development should be gradual and accompanied by efforts to help the poor adapt their institutions to new conditions. When government promotes products for industry, while the poor give priority to products that help meet subsistence and protection needs, market inefficiencies can appear and the poor can be vulnerable when subsidies are removed. For example, growing trees as cash crops has proved to be appropriate primarily for those who had other land for food or cash crops, or those who had off-farm income, not for the very poor.

Another element of market development is to provide a more competitive environment for the provision of goods and services used by the poor. In some cases, this has been accomplished by privatization schemes, such as for the provision of power and water. In such cases where a "natural monopoly" is given to the private sector, the State must still play a role in tightly regulating and supervising the sector. In other cases, healthy competition can be encouraged through allowing private companies to compete with the state sector in providing services used by the poor, such as in water supply, sanitation and energy services. When properly engineered, such programmes can improve the quality of the services the poor receive and lower their costs.

Policy Option: Eliminating Subsidies for the Non-Poor

In many countries, the non-poor receive substantial subsidies from the State. Removing these subsidies can be a source of funds for investment in the resources needed by the poor. For example, many governments give farmers in large-scale government-managed systems free or heavily subsidized water, while farmers managing their own systems can and do pay higher effective rates. Charging rates that more closely approximate the market price of water in many countries would generate millions of dollars of revenue and would lead to more efficient use of water.

Many countries set high import duties and taxes for energy technology and equipment, including those that are very energy efficient. However, many also offer subsidies to conventional energy, often to appease a particular industry or agricultural lobbies. Thus, many efficient energy technologies that could improve energy services and benefit the poor are placed out of their reach.

Furthermore, prices for most conventional energy technologies do not reflect social and environmental externalities. Subsidies are often associated with poor service, such as frequent voltage fluctuations, because energy suppliers find it difficult to generate enough revenue to maintain their equipment properly. If combined with appropriate financing schemes, end-users may be quite willing to use more efficient devices and also pay higher prices in exchange for assured quality of energy services that will lower their total energy consumption. The urban poor often pay more per-unit for energy services and water than the non-poor. Therefore, reducing subsidies for the non-poor and extending higher-quality services to the poor can be financially viable.

Policy Option: Reforming the Planning Procedures

Revising the investment criteria used for planning the State's investments may be needed. The conventional formula for calculating the internal rate of return, or net present value, heavily discounts future benefits and overvalues present consumption. As such, it devalues environmental benefits that might be felt only at some time in the future. "Greening the IRR," (revising the internal rate of return) to place more value on future benefits is one way forward. Similarly, using a "basic needs externalities" approach may be necessary for calculating the returns on investments made for the benefit of the poor. For example, certain basic social services (primary health care, basic education) may be valued at more than what the poor are willing to pay. But they can still be considered viable projects because of the larger benefits that accrue to society from such investments. Finally, investment criteria need to be adjusted to take into account that one dollar of increased income for a poor person is more valuable than one dollar increased income for a rich person. For example, a 7 percent return on an investment for the benefit of the poor may produce more added well-being than, say, a 10 percent return on an investment targeted toward the non-poor. Economists have developed weighted income measures that include such considerations.

Community-based ecosystem planning can help move from ecological poverty to ecosystem health through natural resource regeneration and maintenance of biodiversity, both on land and in aquatic environments. By linking urban biodegradable waste to agriculture and recycling its nutrients, food production can increase and degraded lands can be reclaimed. An urban policy framework can be comprehensive and eco-friendly if it integrates environmental concerns with natural resource management and problems of other sectors (e.g., agriculture and forestry). This requires better understanding of water and nutrient cycles as they pass through communities and households, and a good inventory of the current natural resource base. It also requires a better understanding of resource uses and users (i.e. who has which, and who has access to and control over them). Women's and men's different uses and knowledge of the ecosystem also must be taken into account.

Planning reform also needs a vision across time, and involves bringing in different groups of poor people in longer-term land-use planning efforts to ensure that both their existing use patterns and future needs can be met. Environmental enhancement and poverty eradication strategies also urgently need a spatial vision, so that the solutions for urban-related problems do not cause rural-related problems, and vice versa.

All references for articles included in *Taking Sides: Clashing Views in Sustainability* can be found on the Web at www.mhhe.com/cls.

EXPLORING THE ISSUE

Is Poverty Responsible for Global Environmental Degradation?

Critical Thinking and Reflection

1. What are the causes of environmental degradation? What policies or programs would you advocate that could reduce environmental degradation and improve the quality of the environment?
2. Is there any relationship between the environmental degradation and poverty in developing countries and the affluences of rich countries? If there is, explain how?
3. Discuss the basic elements of globalization and discuss whether it is good or bad for sustainability. How does globalization affect the environmental quality and economy of developing countries?
4. Can better management and policy reduce environmental degradation and poverty in developing countries? If they can, what policies and management strategies work better than other policies?
5. Can developing countries grow their way out of poverty without reducing environmental quality? Does this mean that the environmental Kuznets curve is a correct way to view the relationship between economic growth and environmental quality?

Is There Common Ground?

Developing economies often believe that they need to choose between sustainability and economic development. In some cases this is true. Cheap coal for energy in China and India often leads to air pollution and an increase in greenhouse gases, which leads to global warming. But, for these countries to seek a more sustainable path they would need to consider more expensive alternatives that could negatively impact their economy. Yet, there are many ways that developing countries can and do institute sustainable practices. Collecting rain water, using more natural drainage systems, composting, and recycling are all ways that developing countries can easily maintain a sustainable path. In reality, there are ways that both spur economic development while maintaining sustainability.

Poverty and weak governing institutions can often lead to corruption, which in turn can cause environmental degradation, particularly in developing economies where strong governing institutions are still ascending. Deterioration in natural resources directly affects the economic development as well as the livelihoods of the poor. A common ground is the development of policy at the macro level that connects indicators of both human and economic development, that is, education for women and incentives to farmers to maintain the environment and alleviate poverty.

Internet References

Anton, J. Danillo. (1993) *Thirsty cities urban environments and water supply in Latin America.* Online book available at International Development Research Center. Retrieved on November 15, 2010 from www.idrc.ca/en/ev-9313-201-1-DO_TOPIC.html (see chapter 3).

A book posted on the Internet that investigates environmental issues in many cities of Latin America and Caribbean island. It shows environmental degradation and inadequate policies of Latin America and Caribbean (LAC) regions. Environmental policies are decided in a bureaucratic way with no public participation.

International Fund for Agricultural Development. (1995, Nov 21) Combating environmental degradation. *International Fund for Agricultural Development. Conference on Hunger and Poverty.* Retrieved on November 4, 2010 from www.ifad.org/events/past/hunger/envir.html

This article is posted on the Internet to show how the poor are the vulnerable community in marginal areas.

United Nations Conference on Environment and Development (UNCED). Retrieved on November 1, 2010 from www.un.org/Docs/SG/environ.htm#environ.htm

Full implementation of Agenda 21 and other outcomes of United Nations Conference on Environment and Development (UNCED).

Woodrow Wilson International Center for Scholars. (20011, Jan 30). Retrieved on November 2, 2010 from www.wilsoncenter.org/index.cfm?fuseaction=events.event_summary&event_id=7253

An article posted on the Internet that examines U.S. policy over the developing world and its concern over the rapid population growth about environmental change and human security in developed and developing countries.

ISSUE 7

Is Limiting Consumption Rather Than Limiting Population the Key to Sustainability?

YES: Robert W. Kates, from "Population and Consumption: What We Know, What We Need to Know," *Environment* (April 2000)

NO: J. Anthony Cassils, from "Overpopulation, Sustainable Development, and Security: Developing an Integrated Strategy," *Population and Environment* (January 2004)

Learning Outcomes

After reading this issue, you should be able to:

- Understand the connection between consumption and environmental impact and change.
- Discuss the basic elements of the I-PAT model.
- Discuss the meaning of the global disparity of consumption and differences in ecological footprint.
- Discuss why population has been studied.

ISSUE SUMMARY

YES: Robert W. Kates is an American geographer and independent scholar in Trenton, Maine, and university professor (emeritus) at Brown University. He believes that consumption is more challenging to sustainability than population but more difficult to study because of its varied meanings.

NO: J. Anthony Cassils, a writer and an activist on population issues for the Population Institute of Canada, states that "nothing threatens the future of our species as much as overpopulation," and advocates a comprehensive strategy to address overpopulation.

Both consumption and population impact natural resource use and ecosystem stability. Increased global population will, of course, lead to greater use of natural resources and energy. But not all populations share the same level of resource use. For instance, Jared Diamond (2008) has pointed out that the average rates at which people consume resources like oil and metals, and produce wastes like plastics and greenhouse gases, are about 32 times higher in North America, Western Europe, Japan, and Australia than in the developing world. The world population in 2011 will break the 7 billion mark. If we estimate that 2 billion of these people consume at the rate of the developed world (whether population in the developed world or affluent population in developing countries), then their contribution to world population is actually more like 64 billion. If everyone lived like a citizen in a developed country, then the world's consuming population would be 224 billion rather than 7 billion. Clearly, the resource base and sustainability of global ecosystems would be stressed.

Population growth in developing countries is much faster than in the developed countries. Countries like the Philippines and Nigeria are rapidly growing, while northern and western European countries are close to zero population growth. Policymakers in Western countries often point to the "population explosion" in developing countries and concentrate on population control policies. Leaders in developed countries de-emphasize the need for population controls and, instead, focus on policies that can slow the West's consumption of materials and energy. There is a significant difference in the way both regions, the developed and the developing, view this issue. This write-up seeks to explore that divide.

Mahatma Gandhi once stated that "the world has enough for everyone's need, but not for everyone's greed." As a visionary and world leader, he saw the relationship between growing consumption patterns, pollution, and global environmental crisis. The Union of Concerned Scientists noted how the manufacture and growing use of automobiles have led to increased air pollution through gasoline consumption. As the world becomes more affluent, eating habits change and people add more animal protein to their diets. The growing consumption of meat and poultry requires large amounts of water and land, particularly to grow the crops necessary for animal feed. This exponentially increases the demand for agricultural land. Consumption of grains, fruits, and vegetables increase domestic water use by 30 percent. The application of fertilizers and pesticides for increased agricultural production leads to both surface and groundwater contamination. Use of household appliances, lighting, home heating, hot water, and air-conditioning leads to use of fossil fuels, thereby creating air pollution. Home construction requires land and wood, resulting in depletion of these resources leading to habitat destruction. Household water and sewage are a cause of 11 percent of domestic water pollution.

The world is now beginning to face stresses in the sustainability of key resources, such as fishing, as consumption levels have increased (Jorgenson, 2011). In a study by Patterson et al. (2007), the authors discuss how tourism in Italy has been affected by changes in consumer behavior leading to increased material throughput. Bagliani et al. (2008) utilize ecological footprint data for 141 countries to show how national environmental policies that aim to reduce consumption result in shifting environmental pressures from one nation to another with no net benefit in the protection of nature. A recent report (from the World Wildlife Fund, Zoological Society of London, and Global Footprint

Network) concludes that we will need two earths by 2030 at the present rate of consumption of natural resources. The global ecological footprint has doubled since the mid-1960s, driven by an 11-fold increase in carbon footprint. There has been a 30 percent loss of species in the last 40 years and 60 percent of this loss has been in the tropics (www.treehugger.com).

There is a general consensus that if current consumption patterns continue, it will put a strain on the earth's natural resources and social stability (www.globalissues.org). But the study of consumption is a difficult undertaking, and to create policies to manage it can be even more difficult. For instance, what is the relationship between lifestyle and consumption? Can people have a high-quality lifestyle while minimizing their consumption and ecological footprint? What incentives can be adopted that can reduce consumption?

The notion of population growth and sustainability has been a constant. The relationship is more direct and the policies developed easier to administer. Hence, the issue of overpopulation has historically constituted a central sustainability issue. Many natural resources are finite, arable land is limited, and freshwater resources relative to population growth are stressed. Global sustainability is directly affected by increases in world population. Policies to manage world population are already in place and the positive results from these policies on the environment can be quantitatively measured. Although policies limiting consumption can directly affect economic growth, population control measures have fewer critics and can actually encourage economic growth as less public financial capital needs to be directed as social welfare and, instead, can be redirected into economic innovation.

The United Nations has projected that world population will stabilize at 9.1 billion in 2050 (*Mother Jones* magazine, 2010). To sustain the world's demand for natural resources, global population growth needs to be managed. Problems of environmental degradation caused by exploding population are dynamic and nonlinear, not passive and linear (Harte, 2007). For instance, the I-PAT equation (impacts equals population, affluence, and technology) enunciated by Ehrlich and Holdren in 1971 is misleading if taken too literally. This equation represents linearity, whereas the factors involved can be multiplied significantly by the increase in population. For instance, if population doubles in size, impacts are doubled in magnitude (Harte, 2007).

International family planning has been suggested as one of the ways to curtail growing global population. Developing countries are responsible for 99 percent of global population growth and one way to slow this growth is to provide incentives for people to have smaller families. Potts (2007) quotes from a summit held with the world's renowned scientists in 1994 that "if current predictions of population growth prove accurate and patterns of human activity on the planet remain unchanged, science and technology may not be able to prevent irreversible degradation of the natural environment and continued poverty for much of the world" (Graham-Smith, 1994). Today we are seeing that much of this prediction has come true and is a major issue addressed by advocates for sustainability policies and practices.

Global environmental change is a confounding mix of a lot of social and environmental changes. Tong et al. (2007) note that increasing environmental damage due to increased ecological footprint per capita leads to impacts in human health. These impacts are both direct and indirect. An example of a direct impact is exposure to ultraviolet radiation, which causes skin cancer through depletion of the ozone layer. An indirect impact is the concentration of regional aerosol pollutants due to warmer air and disruption of ecosystems that leads to outbreaks of infectious diseases. Besides loss of biodiversity, excess environmental pollution impacts human health through exposure to polluted water and air, lack of efficient sewage facilities (especially in overpopulated regions), and groundwater depletion (Ebi and Gamble, 2005; Tong, 2000).

Population is a well-studied area compared with consumption. Measures intended to reduce world population, such as more education for women, contraception, and family planning, are already in place. Measures to reduce global consumption are more difficult to implement. But this does not mean that consumption is less important for sustainability to address.

The article for the YES position, that consumption is the key to sustainability, is presented by noted sustainability science professor Robert Kates. He discusses how the I-PAT equation relates to overpopulation and consumption. He points out that demographics or population science is a well-established scientific discipline that encompasses a wide range of issues, that is, mortality, longevity, demographic transition, and so forth. The study of consumption is new and untested, and no consensus even exists that can define it. It means different things to different people. He concludes by explaining that population is simpler and easier to study, and a consensus exists about its trends, policies, and terms. Consumption is harder to quantify, more difficult to study, and more threatening for sustainability.

The article for the NO position, that population is a central issue for sustainability, is presented by J. Anthony Cassils. He addresses the need to curb overpopulation and how it relates to ecological damage, and growing insecurity, which is a major obstacle for sustainability. He provides an overview of overpopulation and growth of interest in sustainable development. He shows the relationship between overpopulation and sustainable development on national and international issues of security. He concludes with recommendations for a transition to a sustainable future.

Robert W. Kates

Population and Consumption: What We Know, What We Need to Know

Thirty years ago, as Earth Day dawned, three wise men recognized three proximate causes of environmental degradation yet spent half a decade or more arguing their relative importance. In this classic environmentalist feud between Barry Commoner on one side and Paul Ehrlich and John Holdren on the other, all three recognized that growth in population, affluence, and technology were jointly responsible for environmental problems, but they strongly differed about their relative importance. Commoner asserted that technology and the economic system that produced it were primarily responsible.[1] Ehrlich and Holdren asserted the importance of all three drivers: population, affluence, and technology. But given Ehrlich's writings on population,[2] the differences were often, albeit incorrectly, described as an argument over whether population or technology was responsible for the environmental crisis.

Now, 30 years later, a general consensus among scientists posits that growth in population, affluence, and technology are jointly responsible for environmental problems. This has become enshrined in a useful, albeit overly simplified, identity known as I-PAT, first published by Ehrlich and Holdren in *Environment* in 1972[3] in response to the more limited version by Commoner that had appeared earlier in *Environment* and in his famous book *The Closing Circle*.[4] In this identity, various forms of environmental or resource impacts (I) equals population (P) times affluence (A) (usually income per capita) times the impacts per unit of income as determined by technology (T) and the institutions, that use it. Academic debate has now shifted from the greater or lesser importance of each of these driving forces of environmental degradation or resource depletion to debate about their interaction and the ultimate forces that drive them.

However, in the wider global realm, the debate about who or what is responsible for environmental degradation lives on. Today, many Earth Days later, international debates over such major concerns as biodiversity, climate change, or sustainable development address the population and the affluence terms of Holdrens' and Ehrlich's identity, specifically focusing on the character of consumption that affluence permits. The concern with technology is more complicated because it is now widely recognized that while technology can be a problem, it can be a solution as well. The development and use of more environmentally benign and friendly technologies in industrialized countries have slowed the growth of many of the most pernicious forms of pollution that originally drew Commoner's attention and still dominate Earth Day concerns.

A recent report from the National Research Council captures one view of the current public debate, and it begins as follows:

> For over two decades, the same frustrating exchange has been repeated countless times in international policy circles. A government official or scientist from a wealthy country would make the following argument: The world is threatened with environmental disaster because of the depletion of natural resources (or climate change or the loss of biodiversity), and it cannot continue for long to support its rapidly growing population. To preserve the environment for future generations, we need to move quickly to control global population growth, and we must concentrate the effort on the world's poorer countries, where the vast majority of population growth is occurring.

Government officials and scientists from low-income countries would typically respond:

> If the world is facing environmental disaster, it is not the fault of the poor, who use few resources. The fault must lie with the world's wealthy countries, where people consume the great bulk of the world's natural resources and energy and cause the great bulk of its environmental degradation. We need to curtail overconsumption in the rich countries which use far more than their fair share, both to preserve the environment and to allow the poorest people on earth to achieve an acceptable standard of living.[5]

It would be helpful, as in all such classic disputes, to begin by laying out what is known about the relative responsibilities of both population and consumption for the environmental crisis, and what might need to be known to address them. However, there is a profound asymmetry that must fuel the frustration of the developing countries' politicians and scientists: namely, how much people know about population and how little they know about consumption. Thus, this article begins by examining these differences in knowledge and action and concludes with the alternative actions needed to go from more to enough in both population and consumption.[6]

Population

What population is and how it grows is well understood even if all the forces driving it are not. Population begins with people and their key events of birth, death, and location. At the margins, there is some debate over when life begins and ends or whether residence is temporary or permanent, but little debate in between. Thus, change in the world's population or any place is the simple arithmetic of adding births, subtracting deaths, adding immigrants, and subtracting outmigrants. While whole subfields of demography are devoted to the arcane details of these additions and subtractions, the error in estimates of population for almost all places is probably within 20 percent and for countries with modern statistical services, under 3 percent—better estimates than for any other living things and for most other environmental concerns.

Current world population is more than six billion people, growing at a rate of 1.3 percent per year. The peak annual growth rate in all history—about 2.1 percent—occurred in the early 1960s, and the peak population increase of around 87 million per year occurred in the late 1980s. About 80 percent or 4.8 billion people live in the less developed areas of the world, with 1.2 billion living in industrialized countries. Population is now projected by the United Nations (UN) to be 8.9 billion in 2050, according to its medium fertility assumption, the one usually considered most likely, or as high as 10.6 billion or as low as 7.3 billion.[7]

A general description of how birth rates and death rates are changing over time is a process called the demographic transition.[8] It was first studied in the context of Europe, where in the space of two centuries, societies went from a condition of high births and high deaths to the current situation of low births and low deaths. In such a transition, deaths decline more rapidly than births, and in that gap, population grows rapidly but eventually stabilizes as the birth decline matches or even exceeds the death decline. Although the general description of the transition is widely accepted, much is debated about its cause and details.

The world is now in the midst of a global transition that, unlike the European transition, is much more rapid. Both births and deaths have dropped faster than experts expected and history foreshadowed. It took 100 years for deaths to drop in Europe compared to the drop in 30 years in the developing world. Three is the current global average births per woman of reproductive age. This number is more than halfway between the average of five children born to each woman at the post World War II peak of population growth and the average of 2.1 births required to achieve eventual zero population growth.[9] The death transition is more advanced, with life expectancy currently at 64 years. This represents three-quarters of the transition between a life expectancy of 40 years to one of 75 years. The current rates of decline in births outpace the estimates of the demographers, the UN having reduced its latest medium expectation of global population in 2050 to 8.9 billion, a reduction of almost 10 percent from its projection in 1994.

Demographers debate the causes of this rapid birth decline. But even with such differences, it is possible to break down the projected growth of the next century and to identify policies that would reduce projected populations even further. John Bongaarts of the Population Council has decomposed the projected developing country growth into three parts and, with his colleague Judith Bruce, has envisioned policies that would encourage further and more rapid decline.[10] The first part is unwanted fertility, making available the methods and materials for contraception to the 120 million married women (and the many more unmarried women) in developing countries who in survey research say they either want fewer children or want to space them better. A basic strategy for doing so links voluntary family planning with other reproductive and child health services.

Yet in many parts of the world, the desired number of children is too high for a stabilized population. Bongaarts would reduce this desire for large families by changing the costs and benefits of childrearing so that more parents would recognize the value of smaller families while simultaneously increasing their investment in children. A basic strategy for doing so accelerates three trends that have been shown to lead to lower desired family size: the survival of children, their education, and improvement in the economic, social, and legal status for girls and women.

However, even if fertility could immediately be brought down to the replacement level of two surviving children per woman, population growth would continue for many years in most developing countries because so many more young people of reproductive age exist. So Bongaarts would slow this momentum of population growth by increasing the age of childbearing, primarily by improving secondary education opportunity for girls and by addressing such neglected issues as adolescent sexuality and reproductive behavior.

How much further could population be reduced? Bongaarts provides the outer limits. The population of the developing world (using older projections) was expected to reach 10.2 billion by 2100. In theory, Bongaarts found that meeting the unmet need for contraception could reduce this total by about 2 billion. Bringing down desired family size to replacement fertility would reduce the population a billion more, with the remaining growth—from 4.5 billion today to 7.3 billion in 2100—due to population momentum. In practice, however, a recent U.S. National Academy of Sciences report concluded that a 10 percent reduction is both realistic and attainable and could lead to a lessening in projected population numbers by 2050 of upwards of a billion fewer people.[11]

Consumption

In contrast to population, where people and their births and deaths are relatively well-defined biological events, there is no consensus as to what consumption includes. Paul Stern of the National Research Council has described the different ways physics, economics, ecology, and sociology view consumption.[12] For physicists, matter and energy cannot be consumed, so consumption is conceived as transformations of matter and energy with increased entropy. For economists, consumption is spending on consumer goods and services and thus distinguished from their production and distribution. For ecologists, consumption is obtaining energy and nutrients by eating something else, mostly green plants or other consumers of green plants. And for some sociologists, consumption is a status symbol—keeping up with the Joneses—when individuals and households use their incomes to increase their social status through certain kinds of purchases.

In 1977, the councils of the Royal Society of London and the U.S. National Academy of Sciences issued a joint statement on consumption, having previously done so on population. They chose a variant of the physicist's definition:

> Consumption is the human transformation of materials and energy. Consumption is of concern to the extent that it makes the transformed materials or energy less available for future use, or negatively impacts biophysical systems in such a way as to threaten human health, welfare, or other things people value.[13]

On the one hand, this society/academy view is more holistic and fundamental than the other definitions; on the other hand, it is more focused, turning attention to the environmentally damaging. This article uses it as a working definition with one modification, *the addition of information to energy and matter, thus completing the triad of the biophysical and ecological basics that support life.*

In contrast to population, only limited data and concepts on the transformation of energy, materials, and information exist.[14] There is relatively good global knowledge of energy transformations due in part to the common units of conversion between different technologies. Between 1950 and today, global energy production and use increased more than fourfold.[15] For material transformations, there are no aggregate data in common units on a global basis, only for some specific classes of materials including materials for energy production, construction, industrial minerals and metals, agricultural crops, and water.[16] Calculations of material use by volume, mass, or value lead to different trends.

Trend data for per capita use of physical structure materials (construction and industrial minerals, metals, and forestry products) in the United States are relatively complete. They show an inverted S shaped (logistic) growth pattern: modest doubling between 1900 and the depression of the 1930s (from two to four metric tons), followed by a steep quintupling with economic recovery until the early 1970s (from two to eleven tons), followed by a leveling off since then with fluctuations related to economic downturns (see Figure 1).[17] An aggregate analysis of all current material production and consumption in the United States averages more than 60 kilos per person per day (excluding water). Most of this material flow is split between energy and related products (38 percent) and minerals for construction (37 percent), with the remainder as

---

WHAT IS CONSUMPTION?

Physicist: "What happens when you transform matter/energy"
Ecologist: "What big fish do to little fish"
Economist: "What consumers do with their money"
Sociologist: "What you do to keep up with the Joneses"

Figure 1

Consumption of Physical Structure Materials in the United States, 1900–1991

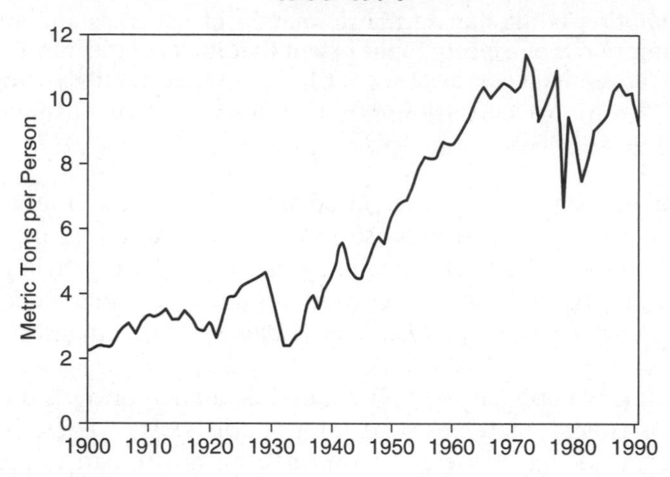

Source: I. Wernick, "Consuming Materials. The American Way," *Technological Forecasting and Social Change,* 53 (1996): 114.

industrial minerals (5 percent), metals (2 percent), products of fields (12 percent), and forest (5 percent).[18]

A massive effort is under way to catalog biological (genetic) information and to sequence the genomes of microbes, worms, plants, mice, and people. In contrast to the molecular detail, the number and diversity of organisms is unknown, but a conservative estimate places the number of species on the order of 10 million, of which only one-tenth have been described.[19] Although there is much interest and many anecdotes, neither concepts nor data are available on most cultural information. For example, the number of languages in the world continues to decline while the number of messages expands exponentially.

Trends and projections in agriculture, energy, and economy can serve as surrogates for more detailed data on energy and material transformation.[20] From 1950 to the early 1990s, world population more than doubled (2.2 times), food as measured by grain production almost tripled (2.7 times), energy more than quadrupled (4.4 times), and the economy quintupled (5.1 times). This 43-year record is similar to a current 55-year projection (1995–2050) that assumes the continuation of current trends or, as some would note, "business as usual." In this 55-year projection, growth in half again of population (1.6 times) finds almost a doubling of agriculture (1.8 times), more than twice as much energy used (2.4 times), and a quadrupling of the economy (4.3 times).[21]

Thus, both history and future scenarios predict growth rates of consumption well beyond population. An attractive similarity exists between a demographic transition that moves over time from high births and high deaths to low births and low deaths with an energy, materials, and information transition. In this transition, societies will use increasing amounts of energy and materials as consumption increases, but over time the energy and materials input per unit of consumption decrease and information substitutes for more material and energy inputs.

Some encouraging signs surface for such a transition in both energy and materials, and these have been variously labeled as decarbonization and dematerialization.[22] For more than a century, the amount of carbon per unit of energy produced has been decreasing. Over a shorter period, the amount of energy used to produce a unit of production has also steadily declined. There is also evidence for dematerialization, using fewer materials for a unit of production, but only for industrialized countries and for some specific materials. Overall, improvements in technology and substitution of information for energy and materials will continue to increase energy efficiency (including decarbonization) and dematerialization per unit of product or service. Thus, over time, less energy and materials will be needed to make specific things. At the same time, the demand for products and services continues to increase, and the overall consumption of energy and most materials more than offsets these efficiency and productivity gains.

What to Do About Consumption

While quantitative analysis of consumption is just beginning, three questions suggest a direction for reducing environmentally damaging and resource-depleting consumption. The first asks: *When is more too much for the life-support systems of the natural world and the social infrastructure of human society?* Not all the projected growth in consumption may be resource-depleting—"less available for future use"—or environmentally damaging in a way that "negatively impacts biophysical systems to threaten human health, welfare, or other things people value."[23] Yet almost any human-induced transformations turn out to be either or both resource-depleting or damaging to some valued environmental component. For example, a few years ago, a series of eight energy controversies in Maine were related to coal, nuclear, natural gas, hydroelectric, biomass, and wind generating sources, as well as to various energy policies. In all the controversies, competing sides, often more than two, emphasized environmental benefits to support their choice and attributed environmental damage to the other alternatives.

Despite this complexity, it is possible to rank energy sources by the varied and multiple risks they pose and, for those concerned, to choose which risks they wish to minimize and which they are more willing to accept. There is now almost 30 years of experience with the theory and methods of risk assessment and 10 years of experience with the identification and setting of environmental priorities. While there is still no readily accepted methodology for separating resource-depleting or environmentally damaging consumption from general consumption or for identifying harmful transformations from those that are benign, one can separate consumption into more or less damaging and depleting classes and *shift* consumption to the less harmful class. It is possible to *substitute* less damaging and depleting energy and materials for more damaging ones. There is growing experience with encouraging substitution and its difficulties: renewables for nonrenewables, toxics with fewer toxics, ozone-depleting chemicals for more benign substitutes, natural gas for coal, and so forth.

The second question, *Can we do more with less?,* addresses the supply side of consumption. Beyond substitution, shrinking the energy and material transformations required per unit of consumption is probably the most effective current means for reducing environmentally damaging consumption. In the 1997 book, *Stuff: The Secret Lives of Everyday Things,* John Ryan and Alan Durning of Northwest Environment Watch trace the complex origins, materials, production, and transport of such everyday things as coffee, newspapers, cars, and computers and highlight the complexity of reengineering such products and reorganizing their production and distribution.[24]

Yet there is growing experience with the three Rs of consumption shrinkage: reduce, recycle, reuse. These have now been strengthened by a growing science, technology, and practice of industrial ecology that seeks to learn from nature's ecology to reuse everything. These efforts will only increase the existing favorable trends in the efficiency of energy and material usage. Such a potential led the Intergovernmental Panel on Climate Change to conclude that it was possible, using current best practice technology, to reduce energy use by 30 percent in the short run and 50–60 percent in the long run.[25] Perhaps most important in the long run, but possibly least studied, is the potential for and value of substituting information for energy and materials. Energy and materials per unit of consumption are going down, in part because more and more consumption consists of information.

The third question addresses the demand side of consumption—*When is more enough?*[26] Is it possible to reduce consumption by more satisfaction with what people already have, by *satiation,* no more needing more because there is enough, and by *sublimation,* having more satisfaction with less to achieve some greater good? This is the least explored area of consumption and the most difficult. There are, of course, many signs of *satiation* for some goods. For example, people in the industrialized world no longer buy additional refrigerators (except in newly formed households) but only replace them. Moreover, the quality of refrigerators has so improved that a 20-year or more life span is commonplace. The financial pages include frequent stories of the plight of this industry or corporation whose markets are saturated and whose products no longer show the annual growth equated with profits and progress. Such enterprises are frequently viewed as failures of marketing or entrepreneurship rather than successes in meeting human needs sufficiently and efficiently. Is it possible to reverse such views, to create a standard of satiation, a satisfaction in a need well met?

Can people have more satisfaction with what they already have by using it more intensely and having the time to do so? Economist Juliet Schor tells of some overworked Americans who would willingly exchange time for money, time to spend with family and using what they already have, but who are constrained by an uncooperative employment structure.[27] Proposed U.S. legislation would permit the trading of overtime for such compensatory time off, a step in this direction. *Sublimation,* according to the dictionary, is the diversion of energy from an immediate goal to a higher social, moral, or aesthetic purpose. Can people be more satisfied with less satisfaction derived from the diversion of immediate consumption for the satisfaction of a smaller ecological footprint?[28] An emergent research field grapples with how to encourage consumer behavior that will lead to change in environmentally damaging consumption.[29]

A small but growing "simplicity" movement tries to fashion new images of "living the good life."[30] Such movements may never much reduce the burdens of consumption, but they facilitate by example and experiment other less-demanding alternatives. Peter Menzel's remarkable photo essay of the material goods of some 30 households from around the world is powerful testimony to the great variety and inequality of possessions amidst the existence of alternative life styles.[31] Can a standard of "more is enough" be linked to an ethic of "enough for all"? One of the great discoveries of childhood is that eating lunch does not feed the starving children of some far-off place. But increasingly, in sharing the global commons, people flirt with mechanisms that hint at such—a rationing system for the remaining chlorofluorocarbons, trading systems for reducing emissions, rewards for preserving species, or allowances for using available resources.

A recent compilation of essays, *Consuming Desires: Consumption, Culture, and the Pursuit of Happiness,*[32] explores many of these essential issues. These elegant essays by 14 well-known writers and academics ask the fundamental question of why more never seems to be enough and why satiation and sublimation are so difficult in a culture of consumption. Indeed, how is the culture of consumption different for mainstream America, women, inner-city children, South Asian immigrants, or newly industrializing countries?

Why We Know and Don't Know

In an imagined dialog between rich and poor countries, with each side listening carefully to the other, they might ask themselves just what they actually know about population and consumption. Struck with the asymmetry described above, they might then ask: "Why do we know so much more about population than consumption?"

The answer would be that population is simpler, easier to study, and a consensus exists about terms, trends, even policies. Consumption is harder, with no consensus as to what it is, and with few studies except in the fields of marketing and advertising. But the consensus that exists about population comes from substantial research and study, much of it funded by governments and groups in rich countries, whose asymmetric concern readily identifies the troubling fertility behavior of others and only reluctantly considers their own consumption behavior. So while consumption is harder, it is surely studied less (see Table 1).

The asymmetry of concern is not very flattering to people in developing countries. Anglo-Saxon tradition has a long history of dominant thought holding the poor responsible for their condition—they have too many children—and

Table 1

A Comparison of Population and Consumption

Population	Consumption
Simpler, easier to study	More complex
Well-funded research	Unfunded, except marketing
Consensus terms, trends	Uncertain terms, trends
Consensus policies	Threatening policies

Source: Robert W. Kates.

an even longer tradition of urban civilization feeling besieged by the barbarians at their gates. But whatever the origins of the asymmetry, its persistence does no one a service. Indeed, the stylized debate of population versus consumption reflects neither popular understanding nor scientific insight. Yet lurking somewhere beneath the surface concerns lies a deeper fear.

Consumption is more threatening, and despite the North–South rhetoric, it is threatening to all. In both rich and poor countries alike, making and selling things to each other, including unnecessary things, is the essence of the economic system. No longer challenged by socialism, global capitalism seems inherently based on growth—growth of both consumers and their consumption. To study consumption in this light is to risk concluding that a transition to sustainability might require profound changes in the making and selling of things and in the opportunities that this provides. To draw such conclusions, in the absence of convincing alternative visions, is fearful and to be avoided.

What We Need to Know and Do

In conclusion, returning to the 30-year-old I-PAT identity—a variant of which might be called the Population/Consumption (PC) version—and restating that identity in terms of population and consumption, it would be: $I = P*C/P*I/C$, where I equals environmental degradation and/or resource depletion; P equals the number of people or households; and C equals the transformation of energy, materials, and information (see Figure 2).

With such an identity as a template, and with the goal of reducing environmentally degrading and resource-depleting influences, there are at least seven major directions for research and policy. To reduce the level of impacts per unit of consumption, it is necessary to separate out more damaging consumption and *shift* to less harmful forms, *shrink* the amounts of environmentally damaging energy and materials per unit of consumption, and *substitute* information for energy and materials. To reduce consumption per person or household, it is necessary to *satisfy* more with what is already had, satiate well-met consumption needs, and *sublimate* wants for a greater good. Finally, it is possible to *slow* population growth and then to *stabilize* population numbers as indicated above.

Figure 2

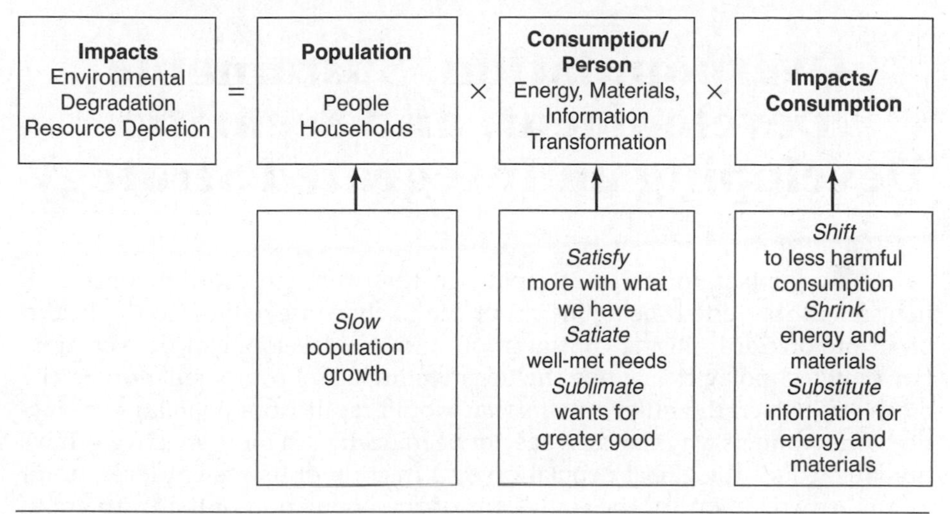

Source: Robert W. Kates.

However, as with all versions of the I-PAT identity, population and consumption in the PC version are only proximate driving forces, and the ultimate forces that drive consumption, the consuming desires, are poorly understood, as are many of the major interventions needed to reduce these proximate driving forces. People know most about slowing population growth, more about shrinking and substituting environmentally damaging consumption, much about shifting to less damaging consumption, and least about satisfaction, satiation, and sublimation. Thus the determinants of consumption and its alternative patterns have been identified as a key understudied topic for an emerging sustainability science by the recent U.S. National Academy of Science study.[33]

But people and society do not need to know more in order to act. They can readily begin to separate out the most serious problems of consumption, shrink its energy and material throughputs, substitute information for energy and materials, create a standard for satiation, sublimate the possession of things for that of the global commons, as well as slow and stabilize population. To go from more to enough is more than enough to do for 30 more Earth Days.

All references for articles included in *Taking Sides: Clashing Views in Sustainability* can be found on the Web at www.mhhe.com/cls.

J. Anthony Cassils

Overpopulation, Sustainable Development, and Security: Developing an Integrated Strategy

. . . As population increased rapidly in the twentieth century, some eminent scientists alerted the public about the likely consequences to the health of all life on Earth. Rising consumption and the development of ever more powerful technologies magnify the negative impact of overpopulation on the biosphere. Given the huge benefits that would result from population reduction, one wonders why the issue has not been dealt with more effectively. If we humans reduce the global population to a fraction of its present level to, for example, two billion[1] or the equivalent of the population of the Earth about 1930, it would provide extraordinary benefits. The quality of life of all people would soar. We would have all the advantages of modern technology but little, if any, environmental deterioration.

The alarm raised by scientists had some impact. Some valuable initiatives have been undertaken in recent decades to reduce fertility rates, but these have been insufficient, for, in recent years, the issue of overpopulation has been pushed to the side by many countries. It is a contentious topic. For thousands of years, we humans have viewed ourselves as struggling for survival against a hostile natural world and consequently many traditional values and institutions favour the growth of human numbers. Most major religions support population growth by opposing birth control and limits to immigration into prosperous countries. Real estate developers want more people because they require housing and cause real estate prices to rise. Banks like rising real estate prices because it makes their mortgage business more secure. Immigration lawyers want more immigrants because it is good for their business. Governments favour an increase in population for it spreads the burden of public debt, and often they support lax immigration laws with the hope of gaining the support of recent immigrants. Meanwhile, Marxists overlook overpopulation and insist on the redistribution of wealth as the global panacea.

These groups all have a vested interest in continual expansion and population growth. They manipulate politicians by bloc voting and intimidate dissenters in the public and the media, placing their own interests above the future well-being of humanity. They work with reflexive speed to take advantage of every opportunity to support population growth. For example, in March 2002, the United Nations Secretariat released a report entitled "The Future of Fertility in Intermediate Fertility Countries." The Report indicated that demographers have changed their assumptions by lowering anticipated fertility rates in many parts of the world and now project that, before 2050, approximately eighty percent of the world's population will have fertility levels below replacement. Advocates of never-ending growth were quick off the mark to reach the media and the predominant reaction to the Report was that the drop in fertility was a crisis. In fact, the real calamity is overpopulation and the resulting environmental degradation. What the UN Report did not emphasize was the bifurcating trend between the least developed countries (LDCs) and the rest of the world. In the LDCs, population growth rates have dropped little or not at all and almost all of the projected growth in global population (from 6.2 billion in 2002 to 8.9 billion in 2050) will occur in those countries.

Most of the developed countries already have fertility rates below replacement. These are important and positive developments. It is true that the population of the world will age as fewer children are born, but this is an essential step to reduce human numbers to a level that the Earth can support over the long term. Most policy makers seem incapable of coming to terms with environmental limits and continue to react to any let-up in growth as unacceptable.

It is important to recognize that the prediction of 8.9 billion people in 2050 is based on just one medium-variant scenario supported by assumptions taking into account a number of factors, such as the young age structure of global population, fertility rates, and life expectancy. This recent UN projection may be too optimistic, reflecting political hopes rather than demographic reality. If the annual rate of increase in global population stays at the current level of about 1.3%, then population would be about 12 billion by 2050. However, the annual growth rate has been trending downwards since the early 1970s when it was over 2%. Some analysts suggest that the UN Report pays insufficient attention to the young age structure of the global population, calculating that if the global fertility rate of two children per female had been reached in the year 2000 (the estimated rate was 2.8 in that year), and stabilized, the world population would peak at 12 billion in about 70 years. China, with a policy of one child per couple, will add about 10 million to their population this coming year. Clearly, there is a broad range of projections of future population levels and much careful conjecture, but most population groups, such as the UN Population Division, the Population Reference Bureau, and The International Institute of Applied Systems Analysis, have medium-variant projections that have the global population growing to about 9 billion by mid-century.

What is notable about these calculations is the omission by demographers of any consideration of carrying capacity (the number of people the global environment can carry sustainably over the medium and long term). The main reason for this serious gap is the difficulty in attaining agreement on the limiting factors, on the evolution of such factors over time, and, if such factors could be reliably predicted, the effects of economic, cultural, and political systems (O'Neill & Balk, 2001). However, environmental limits form a very real constraint on future population growth, whether or not humans can agree on them, and the growing list of environmental crises provides ample warning. The resolution of this issue is urgent but it is beyond the scope of demographers. It will take the marshaling of a broad base of political, economic, and social forces to come to agreement on carrying capacity, and the pace of international progress regarding this issue must accelerate significantly.

There are already many signs of impending limits at the current level of 6.3 billion people. The World Health Organization (WHO, 2002) has reported that there are more than three billion people who are malnourished and lack calories, several vitamins, iron, iodine and other nutrients, making them more susceptible to numerous diseases. This is the largest number of malnourished people ever in human history. The WHO reports a significant increase in various diseases; for example, more than 2.4 billion are infected with malaria and 2 billion each with TB and helminths. Masses of impoverished and malnourished people might serve as ideal incubators of new viruses. The outlook is rendered more grim with the conjunction of key trends, such as the looming shortages of fresh water in most of the world, the imminent peaking of the global production of conventional oil and gas, the collapse of major fisheries, and the loss of prime agricultural land to erosion and urbanization.

The prospect of an almost fifty percent increase in the global population over the next fifty years is nightmarish. This increase is equivalent to the entire population of the Earth in 1960. Almost all of that increase will occur in the least developed countries that are unable to meet the needs of their present population. The impact will include more pollution, an acceleration of global warming, and more severe damage to fisheries, forests, fresh water supplies, fertile soil, and biodiversity. Migration pressures, and ethnic and religious conflict will increase. Most of the population in both rich and poor countries will want to consume more with devastating effects on ecosystems already reeling under current human demands.

The international community needs to come to an unequivocal agreement that a significant reduction of human population is a desirable goal. All steps should be taken by individual nations to make it happen quickly and humanely. The Cairo conference on population in 1994 was a tentative but totally inadequate step in that direction. In this time of crisis, governments will need to show real leadership by designing institutions that help humanity shrink its way to sustainable prosperity. Countries that are in the lead in reducing their populations should not give in to the advocates of growth by allowing massive immigration. This rewards those who multiply irresponsibly. The various peoples of the Earth will need to assume responsibility to restore their respective regions of origin into lands of hope. In an overcrowded world, mass migration is no longer a reasonable option to address overpopulation. The problems must be resolved in situ.

Although overpopulation is a global problem, solutions will be implemented by individual nation states. To realize a coherent global approach, the population issue should be placed at the core of policy in every country of the world and appropriate remedies adopted to stabilize and reduce human population. The immediate issue is "How can this be achieved?"

This paper begins a process to develop a strategy to address that question. At this stage, it is expected that the strategy will grow and evolve as it benefits from the contributions of individuals and organizations.

Causes and Implications of Human Overpopulation

High intelligence has increased the capacity of the human species to acquire resources and to diminish the impact of many population-limiting factors, such as disease and famine. These capabilities have allowed humans to expand their numbers enormously, from about five million in 6000 BCE (Ehrlich, 1968) to over six billion today. In the words of E. O. Wilson (2002): "The pattern of human population growth in the 20th century was more bacterial than primate. When Homo sapiens passed the six-billion mark we had already exceeded by perhaps as much as 100 times the biomass of any large animal species that ever existed on the land. We and the rest of life cannot afford another 100 years like that." The impact on both the physical world and other life forms has been devastating and will continue far into the future.

It is obvious that the success of the human species in expanding its numbers and accessing resources has not been matched by a change in outlook. We have failed to shift our focus from the short to the long term. It is rare that we even acknowledge the influence of our basic instincts on ethics, decision-making, and public policy.

In the past fifty years, there have been major advances in our knowledge of biology and ecology. The best of science and reason suggests that human demands upon the web of life on Earth have exceeded sustainable levels. Meanwhile, human population and expectations continue to rise, making chaos and collapse very likely unless corrective action is taken promptly. Unfortunately, most people, including policy makers, overlook the important message of ecology. The health of the environment is viewed as just one of many political issues rather than the essential underpinning of all life. The ecological reality is that human life and the economy form part of and depend upon a web of many interdependent species. The web is based on countless microorganisms in the soil and water and on plants that, through photosynthesis, directly or indirectly feed all other levels of the food chain. We tend to overlook the importance of microorganisms because we cannot see them. As for the energy produced by terrestrial photosynthesis, the human species consumes, at an estimated 50% of the total (Pimentel et al., 1999), a hugely disproportionate share. It is estimated that the amount of energy consumed annually in the form of fossil fuels in the United States exceeds by fifty percent the total solar energy captured by all the plants in that country (Heinberg, 2003). The expansion of human activities appropriates an ever-larger fraction of the Earth's surface, and this is having a negative impact on biodiversity. Not only do we directly displace other species by occupying or destroying their habitat, we extinguish them, including many of the microorganisms at the foundation of all life, with our wastes and chemical poisons. Current human practices and beliefs are on a collision course with the life support system on Earth. With the short-term focus of daily human activities, the profound implications of this predicament are not given sufficient attention. Old beliefs impede the acceptance of new information and delay the implementation of changes that are essential to render human activities sustainable.

Modern communications and transportation have created an unprecedented global awareness, but the emphasis has been on global exploitation rather than on taking better care of the planet. Individuals, both corporate and human, have been quick to take advantage of opportunities. Transnational corporations comb the world looking for cheaper labour, natural resources, and larger markets, while individual people look for advantages in other countries and migrate in unprecedented numbers. Therefore, transnational corporations and economic refugees share a common interest in relatively open national borders to provide more opportunities for increased profits and consumption. Despite the popular admonition to "think globally and act locally," any nation that acts with foresight to curtail population and protect its environment, thereby creating an area of order in an increasingly chaotic world, will likely attract more international corporate activity and face enormous pressure to allow the entry of people from less ordered regions. Pressure tactics will include demands for free trade and accusations of racism for restrictions on immigration. The net result of such tactics, if successful, is to accelerate the unraveling of the web of life worldwide.

Like locusts and rabbits, we have entered a plague cycle that will end with the collapse of the food supply when environmental constraints can no longer be attenuated by technological interventions. Moreover, our technological capacity for intervention will weaken rapidly as oil and gas become more scarce over the next half-century. Local human population crashes resulting from the depletion of natural resources have occurred previously (e.g., Easter Island), but human activities have never before caused so much damage to virtually every part of the planet at the same time. The coming population crash will be more global in scale, but it will be far from equally shared across the planet.

Urban areas, especially the shantytowns of overcrowded mega-cities, will be the most adversely affected. Their dense population and woefully inadequate sanitation provide the ideal breeding ground for diseases new and old—diseases that will, sooner or later, be carried to all parts of the world through migration and tourism. Health systems in developed countries may not be able to adequately deal with the impact. The poorest parts of the world, where population growth is still rapid, will continue to suffer the most and face decreased life expectancy through resource depletion, conflict, and disease. The downward-spiraling situation in Africa is the most graphic illustration of the reality of environmental constraints. It is popular in the current intellectual climate to pretend that all would be well if only those in the developed regions would consume less. However, this ignores the fact that it is primarily population growth and its concomitant deforestation, erosion, and desertification that, in the poorer countries, destroys the ecological underpinnings of their subsistence way of life. This unfortunate situation is exacerbated by growing international demand for inexpensive natural resources.

Most people overlook the inevitability of the human encounter with global environmental limits. The press has done little to help people understand the "whole system" reality of the population/environmental crisis. The right-wing press does not begin to acknowledge that the economy is a wholly owned subsidiary of the environment. As far as it is concerned, the resources of the planet are infinite and the loss of a few million species is of no great concern. The left-wing press embraces the "it's not population, it's consumption" ideology, as if the two were not inextricably linked. From the perspective of many on the left, the resource pie is big enough for any number of people provided these people divide it into equal pieces. These positions reflect insufficient understanding of biological capacities and human nature.

The major challenge for humanity in the twenty-first century is to learn to live within the web of life on Earth without destroying it. This will be a difficult undertaking for a species with a misplaced sense of its own importance. Only in the past few hundred years have most humans come to accept that the Earth (and by inference humanity) is not the centre of the Universe.

Any attempt to change ethics challenges long-standing traditions. Codes of ethics are found in all religions, some of which have existed for millennia, and to go against them—or even to challenge some of their tenets—appears to many as tantamount to opposing the will of some form of universal power. Regardless of perception, however, ethics evolve over time, for they are practical beliefs that respond, however slowly, to changing circumstances. New information does alter the way we see the world. At present, most new insights come from science.

The human psyche has a need for a sense of purpose. Humans can be thought of as comprising their physical selves and their symbolic selves. The physical self requires food, shelter, sex, and entertainment. The symbolic self is more complicated. It comprises the way that we perceive ourselves and includes beliefs and values, and identification with family, tribe, race, religion, and nationality. The allegiance to the symbolic self is difficult to alter for it is reinforced by the great human fear of death. The identification with symbols provides a sense of being able to transcend the limitations of human life by various means, such as belief in an afterlife, seeking enlightenment through adherence to a religion, and the continuation of one's culture, language, and belief system. Having children, in a sense the continuation of the physical self, often takes on symbolic importance. Asking a person to change beliefs and attitudes about procreation can threaten the symbolic self, triggering fears of death, and frequently, violent reactions.

It is essential to encourage people to identify their symbolic selves with the web of life. If people make it their core purpose to sustain the health of the web of life, then, in a most practical sense, they will contribute to the survival of humanity. At a more spiritual level, the continuation of all life gives them a kind of immortality. When humans make the leap in understanding that we are just one species in an interdependent web, a substantial shift of ethics must follow. The totality of life becomes sacred, not just human life. If human numbers and demands grow too quickly to the point where they endanger the health of ecosystems, as is happening now, then human life becomes relatively less valuable. Like all things, scarcity increases value. Excessive human numbers drive down average wages and quality of life and drive up the cost of the necessities of life. This debasement of the value of human life, while not publicly acknowledged, is occurring in many countries.

Heeding the Warnings of Science and Reason

A battle is being waged between scholars who provide new information and established interests threatened by it. This struggle has persisted for centuries. Throughout history, many thoughtful people have warned of the potential for overpopulation, including Aristotle, Plato, Benjamin Franklin, and David Hume. But it was Thomas Malthus who crystallized the views of these writers in 1798 when he brought out his *Essay on the Principle of Population.* The purpose of the essay was to give a reasoned argument against the theories regarding the perfectibility of society put forward by Rousseau and his school. Malthus maintained that such optimistic hopes are rendered useless by the natural tendency of population to increase faster than the means of subsistence. He pointed out that all forms of life were so prolific that if they were allowed to multiply without restraint, they would fill millions of worlds in the course of a few thousand years. This point was picked up by Charles Darwin who saw that natural selection was the inevitable result of the rapid increase of all organic beings, for such increase necessarily leads to a struggle for existence.

The ideas put forward by Malthus proved threatening to many who have a vested interest in a steadily expanding human population. They charge that the fundamental idea of Malthus has been disproved since the global population has increased more than sixfold in the past two hundred years and food production has kept up with the aid of modern technologies. In the short term, the leap in agricultural production called the "green revolution" seemed to imply that Malthus was mistaken. However, the great environmental deterioration that intensive agriculture has precipitated (in terms of desertification of overused land, salination and erosion of arable land, depleted aquifers, and loss of wildlands, including forests, to agriculture) as well as the rapid depletion of other food resources, such as fisheries, seem to reinforce his analysis. Opponents have connected Malthus to draconian measures that he never proposed. Malthus was a gentle man who concluded his analysis of population with the temperate advice that a man should not marry and have children until he could support them.

Of all the economic development that the human race has ever experienced, over half has occurred since 1950 with devastating effects in the life system of Earth, including the rapid extinction of species and global warming. Many eminent scientists have documented this rapid deterioration, and some have identified overpopulation as the main underlying cause, for example, Paul Ehrlich in *The Population Bomb* published in 1968. As with most forecasts, some of Ehrlich's predictions were off on the timing. The green revolution exceeded his expectations, but the overall integrity of his arguments remains intact. Mass starvation by the mid-1970s did not occur, but now, only thirty years later, the United Nations estimate that half the world, that is 3 billion people, are suffering from malnutrition, and, in 2001, 10 million children died from easily preventable diseases and malnutrition. Many fisheries have collapsed. A steadily increasing amount of farmland becomes depleted or lost to urban expansion. A recent UN Report (2003) states that the global shortage of fresh water is rapidly worsening. Massive environmental deterioration and global warming provide the potential for catastrophe.

Because of its economic clout, the United States has been looked to as a source of leadership on this issue. In the late 1960s and the 1970s, the Government of the United States took the initiative leading to the World Population Plan of Action adopted by a consensus of 137 countries in 1974 at Bucharest. In the same year, the United States developed its strategy to address overpopulation under the heading "National Security Study Memorandum 200: Implications of Worldwide Population Growth for US Security and Overseas Interests." This was and is a superb strategy but it was never implemented. It was put aside under pressure from essentially the same special interests that opposed the ideas of Malthus two hundred years before (Mumford, 1996).

Opponents of population stabilization and reduction have been surprisingly effective and ruthless in their opposition. They are politically powerful and well-funded. This opposition is of particular significance in the US, which has a de facto role of global leadership.

Traditional religions look to childbirth to increase their congregations and political power. Some religious fundamentalists object to the use of contraception and abortion as contrary to the "right to life." Others attack population stabilization and reduction from the perspective of human rights and advocate the free flow of international migration regardless of the environmental and social implications of such trends. Those who oppose population stabilization on the basis of the "right to life" or human rights do not recognize the interconnection of all life. There is no "right to life." By destroying the ecosphere, we destroy the foundation of all human rights.

Many companies view population growth as a source of cheap labour and growing demand, and disregard the longer-term consequences. Much of business is locked into a cycle of rapacity with the more aggressive making the most money, which is then used to buy compliant public policies and fund campaigns to fill key political positions. Some business interests have created misinformation to undermine the impact of new insights from science. Clearly, they place the short-term viability of their respective institutions above the long-term survival of humanity and of all life. They have made the best use of telecommunications to drive their messages home. Often they own the networks and newspapers which they use to convey their message.

Since 1974, the global population has risen from about 3.4 billion to about 6.3 billion. The increase might have been half that if the strategies of the mid-1970s had been implemented. One of the first steps in reinvigorating political initiatives to stabilize and reduce human population will be to reveal the surreptitious actions of special interests that undermined the excellent policy proposals of the mid-1970s, thereby increasing the threat to world order and indeed to all life on Earth. These special interests have to be made accountable for their contribution to growing human hardship due to overpopulation.

Sustainable Development

The term "sustainable development" surfaced in environment-related discussions in the late 1970s and early 1980s but did not become widely known until the release of the report of the World Commission on Environment and Development (the Brundtland Report) in 1987.

The most common definition of sustainable development derived from the Report is "development that meets the needs of the present without compromising the ability of future generations to meet their own needs." While this definition has apparent simplicity, it fails to make clear that a vital variable in the equation is population: most scientists who have studied the matter think that global population would need to be around two billion for all people to sustainably enjoy a modest European quality of life. It is this failure to even allude to population limits, which has led most scientists to dismiss the term "sustainable development" as a contradiction in terms. . . .

While the concept of sustainable development is a good one, the strategy set out in the Brundtland Report in 1987 for achieving it is probably unattainable. The Brundtland path to sustainable development is predicated in the demographic transition theory, . . . demographic developments in many countries do not coincide with these predictions. The demographic transition theory does not merit the appellation "theory." Even if demographic transition could be triggered in all countries by economic growth, it is unlikely that the growth required to trigger it could be achieved given the massive increase of population and looming shortages of natural resources, such as fossil fuels and fresh water.

A surer path to prosperity is to stabilize and reduce population without delay. Falling birth rates have been a factor in the emerging economies of China, Thailand, and South Korea, among others. Insistence of the infallibility of the empirically unsupported demographic transition theory comes, in part, from opponents of effective measures for population stabilization and reduction and from some leaders of poorer countries in the advantageous position of saying to wealthier ones, "give us prosperity and our population growth will diminish," while doing little or nothing to control their numbers. One of the greatest flaws in the Brundtland Report is its failure to address the urgent need for a huge increase in funding for universal family planning and birth control and the removal of any incentives rewarding population growth. To its credit, the Brundtland Report did advocate more equality, education, and opportunities for women and this will contribute significantly to lower birth rates over the long term. . . .

The Concept of Security

Security has many aspects: personal security in an orderly society; social security in terms of public provision to citizens of some financial support and services; national security; and international security. . . .

Population growth contributes to congestion, pollution, and the overexploitation of natural resources, thereby reducing the freedoms of individuals in democratic countries. Environmental degradation adds to the sense of personal vulnerability with a growing number of incidents, such as respiratory and gastrointestinal diseases and neural damage, resulting from air pollution and contamination of fresh water by pollution and toxic waste.

Confronted with the growing threats posed by overpopulation and accelerating scarcity, an almost immediate response is to affirm human rights. However, this very affirmation flies in the face of reality if it ignores the environmental deterioration that makes it increasingly difficult if not impossible to fulfill these rights.

The projected rapid growth of population, most of it in the poorest regions of the Earth, will exacerbate regional disparities, fuel resentment, and possibly give rise to more terrorism. Many of the poor will try to migrate to more prosperous countries which will resist more actively what might easily become an uncontrolled invasion. Mass migration is a phenomenon closely associated with the human expansion over the past ten thousand years. It is tied to the instinct of the hunter-gatherer that if food becomes scarce in one region, population migrates to a region of greater abundance. However, in an overcrowded world this does not work. Mass migration merely spreads chaos and misery throughout the globe. A profound change is needed. The era of mass migration is coming to an end. It is impossible that even a significant proportion of the current global annual population expansion of about seventy-nine million a year (Population Reference Bureau, 2003), almost all of which is occurring in undeveloped nations, could be incorporated into developed nations. Moreover, were this to be attempted, it would overload the world's carbon sinks even more disastrously than is occurring at present. Most of the developed nations are already heavily populated and much of their prosperity depends on abundant energy which may become much more expensive sooner than most people expect.

The benefits of a smaller global population are so promising that one wonders why there is any hesitation to reduce human numbers by making use of many new and benign contraceptive technologies. What is needed is a commitment to act on the best information firmly anchored in science and reason, a global surge of information sharing, and political will. The agendas of those still promoting population growth must be clearly exposed. . . .

All references for articles included in *Taking Sides: Clashing Views in Sustainability* can be found on the Web at www.mhhe.com/cls.

EXPLORING THE ISSUE

Is Limiting Consumption Rather Than Limiting Population the Key to Sustainability?

Critical Thinking and Reflection

1. Are population, technological advancements, and affluence level easily quantifiable?
2. What are the limitations possessed in conducting studies related to affluence and technological advancements besides lack of quantification?
3. Does having international conferences really solve the issue of sustainable science?
4. Will the population ever stabilize at a point that will not impact the earth's carrying capacity?
5. How precise and accurate is "ecological footpriniting" as a tool to address the growing stress on the environment and can more such tools be developed to study the same?

Is There Common Ground?

Writers from Thomas Malthus to Paul Ehrlich have discussed the relationship between population growth and resource use. The larger the population the more stress placed on natural resource use, causing the "limits of growth." Malthus discussed food supply and how human population grows exponentially while food supply grows arithmetically. The neo-Malthusians have emphasized population control as the chief mechanism necessary for achieving sustainability. The technological optimists, on the other hand, believe that human intelligence will allow the human species to develop sustainable solutions that do not necessarily mean a reduction in global population.

Ecological footprint is a concept that is more related to consumption than to population. Richer populations have a higher ecological footprint than poorer societies since they use more energy and materials to achieve their affluent lifestyle. But population is, of course, a key variable in ecological footprint since larger populations will ultimately lead to greater consumption. As can be seen, both concepts, population and consumption, are interrelated. Global consumption will be on the rise as more people in developing countries move into the middle class. Whether or not the stabilization of world population growth is enough to stabilize the global ecological footprint is a question that needs to be addressed and answered. The challenge of sustainability will be to seek equilibrium between population and consumption.

Internet References

Diamond, Jared. (2008). "What's Your Consumption Factor?" *New York Times*, January 2008.

Ebi KL, Gamble JL. (2005). Summary of a workshop on the development of health models and scenarios: strategies for the future. *Environmental Health Perspectives*, pp. 335–338.

Graham-Smith, F. (1994). *Population—the complex reality: A report of the population summit of the world's scientific academies*. London: The Royal Society.

Harte, J. (2007). Human population as a dynamic factor in environmental degradation. *Popul Environ*, pp. 223–236.

Jorgenson, A. K., & B, C. (2011). Societies consuming nature: A panel study of the ecological footprints. *Social Science Research*, pp. 226–244.

Patterson, T. M., Niccolucci, V., & Bastianoni, S. (2007). Beyond "more is better": Ecological footprint accounting for tourism and consumption in Val di Merse, Italy. *Ecological economics*, pp. 747–756.

Potts, M. (2007). Population and environment in the twenty-first century. *Popul Environ*, pp. 204–211.

Tong, S., & Soskolne, C. L. (2007). Global Environmental Change and Population Health: Progress and Challenges. *EcoHealth*, pp. 352–362.

This website discusses all the scientific and social issues that affect the well-being of the planet, which directly affects human welfare. It posts reaction articles on issues such as energy security, global finance, and so forth. This keeps us up-to-date on all the leading global issues and hence helps us make wise decisions about our existence and habits.

http://www.globalissues.org/

Center for Biological Diversity links human welfare to nature. It explains how mankind should exist with mutual symbiosis with the plants and animals of the biota, so that the diversity is maintained.

http://www.biologicaldiversity.org/campaigns/overpopulation/index.html

Website of CASSE—Center for Advancement of the Steady State Economy. CASSE discusses the relationship between economic growth and environment.

http://steadystate.org/

Discusses the impact of overpopulation and overconsumption. It studies how sustainable development can be achieved by addressing these global issues.

http://www.overpopulation.org/

This website is a media outlet publishing blogs, daily newsletters, and regular updates on Twitter and Facebook pages. It basically helps a person inculcate the discipline of leading a "green" life and also provides the resources for doing so.

http://www.treehugger.com/

ISSUE 8

Can Technology Deliver Global Sustainability?

YES: Joanna I. Lewis, from "Technology Acquisition and Innovation in the Developing World: Wind Turbine Development in China and India," *Studies in Comparative International Development* (November/December 2007)

NO: Alan Colin Brent and David E. Rogers, from "Renewable Rural Electrification: Sustainability Assessment of Mini-hybrid Off-grid Technological Systems in the African Context," *Renewable Energy* (2010)

Learning Outcomes

After reading this issue, you should be able to:

- Gain an understanding of the challenges and opportunities for implanting alternative energy on a large scale in emerging nations.
- Discuss the importance of the precautionary principle in assessing new technologies.
- Describe new technologies' potential for equitable and ethical access.
- Understand how the debate on growth versus degrowth involves technology.
- Define geoengineering, leap-frogging, genetically engineered crops, and desalinization.

ISSUE SUMMARY

YES: Joanna Lewis, a professor of Science, Technology and International Affairs at Georgetown University's Edmund A. Walsh School of Foreign Service, discusses how technological "leapfrogging" in emerging economies can "address concerns about rising greenhouse gases." She explores the role that technology transfer holds in accelerating wind power in India and China.

NO: Alan Colin Brent and David E. Rogers, engineers from South Africa's University of Pretoria and leaders in sustainable energy futures, conclude that alternative energy technology cannot always be easily implemented and that policy must consider social and cultural factors and involve multiple stakeholders.

The role of technology and technological systems toward achieving a more sustainable future is a hotly debated topic. Supporters of technological solutions, referred to as Techno-optimists, believe that green technology should play a prominent role in social, economic, ecological, and political policy and that it is the primary hope for reduction of greenhouse gases, while technological critics, referred to as Techno-skeptics, see that reliance on technology is not prudent and that the focus should instead be on lifestyle change and a paradigm shift to a ecologically based sustainability. Both technological optimists and technological skeptics have found some common ground in the belief that change is necessary; however, they differ drastically on the path that should be taken. One major rift is that Techno-optimists attempt to mold green technology into existing economic models, such as the globalized, neo-liberal capitalist paradigm, while Techno-skeptics see the need for local networks to govern natural resource commons. This section covers some of the major technological innovations, including geo-engineering and genetic modification of crops (GM), and features two articles from experts on alternative energy.

Undoubtedly, it is a pivotal moment in human history, and the choices of today will have crucial ecological consequences into the next century. While technological innovation and efficiency offers one approach to sustainability, it should be carefully considered. Technological solutions often have issues of availability, feasibility, and ethics. Technological review should address several considerations:

- Is the technology available, and if not, then what are the best alternatives?
- Is the technology cost-effective not only in terms of traditional economics but also in relation to the value of ecosystem services?
- What are the ethical concerns if this technology becomes integrated into widespread use? Ethical issues can include: health, privacy, democratic involvement, and social and cultural changes.

Those who favor the technological approach tend to minimize the finite quality of resources and hope to extend limits with creative design and engineering such as carbon sequestration, the capture and storage of carbon deep within the Earth. Techno-optimists cite successful innovations, such as hybrid or alternative fuel vehicles and emphasize the capacity of human intelligence to overcome any challenge. They seek ways to repair environmental damage, while continuing material advancement and economic growth by advocating such solutions as the building of sustainable cities. Agriculturally, they favor large-scale, genetically modified crops that offer solutions to the global demand for food and emphasize the green aspects of GMO crops that require less pesticides and herbicides.

Another technological solution that the Techno-optimists support is geo-engineering. To mitigate climate change, geo-engineering, the use of technology to alter climate and weather, is increasingly advocated as a viable option (Crutzen, 2006, Bailey, 2010, Wigley, 2006). Some geo-engineering advocates say that present policies are not sufficient to substantially reduce CO_2 emissions. They project an upward movement of CO_2 and that geo-engineering is necessary to adjust to a warmer planet and its global consequences, which include extreme weather events, flooding, sea-level rise, etc. Their view is partially based on doubts about how rapidly conversion from fossil-fuel–based vehicles to those using renewable energy fuels can be undertaken on a global level. Fresh water availability also promises to be another area for technological innovation. They see technological solutions to future fresh water problems solved through de-salinization technology.

Genetically engineered crops (GM) are seen as a way to meet the food security threat for an increasing population (Lipton, 2000). While acknowledging that GM is still very controversial, Phipps and Park (2000) claim that it "may make further marked reductions in global pesticide use." Techno-optimists tend to overlook the loss of biodiversity, the increase in mono-culture, and the need for further research into health effects, and concentrate on the alleviation of poverty in developing nations (Srivastava, 2009). Certainly, these issues have polarized the debate and each side has cultural values that inform the discourse.

Some cities have certification programs to encourage the use of sustainable technologies, that is, wind-, solar-, geo-thermal, wave-power, etc. Leadership in Energy and Environmental Design (LEED) is one such program that advocates the use of green building technology and technological solutions to mitigate some of the unsustainable characteristics of urbanizations. Also, the belief in modern information technology to create "smart cities" is currently gaining great exposure. But, critics point out that technology can only minimize ecological system impacts since cities are inherently unsustainable because they are heterotrophic or derive move of their resources from outside the ecosystem. Techno-optimists believe that technology will be able to "leap-frog" historical evolution and provide sustainable solutions to major global problems (Lewis, 2007; Quimen & Zhang, 2010).

The critics of technology, the Techno-skeptics, tend to view technological innovation with caution. Some point to how technology has deconstructed human values (Zerzan, 2005). Others discuss how technology fosters economic growth and consumption and that a better approach is to consider de-growth (Levallois, 2010). As early as 1812, the British Luddites, handloom weavers, broke their looms and machinery to resist mechanization in the workplace and some Techno-skeptics have been labeled as neo-Luddites ever since. Also, romantics and transcendentalists such as Henry David Thoreau in *Walden* (1854) noted the increasing materialism, complexity, and misery associated with the Industrial Revolution and advocated a return to nature and preservation of habitat.

Techno-skeptics see technology as potentially unethical because it limits choice and perpetuates itself. Jacques Ellul (1912–1994), like many Techno-skeptics, saw technology as having agency, the ability to affect systems (*The Technological Society* (*La technique ou l'enjeu du siècle.*), 1954). Today, many Techno-skeptics advocate a deep green philosophy. They maintain that personal lifestyle changes away from materialism to ecological self-sufficiency and an emphasis on the local community and appropriate technology (Schumacher's "small is beautiful" philosophy) will produce a more enriching quality of life and a sustainable world (Neisheim et al., 2006; Schumacher, 1973). They favor organic farming over agrobusiness, and support the precautionary principle in regard to nano-technology and genetically modified crops (Dhillion et al., 2010; Johnson, 2007). Interestingly, Europeans have a greater aversion to GM crops and nano-technology than do Americans (Gaskell et al., 2005). Techno-skeptics support low-tech or conservation-based solutions to problems rather than the use of large-scale capital projects like hydro-electric power and dams.

The YES selection, that technology can deliver global sustainability, is presented by Joanna Lewis, an international technology policy scholar and university professor. She discusses the challenges for developing nations in seeking to expand their domestic alternative energy production. She discusses the potential for developing countries to "leap-frog" through technological breakthrough by analyzing the best practices in the wind-power market in China. The author relies on the assumption that multinational corporations should play a role in technology transfer, while acknowledging some of the challenges that they face. The selection provides an example of an optimistic assessment of technological innovation, and assesses the policies and conditions that make it possible.

The NO selection is presented by two engineers from South Africa, Brent and Rogers. They conducted a case study of a hybrid alternative energy project commissioned by the South African Department of Minerals and Energy (DME) to provide off the grid energy for poverty reduction. The project failed to meet UN Millennium sustainability goals and was eventually abandoned. The question that they addressed was why did a seemingly well-designed solar and wind project to create prosperity in a rural area fall short of its goals? Their study did not assume a problem with the technology but instead addressed the social and economic limitations of the project. They concluded that the project failed because it emphasized a technological solution instead of a total solution which would have taken into consideration the culture and social conditions of the host community. Policy formation and implementation were not transparent and did not involve local stakeholders. Cost controls were lacking and the community had higher expectations of the technology than the project could deliver. Hence, the project created distrust. They concluded that project leaders had little knowledge of the local community and could not effectively communicate their message that this project could benefit the community both socially and economically. Hence, they admonished that for projects to be successful, leaders must plan to involve the local community more in decisions and not just rely on the technology.

YES 

Joanna I. Lewis

Technology Acquisition and Innovation in the Developing World: Wind Turbine Development in China and India

. . . The energy development pathways of China and India are a frequent topic of international attention. In the climate change arena, the current and future energy growth trajectories of the two countries raise concerns about rising greenhouse gas emissions. China has recently surpassed the United States as the largest national emitter of greenhouse gases, and India will soon surpass Russia to become the fourth largest emitter after the European Union. China and India use coal to fuel most of their electricity generation, and both countries have plans to expand their coal power capacity considerably in the coming decade. For these reasons, China and India are perhaps two of the least likely places one might expect to find a burgeoning wind power industry.

While there are many potential benefits to local wind manufacturing, there are also significant barriers to entry into an industry containing companies that have been manufacturing wind turbines for more than 20 years. In developing countries, limited indigenous technical capacity and quality control makes entry even more difficult. International technology transfers can be a solution, although leading companies in this industry are unlikely to license proprietary information to companies that could become competitors. This could be even riskier for technology transferred from developed to developing countries, where an identical but cheaper turbine potentially could be manufactured.

Nevertheless, India and China are both home to firms among the global top-10 leading wind turbine manufacturing companies. India currently leads the developing world in the manufacturing of utility-scale (multikilowatt) wind turbines, and China is close behind. Initiatives by domestic firms, supported by national policies to promote renewable energy development, are at the core of wind power innovation in both countries. . . .

Theorizing Developing-Country Technology Innovation

This study analyzes empirical cases of successful energy technology "leapfrogging." Energy leapfrogging has been described as a strategy for developing countries to shift away from an energy development path that relies on traditional energy sources, such as fossil fuels, and onto a new path that incorporates the broad utilization of advanced energy technologies—generally those that have been developed within more industrially advanced countries. As a means of climate change mitigation, observers have argued that developing countries need not adopt the dirty technologies of the past—rather, they can "leapfrog" over them, opting instead for modern, clean technologies as an integral part of capacity additions (Goldemberg 1998).

Promoters of the leapfrogging concept generally give the impression that leapfrogging is feasible, provided that it meets some basic conditions, including strong incentives for firms to reduce their environmental impacts, and the participation of transnational corporations in the development process (Perkins 2003). As some studies have illuminated, this optimistic picture of leapfrogging is not generally supported by empirical studies, and therefore meeting the objectives of clean development in industrializing countries will doubtlessly prove more complex and challenging than many would lead us to believe (Gallagher 2006; Van De Vegte 2005). This study weaves together insights from a range of literatures to identify the facilitating conditions for energy technology leapfrogging in developing-country contexts. It draws both from general theories on technology transfer and from the ecountry-specific literatures on technology transfer to China and India.

Technology Transfer

International technology transfer typically refers to the transfer of technology from industrialized to developing countries.[1] The mechanism of transfer is either private-sector arrangements such as foreign direct investment (FDI), licensing, and joint ventures, or bi- or multilateral technology agreements among governments. In the wind technology domain, countries that were not part of the group of early wind turbine innovators—namely Denmark, The Netherlands, Germany, and the United States—have used different strategies to foster the development of their own domestic large wind turbine manufacturing companies. A common strategy has been to obtain a technology transfer from a company that has already developed advanced wind turbine technology. This can be done through a licensing agreement, or developed through collaborative research and development (R&D). Another model includes establishing joint-venture partnerships between foreign and domestic companies in which a technology license is usually transferred. In a third form of technology transfer, no license is transferred and the know-how and intellectual property associated with the technology remains primarily in the hands of foreign firms. For example, if a foreign-owned firm locally manufactures its wind turbines in China, this may constitute an incomplete transfer: the hardware (technology) without the software (intellectual property rights and know-how). In this situation, the technology physically has been transferred, and the foreign firms may facilitate learning by employing local workers, but it is unlikely that much expertise will end up in the hands of domestic firms.

There is a substantial literature describing the challenges associated with technology transfer in general (Hirschman 1967; Rosenberg and Frischtak 1985; Kranzberg 1986; Goulet 1989; Reddy and Zhao 1990; Mowery and Oxley 1995; Brooks 1995; Nicholson 2000; IPCC 2000), and a growing literature concerning technology transfer challenges that are specific to China (US OTA 1987; Mansfield 1994; Guerin 2001; Miesing et al. 2003) and to India (Kathuria 2002; Vishwasrao and Bosshardt 2001; Kinge 2005; Kumar and Jain 2003). Two challenges to successful technology transfer particularly relevant to India and China are the domestic policy environment and firms' ability to acquire new knowledge.

National Innovation Systems

A complete leapfrogging strategy must include innovation at the receiver's side, which leads to a need to investigate the enabling conditions for technological innovation nationally. The larger domestic context in which the innovative activity is taking place, or what some refer to as the "national innovation system," is likely an important determinant of the ultimate the success of a technology transfer, particularly concerning a country's ability to adopt an externally sourced technology and apply it internally (or its "absorptive capacity"). This transfer includes the policy structures and institutions that contribute to the development and diffusion of new technologies (Lundvall 1992; Metcalfe 1995; Johnson and Jacobsson 2003). Studies have emphasized how the organization and distribution of innovation-related activities often differ fundamentally between developed and developing countries (Liu and White 2001), or as distinguished by regional characteristics (Freeman 1995), with some similarities among the Asian "late industrializing countries" (Amsden 2001).

In the field of innovation economics, there is a growing characterization of how transnational firms may work outside of national innovation systems, and may eventually render them obsolete (Patel 1995). The rise of the multinational corporation with global presence has created a new model for innovation through the global generation of technology. As described by Archibugi and Michie (1995), multinational firms now take advantage of global experience to shape their innovative activity, rather than relying on a national innovation system. They described this as the third in a three-stage process of "technological globalization" which began with global exploitation (firms accessing global markets), then transitioned to global technological collaboration (international technology transfers), and has evolved to a stage in which technological innovation is conducted within the global network of the multinational firm. This process of transition suggests that a conducive national innovation context is necessary but not sufficient for successful technology transfer. Firms themselves must play an active role in the acquisition of new technology and know-how.

Learning Networks—Regional and Global

Regional network-based industrial systems have been identified in many parts of the world and in many historical periods. Characterized as horizontal networks of firms in which producers deepen their own capabilities by engaging in close, nonexclusive relations with other specialists in their field (Saxenian 1994), or "learning by interacting," learning networks have likely played a large role in the development of wind turbine technology over time. The wind industry—characterized by its small number of firms, highly specialized technology, and geographically specific hubs of innovation (often near wind development locations)—is likely to exhibit many of the characteristics of the regional learning networks that have been observed in other industries and locales. Studies have hypothesized that learning networks are a crucial determinant in a firm's ability to obtain success with a new technology (Van Est 1999; Kamp et al. 2004; Karnoe 1990). In the innovation systems of Denmark, where "the focus was on knowledge transfer between turbine producers, turbine owners and researchers . . . conditions for learning by interacting were optimal; in this way, wind turbines were successfully, though slowly, scaled up and improved" (Kamp et al. 2004). In contrast, the U.S. wind industry has been characterized by a lack of collaboration, and actions taken by firms to impede information flow among firms, that "inhibited the transfer of hard-won experience" (Gipe 1995). . . .

Wind Power Development in China and India

Wind in the National Energy System

India and China have excellent wind resources, and the promise of years of wind turbine sales has kept overseas turbine manufacturers closely involved in all three markets.[2] Yet fundamental risks in the Indian and Chinese markets remain, making some investors reluctant to enter. In addition, both countries have been undergoing varying degrees of power sector reforms, and the full impact of such reforms on wind power development is still uncertain.

China's vast electricity sector faces critical challenges surrounding regional resource disparity, and the growing complexity of coordinating supply with demand. Although China currently ranks second in the world in both installed generating capacity and annual power generation, per capita electricity consumption is only 10–15% that of developed countries, thus is expected to increase substantially as China's economy grows. Growth rates for new electricity generation capacity have increased substantially since 2003, and 2006 was the largest capacity addition yet. Most of this new capacity is from coal power plants (about 80%). This continued reliance on coal not only has tremendous local and global environmental implications, but also poses a huge technical challenge for a sector plagued by inefficient state-owned enterprises, aging capacity and transportation bottlenecks. Despite constant discussions of a national, interconnected power grid, China still relies on often unconnected, regional grids that vary in quality and reliability throughout the country. China's booming economic development in its eastern coastal regions has resulted in supply disruptions and boom and bust cycles in planning for power sector capacity additions.

Wind is still a small share of China's total electricity generation—less than 1% nationally, with about 2.66 TWh produced in 2005. The highest saturation of wind power is found in the Inner Mongolia and Xinjiang Autonomous Regions due to the relatively small total power demand and relatively large amount of installed wind power capacity in these provinces. Despite excellent wind resources, wind power's total percent contribution to electricity generation will probably remain low, particularly in the eastern coastal provinces. Even though wind power capacity is growing very quickly, capacity in other generation technologies is growing more. Wind energy could still play a crucial role in meeting demand, especially in eastern coastal China where wind resources are abundant, and electricity demand has seen unprecedented growth rates in the last few years.

India is in a similar situation to China, where electricity demand is growing rapidly, and most of this demand is being met with domestic coal resources. The demand for electricity is projected to more than double in the next 10 years, and projections are even higher if economic growth rates increase.

Less than 2% of India's electricity generation comes from wind power (with about 7.66 TWh produced in 2005), despite record growth rates in new wind capacity of 40% annually for the past 3 years. India had 137.5 GW of installed electricity capacity at the end of 2005, of which wind comprised 3%. The vast majority of India's wind power is located in two states with relatively aggressive wind power support policies: Tamil Nadu (with over 51%) and Maharashtra (with 20.5%). Tamil Nadu is also a wind turbine manufacturing base, home to both Indian and Danish system and component manufacturers.

Despite aggressive forecasts for future development and substantial wind resources, future growth may be limited by a lack of transmission capacity (particularly in Tamil Nadu), and by voltage and reliability problems with the Indian power grid. Problems with inaccurate resource data, poor installation practices, and poor power plant performance have also slowed wind power development in India.

National Policy Support for Wind Power

China and India have undertaken extensive policy support schemes to catalyze wind power development, in turn creating market opportunities for their own domestic turbine manufacturers. While some policies aim to create a demand for wind power, others specifically aim to promote the development of local wind power technology industries, as discussed below.

India has been an active supporter of wind development since the 1990s, and has a government ministry exclusively devoted to renewable energy promotion: the Ministry for Non-Conventional Energy Sources (MNES). However, India's policy support has been somewhat unstable over the years, which led to uneven wind development in the 1990s.

Recent years have seen the market rebound, driven in part by more policy stability and more aggressive support mechanisms. For example, India's

Electricity Act of 2003 requires all state-level energy regulatory commissions to encourage electricity distributors to procure a specified minimum percentage of power generation from renewable energy sources. Many states therefore have aggressive renewable energy targets and policy support mechanisms in place; for example, the Karnataka Energy Regulatory Commission has stipulated a minimum of 5% and maximum of 10% of electricity from renewables, and the Madhya Pradesh Energy Commission has stipulated 0.5% of electricity from wind power by 2007. The state government of Maharashtra has implemented a feed-in tariff for wind electricity, which means wind power producers are guaranteed a long term contract for their power at a subsidized price that declines over time. Maharashtra has also imposed a small, per unit charge on commercial and industrial users to be used in support of nonconventional energy projects. In Gujarat, the government has signed agreements with Suzlon, NEG Micon (now Vestas), Enercon, and NEPC India to develop wind farms on a build–operate–transfer (BOT) basis, with each manufacturer given land for the installation of between 200 and 400 MW in the Kutch, Jamnagar, Rajkot, and Bhavnagar districts (WPM March 2004:57).

Early wind policy in India included the National Guidelines for Clearance of Wind Power Projects implemented in July 1995 (and further refined in June 1996), which mandated that all state electricity boards take the necessary measures to ensure grid compatibility with planned wind developments. Financial incentives were offered as well, with 100% depreciation of wind equipment allowed in the first year of project installation, along with a 5-year tax holiday (Rajsekhar et al. 1999).

More recent Indian polices have been directed at encouraging local wind turbine manufacturing. For example, the government has set customs and excise duties that favor importing wind turbine components over complete machines. India has also developed a national certification program for wind turbines administered by MNES and based on international testing and certification standards to support the development of domestic turbine manufacturers. By 2006, India had reached 6,228 MW of installations, surpassing its long-held target of 5,000 MW of wind capacity by 2012 several years early (WPM March 2006:44).

The development of large-scale wind farms in China has been invigorated in the past few years with the introduction of the government's wind concession program. In this program, government-selected sites are auctioned off through a competitive bidding process to potential wind project developers (Wind Concession Group 2003). To promote the use of locally manufactured wind turbines, a local content requirement has been placed on the developers of the concession projects mandating that 70% of the turbine content used be made in China.[3] Approximately 4,000 MW of new wind power capacity is expected to be installed through the concession process by 2010. The wind concession model was further promoted by China's 2005 National Renewable Energy Law and in the subsequent power pricing regulations that were released in 2006. The Renewable Energy Law stipulates that concession-based pricing be used for the majority of wind power development in China, although in some cases negotiated feed-in tariff or other fixed price contracts are being agreed to for discrete projects.

The 70% local content regulation essentially has forced wind turbine manufacturers wishing to sell to the Chinese market to establish China-based manufacturing facilities to locally source its turbines. Yet many of the foreign-owned companies that have done this have opted to not partner with Chinese-owned companies except for some smaller components, and have not transferred know-how or intellectual property rights as a license. Although Chinese-owned technology (in addition to China-produced content) is not an official requirement for selecting wind concession projects, Chinese-owned manufacturers like Goldwind have dominated the selection process thus far (WPM November, 2006:27).

Several other government programs have encouraged wind turbine manufacturing in China, including the 1997 "Ride the Wind Program" that established two Sino-foreign joint ventures to manufacture wind turbines with limited success (MOST et al. 2002). In addition, the Ministry of Science and Technology (MOST) has subsidized wind energy R&D expenditures at varied levels over time, beginning most notably in 1996 with the establishment of a renewable energy fund (MOST et al. 2002; Liu et al. 2002). MOST had also supported the development of megawatt-size wind turbines, including technologies for variable pitch rotors and variable speed generators, as part of the "863 Wind Program" under the Tenth Five-Year Plan (2001–2005). China has a national target to achieve 30 GW of wind power by 2020, and is well on its way toward meeting this ambitious goal. . . .

National Innovation and Policy Contexts

Both China and India have excellent wind resources and aggressive, long-term government commitments to promote wind energy development. Both countries have been supporting wind energy for more than 10 years through a variety of policy mechanisms. Some of the early support mechanisms in China and India, in particular, led to market instability as developers were faced with regulatory uncertainty, especially concerning pricing structures for wind power. In the early years of wind development in China and India, difficulties also resulted from a lack of good wind resource data, and a lack of information about technology performance stemming from little or no national certification and testing.

Policy reforms in the electric power sectors of both countries, combined with national legislation to promote renewable energy, has led to a series of regional renewable energy development targets in India, national targets in China, and additional financial support mechanisms for wind in particular. There are two key differences in the policy support mechanisms currently used in China and India: (1) China's recent reliance on local content requirements to encourage locally sourced wind turbines, which does not exist in India, and (2) India's use of a fixed tariff price for wind power, versus China's reliance on competitive bidding to set the price for most of its wind projects.

The local content requirement in China has encouraged several foreign-owned companies to shift their manufacturing to China. Yet the primary beneficiary of this policy to date has been the Chinese-owned turbine manufacturer that could meet the local content requirements before the other companies could—namely Goldwind. In regard to pricing policies, several Indian states have adopted feed-in tariffs for wind electricity, while such a policy has yet to be adopted wide-scale in China.[4] Many studies have cited feed-in tariffs as the most effective policy mechanism for promoting wind power development due to the stability and regulatory certainty it provides (Mitchell et al. 2006; Sijm 2002; Lewis and Wiser 2007). The difference in pricing policies are most likely to affect the stability of each country's domestic wind market, and it appears that India's fixed tariff structure has been most successful in providing this stability (reflected in its large annual capacity additions in recent years). Since each country has used both forms of pricing structure over time, it is difficult to draw a direct relationship between the growth occurring in India and China and their respective pricing policies.

Although both countries have manipulated customs duties and related taxes to promote the use of domestically manufactured wind turbines or components, India has generally been much more hands-off than China in promoting local wind turbine manufacturing by not mandating the use of local content in domestically installed wind turbines. India's local manufacturing industry seems to have emerged organically as companies shifted their facilities to India to meet the local market demand. China has not experienced the same magnitude of annual capacity additions as India, and is still several thousand megawatts behind India in terms of total installed capacity. . . .

Notes

1. Although less common, there is a growing recognition of the potential of "south–south" technology transfer (technology transfer between developing countries.)
2. Both countries have vast wind resource potential; China has an estimated 1,000,000 MW of total exploitable wind resources, including about 250,000 MW on land and 750,000 MW offshore, although the amount that is technically and economically viable may closer to 300,000 MW overall (SDPC 2000). Estimates of India's wind resources are less readily available; one estimate puts the range from 20,000 to 45,000 MW (WEC 2001) though this range is likely quite conservative.
3. Local content is generally calculated according to cost, therefore 70% local content represents domestically produced components totaling 70% of the wind turbine cost.
4. Despite some indications that the Chinese government would implement a nation-wide feed-in tariff program, it instead based the pricing structure for wind power on a competitive bidding model in the 2006 pricing regulations of the renewable energy law (WPM, February 2006:25). Certain provinces (namely Guangdong) and projects have continued to utilize fixed feed-in tariffs.

All references for articles included in *Taking Sides: Clashing Views in Sustainability* can be found on the Web at www.mhhe.com/cls.

**Alan Colin Brent and
David E. Rogers**

Renewable Rural Electrification: Sustainability Assessment of Mini-hybrid Off-grid Technological Systems in the African Context

. . . The South African governance system is developing national measures of sustainability. For example, the Millennium Development Goals are pursued to reduce widespread poverty by 2015 [1]. The post Kyoto 2012 commitments to low-carbon technologies to mitigate the effects of climate change are based on renewable energies, which are to be supported by a carbon tax [2]. In terms of mitigation, the application of (energy) technological innovation to meet the objectives of sustainable development and the conditions for sustainability has been stressed [3–5]; a model, based on the principles of sustainability science (see Table 1), has been developed that can be used to assess the sustainability of such technologies (see Figure 1) [5]. The model integrates:

- A life cycle perspective [4] and systems thinking, i.e. systems provide feedback loops and are self-correcting [6].
- Learning methods for the management of information in the paradigm of sustainable development [5].
- Conditions for sustainability to reduce the complexity of systems by clarifying the magnitude of cause and effect on systems, so that priorities can be allocated [5].
- Technology innovation and what is feasible within constraints of time, finances and institutions [3,4].

The model to prioritise assessable sustainability indicators for renewable energy systems initiates with a comprehensive set of sustainable development indicators that are deemed appropriate for the context of integrated renewable energy technological systems under investigation. Only those indicators that are controllable by decision-makers in the context of an integrated technological system, and specifically those that are expected, by the technological

Table 1

Specific theories of the emerging field of sustainability science that relate to sustainability performance indicators for technological systems [5]

Theory	In the context of sustainability science	In the context of performance indicators of technologies
Trans-disciplinarity	The result of a coordination of disciplines such as science and laws of nature; technology and what is achievable; law and politics and what is acceptable to social systems; and ethics of what is right and wrong beyond the bounds of society.	Where: "successful transformation of technologies into marketable commodities requires knowledge and skills from a variety of different specialist fields of science and engineering."
Resilience	A system's ability to bounce back to a reference state after a disturbance and the capacity to maintain characteristic structures and functions despite the disturbance. Where: "ecological resilience is the amount of disturbance that a system can absorb before it changes state. Ecological resilience is based on the demonstrated property of alternative stable states in ecological systems. Engineering resilience implies only one stable state (and global equilibrium)." Further: "a resilient ecosystem can withstand shocks and rebuilds itself when necessary. Resilience in social systems has the added capacity of humans to anticipate and plan for the future." Resilience is conferred in human and ecological systems by adaptive capacity.	The resistance and robustness of an integrated system against surprises, which includes risk-based measures and precautionary regulations; the capacity to buffer change, learn and develop.
Complexity	From a biology perspective: "that understanding of how the parts of a biological system—genes or molecules—interact is just as important as understanding the parts themselves." From a natural systems perspective: "complex interactions of natural systems that are not chaotic." Furthermore, the growing appreciation of the need to work with affected stakeholders to understand the full range of aspects of any particular system.	Deals with the study of complex systems, i.e. is composed of many interacting elements that interact in complex ways; and the ability to model complex interaction structures with few parameters.
Adaptive management	Or adaptive resource management (ARM) is an iterative process of optimal decision-making in the face of uncertainty, with an aim to reducing that uncertainty over time via system monitoring.	
Adaptive capacity	"As applied to human social systems, the adaptive capacity is determined by: • The ability of institutions and networks to learn, and store knowledge and experience. • Creative flexibility in decision-making and problem solving. • The existence of power structures that are responsive and consider the needs of all stakeholders. Adaptive capacity is associated with r and K selection strategies in ecology and with a movement from explosive positive feedback to sustainable negative feedback loops in social systems and technologies."	

Figure 1

Model to Achieve Prioritised Assessable Sustainable Performance Indicators for Technological Systems [5]

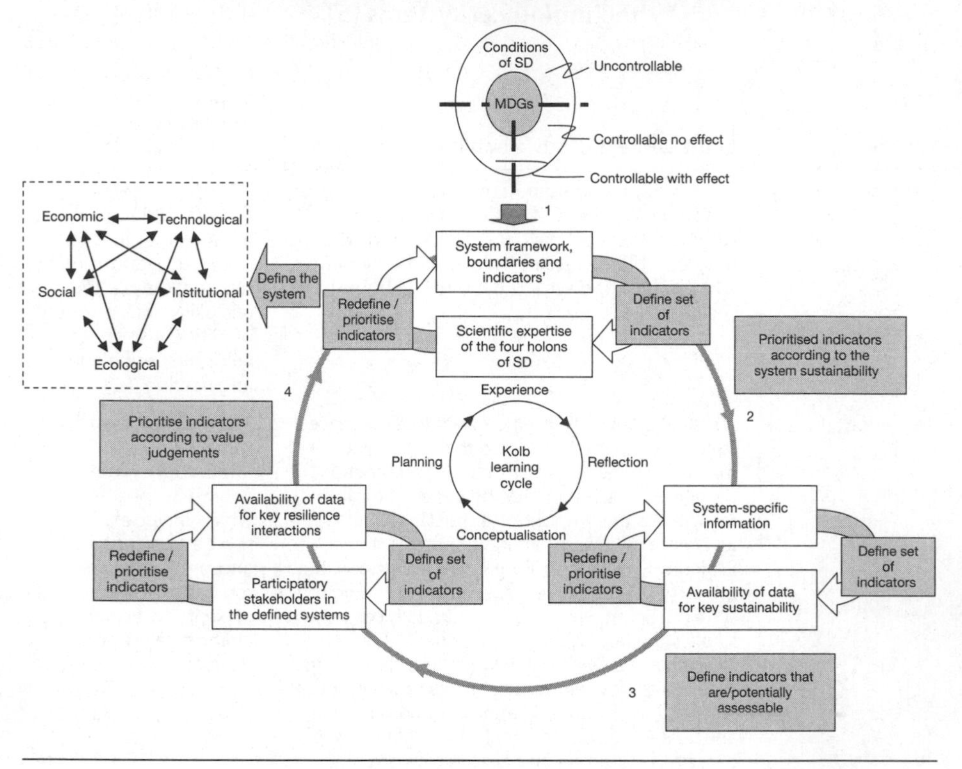

sub-system analysts, to be effected through the implementation of the technological sub-system, are considered further (#1 in Figure 1).

The remainder of the approach is based on the Kolb learning cycle of experience, reflection, conceptualisation and planning [7]. First, the expertise of the technological sub-system analysts, with the expertise of the sustainability of the economic, environmental, institutional and social sub-systems, also termed holons [8], are used interchangeably through a sub-learning cycle to [5]:

- Define a specific system, as a framework, in terms of technology–economic–social–ecological–institution interactions, including the boundaries of the system, and important resilience considerations; and
- Establish a hierarchy of controllable indicators that may be affected in terms of their respective importance to ensure the sustainability, as defined by the concepts of Table 1, of the investigated technology–economic–social–ecological–institution system.

The outcome is an initial set of prioritised indicators for each of the technology, economic, social, ecological, and institutional sub-systems or holons according to the overall system sustainability, as perceived by the sustainability expertises (#2 of Figure 1). The technology holon analysts then re-evaluate, through a number of sub-cycles, the site-specific information to determine which indicators are, potentially, assessable for the specific technological system under investigation (#3 of Figure 1). Thereafter, the different stakeholders of the technology–economic–social–ecological–institution system are engaged to highlight the key aspects of the integrated system to prioritise the indicators and identify aspects of the overall system that may not have been included in the initial set of indicators (#4 of Figure 1). Further learning cycles (2 → 3 → 4 → 2) are utilised to facilitate the transdisciplinarity prioritisation of the key set of indicators. Finally, and considering the market uptake of innovation [9], multiple technology–economic–social–ecological–institution systems at regional, national and international levels may result in different sets of prioritised assessable indicators through a continuous learning process.

The main objective of the investigation summarised in this paper was to apply the introduced model on a rural mini-hybrid off-grid electrification system to determine the sustainability performance of such systems in the African context. Thereby policy makers may be informed as to the key aspects that drive the sustainability of mini-hybrid off-grid renewable energy systems.

Application of the Model on an Implemented Renewable Energy Technological System

Supply of energy for basic needs is an assumption for sustainable development of the South African National Department of Minerals and Energy (DME) [10]. Household electrification and an energy grant of R 55 (~€5) per household per month are administered to local municipalities by the Department of Provincial and Local Government (DPLG). In rural areas up to 84% of households can qualify for this grant [11]. In 2003 the DME embarked on a renewable energy project in the OR Tambo municipality Lucingweni Village in the Eastern Cape Province, which was used to test the viability of renewable energy for locations not accessible to the national grid; the introduced model (of Section 1) was applied to the Lucingweni case study [3,5].

Scope of the Study

The boundaries of the case study were set at the borders of Lucingweni Village with its four neighbouring villages and a nature reserve; the details of the case are described elsewhere [3,12]. The time period for the case study was from September 2004 to January 2007. The boundaries and key elements have been described for the following sub-systems [3]:

- Socio-political—the five villages and the region that is controlled by a traditional, cultural government system.
- Socio-ecological—the area used by the villagers of Lucingweni for their ecological services.
- Socio-economic—the same as the socio-political sub-system with the nature reserve and an associated tourist camp that is a source

of employment, including the economic services that are provided as part of the non-traditional government system, i.e. a clinic and school, through the Eastern Cape Parks Board of the South African government.

- Technological—the area to which the power lines are extended. This is a subset of the Lucingweni village.

Flows Across Boundaries

Productive capacity in the Village is in agriculture. Trade and financial transactions across borders are therefore for production in the village and remittances from government grants, and migrant workers. Energy flows across the boundaries are for fossil liquid fuels for transport, cooking, lighting, and refrigeration; and biomass for heat and cooking.

Sources and Quantities of Renewable Energy

The useful energy that can be provided by the six wind turbines (6 kW-peak) and 540 solar panels (0.113 kW-peak) of the mini-hybrid off-grid technological sub-system is determined from the available wind and sun at the coordinates (Latitude 31.825 S and Longitude 29.254 E). The strongest local wind is located on the edge of an escarpment, and polycrystalline Si collectors are located adjacent to the wind turbines. Table 2 shows the available wind and sun energy per day. This daily energy takes up on average an estimated 25% of the maximum capacity of the wind turbines, and 19% of the photovoltaic cells.

Electrical System Conversion Efficiencies

The flow of energy through the electrical system is shown schematically in Figure 2. The amount of useful energy that can be obtained at the household connections can be determined from the input energy from the turbine and

Table 2

Projected Average Wind and Sun Energy, and Capacity Factors

Wind	Wind velocity	6.32	m/s (10 year average)
	Turbine output	9.00	kW
	Output/day	147	kWh/day
	Capacity factor	25	% – Output power/peak power
Sun	Solar radiance	4.67	kWh/m2/day (10 year average)
		3.48	hrs full sun/day
	Efficiency Si PV	11	% – Output power/input power
	Output/day	190	kWh/day
	Capacity factor	19	% – Output power/peak power

Figure 2

Energy Flows from Generators to Users

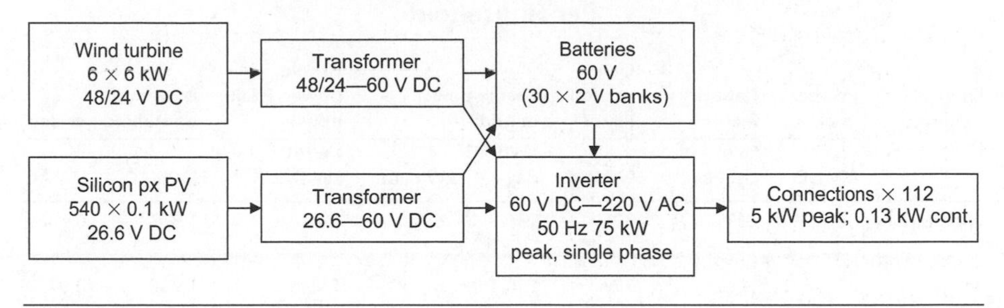

Table 3

Conversion Efficiency of Distribution System Components, Storage, DC/AC Conversion and AC Distribution.

Transformer	99%
Battery	85%
Battery temperature derating	97%
Inverter efficiency	85%
Power conditioning	99%
Line losses	99%
Sum of energy losses	32%

the photo-voltaic and subtracting the energy losses in each of the components of the 220 AC 50 hz distribution system. The energy losses of each component in the system are estimated in Table 3.

112 households were connected to the system. The useful energy from the 97 kW peak system that is available at the 112 household connections is about 125 watts continuous (see Table 4). This provides energy per household connection of just over 3 kWh per month.

Results: Prioritised Sustainability Indicators for the System

The assumptions of three main sustainability paradigms were used as the starting point for identification of the elements in each of the sub-systems (#1 of Figure 1). These were the United Nations Millennium Development Goals (MDGs) to which South Africa has subscribed; the World Commission on Environment and Development, i.e. the Brundtland Report, which has provided the first and only global consensus on conditions for sustainability [5]; and

Table 4

Net System Power Availability for the 97 kW DC System Is 125 W AC Per Household

Energy generators	Power peak	Capacity factor[a]	Con-version losses	Usable power in grid[b]		Usable power/Peak power	Power per household per day	
	kW DC	% of kW DC Peak	% kWp	kWh AC/ day	kW cont.	kW-AC/ kW-DC	kWh AC	kW AC
Wind turbines	36	25%	32%	146.9	6	17%	1.311	0.055
Silicon photo-voltaic	61	19%	32%	190.4	8	13%	1.700	0.071
Total wind and solar	97	—	—	337.3	14	15%	3.012	0.125

[a] Capacity factor = % equivalent of time that the renewable energy converter operates at peak capacity over 10 years sun and wind conditions with equipment operating at delivery specifications.
[b] Usable power estimate has a zero down time, i.e., batteries supply power during maintenance and the supply is greater than demand.

Figure 3

Technological Systems Aim to Contribute to Short-term Millennium Development Goals (MDGs) as a Subset of the Long-term Conditions of Sustainable Development (SD) [5]

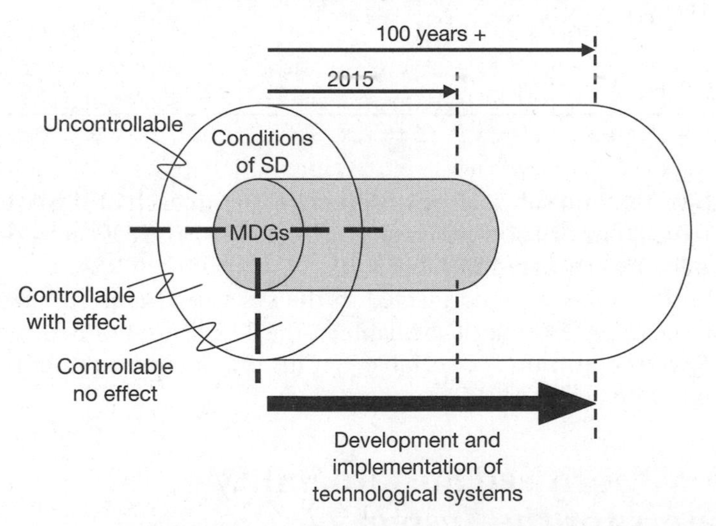

the Stern Review [13], which has provided the most widely accepted techno-economic model for mitigation against climate change. The measured and assessable indicators specified in these paradigms have been used as the initial elements (Figure 3). Additional sustainability and sustainable development indicators were identified through the described learning model (#2 of

Figure 1). Prioritisation was included in the model, and this distinguished between those elements which were deemed uncontrollable, and controllable with, and with no effect (see Figure 3). Expert knowledge was obtained by way of review of initial sustainability mapping of the sub-systems with the University of Pretoria's Departments of Economics, Law and Governance, and Ecology; the Competency Areas of the South African Council for Scientific and Industrial Research (CSIR) in Energy Processing, Energy Infrastructure, and Resource Based Sustainable Development; the DME's Directorate of Renewable Energy; and the National Energy Regulator of South Africa (NERSA) in terms of project management (#3 of Figure 1). Interviews were conducted with the Local Municipality Manager and Council, the Ward Councillor, the Ward Council Committee, the traditional Head Man of Lucingweni Village, and the Headman's Committee (#4 of Figure 1). Technology inputs were supplied by the technology contractor, and its network of technology suppliers; the District Municipality; the national electricity utility (Eskom); and adjacent Eskom grid consumers (#4 of Figure 1).

The conditions for sustainability were prioritised through the engagement with holon expertise and other stakeholders of the system, as described in Section 1. The prioritised set of assessable indicators for the renewable energy system is given in Table 5; details of all the prioritised indicators are provided elsewhere [3,5,9].

Discussion: Outcomes of the Sustainability Assessment

The most important aspects were identified as the economic beneficiation from the technological intervention as expected by the community, and the community ownership of the technological system as expected by the lead implementing agency. From an economic and institutional perspective the community expected that they would receive a similar service, and performance, as provided by the national electricity grid. However, the capacity and reliability of the technological system proved insufficient to meet these expectations.

Demand versus Production

Observation from the site visit was that average demand exceeded average generation capacity of 3 kWh per day per household connection.

Energy Charges and Demand in the Case Study Area

A reason for the high demand for electricity in the region can be seen from the charges for energy in the adjacent areas. Electricity has the lowest charge by a factor of 2–3. For this reason electricity is the energy carrier of choice for high energy services, i.e. cooking and refrigeration. The national electricity utility [14] advises that electricity demand doubles soon after installation when people want stoves and refrigerators. In the Lucingweni Village, household connections were provided with a 20 A trip switch, but behaviours resulted in many of the households bypassing this trip switch.

Table 5

Overall Prioritised Assessable Sustainability Indicators for the Renewable Energy Technological Systems [3,5,9]

Indicator performance			Changes due to technological intervention		Unit	Remark
Holon	Priority	Indicator	Designed for	Outcome after		
Economic	A	PPP	None	None	US$/head/day	Purchase Power Parity; international benchmark of ability to meet basic needs with available resources.
	A	Gini	None	None	% Income lowest quartile	Gini (share of poorest quintile in national consumption).
	A	Health	None	None	10 years of adult working life	World Bank model of health of adults for productivity; 0.4% productivity per 10 years life expectancy.
	B	Education	Some	Some	Years education working adults	World Bank model of education of adults for productivity; 0.5% productivity per year at school.
	C	Access to basic services	Some	Some	No units	Basic services are required for productivity.
	B	Positive return on energy investments	Some	None	% Return	Energy output of system > factor of energy cost of inputs; to ensure viable energy supplies.
	D	Afford-ability energy	Yes	None	% Of income/ disposable resources	Energy cost for users is affordable.
Institu-tional	A	Allocation and control of resources	Some	None	Contracts	Allocation and control of resources. This is the indigent grant system that is controlled by the responsible authority.
	B	Legal protection for controls	None	None	Contracts/ working services	Legal protection to controls for resources. This is via contracts between the suppliers and the users.
	C	Access to credit	None	None	% Of assets	This is via financial institutions that can use the assets as collateral for loans.
	C	Post Kyoto CO_2 eq. targets	None	None	Tonnes CO_2 eq.	Post Kyoto targets for land use. Deforestation rates should be reduced.

Indicator performance			Changes due to technological intervention		Unit	Remark
Holon	Priority	Indicator	Designed for	Outcome after		
	D	Access to basic resources	Yes	3 months	National standards	Access to basic resources is guaranteed by the constitution, water and energy. Includes energy, clean water and sanitation.
Ecology	A	Biological community diversity	None	Some	Acceptable trend	Resilience of ecosystem is indicated by trends in indicator populations for ecosystem type.
	B	Soil type maintenance (fertility)	None	None	Acceptable trend	Resilience of ecosystem is indicated by trends in soil characteristics for soil type.
	A	Available natural energy resource	Yes	Some	% Of need	Natural resources must be available for conversion and the excess should reflect the efficiency and the need for stable supply.
Sociology	A	Jobs (ability to get food)	None	Not direct	Hours of saleable production work	Best indicator of ability to self support for basic needs.
	B	Nutrition	None	Not direct	Stunting of children	Best indicator of food quality that affects productivity and ability to learn.
	B	Life expectancy	None	Not direct	Years	Best overall measure of resilience of social systems is average life expectancy.
	C	Literacy	Yes	Yes	Standard literacy test	Best overall indicator of ability of humans to improve productivity.
Technology	E	Increased productivity	None	None	% Increase in production	Ability of energy system to assist production, e.g. electrical energy for means of production.

Municipalities are authorized to derive revenue from the sale of electricity. Municipal electricity charges are typically made up of a municipal levy of R 0.23/kWh plus the national utility's supply charge of R 0.16/kWh (see Table 6); 10 South African Rands (R) were equal to about 1 Euro (€). The site visit found that electricity sales can be the single largest source of revenue for South African municipalities (see Table 7).

Municipality Subsidy from the DME

As indicated in Section 2, the DME free basic alternative energy policy for off-grid support to indigent households is administered by DPLG. At the national

Table 6

Energy Charges in the OR Tambo District Municipality (March 2007)

Energy carrier	LPG	Diesel	Paraffin	OR Tambo DM rural electricity
MJ/kg	48.55	38.1	37.00	
R/kg	19.00	6.58	7.39	
R/MJ	0.39	0.17	0.20	0.11
R/kWh				0.39

10 South African Rands (R) were equal to approximately 1 Euro (€).

Table 7

OR Tambo District Municipality Income from Electricity at a Rural Connection (March 2007)

Units	OR Tambo DM charge	ESKOM national average cost	OR Tambo DM levy
R/MJ	0.11	0.05	0.06
R/kWh	0.39	0.16	0.23

10 South African Rands (R) were equal to approximately 1 Euro (€).

Table 8

Estimated Municipality Monthly Demand Using the DME Basic Grant Funding for 2007

Municipal charge system to DME indigent grant	Cost[a] for electricity	Demand[b] from policy
	R/kWh	kWh AC
Without levy	0.13	359
With levy	0.39	166

[a] Cost includes VAT.
[b] Demand is based on a CPI + 1.5% escalation of the R 55 grant per annum from 2003.

utility's electricity charges in the vicinity, the off-grid support is equivalent to either 166 or 359 kWh per household (Table 8), depending on the charge of the levy by the municipality. The indigent subsidy policy for urban households is 50 kWh per household connection.

Potential Offsets for Renewable Energy Projects

Renewable energy projects attract carbon subsidies from South Africa and international institutions, but for potential project implementers, a determining barrier is often the administrative costs. In order for registration and auditing to be a small fraction of the total project costs, a minimum number of carbon credits are needed. In January 2007 typical incomes that might have been obtained were:

- Tradable Renewable Energy Certificates; R 0.12/kWh which has been reported by DME and this is equivalent to R 15 000/a, i.e. less than € 1500 per annum.
- A DME Renewable Energy Subsidy of 20% of the capital cost, i.e. R 1.04/kWh or less than € 15 000 per annum was available in early 2007.
- NERSA household connection subsidy; at about R 4500 per household this was equivalent to R 1.78/kWh.
- EU Green House Gas emission trading scheme for 2007 ranged between R 0.04 and R 0.40/kWh.

These incomes could provide a total of R 3.34/kWh, but the administrative costs would be most cost effective if only the NERSA and DME renewable energy subsidies were claimed at the start of the project, i.e. R 2.82/kWh.

Electricity Costs for Off-grid Municipal Supply from Wind and Solar Power and the National Grid Coal Power

The cost of electricity supply for the DME renewable energy village project using wind and sun was estimated from projections of energy outputs and the budget of the Lucingweni Village project. In 2007 Rands, the total system cost was R 7.76/kWh for 119 000 kWh/a. The national utility cost for 2007 was R 0.16/kWh, or approximately 50 times lower. Reasons for this are attributable to:

- Energy conversion losses between the source of electricity and the consumer are higher. About 30% is lost by battery storage and DC to AC conversion (see Tables 3 and 4). In comparison the transmission losses in the national grid are expected to be up to 10%.
- Capital costs for the battery storage make up 40% of the total costs. The pumped storage in the national grid is estimated at 1% of the capital cost.
- Capital costs of renewable electricity generators have low capacity factors, i.e. 25% and 19%. These compare unfavourably with coal fired power stations that operate at 87% of maximum rated output.
- The national utility's capital costs are typically based on old and depreciated plant. New renewable energy generators have yet to be written off.
- Connection costs in rural areas can be subsidized by NERSA in a once-off payment and are therefore not included in the tariff of the national utility.

The DME subsidy system operating in 2007 provided for a once-off subsidy of 20% [12]. This is equivalent to a grant of approximately R 1/kWh. The inclusion of the DME and NERSA subsidies results in a Renewable Village Energy cost of about R 5/kWh. Therefore, the R 55 household energy grant from DME, provides for a maximum of 11 kWh/month. It is reasonable to expect that the traditional leader and the Nyandeni Local Municipality, and the DME, would prefer a quantity closer to the indigent allocation of 50 kWh to national grid connections.

Alternative Technological Solutions for Sustainable Development

The DME needs a more economical electricity supply if the indigent grant subsidy scheme is to be used for renewable energy off-grid applications. Different alternative technologies options were subsequently considered.

Village Grid Energy Storage in Lead-Acid Batteries or Diesel?

The DME renewable energy village has battery capacity for storage for up to 100 h of windless and overcast days, i.e. for approximately 1400 kWh. This storage can be provided by about 500 l of diesel. The cost of a 75 kW peak diesel generator was approximately R 200000, or € 20000, and included a fuel tank. In comparison, lead-acid battery bank capital cost was R 3000000 [12]. While the running costs of diesel were higher they were not high enough in January 2007 (R 5.6/l) to make diesel unaffordable (R 2.85/kWh) compared to renewable energy (R 7.76/kWh). This is still an order of magnitude more expensive than the national utility charge, but improves affordability of a stand alone rural village grid by a factor of 2.7.

Village Grid Storage or National Grid Storage?

The national grid has a smaller differential between peak and average demand (see Table 9) and a lower portion of the supply from pump storage (248 MW and 0.9% of supply capacity) compared to the lead-acid batteries (1400 kWh and 100% of the village supply) [12]. Connection to a national grid for renewable energy would have saved storage costs of R 3.15/kWh [12].

Low Cost Extension of the National Grid?

Technological innovations from the national electrification programme have been attributed to user based standards, rather than supplier based standards [15]. These standards enable uniformity in procurement and national uptake of successful interventions, and shorter times on fault corrections

Table 9

Peak and Average Demand for the Village Grid and the National Grid

Demand	Mini-grid (KWh)	National grid (kWh)
Peak	53	32000000
Average	14	22000000
Ratio	3.7	1.5

during implementation. These innovations included prepaid meters and low cost grid extension; Single Wire Earth Return (SWER) technology replaced both three phase and single phase grid extension.

A national utility cost for a 5000 kW line extension of 13 km from the adjacent Mdumbi Village to Lucingweni Village, using a standard grid controller would have cost about R 40000 per km or about R 0.91/kWh [12]. Normal grid extensions required in rural areas are limited by bulk infrastructure capacity, but in the case of a small additional load (14 kW) this is not a restriction. The extension would provide 50 kWh per household within the DME grant. If this project was funded as a stand alone project the local municipality would, however, have to forgo approximately R 0.23/kWh income on sales (see Table 7).

Comparison with Targets for South African Low-Carbon Technologies

South Africa's response to the high cost of carbon, climate mitigation, and energy shortages has so far been to commit to a long-term policy of power expansion based on renewable energy and nuclear power with high carbon tax [2]. The first response by the national utility is the largest project proposal in South African history, and the latest indications are that the cost for electricity will be in the range of R 1/kWh [16] assuming a growth of consumption at 6% per annum; there is high uncertainty in the cost, with estimates increasing rather than falling. This can be compared with the social cost of carbon, which the Stern Review [13] has estimated to be in the range US$ 85–US# 25 per tonne of CO_2. This is equivalent to additional costs between R 0.7/kWh and R 2.6/kWh, for the South African national grid that relies on coal with and average energy content of 21 MJ/kg and ash at 31% with power station efficiency of 34%.

As this Lucingweni Village case study shows, renewable energy is more practical when connected to a large grid. For 2007 prices the village wind electricity is about R 1.70 R/kWh and solar PV about R 2.45/kWh. These costs are the same range as the latest new costs of coal and nuclear electricity added to the lower social cost of carbon, i.e. R 2/kWh. Renewable energy requires a carbon tax to bring into cost competitiveness with coal. A summary of the costs for three options in Table 10 shows that renewable energy is not affordable for local municipalities.

Table 10

Comparison of Options

Option		R/MJ	R/kWh
National grid/Eskom	ESKOM national average cost	0.05	0.16
	OR Tambo DM levy	0.06	0.23
	OR Tambo DM charge	0.11	0.39
Village grid	Renewable energy village	2.02	7.76
National grid/Renewable	Wind and solar with NERSA and DME subsidies	0.48	1.83

Nontechnical and Institutional Issues

The disregard at the design stage for almost all of the nontechnical aspects has further resulted in an overall unsustainable system. The uncontrolled connections by the community resulted in system overload, disputes between all parties, and disconnections of power by the generator; the system stopped operating continually within one year of commissioning. Overall the management of the technological intervention did not improve the conditions of the social sub-system in the rural village or meet any of the performance aspects raised by the stakeholders. The result was the breakdown of trust between the traditional societal structures and the formal government structures, and the technology developers. The case study therefore emphasizes that in the pre-feasibility and feasibility phase of the technology life cycle a holistic understanding of energy needs and other expectations is crucial. If an integrated system is addressed as a whole the overall resilience and adaptive capacity of all the sustainability aspects can be improved. Also, the system design needs to accentuate strategies for the technological intervention to ensure adaptive management of the integrated system in society.

The case study highlights the importance of the principles of sustainability science (see Table 1) to design and manage renewable energy technologies [9]:

- Transdisciplinarity. The different perspectives of experts and stakeholders on the aspects of sustainability are essential for the design stage. Thereby, technology designers can acquire a practical integrated understanding of the most important aspects and obtain agreement on the most important performance indicators for a type of technological intervention.
- Resilience. A key aspect to the sustainability of the integrated system is the trust between society and institutions, and technology developers and implementers. A breakdown of trust will result in society not accepting and adopting the technology intervention. Depending on the context, ecological and economic aspects may determine the resilience of the overall system to the technological intervention, e.g. the capacity of natural resources, and affordability.
- Complexity. Interactions between and within human and natural systems can result in misunderstanding and a mismatch between expectations and bio-physical and economic capacities. This complexity is likely to be poorly understood initially, and therefore deductive rather than inductive learning should direct technological design and intervention. Especially behavioural changes in the socio-economic, and the implications thereof for ecological systems, have high uncertainty.
- Adaptive management. Renewable energy for electricity generation is relatively new to remote areas of developing countries. The management of a technology during and after intervention requires technical skills and understanding of equipment performance and economic benefits that are not readily available in this traditional context. Traditional social structures that need to support the technological intervention must be engaged to deal with the adaptive responses to changes in social values and eco-services.

- Adaptive capacity. The ability of the stakeholders to agree to experiment with alternatives to mitigate problems with sustainability aspects highlights the potential ability of the social system to learn and adapt to a technological intervention within the carrying capacity of the ecological systems and the technological capacity of the society over time. Renewable energy interventions should therefore provide flexibility for stakeholders to adapt to sustainability aspects within the constraints of the applicable social and institutional systems.

Implications for Policy-making to Promote Renewable Energy Technologies for Off-grid Applications

The literature frequently makes recommendations to governments about their responsibilities and the policies they should implement for long-term sustainable development [17]. However, because social-ecological systems are self-organizing their evolution rarely follows the paths intended by governments [18]. Governments are not free to invest or establish institutions at will, but must take account of the political influence of all stakeholders to promote sustainable technology–economic–social–ecological–institution systems. The capacity of such systems to self-organize is the foundation of their resilience. Rebuilding this capacity at times requires access to external resources. Excessive subsidization can, however, reduce capacity. Cross-scale subsidization should end when self-organization becomes apparent, because cross-scale subsidization can increase the vulnerability of the broader system. A long-term perspective is essential, i.e. cross-scale relationships should in the long-term be mutually sustaining, neither exploitative from above nor parasitic from below [18]. Therein lays the challenge for policy-making related to the promotion of sustainable and adaptable renewable energy technologies in social-ecological systems, especially in remote areas.

Conclusions and Recommendations

The investigation summarised in this paper set out to assess the sustainability of renewable energy technologies for off-grid applications, by applying an introduced learning model. The investigation focused on a rural village in the Eastern Cape Province of South Africa where a renewable energy system was implemented.

The complex interactions between the technological, economic, social, ecological, and institutional sub-system were demonstrated through the case study. The vulnerability of the overall system to issues such as trust and ownership was highlighted. Such issues emphasize that transdisciplinarity understanding is required by renewable energy technology designers to reduce uncertainty and improve the sustainability of technological interventions. Apart from technical aspects a holistic understanding of energy needs and implications, where a technology is to be introduced, is essential. The understating of implications or changes in the integrated system over time, in turn, could identify adaptive strategies for the management of renewable energy technologies. The learning capacity of cultures in specific contexts, especially, is vital for the planning and decision-making of renewable energy systems. With such increased understanding it is envisaged that the sustainability performances of renewable energy technological interventions may be improved during the design stages, i.e. during the pre-feasibility and feasibility phases, and in the uptake stages, i.e. the transfer and adoption phases, of the technology life cycle.

The renewable energy system of the case study was found to be unsustainable. In essence renewable energy for off-grid rural electrification does not meet the South African Millennium Development Goals commitments for poverty reduction, because the return in productivity is uncertain and the cost is too high for the institutional support from the National Department of Minerals and Energy (DME). The failure of the integrated systems was further found to be attributed to:

- The complexity of the social-institutional sub-system, which resulted in uncertainty for project planners and system designers; and
- The lack of resilience of the technological system to demands from the social, economic and institutional sub-systems.

If renewable energy in remote off-grid areas is to be linked to sustainable development goals then national grid connection is required. . . .

All references for articles included in *Taking Sides: Clashing Views in Sustainability* can be found on the Web at www.mhhe.com/cls.

EXPLORING THE ISSUE

Can Technology Deliver Global Sustainability?

Critical Thinking and Reflection

1. What are some alternatives to globalization and corporate investment in promoting leapfrogging?
2. How can low-impact technologies be encouraged and developed?
3. Does technological innovation require an international policy-making body?
4. Is climate change mitigation a matter of national defense?
5. What are the potential dangers of carbon sequestration, de-salinization?

Is There Common Ground?

Both camps agree that reduction of carbon emissions is a crucial concern for creating a sustainable future. The issue of poverty alleviation and equity of resource allocation in the face of growing energy demand and finite natural resources will certainly require broad-based cooperation. The ultimate drivers of sustainability are human choices and the policies and actions that result. Although, the role of technology is debated, both optimists and skeptics see hope in human innovation and adaptability. The youth of today—having experienced many technological changes, and also much ecological destruction, in their lifetime—seem willing to take a new path even if it means having "less stuff" and more local focus.

Internet References

Tree-hugger.com

This site is a great resource for balanced news articles and commentary. There is a recent article about the recent UN geo-engineering moratorium and how it doesn't completely ban geo-engineering. Here is a link to their page that has some exciting and innovative breakthroughs in sustainable technology, like solar film night in Kenya and other applications.

www.treehugger.com/science technology/?campaign=th_nav_scitech

Natural Resources Defense Council-LEED Building Resource Page

The NRDC has an informative website and its page on LEED certified building certification is informative on the latest technologies that are available. It is not too technical so the average person can understand the process.

www.nrdc.org/buildinggreen/leed.asp

CulturalSurvival.Org

This is a website about the struggle for indigenous land, language, and culture. It has good articles about topics regarding technology's effects on the least powerful group of people in the world. Topics include: the effects of large dam projects for indigenous people such as relocation, etc.

www.culturalsurvival.org

NewYorkSkywatch.org

New York Metropolitan area residents discuss geo-engineering. There are links to other websites and supportive documentation of worldwide geo-engineering. It is updated on a frequent basis. Very intriguing videos of supposed "chem-trails" are accompanied by identified and "unidentified" pictures of planes spraying what looks to be chemicals. As can be seen from the various regional links-there is a huge movement to stop geo-engineering.

http://newyorkskywatch.com/

Scitizen.com

This website brings science, technology, and society together with interdisciplinary articles. Here is one link to an article about the benefits of nano-technology to sustainability—the public can comment in some of the forums.

http://scitizen.com/nanoscience/nanotechnology-and-sustainability_a-5-750.html

Internet References . . .

Economy and Ecosystem Services

This site contains analytical information regarding the economic aspects of ecosystem services.

http://www.teebweb.org/

Economy and Valuation of Ecosystem Services

This website discusses the benefits of ecosystem services from an economic perspective, and how degradation of ecosystems can have economic costs.

http://www.conservationcommons.net/

Climate Change—Cap and Trade

The Chicago Climate Exchange offers the only voluntary cap-and-trade program in the United States.

http://www.chicagoclimatex.com/content.jsf?id=821

Carbon Trading

The European Commission carbon trading system is the first instituted cap-and-trade program. Background on how the system was implemented is provided.

http://ec.europa.eu/clima/policies/ets/index_en.htm

Free-Market Sustainability

The Competitive Enterprise Institute is a conservative free-market think tank dedicated toward disproving global warming and promoting limited government.

http://www.cei.org/

Green Building

The Green Globes system is a revolutionary building environmental design and management tool. It delivers an online assessment protocol, rating system, and guidance for green building design, operation, and management.

http://www.greenglobes.com

World Water Council

This council is a beneficial resource for obtaining information on water issues.

http://www.worldwatercouncil.org/

Natural Resources Defense Council (NRDC)

The NRDC has an informative website and its page on LEED-certified building certification is informative on the latest technologies that are available.

http://www.nrdc.org/buildinggreen/leed.asp

UNIT 3

Policy

For sustainability to become operational, governments need policy that will produce incentives for people and society to change unsustainable behavior. Two essential approaches can be considered. First, governmental leadership is needed to develop regulations and policy that promote conservation, a change to renewable energies, and encouragement of sustainable practices. An example of this approach would be for government to establish regulations imposing limits on the emission of carbon dioxide as a way to deal with climate change. For instance, the U.S. Environmental Protection Agency would establish a standard limiting carbon emissions on utilities. Another policy approach would be to rely more on the market to promote sustainable behavior. This second approach emphasizes the use of subsidies and tax incentives to change behavior. Today, renewable energy development is promoted with these incentives. Since renewable energies today, such as solar, cannot compete cost-wise with fossil fuel–based energy, this is a way for government to even the playing field. The problem with this approach is that government needs to continually fund this approach, which means that it is continually a political issue. Other market-based approaches are carbon taxes and cap-and-trade policies. Although allowing the market to operate, these policies require governmental leadership in development and implementation.

There are a number of ways to affect good policy for sustainability. One way is to affix a monetary value on ecosystem services. The belief is that people will value nature if a monetary cost is associated with their decisions to change it. Another issue is related to land-use control and the loss of valuable land resources to urbanization. The battle here is over policies that encourage smart growth or the use of governmental zoning and planning regulations, which curtail the freedom of ownership of land owners. And lastly, a main question to discuss is whether the private sector can provide more efficiency and lower costs to deliver services than government. This is a particularly sensitive issue when it comes to water.

- Is Monetizing Ecosystem Services Essential for Sustainability?
- Does the Market Work Better Than Government at Transitioning to Sustainability?
- Does Sustainable Urban Development Require More Policy Innovation and Planning?
- Should Water Be Privatized?

ISSUE 9

Is Monetizing Ecosystem Services Essential for Sustainability?

YES: Stephen Polasky, from "What's Nature Done for You Lately: Measuring the Value of Ecosystem Services," *Choices* (2nd Quarter, 2008)

NO: Clive L. Spash, from "How Much Is That Ecosystem in the Window? The One with the Bio-Diverse Trail," *Environmental Values* (May 2008)

Learning Outcomes

After reading this issue, you should be able to:

- Understand the evolution of the term "ecosystem services."
- Describe the moral arguments both against and for economic valuation of nature.
- Discuss the limitations of market-based systems for valuation of ecosystem services.
- Describe the criticisms raised against survey-based methodologies.
- Describe the issues involved in coupling poverty alleviation and conservation.
- Understand the practical uses of monetization of ecosystem services.

ISSUE SUMMARY

YES: Writer Stephen Polasky presents the argument why putting a monetary value on ecosystem services will improve decision making by clearly illustrating the consequences of alternative choices.

NO: European professor and economist Clive L. Spash questions the model of human motivation and behavior underlying orthodox economics and its use in ecosystem valuation and states that ecologists and conservation biologists who use it fail in their awareness of the political and ideological system within which it is embedded.

The term "ecosystem services" was first used in the 1980s as a tool to focus attention on the complex workings of the natural world, and how fundamental nature was to human survival and well-being (Gomez-Baggethun et al., 2010). The concept had the purpose of highlighting the relationship between ecosystem degradation and sustainability. In 1997, with a publication in the journal *Nature* of an article titled "The Value of the World's Ecosystem Services and Natural Capital," Costanza et al. (1997) formally equated monetization of ecosystem services with sustainability. This approach brought with it the idea that if monetary values could be placed on ecosystem functions, concrete incentives could be designed to promote sustainable use of ecosystems.

Monetization of ecosystem services, however, has its opponents. One of the most common arguments given in opposition for the exclusive use of monetization of ecosystem services is that markets by their nature favor short-term consumption of goods and services (McCauley, 2006). Such a short-term view of nature could be catastrophic, as has been seen with the collapse of fisheries. Viewing nature solely in monetary terms could hasten similar collapses of other services. The detrimental aspect of a shortsighted view of ecosystem services is related to another argument used against monetization of nature. This argument states that ecosystems are highly complex, nonlinear, and interconnected and so an attempt to value any one component will never be able to capture the overall, true value of the entire system (Kosoy and Corbera, 2010). In that case, some components, which may be vital, will be undervalued and deemed unnecessary for conservation purposes. If these components then are allowed to be degraded, that could disrupt the entire ecosystem and those services that had been deemed valuable. Furthermore, because ecosystems are interrelated, any disruption in one ecosystem could have negative effects on other ecosystems, thereby destabilizing an entire region.

Opponents of monetization claim that in general we as humans do not have sufficient information to truly understand what is needed to allow an ecosystem to properly function (Spash, 2008). This raises difficulties in trying to determine the truly "critical" aspects of nature when trying to monetize those aspects. It is because of these difficulties that the opponents of monetization heavily criticize survey-based methodologies used to determine the value of natural resources (Edward-Jones, 2006; Farley, 2008). The public tends to focus on those species that are "iconic" and that are better advertised through conservation campaigns (Martin-Lopez et al., 2008); how can they be asked to place a value on more obscure ecosystem aspects about which they know very little?

Those against using monetization also often take issue with the fact that conservation has been coupled with poverty alleviation (Foreman, 2006; Chan et al., 2007). The poor are often not part of global markets in the first place, nor do they have the luxury of placing value on ecosystem services outside of what they need to survive, so market techniques are often completely inadequate for addressing issues of poverty (Edward-Jones, 2006; Farley, 2008; McCauley, 2006). Even when there are conservation efforts in which they could benefit, such as ecotourism, most of the profits go to private companies that are not based locally (Spash, 2008). Moreover, poverty alleviation is seen by some to at times actually be counterproductive to conservation. The introduction of the Nile perch into Lake Victoria is a classic example. The Nile perch boosted the economy of the locals, catapulting them into the global market, but also led to a staggering loss of local biodiversity in the lake and accelerated rates of deforestation (Pringle, 2005). Finally, most monetization opponents claim that an economic aspect could very well be detrimental to conservation in the long run. Assigning a monetary value to nature could be seen as the equivalent of "bribing" people to care, and this could corrupt the relationship between humans and nature, actually leading to a devaluation of nature (McCauley, 2006; Kosoy and Corbera, 2010). Those who dislike the monetization idea are in favor of highlighting the intrinsic value of nature, or nature for its own sake. Ecological ethics and morals should be used to inform policy and management decisions with regards to ecosystems (Foreman, 2006).

Support for monetization of ecosystem services is quite strong (Spash, 2008). At the heart of this position lies the belief that money is what drives everything. Unless some economic incentive can be presented for nature conservation, most people will opt to take the cheaper route, which often results in degradation of nature (McNeely, 2007; Polasky, 2008; Goldman, 2010). Moreover, in agreement with Garrett Hardin's "Tragedy of the Commons" (1968), the proponents of monetization believe that people will "free ride" on natural resource use if there is no cost associated with that use (Lant et al., 2008). This leads to an under-provisioning of that resource for some users and an overall degradation of that use. If a monetary cost can be associated with resource use and degradation, human beings will be more careful regarding the extent and nature of environmental degradation. When degradation does occur, monetary valuation can be used to assess exactly how much damage has occurred and assign responsibility for that damage. Indeed, this methodology has already been put into place in numerous courts around the world (Edward-Jones, 2006; Mendelsohn and Olmstead, 2009).

One of the most common arguments used in favor of ecosystem service monetization is that since there are limited funds for conserving natural resources, there must be a way of comparing policy and management alternatives (Maguire and Justus, 2008; Polasky, 2008; Nelson, 2009; Goldman, 2010). The idea of intrinsic value championed by the opponents of monetization is a difficult bargaining tool and economics is seen as much more conducive to the decision-making process. One needs only look at the Endangered Species Act to understand this. Many species are "wait-listed" and not given protection because of insufficient funding; those that are lucky enough to make it on the list are deemed "more important" than others for one reason or another.

Although the opponents of monetization stress moral arguments to conserve nature, those in favor of monetization stress the moral argument that no one should go hungry (Edward-Jones, 2006; Maguire and Justus, 2008). Associating monetary benefits with conservation, then, is an efficient way to alleviate poverty. The poor often have no choice but to degrade resources to sustain their livelihoods. As long as payment schemes make sure to include locals, monetization of ecosystem services could be a great benefit to local economies. Proponents of this position cite myriad examples where this has occurred. Conservation of carnivores in Africa and Asia has benefited from monetization, for example. Although carnivores in these locations often have a negative impact on livelihoods through predation of livestock, locals are refraining from eliminating them due to efforts to provide payment for their protection or protection of their native prey (Nelson, 2009). Those in favor of monetization recognize that ecosystems are complex and that we may never understand the interactions between all components, but that should not prevent more research into how to value ecosystems for conservation purposes and to improve human well-being (Hodgson et al., 2007; McNeely, 2007; Nelson, 2009).

The article for the YES position, which favors monetization, is presented by writer Stephen Polasky. He claims that the provisioning of ecosystem services is seriously flawed due to the lack of incentives for ecosystem protection; economists can help correct that flaw by placing a monetary value on ecosystem services and creating mechanisms that focus on economic incentives for policy and management decisions. He discusses various techniques to measure the value of ecosystem services and claims that despite criticisms, difficulties, and incomplete information, cost-benefit analyses are the best way to decide between management and policy alternatives for natural resources.

The article for the NO position, critical of monetization, is presented by university professor and economist Clive L. Spash. He points out the flaws in some of the more common justifications behind the ecosystems services approach. The author notes that proponents of monetization claim that monetary value is merely one input for decision-making activities aimed at conservation. But the author points out that the other inputs and the decision-making activities themselves are never clearly defined. Spash also attacks the idea that sees free-market values as the best way to allocate resources and views resource use more of a condition imposed by economic elites. And finally, the author brings up how adding a monetary component to either a sexual relationship or a relationship between parents and offspring transforms those relationships, and such is the case with nature. Because of this, the author argues for an understanding of ethics and human motivations in the interactions with ecosystems.

Stephen Polasky

What's Nature Done for You Lately: Measuring the Value of Ecosystem Services

The natural world generates a range of valuable goods and services that support human well-being. These goods and services, collectively called ecosystem services, are typically provided free of charge and often have characteristics of public goods. Like other public goods, ecosystem services will not be provided optimally by aggregating the decisions of individuals motivated by self-interest. For example, an individual farmer gains the benefits of increased yields from the application of nitrogen fertilizer but often bears an insignificant portion of the costs from additional release of nitrous oxide, which is a powerful greenhouse gas, increased air pollution from emissions of nitrogen oxides and ammonia, and increased water pollution from release of nitrates into ground or surface water. In such cases, the sum of individual actions may result in the disruption of the flow of valuable ecosystem services thereby making all individuals collectively worse off. Even in cases where ecosystem services provide localized benefits, if individuals are not aware of the consequences of their actions they may still take actions that unknowingly damage ecosystem services on which their long-term welfare depends.

The presence of both incentive problems and information problems means that ecosystem services are often not provided efficiently. There is an important role for economists to play in improving the provision of ecosystem services, which includes understanding how management choices affect ecosystems and the services they provide, understanding of the relative value of ecosystem services to different groups in society, and designing appropriate incentive mechanisms for the efficient provision of ecosystem services.

The recent focus on ecosystem services grew out of efforts, led primarily by ecologists, to highlight the importance of ecosystems and the natural world to human welfare. Just over a decade ago, the publication of *Nature's Services: Societal Dependence on Natural Ecosystems* (Daily 1997) and a controversial article published in the journal *Nature* entitled *The Value of the World's Ecosystem*

Polasky, Stephen. From *Choices: The Magazine of Food, Farm, and Resource Issues,* 2nd Quarter 2008, pp. 42–46.

Services and Natural Capital (Costanza et al. 1997) brought significant attention and research focus to assessing ecosystem services. The *Millennium Ecosystem Assessment,* a major international research effort to summarize the current condition and potential future trajectories of the world's ecosystems and biodiversity, used ecosystem services as its major organizing principle and emphasized the link between ecosystems and human well-being (MEA 2005). Major research efforts on ecosystem services are underway in government agencies such as the U.S. Environmental Protection Agency, international organizations such as the World Bank, and nongovernmental organizations such as The Nature Conservancy and World Wildlife Fund. Many of these efforts are being led by natural scientists and there is a compelling need for greater economic input.

Economists have much to contribute to research on ecosystem services. In fact, properly understood the research agenda on ecosystem services is a continuation of a long-standing set of research objectives in agricultural, resource, and environmental economics. Agricultural economists know that soil and climate are necessary inputs to the production of agricultural crops and have studied production functions and agricultural profitability under a wide variety of circumstances. Resource economists know that natural resources (oil, minerals, timber, and fish) contribute to a wide range of intermediate and final products and have studied optimal harvesting and inefficiencies caused by open access. Environmental economists know that people value the environment directly even where there is no market and have developed tools of nonmarket valuation to analyze such things as the value of a scenic vista or clean air. In fact, in the 1970s economists set out a research agenda to measure "the value of services that natural areas provide" (Krutilla and Fisher 1975, p. 12). The "new" topic of measuring the value of ecosystem services can build from a large existing base of prior research on the value of agricultural production (Beattie and Taylor 1985), bioeconomic modeling of fisheries and other renewable resources (Clark 1990), nonrenewable resources (Dasgupta and Heal 1979), and nonmarket valuation of environmental amenities (Freeman 1993).

A Research Agenda for Economists on Ecosystem Services

What is needed now is to bring the full set of economic tools and expertise to bear on the analysis of ecosystem services. To do this, economists will need to engage with ecologists as well as other natural and social scientists. In measuring, valuing, and providing proper incentives for the provision of ecosystem services, economics is necessary but not sufficient. Knowledge of ecosystems and how they are altered by human actions, which is more in the domain of natural sciences, is also necessary but not sufficient. In research on ecosystem services, integrating both economics and natural science is essential. In what follows, I briefly describe a research agenda and a set of challenges for economists in addressing issues related to ecosystem services. Challenges for economists exist [in both] developing new applications and analysis as well as more effectively integrating with other disciplines.

Measuring the value of ecosystem services and providing an efficient level of provision of these services requires tackling three main tasks:

- Provision of ecosystem services ("ecological production functions")
- Value of ecosystem services ("valuation")
- Designing policies for efficient provision of ecosystem services ("incentives")

I briefly discuss each of these three tasks in the following sections.

The Provision of Ecosystem Services: The Ecological Production Function

Policy and management actions chosen to accomplish certain objectives, such as increasing the yield of agricultural commodities or allowing development of industry, often have a range of effects, both intended and unintended, on ecosystems and the services they provide. For example, expanding agricultural land will increase crop production but may also lead to greater release of greenhouse gases and a decline in water quality downstream. Evaluating alternative policy or management actions in terms of ecosystem services involves understanding the full range of consequences the action has on ecosystems and how these consequences translate into changes in the suite of ecosystem services provided. Like a typical production function that predicts output of goods (e.g., crop production) as a function of inputs (e.g., land, fertilizer, water), an ideal "ecological production function" would predict the outputs of a range of ecosystem services given ecosystem structure and function.

Though considerable ecological knowledge exists about the structure and function of ecosystems, the translation to how these contribute to the provision of important ecosystem services is sometimes lacking. Ecological production functions for some services, such as above-ground carbon sequestration in plant material are well understood. But understanding carbon sequestration or release in soils or the net production of other greenhouse gases (e.g., nitrous oxide or methane) is less predictable. Sequestration or release of greenhouse gases in soil is a complex function that depends on whether chemical reactions are aerobic (with air) or anaerobic (without air), temperature, soil water content, [and] the presence of various organic compounds and minerals.

In general, estimating the provision of the complete range of ecosystem services from any particular ecosystem is beyond our ability at present (NRC 2005). Key limitations that prevent complete understanding of ecological production functions include imprecise understanding of ecological processes, complex interaction among ecosystem processes, and lack of data.

Despite these limitations, ecological understanding is often sufficient to provide reasonable estimates of many important ecosystem services. The intense interest focused on ecosystem services at present is also helping to advance our understanding of ecological production functions for important services. In fact, framing issues in terms of ecosystem services has helped to redirect ecological research creating more rapid progress and easier links between ecological and economic analysis.

The Value of Ecosystem Services: Market and Nonmarket Valuation

The provision of ecosystem services yields outcomes in terms of physical units (e.g., bushels of crops, tons of carbon sequestered, [and] concentrations of nitrate in water). But comparing outcomes of alternative management options is difficult when there are impacts on multiple ecosystem services and when each service is measured in its own physical units. Is a management option that increases crop yields but also results in increased carbon release and decreased water quality beneficial for society? The answer to this question depends on how one views the trade-offs between various services. In a standard economic problem, economists compare consumption bundles that might differ in many dimensions by converting the measures to a common metric of value measured in monetary terms. The same conversion to a common metric of value can be done with ecosystem services through the application of market and nonmarket valuation techniques.

Some ecosystem services result in outputs of marketed commodities (e.g., agricultural crops, commercial fisheries, [and] timber) making valuation relatively straightforward. The analysis of the value of these ecosystem services only requires the application of standard tools of market analysis to assess the change in consumer and producer welfare with a change in the provision of ecosystem services. Ecosystem services that provide a necessary input to the output of a marketed commodity can be analyzed in a similar fashion. For example, the value of pollination services can be assessed by looking at the change in the quantity and quality of crop production when pollinators are present versus when they are absent. The only danger in analyzing the value of ecosystem services that are inputs to the production of other ecosystem services (e.g., pollination for crop production) is that one cannot count both the value of the input and the value of output at the same time because this would result in double-counting.

Most ecosystem services, however, are public goods that are not traded in markets. As mentioned above, the lack of markets is one of the main reasons for concern over the inadequate provision of ecosystem services. For such ecosystem services, nonmarket valuation methods (revealed preference, stated preference) are needed. The value of some nonmarket ecosystem services has been well studied by economists. For example, there are numerous applications of random utility models to assess the value of outdoor recreation (hunting, fishing, bird watching, backpacking), and numerous applications of the hedonic property price model to assess the value of various environmental amenities (access to open space, access to water resources, local air quality). The strengths and weaknesses of applying both revealed and stated preference methods to value aspects of the environment are well understood and a number of excellent summaries of this literature exist (e.g., Freeman 1993,

Champ, Boyle and Brown 2003, Haab and McConnell 2003). Though estimating nonmarket values can be challenging, valuing ecosystem services is not inherently more difficult than applying nonmarket valuation to other areas of environmental economics. In fact, many things that are now called ecosystem services are things for which economists have routinely applied nonmarket valuation techniques.

Some prominent examples of the value of ecosystem services have been derived using replacement cost, i.e., what would it cost to replace a naturally provided ecosystem service with a human-engineered alternative. For example, the value of providing clean drinking water to New York City by protecting watersheds in the Catskills has been estimated to be worth $6–8 billion dollars because this is the cost of building and operating a water filtration plant (Chichilnisky and Heal 1998). Though popular, especially with non-economists in part because it is easier to understand than methods to estimate willingness-to-pay, the replacement cost approach should be used with caution. Costs are not the same thing as benefits and estimates of cost can only be used to give an estimate of the value of ecosystem services under certain conditions: i) there are alternatives to provide the service and ii) people would be willing to pay the cost of the alternative if the ecosystem service is not available (Shabman and Batie 1978).

What the Millennium Ecosystem Assessment labeled "cultural services," which includes aesthetic and spiritual values, can be quite important and is perhaps the most difficult type of value to assess using economic tools. Critics of economic valuation of the cultural or spiritual significance of nature raise both practical and philosophical objections. For some non-economists, attempting to "put a price on nature" is deeply troubling (e.g. Sagoff 1988). One critique of the ecosystem services approach is that conservationists should use ethical arguments based on moral principles: "Nature has an intrinsic value that makes it priceless, and that is reason enough to protect it." (McCauley 2006, p. 28) Most economists including me find it hard to apply arguments about "intrinsic value" to typical policy and management questions. For example, should we view decisions by farmers to convert a wetland to an agricultural field, or to increase the amount of fertilizer application, each of which will have an impact on an ecosystem, as a moral issue with clear right and wrong? These types of decisions seem better suited to weighing the full set of costs and benefits rather than being subject to moral absolutes.

Setting aside the philosophical debate, practical difficulties in assessing value in a manner that will be viewed as objective, authoritative, and accurate is difficult for some ecosystem services like cultural services. This difficulty may argue for simply providing information about potential trade-offs among services without attempting to measure all services in the same monetary metric. For example, Polasky et al. (2008) derive a production possibility frontier showing trade-offs between feasible combinations of the value of commodities produced measured in dollars and species conservation measured in biological units. This approach illustrates the consequences of alternative land use decisions but avoids the difficult task of putting a dollar value on species conservation. It is then up to the decision-making process to make value judgments about the relative value of species conservation versus commodity production and choose which land use alternative is most preferred.

Valuation of ecosystem services is likely to become more important in the future. With improvements in our understanding of ecological production functions there is greater understanding of the impacts of human actions on ecosystems and the consequences these impacts have on the provision of a suite of valuable ecosystem services. Application of valuation methods can help illuminate what policy or management options generate the greatest social welfare.

Policies and Institutions for Efficient Provision of Ecosystem Services

Though there are many interesting and worthwhile scientific questions to pursue, the prime motivation for assessing the value of ecosystem services is practical. Understanding the full consequences of policy or management decisions and comparing the net benefits to society of alternative choices can result in better policy and management decisions for use of land, water, and natural resources. The title of a National Research Council report on valuing ecosystem services sums it up nicely: *Valuing ecosystem services: towards better environmental decision-making.* Integrating ecological and economic analysis to value ecosystem services can improve decision making by clearly illustrating the consequences of alternative choices.

Information on ecological production functions and on values will almost surely be incomplete. Such incomplete information, however, should not paralyze decision making. In some cases, enough information will be available to make good decisions. In the Catskills watershed example, watershed protection could be justified on the basis of avoiding building a filtration plant, making it unnecessary to know the value of other ecosystem services. In other cases, decision makers may have to make choices based on the best available information, with an eye to learning and adjusting policy or management based on new information ("adaptive management").

The supply of ecosystem services is often influenced by a different set of individuals than those who benefit from the provision of these services. For example, the farmer who maintains wetlands and limits fertilizer application provides benefits of cleaner water and lower probability of flooding to individuals who live downstream. The mismatch between those who influence the supply of services and those who benefit from services gives rise to a classic externality problem. Numerous potential solutions have been proposed for internalizing externalities, including payments for ecosystem services, tradable development rights, taxes on activities that result in damages to services, or some form of direct regulation (e.g., zoning laws, restrictions on actions that harm endangered species). Research that studies the incentive properties of these approaches and empirical analysis of results of implementation should be a high priority.

In the end, more efficient provision of ecosystem services will require that society overcome both information and incentive problems. The challenge for economists in the first case is to be able to work closely with natural scientists to build understanding of ecological production functions and to apply appropriate valuation methods. The challenge in the second case is to design policies simple enough to be implemented yet sophisticated enough to do justice to the underlying biophysical and socioeconomic complexities involved. These are important tasks and the sooner and more fully that economists tackle them the better.

Notes

1. Carbon taxes can apply to carbon emissions only or to a broader array of greenhouse gases. In this paper, we will use the term "carbon tax" to apply to a tax on some or all greenhouse gases.
2. We set aside here the distributional implications of climate change itself.

All references for articles included in *Taking Sides: Clashing Views in Sustainability* can be found on the Web at www.mhhe.com/cls.

Clive L. Spash

How Much Is that Ecosystem in the Window? The One with the Bio-Diverse Trail

Introduction

There has been an observable increase in the desire, especially of conservation biologists and ecologists, for concepts such as biodiversity and ecosystems functions to be expressed as part of a mainstream economic philosophy of value (McCauley, 2006). In recent years the need for monetary valuation of ecosystems has been voiced internationally. In 2005 the National Research Council (NRC) in the USA published a report on the subject commissioned from six economists, four ecologists and one philosopher; aiming for 'better environmental decision making' they adopted a narrow 'total economic value' approach (Heal et al., 2005). In 2007 the G8 and five other industrialising nations proposed a global cost-benefit assessment of biodiversity loss called the 'Potsdam Initiative—Biological Diversity 2010.' Under a clause entitled 'The economic significance of the global loss of biological diversity,' the parties stated:

> In a global study we will initiate the process of analysing the global economic benefit of biological diversity, the costs of the loss of biodiversity and the failure to take protective measures versus the costs of effective conservation.

This study aims to emulate Stern's climate change report, but apparently has neglected critiques of that report and other such global environmental cost-benefit analyses (see Spash, 2002a; 2007a; 2007b). Some published studies—notably led by non-economists and appearing in natural science journals—claim to have already estimated the monetary value of the World's ecosystems (Costanza et al., 1997) and all remaining wild Nature (Balmford et al., 2002). The main approach consists of averaging and summing values from various contingent valuation method studies. Groups in the USA (e.g., Batker et al., 2005), and elsewhere, are trying to formalise such ecosystem services 'valuation' for inclusion in public policy decision processes.

One major thrust of all this work is linked to a general movement called 'benefit transfer' which aims for common use in policy of values taken from original monetary valuation studies but applied to represent the value of other sites, entities or environmental changes, as political necessity dictates (Abt Associates Inc, 2005). For example, the mean willingness to pay for wetland 'goods and services' of UK respondents to a stated preference survey may be averaged to a per hectare value and transferred to North America, or visa versa. More sophisticated approaches try to use a transfer function, although these are almost impossible to employ with any validity, due to the prevalence of noneconomic and socio-psychological factors for which there is a lack of data across populations (Brouwer and Spanninks, 1999). Regardless of sophistication, these transfers suffer from serious practical and methodological problems (Spash and Vatn, 2006), but are defended as pragmatic. The numbers are attractive because they appear simple to derive and seem to place a 'market value' on a wide scope and scale of things. However, the stated preference methods (i.e., contingent valuation and choice experiments) which mainly underlie these value transfers are themselves deceptively simple and the numbers they produce may not be all that is claimed, even before such transfers distort them out of context. The problems can be taken on two levels. There are the concerns of economists over such things as the use of statistical techniques, cross validation, incentive-compatible mechanisms, strategic behaviour, information provision, survey design and treatment of different bid categories (Spash, 2008a). Then there are the concerns of political scientists, applied philosophers and heterodox economists, amongst others, over the role and meaning of the mainstream economic approach. Thus, Holland (1997: 484) notes the difference between internal critiques, where refinements and scientific advance are assumed an adequate response, and principled arguments, which point toward the need for alternative approaches. He identifies six principled arguments in the literature which question the applicability of environmental cost-benefit analysis (CBA) and stated preferences in particular: (i) a variety of ethical commitments exist which cannot be made commensurable; (ii) methodological individualism inadequately addresses the collective values which constitute environmental goods and bads; (iii) the market approach to value elicitation is incapable of recognising certain values and precludes their expression; (iv) social context is inadequately addressed because the method is too abstract; (v) the process of economic valuation suppresses articulation and active thinking by assuming values are pre-formed; (vi) environmental values falling within the domain of political action are inappropriately addressed as preferences. These principled arguments appear to fall into two broad categories. First are those concerning what constitutes environmental values, raising such issues as incommensurability, pluralism vs. monism, community vs. individualism, utilitarianism vs. deontology, objective truth vs. subjective judgement. Second are those relating to the appropriate process whereby values should be expressed, namely markets vs. politics, group vs. individual, hypothetical vs. actual, reflective deliberation vs. instant reaction, and who such processes should represent (experts, vested interests, public) and how (statistically, politically). The first set of arguments inevitably feed into the second, while, especially for an empirically based science, application of the second should influence and inform the first. That is, belief in say monism leads to design of processes using a single numeraire, but when incommensurable and plural values arise and are recognised in the value-articulating process these should bring into question the belief in and relevance of monism.

A contention of this paper is that there is a stark disconnect between environmental values as constituted in mainstream economics and as recognised by wider society and other disciplines. Yet some of those other disciplines, such as ecology, are actually employing a broadly defined economic approach in the apparent belief that this is a pragmatic solution to the neglect of their principled concerns over the loss of wild Nature and biodiversity. The ecologists' pragmatic argument in support of ecosystem valuation is critically appraised in the next section, along with those arising in politics and from orthodox economics. One result, pursued in the ensuing section, is to question the model of human motivation and behaviour underlying orthodox economics and to point to alternative, more empirically accurate, models. This leads, in the final section, to recognition of the need for variety in value-articulating institutions: so enabling a more comprehensive and complex picture of how humans value the environment. In particular, respecting plural values brings out the contrast between a process of judgment through deliberation and an appeal to instantaneously stated preferences. The potential role for monetary calculations is not excluded from the former, but put in a very different light from use in the latter and by those currently valuing ecosystems as goods and services.

Conclusion

Modern economics professes to be an empirically based science but seems to defend highly abstract theory over empirical fact. The ideological position held by mainstream economists concerning what constitutes value in society is used to reinforce a specific model of political economy. Ecologists and conservation biologists then appear to be rather naively attempting to employ the economic value approach without showing much awareness of the political and ideological system within which it is embedded. The problem then is that values which fail to fall within the ideological constructs must be ignored, excluded or transformed into those which do. The articulation of those values requires the design and implementation of specific institutions which control and manipulate the type and range of values allowed into the decision-making process. That different institutional processes result in different values being articulated seems poorly understood by both economists and ecologists trying to value ecosystems as goods and services.

The idea of monetary valuation work is not just to show a value exists but that it is tangible in economic terms. The belief then is that this will affect human behaviour because 'if only people knew how much money biodiversity and ecosystems services were worth they would do something about preserving them.' One response from ecologists is to produce lists of what ecosystems 'goods and services' people should value, in their opinion. Any stated preference survey using these lists then needs to make a case explaining why the respondent should value things taken from the list. Realising people may not value and/or understand the listed items has led some to transfer values from wherever appears convenient and use their own judgment to calculate 'total values' for ecosystems and/or their services. Yet there seems inadequate comprehension, or at least discussion, of the arbitrary nature of any outcomes from this approach, its biases and limitations.

That judgment is required is not *per se* the problem. The problem is how judgment is concealed and used to frame public policy. The approach to ecosystem services valuation encapsulates an implicit model of both human behaviour and the relevant decision process for addressing environmental problems. The standard justifications for this are embedded in support for or acceptance of the dominance of market systems. This ignores the many ways in which humans operate outside such systems and without being psychological egoists whose only concern is their wallet. Taking human motivation into account is necessary to address why ecosystems are being destroyed and biodiversity lost. This implies something more is required than valuation studies allocating numbers to ecosystems in the pretence they are goods and services which can be mentally placed amongst items in a shop window.

EXPLORING THE ISSUE

Is Monetizing Ecosystem Services Essential for Sustainability?

Critical Thinking and Reflection

1. How has the concept of "ecosystem services" changed since it was first introduced in the 1980s?
2. Many people claim that conservation of ecosystems is a moral question. What kind of moral issues are involved? Is morality at play for both proponents and opponents of monetization of ecosystem services?
3. What do critics of monetization claim are the problems with market-based techniques for valuation of ecosystem services?
4. Survey-based methodologies for assigning value to ecosystem services are also criticized. What are those criticisms?
5. Are there any problems with coupling poverty alleviation and conservation via an economics-based approach?
6. How can monetization of ecosystem services be used in a practical way?

Is There Common Ground?

There does seem to be a common ground on the issue. All parties believe conserving nature is important, for one. And it seems as if all parties believe that conserving nature is not important for merely material benefits. Indeed, even the proponents of monetization believe that nature has intrinsic value for nonuse benefits; they just believe that that value does not have enough bargaining power when it comes to the practical business of convincing stakeholders to invest in nature preservation. Moreover, the opponents of monetization actually do agree that some aspects of ecosystems may benefit from having a monetary value attached to them and that preservation could actually benefit from being associated with profit. What they do not agree on is the extent to which monetization should be used and, in essence, its relative importance. Therefore, it might be a good idea for representatives of each group to discuss what cases may most benefit from monetization and what cases would benefit from an approach driven more by noneconomic ethics. Also, both sides seem to agree that preservation of nature involves certain values regarding what is truly important and what should truly be saved; they also agree that what one culture values may not be identical to what another culture values. Again they are not in agreement over how best to address those values. Rather than continuing to make a case for one standpoint over the other, I believe again that representatives of both groups must meet to discuss how to "practically" ascertain and address the values relevant to nature conservation and how to incorporate those values into environmental management policies.

Internet References

Chan, K.M., et al. (2007). When Agendas Collide: Human Welfare and Biological Conservation. *Conservation Biology,* vol. 21, pp. 59–68.

Costanza, R., et al. (1997). The Value of the World's Ecosystem Services and Natural Capital. *Nature,* vol. 387, pp. 253–260.

Edward-Jones, G. (2006). Ecological Economics and Nature Conservation. In: M. J. Groom, G. K. Meffe, and C. R. Carroll, eds. *Principles of Conservation Biology.* Sunderland, MA: Sinauer Associates.

Farley, J. (2008). The Role of Prices in Conserving Critical Natural Capital. *Conservation Biology,* vol. 22, pp. 1399–1408.

Foreman, D. (2006). Take Back the Conservation Movement. *International Journal of Wilderness,* vol. 12, pp. 4–8.

Goldman, R. (September/October 2010). Ecosystem Services: How People Benefit from Nature. *Environment,* pp. 15–23.

Gomez-Baggethun, E., et al. (2010). The History of Ecosystem Services in Economic Theory and Practice: From Early Notions to Markets and Payment Schemes. *Ecological Economics,* vol. 69, pp. 1209–1218.

Hardin, G. (1968). The Tragedy of the Commons. *Science,* vol. 162, pp. 1243–1248.

Hodgson, S.M., et al. (2007). Getting a Measure of Nature: Cultures and Values in an Ecosystem Services Approach. *Interdisciplinary Science Reviews,* vol. 32, pp. 249–262.

Kosoy, N., Corbera, E. (2010). Payments for Ecosystem Services as Commodity Fetishism. *Ecological Economics,* vol. 69, pp. 1228–1236.

Lant, C.L., Ruhl, J.B., Kraft, S.E. (2008). The Tragedy of Ecosystem Services. *BioScience,* vol. 58, pp. 969–974.

Lovett, J.C., Noel, S. (2008). Valuing Ecosystem Goods and Services. *African Journal of Ecology,* vol. 46, pp. 117–118.

Maguire, L.A., Justus, J. (2008). Why Intrinsic Value Is a Poor Basis for Conservation Decisions. *BioScience,* vol. 58, pp. 910–911.

Martin-Lopez, B., Montes, C., Benayas, J. (2008). "Economic Valuation of Biodiversity Conservation: The Meaning of Numbers. *Conservation Biology,* vol. 22, pp. 624–635.

McCauley, D.J. (2006). Selling Out on Nature. *Nature,* vol. 443, pp. 27–28.

McNeely, J.A. (2007). A Zoological Perspective on Payments for Ecosystem Services. *Integrative Zoology,* vol. 2, pp. 68–78.

Mendelsohn, R., Olmstead, S. (2009). The Economic Valuation of Environmental Amenities and Disamenities: Methods and Applications. *Annual Review of Environment and Resources,* vol. 34, pp. 325–347.

Nelson, F. (2009). Developing Payments for Ecosystem Services Approaches to Carnivore Conservation. *Human Dimensions of Wildlife,* vol. 14, pp. 381–392.

Olmsted, P. (2010). Getting the Price Right. *Alternatives Journal,* vol. 36, pp. 14–16.

This website discusses data, breaking news, and analytical information regarding the economic aspects of ecosystem services. Its aim is to make information more accessible, therefore facilitating scientific and social discussions on the benefits of ecosystem valuation.

http://www.ecosystemmarketplace.com

The aim of this website is to make information accessible for facilitation of discussions. This site is also a pro-monetization site and discusses the benefits of viewing ecosystem services from an economic perspective, as well as how degradation of ecosystems can have dramatic economic costs.

http://www.teebweb.org/

This site is mainly focused on biodiversity conservation and sustainable use of biodiversity. It is a website that aims to facilitate transfer and management of data with the hope that it will make people more aware of the importance of biodiversity.

http://www.conservationcommons.net/

This is the blog for The Nature Conservancy, an international organization that strives to protect "ecologically important lands and waters for nature and people." Readers of this blog get viewpoints on both sides of the ecosystem services debate, as this is a conservation blog and not one established with an economic framework in mind.

http://blog.nature.org/2009/01/value-of-nature/

ISSUE 10

Does the Market Work Better Than Government at Transitioning to Sustainability?

YES: Paul Krugman, from "Green Economics: How We Can Afford to Tackle Climate Change," *The New York Times Magazine* (April 11, 2010, p. 34)

NO: Leigh K. Fletcher, from "Green Construction Costs and Benefits: Is National Regulation Warranted?" *Natural Resources & Environmental* (vol. 24, Summer 2009, p. 18)

Learning Outcomes

After reading this issue, you should be able to:

- Understand the difference between market-based and command-and-control approaches.
- Comprehend the challenges of international climate negotiations.
- Discuss the historical development of a market-based approach.
- Describe the concept and implications of a carbon tariff.
- Discuss the debate of a national building code over a state or local building code.

ISSUE SUMMARY

YES: Noted national economist Paul Krugman provides a history of both market-based and command-and-control (regulatory) approaches in environmental economics and recommends cap-and-trade, carbon taxes, and a carbon tariff as the best market-based approaches to reduce carbon.

NO: Leigh K. Fletcher, who is LEED certified and a lawyer in Tampa, Florida, believes that building codes as a regulatory policy can reduce electricity which would significantly limit carbon since buildings are the largest contributor to electricity consumption.

This issue of which works better at achieving societal goals, government or the free market, is not new. The relationship between this argument and sustainability is interesting to pursue. One of the most important goals of sustainability is to reduce carbon and limit pollution. Which is the best approach to accomplish these goals? Will there be greater success at pursuing a policy of free-market incentives, or does it make more sense to use a more punitive approach through governmental regulations?

The role of regulation in society is a shifting pattern of continual negotiations between different stakeholders. In the case of trying to control carbon dioxide, the negotiation process is complex since air pollution does not respect international boundaries, and is therefore similar to the problem in the article, "The Tragedy of the Commons," by Garret Hardin (1968). In the Hardin article the basic premise is that while individual actors are pursuing their best self-interests, they are simultaneously jeopardizing the long-range interest of the community. While Mr. Hardin's article discusses the management of the commons, the management of carbon dioxide is similar in the sense that individual countries do not want to jeopardize their economy by raising the cost of energy. Emerging countries argue that they have the right to develop their economies since industrialized countries are responsible for the increase in carbon dioxide levels since the Industrial Revolution. Therefore, succinctly stated by Beardsley et al. (2008), "Societies form regulations through the process that seeks to reconcile the often conflicting objectives of governments and stakeholders."

There are two basic ways that free market incentives are built into public policy, either through an emissions trading approach or through a carbon tax. The emissions trading approach was built into the U.S. Air Pollution Act of 1990 and the Kyoto Protocol of 1997. In the Air Pollution Act, a cap-and-trade program was established for sulfur dioxide from coal-burning utilities. In the Kyoto Protocol, a global emissions trading program for carbon was set up by its signatories. The closest thing to a carbon tax in the United States is the tax imposed by both states and the federal government on gasoline.

The United States Environmental Protection Agency (EPA) was established over 40 years ago to protect the American environment through the regulation of standards in such areas as water and air. Congress established these standards and delegated the EPA to enforce them though the power granted to the federal government through the interstate commerce clause of the U.S. Constitution. This command-and-control approach has been largely successful through statistics that show a cleaner American environment. Yet, it did not come easy as business and industry challenged this strong governmental approach by contending that environmental regulations hindered business development and fostered job loss. Over time environmental regulations such as the Clean Air Act and the Clean Water Act were effective in producing an enhanced environment while simultaneously creating jobs. The Clean Air Act saw noted automobile executives such as Lee Iacocca state that such laws would bankrupt the industry. In hindsight, while the auto industry spent a great deal of money fighting the legislation, the result has been better air quality, and at a cost greatly below industries' initial expected cost.

Recently, the EPA has moved toward a market-based approach to regulation. In the 1990 Clean Air Act, a cap-and-trade of sulfur dioxide was permitted. In hindsight, this legislation set a precedent that a market-based approach, other than taxation, could have the desired effect of decreasing air pollution at a lower cost to utilities and users. At the same time, it provided an economic incentive for business to experiment with new technology to reduce pollution. The purpose of the 1990 Clean Air Act Amendment was to rectify the problem of acid rain, which was attributed to Midwestern power plants fueled by coal. At the time that this cap-and-trade policy tool was implemented, opposition politicians argued that governmental command-and-control measures, such as the requirement for the installation of expensive scrubbers, would produce greater success. The idea, which was popular among economists since the 1960s, was that a market-based emissions trading program, a radical departure from traditional command-and-control policy, would provide the necessary incentives to the market for a reduction of pollution. Opponents viewed cap-and-trade as a license to pollute. The legislation bill was bitterly opposed by the electric utilities of the Midwest, which argued that the legislation would have tremendous negative effects on the local and national economy, but eventually passed with bi-partisan support. The results show that compliance costs were grossly overestimated, and according to Kranish (2010), the cost was one-fourth of what the industry had estimated. The strength of the sulfur dioxide cap-and-trade program was that it allowed companies to pursue reduced emissions by implanting whatever technology was cheapest. The success of the cap-and-trade program changed the view of the capacity of the market to reduce environmental pollution, and policymakers began to look at other ways the approach could be implemented.

In Europe, the battle to control carbon dioxide has led to a cap-and-trade program. The European Union Emission Trading Scheme (ETS) regulates over 10,000 organizations that emit approximately 40 percent of the carbon dioxide in Europe. The ETS began in 2005 with three phases that gradually lowers the cap in order to meet the conditions of the Kyoto Protocol. In the first phase from 2005 to 2007 permits were freely allocated by the national governments. Establishing an appropriate price for carbon offset permits proved tricky. The price of carbon permits fluctuated greatly. A number of environmentalists and politicians were disturbed that market traders were in some instances making a profit on the trading of permits. However, the current second phase between 2008 and 2012 has seen the price of carbon permits somewhat stabilized. The importance of having a stable price is vital for industries that have long-term investment portfolios in different energy sources. A stable price allows for economists and business people to accurately conduct a financial analysis such as the net present value (NPV) of different energy investments. The result is that with a stable market and accurately priced carbon permits, the costs associated with operating a coal fired utility plant can be determined. Thus, the economic externality of pollution can be added into operating costs. Also, an even stronger positive development is that by accurately pricing the true cost of carbon, alternative energy sources such as solar, wind, and geothermal become more financially attractive. The net result is that in Europe there has been a switch from coal burning plants into cleaner burning natural gas and the furlough of coal burning plants that are only used in periods of exceptional energy demand. And finally, the free-market approach has shifted power portfolios in Europe producing a boom in the alternative energy market.

The YES selection is presented by noted economist Paul Krugman. He argues for quick action to combat climate change through either a carbon tax or cap-and-trade program. He presents a historical background on each policy tool and states that a carbon tax may be simpler to implement, but indicates that an international cap-and-trade policy would have further benefits such as trading between the United States and China. A second interesting proposal in Krugman's analysis is the call for a carbon tariff on imported goods from countries that do not regulate carbon dioxide. Krugman strongly argues that the United States should set the precedent by establishing a cap-and-trade approach to lead the world. He goes on to argue that countries that do not have a carbon trading policy in place could have a carbon tariff imposed on their goods. It would seem that a carbon tariff would be contrary to a free market, and would be protectionist, but he maintains that a carbon tariff would be legal under the World Trade Organization.

The NO selection is presented by lawyer Leigh Fletcher. Fletcher argues that regulations at the local, state, and national levels are needed in the construction industry since buildings account for 65 percent of electricity consumption (Fletcher, 2009). She believes regulations can be built into zoning and planning laws and can encourage energy conservation and sustainability by reducing global greenhouse gases. Fletcher maintains that since buildings have a long life cycle and use such large amounts of energy, that by developing stronger regulations at all governmental levels, energy demand will be significantly reduced.

Paul Krugman

Green Economics: How We Can Afford to Tackle Climate Change

If you listen to climate scientists—and despite the relentless campaign to discredit their work, you should—it is long past time to do something about emissions of carbon dioxide and other greenhouse gases. If we continue with business as usual, they say, we are facing a rise in global temperatures that will be little short of apocalyptic. And to avoid that apocalypse, we have to wean our economy from the use of fossil fuels, coal above all.

But is it possible to make drastic cuts in greenhouse-gas emissions without destroying our economy?

Like the debate over climate change itself, the debate over climate economics looks very different from the inside than it often does in popular media. The casual reader might have the impression that there are real doubts about whether emissions can be reduced without inflicting severe damage on the economy. In fact, once you filter out the noise generated by special-interest groups, you discover that there is widespread agreement among environmental economists that a market-based program to deal with the threat of climate change—one that limits carbon emissions by putting a price on them—can achieve large results at modest, though not trivial, cost. There is, however, much less agreement on how fast we should move, whether major conservation efforts should start almost immediately or be gradually increased over the course of many decades.

In what follows, I will offer a brief survey of the economics of climate change or, more precisely, the economics of lessening climate change. I'll try to lay out the areas of broad agreement as well as those that remain in major dispute. First, though, a primer in the basic economics of environmental protection.

Environmental Econ 101

If there's a single central insight in economics, it's this: There are mutual gains from transactions between consenting adults. If the going price of widgets is $10 and I buy a widget, it must be because that widget is worth more than $10 to me. If you sell a widget at that price, it must be because it costs you less than $10 to make it. So buying and selling in the widget market works to the benefit of both buyers and sellers. More than that, some careful analysis shows that if there is effective competition in the widget market, so that the price ends up matching the number of widgets people want to buy to the number of widgets other people want to sell, the outcome is to maximize the total gains to producers and consumers. Free markets are "efficient"—which, in economics-speak as opposed to plain English, means that nobody can be made better off without making someone else worse off.

Now, efficiency isn't everything. In particular, there is no reason to assume that free markets will deliver an outcome that we consider fair or just. So the case for market efficiency says nothing about whether we should have, say, some form of guaranteed health insurance, aid to the poor and so forth. But the logic of basic economics says that we should try to achieve social goals through "aftermarket" interventions. That is, we should let markets do their job, making efficient use of the nation's resources, then utilize taxes and transfers to help those whom the market passes by.

But what if a deal between consenting adults imposes costs on people who are not part of the exchange? What if you manufacture a widget and I buy it, to our mutual benefit, but the process of producing that widget involves dumping toxic sludge into other people's drinking water? When there are "negative externalities"—costs that economic actors impose on others without paying a price for their actions—any presumption that the market economy, left to its own devices, will do the right thing goes out the window. So what should we do? Environmental economics is all about answering that question.

One way to deal with negative externalities is to make rules that prohibit or at least limit behavior that imposes especially high costs on others. That's what we did in the first major wave of environmental legislation in the early 1970s: cars were required to meet emission standards for the chemicals that cause smog, factories were required to limit the volume of effluent they dumped into waterways and so on. And this approach yielded results; America's air and water became a lot cleaner in the decades that followed.

But while the direct regulation of activities that cause pollution makes sense in some cases, it is seriously defective in others, because it does not offer any scope for flexibility and creativity. Consider the biggest environmental issue of the 1980s—acid rain. Emissions of sulfur dioxide from power plants, it turned out, tend to combine with water downwind and produce flora- and wildlife-destroying sulfuric acid. In 1977, the government made its first stab at confronting the issue, recommending that all new coal-fired plants have scrubbers to remove sulfur dioxide from their emissions. Imposing a tough standard on all plants was problematic, because retrofitting some older plants would have been extremely expensive. By regulating only new plants, however, the government passed up the opportunity to achieve fairly cheap pollution control at plants that were, in fact, easy to retrofit. Short of a de facto federal takeover of the power industry, with federal officials issuing specific instructions to each plant, how was this conundrum to be resolved?

Enter Arthur Cecil Pigou, an early-20th-century British don, whose 1920 book, "The Economics of Welfare," is generally regarded as the ur-text of environmental economics.

Somewhat surprisingly, given his current status as a godfather of economically sophisticated environmentalism, Pigou didn't actually stress the problem of pollution. Rather than focusing on, say, London's famous fog (actually acrid smog, caused by millions of coal fires), he opened his discussion with an example that must have seemed twee even in 1920, a hypothetical case in which "the game-preserving activities of one occupier involve the overrunning of a neighboring occupier's land by rabbits." But never mind. What Pigou enunciated was a principle: economic activities that impose unrequited costs on other people should not always be banned, but they should be discouraged. And the right way to curb an activity, in most cases, is to put a price on it. So Pigou proposed that people who generate negative externalities should have to pay a fee reflecting the costs they impose on others—what has come to be known as a Pigovian tax. The simplest version of a Pigovian tax is an effluent fee: anyone who dumps pollutants into a river, or emits them into the air, must pay a sum proportional to the amount dumped.

Pigou's analysis lay mostly fallow for almost half a century, as economists spent their time grappling with issues that seemed more pressing, like the Great Depression. But with the rise of environmental regulation, economists dusted off Pigou and began pressing for a "market-based" approach that gives the private sector an incentive, via prices, to limit pollution, as opposed to a "command and control" fix that issues specific instructions in the form of regulations.

The initial reaction by many environmental activists to this idea was hostile, largely on moral grounds. Pollution, they felt, should be treated like a crime rather than something you have the right to do as long as you pay enough money. Moral concerns aside, there was also considerable skepticism about whether market incentives would actually be successful in reducing pollution. Even today, Pigovian taxes as originally envisaged are relatively rare. The most successful example I've been able to find is a Dutch tax on discharges of water containing organic materials.

What has caught on instead is a variant that most economists consider more or less equivalent: a system of tradable emissions permits, aka cap and trade. In this model, a limited number of licenses to emit a specified pollutant, like sulfur dioxide, are issued. A business that wants to create more pollution than it is licensed for can go out and buy additional licenses from other parties; a firm that has more licenses than it intends to use can sell its surplus. This gives everyone an incentive to reduce pollution, because buyers would not have to acquire as many licenses if they can cut back on their emissions, and sellers can unload more licenses if they do the same. In fact, economically, a cap-and-trade system produces the same incentives to reduce pollution as a Pigovian tax, with the price of licenses effectively serving as a tax on pollution.

In practice there are a couple of important differences between cap and trade and a pollution tax. One is that the two systems produce different types of uncertainty. If the government imposes a pollution tax, polluters know what price they will have to pay, but the government does not know how much pollution they will generate. If the government imposes a cap, it knows the amount of pollution, but polluters do not know what the price of emissions will be. Another important difference has to do with government revenue. A pollution tax is, well, a tax, which imposes costs on the private sector while generating revenue for the government. Cap and trade is a bit more complicated. If the government simply auctions off licenses and collects the revenue, then it is just like a tax. Cap and trade, however, often involves handing out licenses to existing players, so the potential revenue goes to industry instead of the government.

Politically speaking, doling out licenses to industry isn't entirely bad, because it offers a way to partly compensate some of the groups whose interests would suffer if a serious climate-change policy were adopted. This can make passing legislation more feasible.

These political considerations probably explain why the solution to the acid-rain predicament took the form of cap and trade and why licenses to pollute were distributed free to power companies. It's also worth noting that the Waxman-Markey bill, a cap-and-trade setup for greenhouse gases that starts by giving out many licenses to industry but puts up a growing number for auction in later years, was actually passed by the House of Representatives last year; it's hard to imagine a broad-based emissions tax doing the same for many years.

That's not to say that emission taxes are a complete nonstarter. Some senators have recently floated a proposal for a sort of hybrid solution, with cap and trade for some parts of the economy and carbon taxes for others—mainly oil and gas. The political logic seems to be that the oil industry thinks consumers won't blame it for higher gas prices if those prices reflect an explicit tax.

In any case, experience suggests that market-based emission controls work. Our recent history with acid rain shows as much. The Clean Air Act of 1990 introduced a cap-and-trade system in which power plants could buy and sell the right to emit sulfur dioxide, leaving it up to individual companies to manage their own business within the new limits. Sure enough, over time sulfur-dioxide emissions from power plants were cut almost in half, at a much lower cost than even optimists expected; electricity prices fell instead of rising. Acid rain did not disappear as a problem, but it was significantly mitigated. The results, it would seem, demonstrated that we can deal with environmental problems when we have to.

So there we have it, right? The emission of carbon dioxide and other greenhouse gases is a classic negative externality—the "biggest market failure the world has ever seen," in the words of Nicholas Stern, the author of a report on the subject for the British government. Textbook economics and real-world experience tell us that we should have policies to discourage activities that generate negative externalities and that it is generally best to rely on a market-based approach.

Climate of Doubt?

This is an article on climate economics, not climate science. But before we get to the economics, it's worth establishing three things about the state of the scientific debate.

The first is that the planet is indeed warming. Weather fluctuates, and as a consequence it's easy enough to point to an unusually warm year in the recent past, note that it's cooler now and claim, "See, the planet is getting cooler, not warmer!" But if you look at the evidence the right way—taking averages over periods long enough to smooth out the fluctuations—the upward trend is unmistakable: each successive decade since the 1970s has been warmer than the one before.

Second, climate models predicted this well in advance, even getting the magnitude of the temperature rise roughly right. While it's relatively easy to cook up an analysis that matches known data, it is much harder to create a model that accurately forecasts the future. So the fact that climate modelers more than 20 years ago successfully predicted the subsequent global warming gives them enormous credibility.

Yet that's not the conclusion you might draw from the many media reports that have focused on matters like hacked e-mail and climate scientists' talking about a "trick" to "hide" an anomalous decline in one data series or expressing their wish to see papers by climate skeptics kept out of research reviews. The truth, however, is that the supposed scandals evaporate on closer examination, revealing only that climate researchers are human beings, too. Yes, scientists try to make their results stand out, but no data were suppressed. Yes, scientists dislike it when work that they think deliberately obfuscates the issues gets published. What else is new? Nothing suggests that there should not continue to be strong support for climate research.

And this brings me to my third point: models based on this research indicate that if we continue adding greenhouse gases to the atmosphere as we have, we will eventually face drastic changes in the climate. Let's be clear. We're not talking about a few more hot days in the summer and a bit less snow in the winter; we're talking about massively disruptive events, like the transformation of the Southwestern United States into a permanent dust bowl over the next few decades.

Now, despite the high credibility of climate modelers, there is still tremendous uncertainty in their long-term forecasts. But as we will see shortly, uncertainty makes the case for action stronger, not weaker. So climate change demands action. Is a cap-and-trade program along the lines of the model used to reduce sulfur dioxide the right way to go?

Serious opposition to cap and trade generally comes in two forms; an argument that more direct action—in particular, a ban on coal-fired power plants—would be more effective and an argument that an emissions tax would be better than emissions trading. (Let's leave aside those who dismiss climate science altogether and oppose any limits on greenhouse-gas emissions, as well as those who oppose the use of any kind of market-based remedy.) There's something to each of these positions, just not as much as their proponents think.

When it comes to direct action, you can make the case that economists love markets not wisely but too well, that they are too ready to assume that changing people's financial incentives fixes every problem. In particular, you can't put a price on something unless you can measure it accurately, and that can be both difficult and expensive. So sometimes it's better simply to lay down some basic rules about what people can and cannot do.

Consider auto emissions, for example. Could we or should we charge each car owner a fee proportional to the emissions from his or her tailpipe? Surely not. You would have to install expensive monitoring equipment on every car, and you would also have to worry about fraud. It's almost certainly better to do what we actually do, which is impose emissions standards on all cars.

Is there a comparable argument to be made for greenhouse-gas emissions? My initial reaction, which I suspect most economists would share, is that the very scale and complexity of the situation requires a market-based solution, whether cap and trade or an emissions tax. After all, greenhouse gases are a direct or indirect byproduct of almost everything produced in a modern economy, from the houses we live in to the cars we drive. Reducing emissions of those gases will require getting people to change their behavior in many different ways, some of them impossible to identify until we have a much better grasp of green technology. So can we really make meaningful progress by telling people specifically what will or will not be permitted? Econ 101 tells us—probably correctly—that the only way to get people to change their behavior appropriately is to put a price on emissions so this cost in turn gets incorporated into everything else in a way that reflects ultimate environmental impacts.

When shoppers go to the grocery store, for example, they will find that fruits and vegetables from farther away have higher prices than local produce, reflecting in part the cost of emission licenses or taxes paid to ship that produce. When businesses decide how much to spend on insulation, they will take into account the costs of heating and air-conditioning that include the price of emissions licenses or taxes for electricity generation. When electric utilities have to choose among energy sources, they will have to take into account the higher license fees or taxes associated with fossil-fuel consumption. And so on down the line. A market-based system would create decentralized incentives to do the right thing, and that's the only way it can be done.

That said, some specific rules may be required. James Hansen, the renowned climate scientist who deserves much of the credit for making global warming an issue in the first place, has argued forcefully that most of the climate-change problem comes down to just one thing, burning coal, and that whatever else we do, we have to shut down coal burning over the next couple decades. My economist's reaction is that a stiff license fee would strongly discourage coal use anyway. But a market-based system might turn out to have loopholes—and their consequences could be dire. So I would advocate supplementing market-based disincentives with direct controls on coal burning.

What about the case for an emissions tax rather than cap and trade? There's no question that a straightforward tax would have many advantages over legislation like Waxman-Markey, which is full of exceptions and special situations. But that's not really a useful comparison: of course an idealized emissions tax looks better than a cap-and-trade system that has already passed the House with all its attendant compromises. The question is whether the emissions tax that could actually be put in place is better than cap and trade. There is no reason to believe that it would be—indeed, there is no reason to believe that a broad-based emissions tax would make it through Congress.

To be fair, Hansen has made an interesting moral argument against cap and trade, one that's much more sophisticated than the old view that it's wrong to let polluters buy the right to pollute. What Hansen draws attention to is the fact that in a cap-and-trade world, acts of individual virtue do not contribute to social goals. If you choose to drive a hybrid car or buy a house with a small carbon footprint, all you are doing is freeing up emissions permits for someone else, which means that you have done nothing to reduce the threat of climate change. He has a point. But altruism cannot effectively deal with climate change. Any serious solution must rely mainly on creating a system that gives everyone a self-interested reason to produce fewer emissions. It's a shame, but climate altruism must take a back seat to the task of getting such a system in place.

The bottom line, then, is that while climate change may be a vastly bigger problem than acid rain, the logic of how to respond to it is much the same. What we need are market incentives for reducing greenhouse-gas emissions—along with some direct controls over coal use—and cap and trade is a reasonable way to create those incentives.

But can we afford to do that? Equally important, can we afford not to?

The Cost of Action

Just as there is a rough consensus among climate modelers about the likely trajectory of temperatures if we do not act to cut the emissions of greenhouse gases, there is a rough consensus among economic modelers about the costs of action. That general opinion may be summed up as follows: Restricting emissions would slow economic growth—but not by much. The Congressional Budget Office, relying on a survey of models, has concluded that Waxman-Markey "would reduce the projected average annual rate of growth of gross domestic product between 2010 and 2050 by 0.03 to 0.09 percentage points." That is, it would trim average annual growth to 2.31 percent, at worst, from 2.4 percent. Over all, the Budget Office concludes, strong climate-change policy would leave the American economy between 1.1 percent and 3.4 percent smaller in 2050 than it would be otherwise.

And what about the world economy? In general, modelers tend to find that climate-change policies would lower global output by a somewhat smaller percentage than the comparable figures for the United States. The main reason is that emerging economies like China currently use energy fairly inefficiently, partly as a result of national policies that have kept the prices of fossil fuels very low, and could thus achieve large energy savings at a modest cost. One recent review of the available estimates put the costs of a very strong climate policy—substantially more aggressive than contemplated in current legislative proposals—at between 1 and 3 percent of gross world product.

Such figures typically come from a model that combines all sorts of engineering and marketplace estimates. These will include, for instance, engineers' best calculations of how much it costs to generate electricity in various ways, from coal, gas and nuclear and solar power at given resource prices. Then estimates will be made, based on historical experience, of how much consumers would cut back their electricity consumption if its price rises. The same process is followed for other kinds of energy, like motor fuel. And the model assumes that everyone makes the best choice given the economic environment—that power generators choose the least expensive means of producing electricity, while consumers conserve energy as long as the money saved by buying less electricity exceeds the cost of using less power in the form either of other spending or loss of convenience. After all this analysis, it's possible to predict how producers and consumers of energy will react to policies that put a price on emissions and how much those reactions will end up costing the economy as a whole.

There are, of course, a number of ways this kind of modeling could be wrong. Many of the underlying estimates are necessarily somewhat speculative; nobody really knows, for instance, what solar power will cost once it finally becomes a large-scale proposition. There is also reason to doubt the assumption that people actually make the right choices: many studies have found that consumers fail to take measures to conserve energy, like improving insulation, even when they could save money by doing so.

But while it's unlikely that these models get everything right, it's a good bet that they overstate rather than understate the economic costs of climate-change action. That is what the experience from the cap-and-trade program for acid rain suggests: costs came in well below initial predictions. And in general, what the models do not and cannot take into account is creativity; surely, faced with an economy in which there are big monetary payoffs for reducing greenhouse-gas emissions, the private sector will come up with ways to limit emissions that are not yet in any model.

What you hear from conservative opponents of a climate-change policy, however, is that any attempt to limit emissions would be economically devastating. The Heritage Foundation, for one, responded to Budget Office estimates on Waxman-Markey with a broadside titled, "C.B.O. Grossly Underestimates Costs of Cap and Trade." The real effects, the foundation said, would be ruinous for families and job creation.

This reaction—this extreme pessimism about the economy's ability to live with cap and trade—is very much at odds with typical conservative rhetoric. After all, modern conservatives express a deep, almost mystical confidence in the effectiveness of market incentives—Ronald Reagan liked to talk about the "magic of the marketplace." They believe that the capitalist system can deal with all kinds of limitations, that technology, say, can easily overcome any constraints on growth posed by limited reserves of oil or other natural resources. And yet now they submit that this same private sector is utterly incapable of coping with a limit on overall emissions, even though such a cap would, from the private sector's point of view, operate very much like a limited supply of a resource, like land. Why don't they believe that the dynamism of capitalism will spur it to find ways to make do in a world of reduced carbon emissions? Why do they think the marketplace loses its magic as soon as market incentives are invoked in favor of conservation?

Clearly, conservatives abandon all faith in the ability of markets to cope with climate-change policy because they don't want government intervention. Their stated pessimism about the cost of climate policy is essentially a political ploy rather than a reasoned economic judgment. The giveaway is the strong tendency of conservative opponents of cap and trade to argue in bad faith. That Heritage Foundation broadside accuses the Congressional Budget Office of making elementary logical errors, but if you actually read the office's report, it's clear that the foundation is willfully misreading it. Conservative politicians have been even more shameless. The National Republican Congressional Committee, for example, issued multiple press releases specifically citing a study from M.I.T. as the basis for a claim that cap and trade would cost $3,100 per household, despite repeated attempts by the study's authors to get out the word that the actual number was only about a quarter as much.

The truth is that there is no credible research suggesting that taking strong action on climate change is beyond the economy's capacity. Even if you do not fully trust the models—and you shouldn't—history and logic both suggest that the models are overestimating, not underestimating, the costs of climate action. We can afford to do something about climate change.

But that's not the same as saying we should. Action will have costs, and these must be compared with the costs of not acting. Before I get to that, however, let me touch on an issue that will become central if we actually do get moving on climate policy: how to get the rest of the world to go along with us.

The China Syndrome

The United States is still the world's largest economy, which makes the country one of the world's largest sources of greenhouse gases. But it's not the largest. China, which burns much more coal per dollar of gross domestic product than the United States does, overtook us by that measure around three years ago. Over all, the advanced countries—the rich man's club comprising Europe, North America and Japan—account for only about half of greenhouse emissions, and that's a fraction that will fall over time. In short, there can't be a solution to climate change unless the rest of the world, emerging economies in particular, participates in a major way.

Inevitably those who resist tackling climate change point to the global nature of emissions as a reason not to act. Emissions limits in America won't accomplish much, they argue, if China and others don't match our effort. And they highlight China's obduracy in the Copenhagen negotiations as evidence that other countries will not cooperate. Indeed, emerging economies feel that they have a right to emit freely without worrying about the consequences—that's what today's rich countries got to do for two centuries. It's just not possible to get global cooperation on climate change, goes the argument, and that means there is no point in taking any action at all.

For those who think that taking action is essential, the right question is how to persuade China and other emerging nations to participate in emissions limits. Carrots, or positive inducements, are one answer. Imagine setting up cap-and-trade systems in China and the United States—but allow international trading in permits, so Chinese and American companies can trade emission rights. By setting overall caps at levels designed to ensure that China sells us a substantial number of permits, we would in effect be paying China to cut its emissions. Since the evidence suggests that the cost of cutting emissions would be lower in China than in the United States, this could be a good deal for everyone.

But what if the Chinese (or the Indians or the Brazilians, etc.) do not want to participate in such a system? Then you need sticks as well as carrots. In particular, you need carbon tariffs.

A carbon tariff would be a tax levied on imported goods proportional to the carbon emitted in the manufacture of those goods. Suppose that China refuses to reduce emissions, while the United States adopts policies that set a price of $100 per ton of carbon emissions. If the United States were to impose such a carbon tariff, any shipment to America of Chinese goods whose production involved emitting a ton of carbon would result in a $100 tax over and above any other duties. Such tariffs, if levied by major players—probably the United States and the European Union—would give noncooperating countries a strong incentive to reconsider their positions.

To the objection that such a policy would be protectionist, a violation of the principles of free trade, one reply is, So? Keeping world markets open is important, but avoiding planetary catastrophe is a lot more important. In any case, however, you can argue that carbon tariffs are well within the rules of normal trade relations. As long as the tariff imposed on the carbon content of imports is comparable to the cost of domestic carbon licenses, the effect is to charge your own consumers a price that reflects the carbon emitted in what they buy, no matter where it is produced. That should be legal under international-trading rules. In fact, even the World Trade Organization, which is charged with policing trade policies, has published a study suggesting that carbon tariffs would pass muster.

Needless to say, the actual business of getting cooperative, worldwide action on climate change would be much more complicated and tendentious than this discussion suggests. Yet the problem is not as intractable as you often hear. If the United States and Europe decide to move on climate policy, they almost certainly would be able to cajole and chivvy the rest of the world into joining the effort. We can do this.

The Costs of Inaction

In public discussion, the climate-change skeptics have clearly been gaining ground over the past couple of years, even though the odds have been looking good lately that 2010 could be the warmest year on record. But climate modelers themselves have grown increasingly pessimistic. What were previously worst-case scenarios have become base-line projections, with a number of organizations doubling their predictions for temperature rise over the course of the 21st century. Underlying this new pessimism is increased concern about feedback effects—for example, the release of methane, a significant greenhouse gas, from seabeds and tundra as the planet warms.

At this point, the projections of climate change, assuming we continue business as usual, cluster around an estimate that average temperatures will be about 9 degrees Fahrenheit higher in 2100 than they were in 2000. That's a lot—equivalent to the difference in average temperatures between New York and central Mississippi. Such a huge change would have to be highly disruptive. And the troubles would not stop there: temperatures would continue to rise.

Furthermore, changes in average temperature will by no means be the whole story. Precipitation patterns will change, with some regions getting much wetter and others much drier. Many modelers also predict more intense storms. Sea levels would rise, with the impact intensified by those storms: coastal flooding, already a major source of natural disasters, would become much more frequent and severe. And there might be drastic changes in the climate of some regions as ocean currents shift. It's always worth bearing in mind that London is at the same latitude as Labrador; without the Gulf Stream, Western Europe would be barely habitable.

While there may be some benefits from a warmer climate, it seems almost certain that upheaval on this scale would make the United States, and the world as a whole, poorer than it would be otherwise. How much poorer? If ours were a preindustrial, primarily agricultural society, extreme climate change would be obviously catastrophic. But we have an advanced economy, the kind that has historically shown great ability to adapt to changed circumstances. If this sounds similar to my argument that the costs of emissions limits would be tolerable, it ought to: the same flexibility that should enable us to deal with a much higher carbon prices should also help us cope with a somewhat higher average temperature.

But there are at least two reasons to take sanguine assessments of the consequences of climate change with a grain of salt. One is that, as I have just pointed out, it's not just a matter of having warmer weather—many of the costs of climate change are likely to result from droughts, flooding and severe storms. The other is that while modern economies may be highly adaptable, the same may not be true of ecosystems. The last time the earth experienced warming at anything like the pace we now expect was during the Paleocene-Eocene Thermal Maximum, about 55 million years ago, when temperatures rose by about 11 degrees Fahrenheit over the course of around 20,000 years (which is a much slower rate than the current pace of warming). That increase was associated with mass extinctions, which, to put it mildly, probably would not be good for living standards.

So how can we put a price tag on the effects of global warming? The most widely quoted estimates, like those in the Dynamic Integrated Model of Climate and the Economy, known as DICE, used by Yale's William Nordhaus and colleagues, depend upon educated guesswork to place a value on the negative effects of global warming in a number of crucial areas, especially agriculture and coastal protection, then try to make some allowance for other possible repercussions. Nordhaus has argued that a global temperature rise of 4.5 degrees Fahrenheit—which used to be the consensus projection for 2100—would reduce gross world product by a bit less than 2 percent. But what would happen if, as a growing number of models suggest, the actual temperature rise is twice as great? Nobody really knows how to make that extrapolation. For what it's worth, Nordhaus's model puts losses from a rise of 9 degrees at about 5 percent of gross world product. Many critics have argued, however, that the cost might be much higher.

Despite the uncertainty, it's tempting to make a direct comparison between the estimated losses and the estimates of what the mitigation policies will cost: climate change will lower gross world product by 5 percent, stopping it will cost 2 percent, so let's go ahead. Unfortunately the reckoning is not that simple for at least four reasons.

First, substantial global warming is already "baked in," as a result of past emissions and because even with a strong climate-change policy the amount of carbon dioxide in the atmosphere is most likely to continue rising for many years. So even if the nations of the world *do manage* to take on climate change, we will still have to pay for earlier inaction. As a result, Nordhaus's loss estimates may overstate the gains from action.

Second, the economic costs from emissions limits would start as soon as the policy went into effect and under most proposals would become substantial within around 20 years. If we don't act, meanwhile, the big costs would probably come late this century (although some things, like the transformation of the American Southwest into a dust bowl, might come much sooner). So how you compare those costs depends on how much you value costs in the distant future relative to costs that materialize much sooner.

Third, and cutting in the opposite direction, if we don't take action, global warming won't stop in 2100: temperatures, and losses, will continue to rise. So if you place a significant weight on the really, really distant future, the case for action is stronger than even the 2100 estimates suggest.

Finally and most important is the matter of uncertainty. We're uncertain about the magnitude of climate change, which is inevitable, because we're talking about reaching levels of carbon dioxide in the atmosphere not seen in millions of years. The recent doubling of many modelers' predictions for 2100 is itself an illustration of the scope of that uncertainty; who knows what revisions may occur in the years ahead. Beyond that, nobody really knows how much damage would result from temperature rises of the kind now considered likely.

You might think that this uncertainty weakens the case for action, but it actually strengthens it. As Harvard's Martin Weitzman has argued in several influential papers, if there is a significant chance of utter catastrophe, that chance—rather than what is most likely to happen—should dominate cost-benefit calculations. And utter catastrophe does look like a realistic possibility, even if it is not the most likely outcome.

Weitzman argues—and I agree—that this risk of catastrophe, rather than the details of cost-benefit calculations, makes the most powerful case for strong climate policy. Current projections of global warming in the absence of action are just too close to the kinds of numbers associated with doomsday scenarios. It would be irresponsible—it's tempting to say criminally irresponsible—not to step back from what could all too easily turn out to be the edge of a cliff.

Still that leaves a big debate about the pace of action.

The Ramp versus the Big Bang

Economists who analyze climate policies agree on some key issues. There is a broad consensus that we need to put a price on carbon emissions, that this price must eventually be very high but that the negative economic effects from this policy will be of manageable size. In other words, we can and should act to limit climate change. But there is a ferocious debate among knowledgeable analysts about timing, about how fast carbon prices should rise to significant levels.

On one side are economists who have been working for many years on so-called integrated-assessment models, which combine models of climate change with models of both the damage from global warming and the costs of cutting emissions. For the most part, the message from these economists is a sort of climate version of St. Augustine's famous prayer, "Give me chastity and continence, but not just now." Thus Nordhaus's DICE model says that the price of carbon emissions should eventually rise to more than $200 a ton, effectively more than quadrupling the cost of coal, but that most of that increase should come late this century, with a much more modest initial fee of around $30 a ton. Nordhaus calls this recommendation for a policy that builds gradually over a long period the "climate-policy ramp."

On the other side are some more recent entrants to the field, who work with similar models but come to different conclusions. Most famously, Nicholas Stern, an economist at the London School of Economics, argued in 2006 for quick, aggressive action to limit emissions, which would most likely imply much higher carbon prices. This alternative position doesn't appear to have a standard name, so let me call it the "climate-policy big bang."

I find it easiest to make sense of the arguments by thinking of policies to reduce carbon emissions as a sort of public investment project: you pay a price now and derive benefits in the form of a less-damaged planet later. And by later, I mean much later; today's emissions will affect the amount of carbon in the atmosphere decades, and possibly centuries, into the future. So if you want to assess whether a given investment in emissions reduction is worth making, you have to estimate the damage that an additional ton of carbon in the atmosphere will do, not just this year but for a century or more to come; and you also have to decide how much weight to place on harm that will take a very long time to materialize.

The policy-ramp advocates argue that the damage done by an additional ton of carbon in the atmosphere is fairly low at current concentrations; the cost will not get really large until there is a lot more carbon dioxide in the air, and that won't happen until late this century. And they argue that costs that far in the future should not have a large influence on policy today. They point to market rates of return, which indicate that investors place only a small weight on the gains or losses they expect in the distant future, and argue that public policies, including climate policies, should do the same.

The big-bang advocates argue that government should take a much longer view than private investors. Stern, in particular, argues that policy makers should give the same weight to future generations' welfare as we give to those now living. Moreover, the proponents of fast action hold that the damage from emissions may be much larger than the policy-ramp analyses suggest, either because global temperatures are more sensitive to greenhouse-gas emissions than previously thought or because the economic damage from a large rise in temperatures is much greater than the guesstimates in the climate-ramp models.

As a professional economist, I find this debate painful. There are smart, well-intentioned people on both sides—some of them, as it happens, old friends and mentors of mine—and each side has scored some major points. Unfortunately, we can't just declare it an honorable draw, because there's a decision to be made.

Personally, I lean toward the big-bang view. Stern's moral argument for loving unborn generations as we love ourselves may be too strong, but there's a compelling case to be made that public policy should take a much longer view than private markets. Even more important, the policy-ramp prescriptions seem far too much like conducting a very risky experiment with the whole planet. Nordhaus's preferred policy, for example, would stabilize the concentration of carbon dioxide in the atmosphere at a level about twice its preindustrial average. In his model, this would have only modest effects on global welfare; but how confident can we be of that? How sure are we that this kind of change in the environment would not lead to catastrophe? Not sure enough, I'd say, particularly because, as noted above, climate modelers have sharply raised their estimates of future warming in just the last couple of years.

So what I end up with is basically Martin Weitzman's argument: it's the nonnegligible probability of utter disaster that should dominate our policy analysis. And that argues for aggressive moves to curb emissions, soon.

The Political Atmosphere

As I've mentioned, the House has already passed Waxman-Markey, a fairly strong bill aimed at reducing greenhouse-gas emissions. It's not as strong as what the big-bang advocates propose, but it appears to move faster than the policy-ramp proposals. But the vote on Waxman-Markey, which was taken last June, revealed a starkly divided Congress. Only 8 Republicans voted in favor of it, while 44 Democrats voted against. And the odds are that it would not pass if it were brought up for a vote today.

Prospects in the Senate, where it takes 60 votes to get most legislation through, are even worse. A number of Democratic senators, representing energy-producing and agricultural states, have come out against cap and trade (modern American agriculture is strongly energy-intensive). In the past, some Republican senators have supported cap and trade. But with partisanship on the rise, most of them have been changing their tune. The most striking about-face has come from John McCain, who played a leading role in promoting cap and trade, introducing a bill broadly similar to Waxman-Markey in 2003. Today McCain lambastes the whole idea as "cap and tax," to the dismay of former aides.

Oh, and a snowy winter on the East Coast of the U.S. has given climate skeptics a field day, even though globally this has been one of the warmest winters on record.

So the immediate prospects for climate action do not look promising, despite an ongoing effort by three senators—John Kerry, Joseph Lieberman and Lindsey Graham—to come up with a compromise proposal. (They plan to introduce legislation later this month.) Yet the issue isn't going away. There's a pretty good chance that the record temperatures the world outside Washington has seen so far this year will continue, depriving climate skeptics of one of their main talking points. And in a more general sense, given the twists and turns of American politics in recent years—since 2005 the conventional wisdom has gone from permanent Republican domination to permanent Democratic domination to God knows what—there has to be a real chance that political support for action on climate change will revive.

If it does, the economic analysis will be ready. We know how to limit greenhouse-gas emissions. We have a good sense of the costs—and they're manageable. All we need now is the political will.

Leigh K. Fletcher

Green Construction Costs and Benefits: Is National Regulation Warranted?

Buildings have a significant impact on the environment. According to the U.S. Green Building Council (USGBC), in the United States alone, buildings account for 65 percent of electricity consumption; 36 percent of energy use; 30 percent of greenhouse gas (GHG) emissions; 30 percent of raw materials use; 30 percent of waste output (136 million tons annually); and 12 percent of potable water consumption. These environmental impacts result in short-term expenditures (utility costs, materials, and construction costs and disposal fees) as well as long-term costs. Although there is a not consensus on dollar amounts, there is general recognition that the long-term costs of global warming will be astronomical.

With growing public awareness of climate change issues and an absent national climate change policy, several federal agencies and state and local governments have adopted policies and ordinances that encourage and, in some instances, mandate green building for certain commercial and residential buildings to reduce GHG emissions and fight global warming. Although there is no single, standardized definition of green building, the U.S. Environmental Protection Agency (EPA) holds that "(g)reen building is the practice of creating structures and using processes that are environmentally responsible and resource-efficient throughout a building's life-cycle from siting to design, construction, operation, maintenance, renovation and deconstruction" www.epa.gov/greenbuilding/pubs/about.htm. Proponents of green building argue that it is a powerful tool for combating climate change because buildings constitute a large segment of the U.S. carbon footprint. Thus, in the long term, green building makes financial sense because reducing GHG emissions will reduce the potential for global warming and its associated costs. However, in the short term, green building increases the costs of construction and sometimes entails paying for certification of "greenness." Thus, although the green building industry is poised to expand in the United States, the immediate financial barriers to green building must be addressed in order for that expansion to occur. State and local governments and regulatory agencies have attempted to offset some of these short-term costs by utilizing incentives adopted as part of regulatory programs. This article surveys the regulatory systems in place now and suggests that implementing changes to the way green buildings are regulated could better offset the increased expense of building green, thereby lowering the long-term costs associated with global warming.

In the broadest terms, already adopted green building programs fall into three categories: requirements that public construction projects be green; incentive programs to encourage voluntary private green development; or mandatory programs requiring green construction for certain types of public and private construction projects. Most often, green building programs have focused on commercial buildings. Recently, however, some local governments have included residential projects within the scope of green building programs. In light of the country's economic distress, the growing recognition of the need to address climate change, and the Obama administration's commitment to making the United States a leader in climate change policy, it is important to consider whether state, local, or federal regulation will be the most effective means of implementing green building policies in the future. This article reviews the current lack of national consensus on green building standards and analyzes the current costs and benefits of green building to identify what issues need to be addressed legislatively to promote green building as part of a broader climate change policy. It then surveys the variety of existing regulations in the United States and analyzes the legal framework in which green building legislation exists to explain why the most effective means of expanding green building initiatives in the United States will probably require a multifaceted legislative approach, including establishment of national standards for the key metrics against which green building performance is measured, federally enforced mandates for reductions in GHGs, as well as continued state and local government regulation and incentivization of green building through modification of state and local taxing, land use, zoning, and building regulations. If adopted, these changes to the regulatory system would balance the financial playing field and, by lowering the effective cost of green building today, would promote wider use of green construction now, ultimately lowering the costs associated with global warming in the future.

Several competing rating systems exist for identifying a building as green. The Leadership in Energy and Environmental Design (LEED) program developed by the nonprofit, nongovernmental USGBC is the most well known of these rating systems. The LEED program evaluates the sustainable features of new commercial construction by giving points in six areas: (1) location and siting, (2) water efficiency, (3) energy and atmosphere, (4) materials and resources, (5) indoor environmental quality, and (6) innovation and design. Point ranges determine how "green" a building is. LEED has four certification levels: certified, silver, gold, and platinum. Higher points equal a higher certification level. The certification process includes registering a project with the USGBC, documenting the use of equipment and materials that have been established by the USGBC as "green," and paying fees for review. Examples of the types of systems and materials that are documented in a LEED review for a commercial building include plumbing, insulation, and heating and cooling systems. LEED has promulgated separate rating and certifications systems for renovation projects, commercial interiors, and residential construction. The Green Building Initiative has developed the Green Globes certification to qualify a building as green. Green Globes evaluates a building's use of energy, the indoor environment, emissions and effluents, resources, environmental management, and water use. Earthcraft certifies homes based on criteria related to site planning, energy-efficient building techniques and equipment, resource-efficient design and materials, waste management, indoor air quality, water conservation, and homebuyer education. The federal ENERGY STAR program is a partnership between the U.S. Department of Energy and EPA that rates appliances and buildings for energy efficiency and is intended to assist commercial businesses and new home buyers with making energy-efficient choices. According to the program's Web site, to earn the ENERGY STAR, a home must meet strict guidelines for energy efficiency set by EPA. These homes are at least 15 percent more energy efficient than homes built to the 2004 International Residential Code (IRC) and include additional energy-saving features that typically make them 20–30 percent more efficient than standard homes. In addition to the certifications of LEED, Green Globes, Earthcraft, and ENERGY STAR, several states and local governments, including Minnesota and North Carolina, have created their own standards for green buildings. Some are based on LEED, and others are based on the jurisdiction's building code.

For residential construction, the National Association of Homebuilders and American National Standards Institute (ANSI) promulgated the National Green Building Standard, a residential green construction guideline that competes with the LEED residential certification program. On January 29, 2009, the International Code Council (ICC) adopted the National Green Building Standard, known as ICC-700, as an American National Standard Code. According to the ICC, the new Standard provides guidance for safe and sustainable building practices for residential construction, including both new and renovated single-family to high-rise residential buildings. This is the first and only green standard that is consistent and coordinated with the ICOs family of I-Codes and standards, which is significant because many jurisdictions require compliance with specific ICC building codes in their regulations. Additionally, numerous smaller state and regional organizations have promulgated green building guides for new home construction.

Although each certification and guideline is unique, all of these competing definitions of what constitutes green construction are premised on the concept that by increasing energy efficiency, water conservation, use of recycled materials and improving air quality in buildings, a building becomes green. However, the presence of multiple competing standards has created challenges for the building industry, particularly for developers who are working in multiple jurisdictions with different requirements for certifying green projects. Some commentators have questioned whether these measures are a true measure of greenness, particularly when, for example, under the LEED new home certification process, single-family homes greater than 12,000 square feet are applying for and obtaining green certification on the basis of technological improvements, such as solar pool heaters and other more mundane improvements. These commentators have argued that true "green construction" means less construction overall, reuse of existing space, smaller buildings, multifamily residential projects, urban infill, and location of projects to promote pedestrian access and public transit. It should be noted that the LEED certification process does consider and award points for many of these criteria; however, none of them are prerequisites to obtaining certification, and there are no size limits for single-family homes qualifying for certification under the LEED system. The lack of consensus on what counts as green has been an impediment to the implementation of green building programs nationwide. In at least two states, legislation supporting green building was derailed as a result of a debate over which rating system or standard to require.

Costs and Benefits of Green Building

Advocates for green building argue that in addition to the global environmental benefits of green building relating to reducing GHG emissions, green building also makes economic sense. The green features that are good for the environment are also good for a building owner's budget. Increased energy efficiency and water conservation result in lower utility bills for the completed building. Increased air quality results in fewer sick days and measurable increases in worker productivity. USGBC data supports these assertions. Recently, another study analyzed CoStar real estate data for the U.S. Class A office space market to compare ENERGY STAR-rated commercial buildings to non-ENERGY STAR-rated buildings. The results indicated that during the years 2004–07, ENERGY STAR buildings in the CoStar database had higher occupancy rates, higher direct rental rates, higher sales price per square foot, and, after 2005, lower cap rates than non-ENERGY STAR buildings. However, the CoStar analysis also noted that in the buildings surveyed, going green resulted in an extra 3–6 percent in construction costs, with higher costs associated with higher levels of LEED certification and varying by region.

Green building is considered a growing sector within the construction industry. The McGraw-Hill "Construction SmartMarket Report on Commercial and Industrial Green Building, Green Trends Driving Market Change" reports that green building projects now constitute more than 5 percent of the construction market and are projected to be 20–25 percent of commercial and institutional building by value or an approximate marketplace of $56–60 billion dollars within five years. In addition, the McGraw-Hill report notes that the market share of the education market is projected to grow to 30 percent within the next five years. Likewise, the "Turner Construction Company 2008 Green Building Market Barometer" reports that 75 percent of the real-estate executives surveyed said that recent developments in the credit markets would not make their companies less likely to construct green buildings. The National Association of Homebuilders reports that there is growing interest in residential green building from consumers as well as greater interest in remodeling homes for greater energy efficiency.

The financial benefits of green building must be considered, especially in light of an incremental increase in construction costs for green commercial buildings, at least on the commercial side, as noted by the CoStar study. LEED-certified green buildings are a minimum of 1–2 percent more expensive to construct than traditional buildings. Another important consideration in determining the true costs and benefits of green buildings is who reaps the benefits. Many of the touted benefits, such as lower energy costs and higher worker productivity, are reaped by the building users or tenants as opposed to the owners unless leases are structured to avoid that result.

Other variable factors in a cost-benefit analysis for green building include the changing costs of green construction, the administrative and regulatory costs associated with green building programs, and the cost of global warming. Over the last ten years, costs of "green" equipment and building components have decreased. For example, the cost of photovoltaic arrays (solar panel technology) has decreased, and availability has increased since its advent in the 1970s. Proponents of green building argue that the more demand there is for green building products the lower the costs for those products will be as green building becomes even more commonplace, the incrementally higher cost of green building should decrease over time. On the other end of the spectrum, the growing scientific consensus is that global warming over the next century will result in more extreme temperatures, increased likelihood of natural disasters, and reduced availability of potable water in many population centers. Advocates argue that green buildings are designed to address these potential issues through adaptive technologies for temperature control, reduced water demand, and resilient building materials. The economic impact of these factors is hard to quantify on a per-building basis.

Due to the variety of U.S. green building programs, the costs of regulating green building is difficult to quantify. However, an often-cited barrier to adopting local green building programs is the inability, due to workload or existing regulations, of local regulators to review plans and/or approve variances to building codes and regulations to permit green building. Likewise, preexisting zoning and land-use policies are often cited as barriers to implementation of green building programs. When projects are delayed because of these types of regulatory barriers, the cost of developing green building projects increases. Another cost factor in green building is how much the regulating governments will absorb for dedicating personnel and resources to the projects.

The bottom line is, despite the benefits, it simply costs more money to build green. A survey of existing regulations demonstrates that several state and local governments have created voluntary incentive programs to offset some of these costs upfront by reducing taxes, fees, and permitting timeframes.

Green Building Programs Today

To date, green building legislation, whether on the state or local level, has typically followed one of three approaches. The first approach, which has been taken by federal agencies, is to mandate that all public construction projects meet a particular LEED certification level. The second approach is to create financial incentives for private developers to build green, such as tax relief, grants, or expedited permitting. The third approach, which has been met with the most opposition, is to require that all new construction projects that exceed a certain square footage, whether private or public, meet a particular LEED standard.

State and local governments have adopted a variety of green building standards. For example, buildings owned by the cities of San Francisco and Oakland, California, that were constructed after their city's ordinances were adopted must obtain LEED-Silver certification. In New York, public-capital projects valued in excess of $2 million must meet the equivalent of a LEED-Silver or LEED-Certified rating. In Florida, all buildings constructed by the Department of Management Services must be built to LEED standards, and all leases entered into by the agency for state government office space must be located in buildings that meet ENERGY STAR standards. Monroe County, Florida, requires adherence to LEED standards for new county buildings and major renovations of greater than 5,000 gross square feet. Maryland's governor issued an executive order calling for all capital projects greater than 5,000 square feet to earn LEED certification, and the Maryland legislature adopted a requirement that for state capital projects, a green building standard, such as LEED Silver, be used. Kentucky, Alabama, Arizona, and Alaska require increases in energy efficiency for public buildings. Both Colorado and Connecticut state governments are developing their own sustainable building standards for public buildings.

On the federal level, many federal government agencies have adopted LEED requirements for public buildings. These include the General Services Administration, NASA, the National Park Service, the Department of State, the Department of Energy, EPA, and every branch of the military.

The second major category of local and state government regulation related to green building entails creating certain incentives for private development that incorporates green design techniques, including outright grants, tax incentives, expedited permitting, reduced development fees, and density bonuses. In addition, in most states, there is some state, local, or utility program or a combination of programs that offers incentives that can be monetized for green building or energy-efficient equipment or features.

Some state and local governments are offering outright grants to offset the costs of certification and the costs of certain equipment. For example, Costa Mesa, California, established a green building incentive program for private development, effective September 5, 2007, through June 30, 2008, which provided fee waivers for all green installations and fee reductions to cover the cost of LEED certification. The El Paso, Texas, Grant Program provides grants for commercial and multifamily, multistory residential projects earning LEED certification. Grant recipients are required to have obtained a certificate of occupancy and have submitted LEED certification demonstrating that ten of the seventeen available points in the Energy & Atmosphere credit category have been earned. Grants are awarded at increasing intervals based on the level of certification achieved by the building. The maximum grant is $200,000 for LEED Platinum for new construction and $400,000 for LEED Platinum for "multistory existing buildings" that are mixed use and have been at least 50 percent vacant for five years. King County, Washington, established a Green Building Grants Program that offers $15,000–$25,000 in grant funding to building owners who meet a minimum of LEED Silver for new construction or major renovation in the county but outside the City of Seattle. Mecklenberg County, North Carolina, amended the county fee ordinance to include the Green Building Rebate Program, offering permit fee rebates to projects with proof of LEED certification. Rebates increase based on the level of certification achieved: 10 percent reductions for LEED Certified, 15 percent for LEED Silver, 20 percent for LEED Gold, and 25 percent for LEED Platinum. Projects with proof of Green Globes certification are also eligible. In Los Angeles, builders and developers can take advantage of the Los Angeles Department of Water and Power Board Green Building Incentive that offers up to $250,000 in financial incentives to assist a building in becoming more green and meeting LEED standards. In Pennsylvania, grant resources include four state funds, including the $20 million Sustainable Energy Fund targeted to provide grants, loans, and "near-equity" investments in energy-efficiency and renewable-energy projects.

Another frequently used incentive is tax credits for developers and purchasers of green buildings. New Mexico, Oregon, and Maryland offer credits at a state level subject to specific requirements. Virginia has declared energy-efficient buildings to be in a separate class of taxation from other real property and permits local governments to levy equal or lesser taxes on energy-efficient buildings. Virginia code defines energy-efficient buildings as meeting the performance standards of LEED, ENERGY STAR, Green Globes, or EarthCraft and provides a sustainable building tax credit for sustainable buildup. However, the total amount of tax credits awarded under the program is capped at an aggregate amount of $5 million for both commercial buildings and residential buildings. New York offers a green building credit to owners of green buildings, as does Maryland. Las Vegas, Nevada, recently revamped its green building tax credit program to reduce benefits after applications for the credit would have had a more than $450 million impact on revenues. Local governments, including Arlington County, Virginia, give a county property tax credit for a duration of ten consecutive years to any commercial building that achieves LEED-NC Silver certification.

The third incentive approach is to authorize expedited permitting and/or reduced regulatory development fees for green projects. Many jurisdictions offer some combination of fee reductions and expedited permitting. For example, Gainesville, Florida's Green Building Program calls for fast-track permitting for building permits. The District of Columbia created a Green Building Expedited Construction Documents Review Program. Costa Mesa adopted a green building incentive program for private development that encouraged green building practices through various incentives, including priority permitting and fee waivers for all green installations and fee reductions to cover the cost of LEED certification. El Paso County, Texas, provides a fast-track building permit incentive and a 50 percent reduction in the cost of building permit fees for private contractors who use LEED. Issaquah, Washington, has adopted a sustainable building and infrastructure policy. Pursuant to the policy, developers intending to use LEED may receive free professional consultation, and projects achieving LEED certification are placed at the head of the building permit review line. The Hawaii state legislature requires counties to give priority processing for all construction or development permits for projects that achieve LEED Silver or the equivalent. North Carolina permits cities and counties to encourage green building practices in their jurisdictions by reducing permitting fees or providing partial fee rebates for construction projects that achieve LEED certification or certification from other rating systems.

Density bonuses are another incentive that is frequently offered to permit increased size or numbers of units or the amount of commercial square footage permitted in a particular zoning district. Similar to bonuses granted in economic development incentive packages, this type of incentive can increase profitability for a developer if the increase in density results in additional marketable product and the market supports the sale of denser product. As demonstrated in the following examples, many of these density bonuses are only available in designated parts of a community, reflecting the community's intent to redevelop a particular area. Acton, Ohio, revised its land development code to allow for a density bonus for buildings achieving LEED certification in the East Acton Village District. Arlington County, Virginia, allows density bonuses for commercial projects and private developments earning LEED Silver certification. To qualify for the bonus, all site-plan applications must include a LEED scorecard and have a LEED accredited professional on the project team. However, projects are not required to obtain certification. All projects in Arlington contribute to a green building fund for countywide education and outreach activities. Contributions are refunded if projects earn LEED certification. Bar Harbor, Maine, amended its code to award a single, additional market-rate dwelling-unit density bonus for construction projects, provided all dwelling units meet LEED standards. This bonus is only available to projects within a Planned Unit Development. The Pittsburgh Code grants a density bonus of an additional 20 percent floor area ratio and an additional variance of 20 percent of the permitted height for all projects that earn LEED for New Construction or LEED for Core and Shell certification. The bonus is available in all nonresidential zoning districts.

The purpose of incentive regulations is to offset some of the financial barriers to green building. Although no comprehensive analysis of the effectiveness of the various incentives has been conducted, it appears that the challenge for governments utilizing incentives to facilitate voluntary green building is adequately matching the perceived value of the incentive to the perceived increase in cost associated with green building. As discussed below, mismatches on incentives and perceived cost have hampered the effectiveness of certain incentive programs.

Many of these incentive-based laws have been adopted since 2003, and, as reflected in the construction industry surveys referenced earlier, appear to have increased awareness of sustainable development opportunities. In most cases, with the notable exception of Las Vegas, Nevada, they have not, however, brought about the large numbers of green construction projects that they were designed to encourage. In certain cases, the programs were underfunded or underpublicized. In other cases, the incentives were targeted for areas that were not marketable for redevelopment. Consequently, some cities are considering alternatives to voluntary programs as a means of increasing green building projects within their jurisdiction.

Mandated green construction for private developments represents a final and most controversial category of green building regulation. Several large cities have adopted mandatory requirements for green buildings. Proponents of these policies advocate they are appropriate due to growing climate-change concerns and sustainable development being more accepted and, in fact, supported by the general public. Furthermore, the cost of green building materials has dropped significantly in the last five years, and more developers, architects, planners, and government officials are familiar with and have the expertise necessary to implement green building techniques. However, this type of regulation has been successfully challenged in Albuquerque, New Mexico, by heating and air conditioning industry groups who alleged that the policies unfairly disadvantaged vendors and suppliers of nongreen building materials.

Notwithstanding the opposition, Boston, Washington, D.C., Los Angeles, and San Francisco have enacted such laws. Most of the laws apply solely to large commercial buildings. The largest city to embrace green building mandates is Boston. In the summer of 2007, the city amended its zoning ordinance to require that all private construction over 50,000 square feet meet minimum LEED criteria. San Francisco's code is arguably the most comprehensive. The standards apply to newly constructed commercial buildings over 5,000 square feet, new residential buildings taller than 75 feet, and building renovations that involve more than 25,000 square feet. In each case, structures are subject to a specific level of certification under LEED standards or other widely accepted green building ratings systems. San Francisco also imposes fees, including impervious surface fees, to offset the impacts of construction on the environment.

Through its Green Points Program, Boulder, Colorado, has become one of the few cities to mandate green building in the residential context. The program requires some combination of recycled materials (e.g., fiber concrete, reclaimed lumber, or recycled roofing materials), green insulation products, energy-efficient windows, radiant floor heating, or other sustainable products in private residential-addition and remodeling projects larger than 500 square feet.

A few smaller towns and cities have also imposed mandatory requirements. The town of Babylon, New York, requires LEED certification for any new construction of commercial buildings, office buildings, industrial buildings, multiple residence, or senior-citizen multiple residence that exceeds 4,000 square feet. The town refunds the certification fees paid to USGBC by the developer when certification is achieved. Portland, Oregon, adopted a mandatory green building program for commercial new construction in January 2009, which provides that buildings that meet the standard will qualify for a waiver of a city GHG emissions fee; all other new commercial construction will have to pay a fee for the projected GHG emissions resulting from operating the nongreen building for fifteen to thirty years. The program was originally proposed in 2008 to include new residential construction; however, after substantial controversy, it was scaled back and, instead of charging a fee, Portland will track the number of green residential projects constructed. If the targeted number of units has not been built by 2012, the city will consider imposing a fee on nongreen residential construction.

Legal Challenges to Green Building Regulations

The local and state government programs described in this article have been adopted within the context of the existing laws governing land use zoning and public health. It is important to note that while 911 mayors have signed onto the U.S. Mayors Climate Protection Agreement, only approximately seventy-five communities have adopted green building initiatives of any sort. To put this in context, there are 38,967 municipalities in the United States.

There are many possible reasons why more local governments that have committed to climate change initiatives have not addressed green building and why thousands more have addressed neither climate change nor green building. Local government officials have resisted changes based on new, unfamiliar technologies and approaches to construction. Revisions to existing laws, such as incorporating green building technologies or performance standards, often require wholesale restructuring of land-use and building codes, which have historically relied on prescriptive rules instead of flexible standards. The work to complete the revisions is time consuming, and enforcing flexible standards is more complex and time consuming than enforcing traditional codes. Typically underfunded and understaffed, local land-use departments may not have the manpower or resources to address green building innovations.

Traditional zoning and design codes further two broad purposes: protecting and enhancing property values and protecting public health, safety, and welfare. Land development and zoning codes also represent something of a historical consensus on community aesthetic standards. Green building technologies, many of which are new, have not been tested or anticipated in zoning and building codes. Consequently, many of the technologies and equipment do not comply with code standards. Perhaps the most frequently disputed and common sustainable technology barred by zoning laws and building codes is solar panels. In the 1970s solar panels were extremely bulky, utilizing metal frames that were highly visible. Solar panels are now thinner and can be incorporated into building and roof design to decrease visibility without compromising performance. Despite these changes, zoning codes, homeowners' covenants and restrictions, and historic preservation policies often prohibit their installation or restrict it in a manner that prohibits functionality. In historic building renovations, builders have been prohibited from changing windows to energy-efficient models because until recently there were limited design options for replicating the original windows' materials, casing, sash width, muntin profile, or color. Landscaping requirements can also conflict with green building techniques. Codes that specify permissible plant and tree palettes frequently do not permit xeriscaping or hardscaping to reduce irrigation water demand. Zoning ordinances frequently also omit newer technologies such as windmills, freestanding solar panels, turbines, fuel cells, and water collectors/cisterns in lists of permitted and prohibited uses, leaving applicants unclear as to whether the technology can be utilized in a project. A related problem is that zoning boards often have no code-based standards to evaluate applications for zoning relief. Instead, the boards engage in ad hoc inquiries leading to inconsistent results, which are often highly influenced by the presence or absence of "not in my backyard" voices opposing the uses.

Despite these challenges, the local governments that have successfully implemented green building programs or policies have been able to do so by modifying or replacing zoning and building codes that were an impediment to green building with codes that either specifically permit green technologies or provide administrative vehicles to obtain approval for the technologies. These communities have acted within the context of zoning and land-use jurisprudence, which has, since the U.S. Supreme Court decided Vittage of Euclid, *Ohio v. Ambler Realty Co.,* 272 U.S. 365 (1926), required adopted zoning ordinances to protect public health, safety, and welfare. Critical to green building ordinances is the causal link between GHG emissions, global climate change, and the threats to public health, safety, and welfare arising therefrom. Findings of public benefit are also essential elements of government-incentive grant programs and expedited permitting policies.

An evolving area of the law concerns federal regulation of GHGs. In *Massachusetts v. EPA,* 111 S.Ct. 1438, 167 L.Ed.2d 248 (2007), the Supreme Court designated carbon dioxide (CO_2) a pollutant under the Clean Air Act definitions to the extent that reports connect the gas to climate change. As such, EPA is required to regulate CO_2 under the Clean Air Act. Categorizing CO_2 as a pollutant makes it easier for local governments to make the causal link between green building regulations and public health, safety, and welfare. However, depending on the shape of federal regulation, some local and state GHG emissions legislation, including green building programs that include impact fees, may be preempted by federal legislation when adopted.

Given this existing regulatory scheme, EPA may once again turn to mandatory local regulations as a means of addressing CO_2 emissions. Such mandatory local ordinances could include anything from green building standards to comprehensive planning requirements that target a reduction in automobile dependence. From a dollars and cents perspective, this increased regulatory burden would transform some of the increased costs associated with green building into the cost of doing business.

Is National Green Building Regulation Called For?

Local and state governments have essentially been functioning as laboratories for green building policy. Green building is being advanced as part of the toolkit for addressing climate change—a global issue. Thus it has its roots in science-based calculations of reduced GHG emissions and stewardship of natural resources. However, in most cases green building policies affect land use, zoning, and building codes—traditionally in the purview of local governments. Local governments typically have neither the budgets nor the personnel to advance the science of climate change but have devoted resources to understanding the impact of green building practices on local economies and the environment. If the United States is going to become a leader in climate change policy, as promised by President Obama, then more local and state governments need to address barriers in existing land use, zoning, and building codes, which impede implementation of climate change policies, including permitting green building programs. This probably will not happen without federal legislation mandating attainment of GHG emissions reductions by state and local governments. The rationale for this legislation is the same as for many of the other key environmental regulations adopted in the last fifty years, including the Clean Air Act and the Clean Water Act: consistent nationwide standards mandating GHG reductions are necessary to protect public health and safety, and without national regulations, the GHG emissions cannot be managed successfully. The science supporting climate change certainly supports that position, as it is the cumulative impact of GHGs that contributes to climate change.

Federal legislation should also resolve the issue of what constitutes green building by defining key metrics of sustainability: energy-efficiency standards for structures, per-person potable water-use standards, renewable-energy requirements, and additional indoor air-quality requirements. Similar to other environmental laws, the federal standards could then be implemented by state and local governments. Adopting metrics rather than a unified green building standard would preserve the ability of local and state governments to tailor solutions to their jurisdictions. However, developers and others would have clear guidance on the minimum requirements for building green. States should be permitted to adopt more stringent guidelines if necessary to attain required emission reductions.

Federal legislation should also make illegal any provisions contained in regulatory codes or in deed restrictions that prohibit xeriscaping and use of sustainable-energy technologies on solely aesthetic grounds. Although such an act has the potential to impair contracts, it would likely be enforceable if determined to be necessary to address a national environmental crisis.

With these measures in place, all state and local governments would be spurred to action to reduce GHG emissions in their communities to attain federal standards. At the same time, determining the details of how to achieve emissions reductions and enforce the federal metrics of sustainability should be left to state and local governments so that the experts in local conditions can shape the type and design of green building projects, how to balance the competing interests of historic preservation and energy efficiency in building renovation, and which types of incentives, if any, are needed to ensure success. Thus, federal law should not preempt the field entirely and should allow the states and local governments to continue to function as innovators in the regulation of green building. If all of these regulatory changes occur, the financial disincentives to green building will be greatly reduced, and the costs of going green will begin to make sense in the short term as well as the long term.

EXPLORING THE ISSUE

Does the Market Work Better Than Government at Transitioning to Sustainability?

Critical Thinking and Reflection

1. What are the political ramifications of enacting a carbon tariff on imported goods from countries that do not have regulations on carbon dioxide?
2. What is the cost of the current inaction of the United States to climate change? Is it possible to accurately measure the costs?
3. Are carbon taxes easier to implement than cap-and-trade legislation?
4. Why is the United States the only country to have not signed the Kyoto Protocol? Why did President Obama not attend the 2010 Cancun climate talks?
5. Are national regulations on building codes justified or are states more inclined to develop such building codes?

Is There Common Ground?

In the United States, there is some "common ground" toward seeking regulation for carbon dioxide. Most economists and scientists believe that global warming is a real threat, and that a market approach is the best solution. Politicians, on the other hand, are in strong disagreement with the Republican Party and industrial state Democrats arguing against any form of increase in electricity costs, and the Democratic Party gradually losing enthusiasm for a cap-and-trade approach. The ironic situation is that it appears that what no party wanted, not even the EPA, is a command-and-control policy. In January 2011, the United States will have a command-and-control policy that has very little support. The expectations are that the Republicans will fight every dollar of appropriations toward the EPA, and it can be expected that any ruling by the EPA will be tied up in the courts for a number of years.

Seeking a balance between political parties for climate change regulation will be even tougher after last November's 2010 mid-term elections. It appears that the Obama administration may have pushed too fast for changes that the American people do not want to fund at this time. While a cap-and-trade policy was politically popular until very recently, the change due to elections will undoubtedly result in political gridlock for the next 2 years, and the projected command-and-control policies to be established by the EPA in January will only end up tying up the court system. The one main question to raise is that can we as a society afford not to do anything.

Internet References

The Story of Stuff is an educational video about the manufacturing of products and the current linear cycle that needs to be shaped into more of an industrial ecology approach. In addition to the original video, the authors have developed a critical movie on cap and trade.

www.storyofstuff.com/capandtrade/

The Greenpeace website suggests that even strong environmental groups are supporting market-based regulations such as cap-and-trade. In this article, the authors feel that the House of Representatives bill was not strong enough, and did not support the science behind climate change.

www.greenpeace.org/usa/en/media-center/news-releases/greenpeace-opposes-waxman-mark/

The Chicago Climate Exchange offers the only voluntary cap-and-trade program in the United States.

www.chicagoclimatex.com/content.jsf?id=821

The European Commission carbon trading system is the first instituted cap-and-trade program. The website provides background on how the system was implemented.

http://ec.europa.eu/clima/p olicies/ets/index_en.htm

The Competitive Enterprise Institute is a conservative free-market think tank dedicated toward disproving global warming and promoting limited government. The website was selected because it presented an extreme view of the current issue.

www.cei.org/

ISSUE 11

Does Sustainable Urban Development Require More Policy Innovation and Planning?

YES: Bruce Katz, from *Smart Growth: The Future of the American Metropolis,* (Center for Analysis of Social Exclusion and Brookings Institution, 2002)

NO: David B. Resnik, from "Urban Sprawl, Smart Growth, and Deliberative Democracy," *American Journal of Public Health* (October 2010)

Learning Outcomes

After reading this issue, you should be able to:

- Answer how and why Americans moved to the suburbs and outline residential patterns that developed.
- Discuss the characteristics of urban sprawl and how smart growth is a planning response.
- Discuss the nature and characteristics of "sustainable communities."
- Understand the relationship between urban sprawl and race.
- Outline how compact cities are more sustainable.
- Discuss how cities can be made more sustainable.

ISSUE SUMMARY

YES: Bruce Katz, of the ESRC Research Center for Analysis of Social Exclusion within the Suntory and Toyota International Centers for Economics and Related Disciplines at the London School of Economics and Political Science, describes how current public policies facilitate the "excessive decentralization" of people and jobs and how smart growth reforms are being enacted, particularly at the state level, to shape new, more urban-friendly growth patterns.

NO: David B. Resnik, a bioethicist and vice-chair of the Institutional Review Board for Human Subjects Research at the National Institute of Environmental Health Sciences, National Institutes of Health, explains why urban sprawl, a model of unsustainable development around the periphery of a city, has a negative effect on human health and the environment. He believes that smart growth is an alternative to the problem of urban sprawl; nevertheless, he argues that smart growth has many disadvantages including a decrease in property values, decrease in the availability of affordable housing, restriction of property owners' use of their land, disruption of existing communities, and a likely increase in sprawl.

Sustainable urban development seeks to maintain environmental integrity over time, while promoting a prosperous economy, and hosting a vibrant, equitable society. Sustainable communities integrate the three pillars of sustainability—social equity, economic capacity, and environmental quality. As an operational concept, Taylor et al. (2009) states that a sustainable community is one that possesses the following characteristics: it emphasizes compact design to ensure a walkable, bikeable, and mass transit-oriented environment; it provides mixed-use development and a wide range of housing opportunities; it maintains sensitive natural environments; it encourages the reduction of greenhouse gas emissions through green building standards, energy efficiency, and renewable energy sources; and it designs for open green space, parks, and natural drainage systems. Although the goals of sustainable urban development might have strong support among urban and regional planners and certain policy makers, nevertheless, the question remains on how to implement them. Should government lead the way and provide a set of incentives and planning guidelines to make sustainable urban development a reality? Or, should market preferences as determined by property developers be the main driver? This issue explores whether public policy and planning should guide urban development sustainability or whether the private marketplace, a measure of people's preferences, should lead, even though these preferences might not coincide with the traditional goals of sustainable communities.

Traditionally, there has been little public control over urban development in the United States. This has led to excessive decentralization or urban sprawl. Urban sprawl is an unsustainable growth pattern for the following reasons: It is of low density, consumes land resources, and is largely uncontrolled; it exists on the periphery of American cities and causes environmental impacts through the fragmentation of natural ecosystems and the loss of habitat. Sprawl communities are characterized by single-family detached homes in locations that lack public transportation and place a heavy reliance on automobile use. Automobile-dependent communities have an impact on public health. Studies have shown that people who live in automobile-dependent communities walk less, weigh more, and are more exposed to other public health risks. Cities built around automobile use release more air pollutants into the atmosphere; can affect water quality through storm-water runoff; and impact natural ecosystems, thereby causing loss of biodiversity. Sprawl places an added cost to infrastructure development (water distribution, sewage, electricity, etc.) as disperse location of homes drives the cost of service delivery up.

Planners and decision makers are looking for ways to make metropolitan areas more sustainable by relying on various policy options such as smart growth. Smart growth promotes higher population densities; mixed land use development combining residential and commercial buildings; population nodes or centers next to public transportation; and preservation of open space. Smart growth solutions foster a greater sense of place and reduce infrastructure costs leading to more affordable housing. Also, smart growth policies can conserve forests and water recharge areas, preserve wetlands, reduce the loss of agricultural land, and foster redeveloping older urban centers.

Although there are advocates of smart growth and sustainable planning strategies, there are also critics. The opponents believe that smart growth policies encroach on individual freedom and are endangering the American dream of owning a single-family detached home in the suburbs. They point out that due to increased regulations the cost of new housing rises while there is a loss of value of existing suburban properties due to the added congestion.

The traditional American dream is to own land, build a home, and be in control of one's destiny. This ideal has motivated wealthier Americans to leave the problems of the central city, with its congestion, grit, and danger, to move to the pristine suburbs. The "flight from the city" became a torrent after World War II. The growing affluence of Americans and the high ownership of automobiles gave the middle class the opportunity to flee the city. This soon created a divide in American society, between black and white and rich and poor. As more people left the city, the tax base of the central cities eroded and conditions grew worse as the infrastructure began to degrade. Federal policies such as the Interstate Highway Act of 1956 provided people the means to leave the city and long-term, low-interest, federally subsidized housing loans allowed people to own their dream house in the suburbs. As cities decayed, agricultural land was consumed in the building of a new urban type, the sprawling horizontal metropolitan city. Each small suburban community, even though it was a part of a larger metropolitan region, used home rule and local zoning to control the land use patterns in their community. Generally, this meant a preponderance of single-family homes and few apartment dwellings. This was another device that kept the outer suburbs more affluent and kept lower-income populations out.

For the latter part of the twentieth century and now the twenty-first century, the suburbs have changed their demographic makeup and their housing options, but not their pattern of decentralization. Suburbs are diverse today, but with the depopulation of the central cities, people have moved even further to the periphery. The environmental consequences of this decentralization are reflected in the loss of agricultural land and the increase in highways and automobile use. The result has been the loss of open space, degradation of urban watersheds, and the need to transport food for longer distances.

The question remains whether or not Americans buy into smart growth, which requires a cultural paradigm shift. The dream of owning a single-family home on a large parcel of land in the suburbs is still appealing. Real estate professionals and developers still see the attraction of this dream in their building promotions. What is interesting to observe is that this dream is beginning to spread internationally. The urban development schemes in some developing countries are trying to emulate American suburbanization with its emphasis of automobile usage and land loss. It shows the great strength in this unsustainable growth pattern globally and how international affluence can spark this international consumption pattern that can threaten ecosystems. But, is this a question of human desire and want? Is sustainability only possible by forcing people to change by regulating human behavior? Can policy makers and planners convince people that smart growth is a desirable living style? This would mean that people would need to live in more compact communities, mix housing and commercial land use, and live in attached housing. Do they want to do this? Can people be attracted back to the central cities?

Attracting the young middle class back to the cities might be a tough sell since the quality of public education and school safety might be an issue. But apartments with extra bedrooms, safe streets, better schools, and more housing affordability can be a substantial lure. Also, compact urbanized communities can be created in the suburbs, based on walking and mass transit. Recycling suburban malls into mixed use housing and commerce is a first step. Growing city centers help create communities. Innovative zoning and land use controls can promote this type of growth. Development needs to be steered into areas with infrastructure in place. Mixed levels of affordable housing can be incorporated into communities so that all may live and work within the community framework.

The article for the YES position, that sustainable urban development does require more policy innovation and planning, is written by Bruce Katz of the London School of Economics and Political Science. He describes how past and current public policies have facilitated the excessive decentralization of people and jobs and how smart growth reforms are being enacted, particularly at the state level, to shape new, more urban-friendly growth patterns. He provides examples of policies and planning tools that can support urban sustainability and smart growth. He addresses the idea that people need to be convinced that living in a more urbanized setting is a desired lifestyle. Cities need to emphasize their advantages, their proximity to activities, services, mass transit, etc. Sustainable development is possible with the right planning incentives and the commitment of government.

The article for the NO position, that sustainable urban development does not require more policy innovation and planning, is presented by bioethicist David B. Resnik. Although he sees the problems associated with suburban sprawl and some positives in smart growth, he points to its disadvantages. These consist of often causing a decrease in property values and restrictive regulations decreasing the availability of affordable housing. He also notes how land use controls limit the use of land by property owners and how smart growth changes can actually disrupt community stability and lead to increased sprawl. People become discontented and vote with their feet. They move even further outward.

Bruce Katz

Smart Growth: The Future of the American Metropolis

. . . In the past few years, widespread frustration with sprawling development patterns has precipitated an explosion in innovative thinking and action across the United States. This new thinking—generally labelled as "smart growth"—contends that the shape and quality of metropolitan growth in America are no longer desirable or sustainable. It asserts that current growth patterns undermine urban economies and broader environmental objectives and exacerbate deep racial, ethnic, and class divisions. It argues that these growth patterns are not inevitable but rather the result of major government policies that distort the market and facilitate the excessive decentralization of people and jobs.

Across the country, smart growth language and rhetoric have become common not only among political, civic, and corporate leaders but also among developers and other participants in the real estate industry. A growing chorus of constituencies—corporations, local elected officials, environmentalists, ordinary citizens—are demanding that the market and the government change the way they do business and take actions to curb sprawl, promote urban reinvestment, and build communities of quality and distinction. Governors and state legislatures are responding by proposing and enacting important reforms in governance, land use, and infrastructure policies. Voters at the ballot box are regularly approving measures to address the consequences of sprawling development patterns.

This is a powerful paradigm shift, a sweeping rethinking of the costs and consequences of metropolitan growth in the country. It offers a compelling vision of how to achieve environmental quality, urban revitalization, economic competitiveness, and even racial and social justice in metropolitan America. . . .

How Metropolitan America Grows

The release of the 2000 US census created an almost euphoric mood among many long-time observers of American cities. New York City topped 8 million people for the first time. Cities left for dead not long ago—Boston, Atlanta, Chicago—registered strong population growth. With visible signs of prosperity in refurbished downtowns, with immigrants spurring neighborhood revitalization,

Katz, Bruce. From *Smart Growth: The Future of the American Metropolis,* (Center for Analysis of Social Exclusion, London School of Economics, 2002) CASE paper 58, July 2002.

[and] with homeownership rates generally going up and poverty and crime rates generally going down, many American cities are enjoying a hard-won optimism. Yet a closer look at the census (and other market trends) shows that the decentralization of economic and residential life, not the renewal of core cities, remains the dominant growth pattern in the United States. America's cities and metropolitan areas are experiencing similar patterns of growth and development—explosive sprawl where farmland once reigned, matched by decline or slower growth in the central cities and older suburbs. Suburban areas, some of which were small towns a few decades ago, are capturing the lion's share of population and employment growth. In the largest metropolitan areas, the rate of population growth for suburbs was twice that of central cities—9.1 percent versus 18 percent—from 1990 to 2000 (Berube, 2002). Suburban growth outpaced city growth irrespective of whether a city's population was falling like Baltimore or staying stable like Kansas City or rising rapidly like Denver. Even sunbelt cities like Phoenix, Dallas, and Houston grew more slowly than their suburbs.

Percentage growth only tells part of the story. More and more people are living, working, shopping, and paying taxes at the farthest edges of metropolitan areas. Atlanta, often touted as a "turnaround city," is a case in point. The central city grew by a respectable 6 percent during the 1990s and gained 22,000 people. Yet its metropolitan area grew by 39 percent during this period and gained 1.1 million people. Incredibly, rural counties dozens of miles from the central business district gained more people—in both percentage and absolute terms—than the city of Atlanta. Both Henry County and Forsyth County, for example, more than doubled their population in the 1990s and now have 119,000 and 98,000 people respectively. As people go, so do jobs. Consequently, the suburbs now dominate employment growth and are no longer just bedroom communities for workers commuting to traditional downtowns. Rather, they are now strong employment centers serving a variety of functions in their regional economies. The American economy is rapidly becoming an "exit ramp" economy, with office, commercial and retail facilities increasingly located along suburban freeways. This is particularly true in leading technology regions like Washington, D.C.; Austin; and Boston where firms like American Online, Dell, and Raytheon have built large exurban campuses far from the city center. As Edward Glaeser and Matthew Kahn have recently demonstrated, employment decentralization has become the norm in American metropolitan areas. Across the largest 100 metro areas, on average, only 22 percent of people work within three miles of the city center and more than 35 percent work more than ten miles from the central core. In cities like Chicago, Atlanta, and Detroit, employment patterns have radically altered, with more than 60 percent of the regional employment now located more than 10 miles from the city center (Glaeser and Kahn, 2001).

The Consequences of Unbalanced Growth

The shape of metropolitan growth in America has disparate impacts on central cities, older suburbs and rapidly growing areas. Central cities, while generally improving, remain home to the nation's very poor. While poverty has declined in central cities, for example, urban poverty rates are still twice as high as suburban poverty rates, 16.4% versus 8.0% in 1999. Cities are also disproportionately home to families whose earnings are above the poverty level, but below median income (Berube and Forman, 2001).

Cities are not just home to too many poor families; they are also home to neighborhoods where poverty is concentrated. From 1970 to 1990, the number of people living in neighborhoods of high poverty—where the poverty rate is 40% or more—nearly doubled from 4.1 million to 8 million. As Paul Jargowsky, John Powell, and others have shown, concentrated poverty is principally an urban (and racial) phenomenon. The implications of concentrated poverty are severe. People in these neighborhoods often face a triple whammy: poor schools, weak job information networks, and scarce jobs. They are more likely to live in female-headed households and have less formal education than residents of other neighborhoods (Jargowsky, 1997).

Yet the consequences of sprawl are not confined to central cites. Metropolitan growth patterns are also transforming the suburbs. Like city neighborhoods, there are a wide range of suburban experiences and realities. Although American popular culture tends to paint a picture of the monolithic "suburb," the reality on the ground is infinitely more complex. . . .

The patterns of extensive growth in some communities and significantly less growth in others are inextricably linked. Poor schools in one jurisdiction push out families and lead to overcrowded schools in other places. A lack of affordable housing in thriving job centers leads to long commutes on crowded freeways for a region's working families. Expensive housing—out of the reach of most households—in many close-in neighborhoods creates pressures to pave over and build on open space in outlying areas, as people decide that they have to move outwards to build a future.

The cumulative impacts of these trends are severe. Many American metropolitan areas are struggling with traffic problems, environmental problems and a wide gap—both spatial and social—between low-income people and jobs.

Traffic congestion has become a way of life in most major metropolitan areas. A recent study by the Texas Transportation Institute of 68 metropolitan areas in the US found that the average annual delay per person was 36 hours or the equivalent of about one workweek of lost time. The total congestion bill in 1999 for these places came to $78 billion, which was the value of 4.5 billion hours of delay and 6.8 billion gallons of excess fuel consumed (Texas Transportation Institute, 2001).

Congestion and auto dependence also affect the pocketbooks of metropolitan residents and commuters. The shape of suburban growth—low-density housing, low-density employment centers—have made residents and commuters completely dependent on the car for all travel needs. Across the country, household spending on transportation has risen substantially. Transportation is now the second largest expense for most American households, consuming on average 18 cents out of every dollar. Only shelter eats up a larger chunk of expenditures (19 cents), with food a distant third (13 cents) (Surface Transportation Policy Project, 2000). The transportation burden disproportionately affects the poor and working poor. Households earning between $12,000 and $23,000 spend 27 cents of every dollar they earn on transportation. For the very poor (households who earn less than $12,000), the transportation burden rises to 36 cents per dollar.

Unbalanced growth and the outward movement of metropolitan areas ha[ve] taken [their] toll on green space. Urbanized land increased by over 47 percent between 1982 and 1997. The pace appears to be quickening. In the five-year period between 1992 and 1997, the pace of development (2.2 million acres a year) was more than 1.5 times that of the previous 10-year period (1.4 million acres a year). What is remarkable is that the growth in urbanized land is occurring even in metropolitan areas that lost population. The Pittsburgh, Pennsylvania, metropolis lost 8 percent of its population between 1982 and 1997; yet its urbanized land area grew by close to 43 percent during this period (Fulton et. al., 2001).

The broader environment suffers as metropolitan economies decentralize and subdivisions replace forestland and prime farmland. Researchers in Atlanta have demonstrated the powerful connection between driving, land use, and air pollution. Rapid and unbalanced growth has also degraded the water quality of rivers and lakes because of polluted runoff from new, environmentally unfriendly developments. Deforestation threatens the health of the urban ecosystem since trees slow stormwater, reduce runoff, and improve air quality (Bradley, Katz, and Liu, 2000).

Finally, unbalanced growth has enormous social implications for low-wage and minority workers. As economies and opportunity decentralize and working poverty concentrates, a "spatial mismatch" has arisen between jobs and people (Pugh, 1998). In suburbs, entry-level jobs abound in manufacturing, wholesale trade and retailing. All offer opportunities for people with limited education and skills, and many pay higher wages than similar positions in the central cities. But persistent residential racial discrimination and a lack of affordable suburban housing effectively cut many inner city minorities off from regional labor markets. Low rates of car ownership and inadequate public transit keep job seekers in the core from reaching the jobs at the fringe. Often, inner city workers, hobbled by poor information networks, do not even know that these jobs exist. . . .

Smart Growth and the New Metropolitan Agenda

An emerging awareness of the costs of sprawl—and the role of government policies in facilitating sprawl—is triggering an intense debate about growth around the country. Elected officials from cities and inner suburbs; downtown corporate, philanthropic, and civic interests; minority and low-income community representatives; environmentalists and land conservationists; slow-growth advocates in the new suburbs; farmers and rural activists; and religious leaders all are realizing that uncoordinated suburban expansion brings needless costs. These constituencies are beginning to define, advocate for, and implement a smart growth agenda at all levels of government. This agenda principally revolves around changing the state "rules of the development game" to slow decentralization, promote urban reinvestment, and promote a new form of development that is mixed use, transit-oriented and pedestrian friendly.

While the building of new coalitions is taking place at all levels, state governments have become the principal targets of reform for many of these coalitions. This reflects the recognition that states have the most extensive impact on growth trends—in part because of their traditional control over issues like land use, governance, and local taxation and in part because of their increased powers in areas like transportation, workforce, housing and welfare policy due to federal devolution. In recent years, support for smart growth reforms has increased markedly among governors and state legislatures. . . .

The smart growth agenda generally consists of five sets of complementary policies. First, states are experimenting with new forms of metropolitan governance to handle such issues as transportation, environmental protection, waste management, cultural amenities, and economic development. Second, they have embraced land use reforms to manage growth at the metropolitan fringe. Third, they are using state resources to preserve tracts of land threatened by sprawl as well as reclaim urban land for productive use. Fourth, they have begun to steer infrastructure investment and other resources to older established areas. Finally, they are considering tax reforms to reduce fiscal disparities between jurisdictions and reduce the competition between jurisdictions for sprawl inducing commercial development (Katz, 1999).

Conclusion

The smart growth movement has the potential to change the landscape of metropolitan America and, in the process, build stronger cities, sustainable regions, and more inclusive communities. It has the power to unite formerly disparate constituencies—environmentalists, land conservationists, farm preservationists, community development advocates, downtown business interests—into a strong, sustainable force for change. It has the ability to build new kinds of political coalitions that cross-parochial borders and move beyond current racial and ethnic divisions. . . .

David B. Resnik

Urban Sprawl, Smart Growth, and Deliberative Democracy

Urban sprawl is an increasingly common feature of the built environment in the United States and other industrialized nations. Although there is considerable evidence that urban sprawl has adverse affects on public health and the environment, policy frameworks designed to combat sprawl—such as smart growth—have proven to be controversial, making implementation difficult.

Smart growth has generated considerable controversy because stakeholders affected by urban planning policies have conflicting interests and divergent moral and political viewpoints. In some of these situations, deliberative democracy—an approach to resolving controversial public-policy questions that emphasizes open, deliberative debate among the affected parties as an alternative to voting—would be a fair and effective way to resolve urban-planning issues.

In the last two decades, public health researchers have demonstrated how the built environment—homes, roads, neighborhoods, workplaces, and other structures and spaces created or modified by people—can affect human health adversely.[1–7] Urban sprawl, a pattern of uncontrolled development around the periphery of a city, is an increasingly common feature of the built environment in the United States and other industrialized nations.[8] Although there is considerable evidence that urban sprawl has adverse environmental impacts and contributes to a variety of health problems—including obesity, diabetes, cardiovascular disease, and respiratory disease[9]—implementation of policies designed to combat sprawl, such as smart growth, has proven to be difficult.[10–17] One of the main difficulties obstructing the implementation of smart-growth policies is the considerable controversy these policies generate. Such controversy is understandable, given the fact that the stakeholders affected by urban-planning policies have conflicting interests and divergent moral and political viewpoints.[18] In some of these situations, deliberative democracy—an approach to resolving controversial public-policy questions that emphasizes open, deliberative debate among the affected parties as an alternative to voting—would be a fair and effective way to resolve urban-planning issues.

Urban Sprawl

Urban sprawl in the United States has its origins in the flight to the suburbs that began in the 1950s. People wanted to live outside of city centers to avoid traffic, noise, crime, and other problems, and to have homes with more square footage and yard space.[8,9]

As suburban areas developed, cities expanded in geographic size faster than they grew in population. This trend has produced large metropolitan areas with low population densities, interconnected by roads. Residents of sprawling cities tend to live in single-family homes and commute to work, school, or other activities by automobile.[8,9] People who live in large metropolitan areas often find it difficult to travel even short distances without using an automobile, because of the remoteness of residential areas and inadequate availability of mass transit, walkways, or bike paths. In 2002, the 10 worst US metropolitan areas for sprawl were Riverside–San Bernardino, CA; Greensboro–Winston-Salem–High Point, NC; Raleigh–Durham, NC; Atlanta, GA; Greenville–Spartanburg, SC; West Palm Beach–Boca Raton–Delray Beach, FL; Bridgeport–Stamford– Norwalk–Danbury, CT; Knoxville, TN; Oxnard–Ventura, CA; and Fort Worth–Arlington, TX.[8]

There is substantial evidence that urban sprawl has negative effects on human health and the environment.[4,7,9,19] An urban development pattern that necessitates automobile use will produce more air pollutants, such as ozone and airborne particulates, than a pattern that includes alternatives to automotive transportation. The relationship between air pollution and respiratory problems, such as asthma and lung cancer, is well documented.[4] Cities built around automobile use also provide fewer opportunities to exercise than cities that make it easy for people to walk or bike to school, work, or other activities.[4] Exercise has been shown to be crucial to many different aspects of health, such as weight control, cardiovascular function, stress management, and so on.[20,21]

Because socioeconomically disadvantaged people in sprawling cities may have less access to exercise opportunities and healthy food than do wealthier people, sprawl may also contribute to health inequalities.[22] Urban sprawl can reduce water quality by increasing the amount of surface runoff, which channels oil and other pollutants into streams and rivers.[4] Poor water quality is associated with a variety of negative health outcomes, including diseases of the gastrointestinal tract, kidney disease, and cancer.[23] In addition to air and water pollution, adverse environmental impacts of sprawl include deforestation and disruption of wildlife habitat.[4]

Smart Growth

Many public health advocates have recommended smart growth as a potential solution to the problem of urban sprawl.[4,7,9,20] Smart growth can be defined as a policy framework that promotes an urban development pattern characterized by high population density, walkable and bikeable neighborhoods, preserved green spaces, mixed-use development (i.e., development projects that include both residential and commercial uses), available mass transit, and limited road construction.[4,7,11] Smart growth was originally conceptualized as an aesthetically pleasing alternative to urban sprawl that would offer residents a high quality of life and the convenience of local amenities,[24] but it also has many potential health benefits, such as diminished air pollution, fewer motor vehicle accidents, lower pedestrian mortality, and increased physical exercise.[4,7] Smart growth is different from the concept of "garden suburbs" because it addresses issues of population density and transportation, not just availability of green space and preservation of agricultural land.[4]

In the 1970s, Portland, Oregon, was the first major city in the United States to establish smart-growth urban planning by limiting urban growth to an area around the inner city.[11] Since the 1990s, many other urban areas have encouraged the development of planned communities in which people can live, shop, work, go to school, worship, and recreate without having to travel great distances by automobile. An example of one of these planned communities is Southern Village, situated on 300 acres south of Chapel Hill, North Carolina. Launched in 1996, Southern Village features apartments, townhouses, single-family homes, and a conveniently located town center with a grocery store, restaurants, shops, a movie theater, a dry cleaner, common areas, offices, health care services, a farmer's market, a day-care center, an elementary school, and a church. Southern Village is a walkable community with sidewalks on both sides of the streets and a 1.3-mile greenway running through the middle of town. Southern Village residents have access to mass transit via Chapel Hill's bus system and can enjoy free outdoor concerts in the common areas. More than 3000 people live in Southern Village.[25]

Urban sprawl has occurred largely because land owners and developers have made choices that promote their own economic and personal interests, which do not necessarily coincide with the public good.[18,25] Many community leaders have found it necessary to engage in centralized urban planning to promote smart growth.[11] Various laws and regulations can help to control land use and development. One of the most useful land-use policy tools is to change zoning laws to promote mixed-use development.[18] Zoning laws that forbid commercial development in residential areas promote sprawl because they require residents to travel greater distances to buy groceries, shop for clothes, and so on. Zoning laws can also be written to encourage high-density development and to require sidewalks and bike lanes.

Another important policy tool for promoting smart growth is to take steps to prevent development outside of a defined urban area, such as forbidding new housing construction on rural land, or setting administrative boundaries for city services, such as water and sewer connections.[18] The government can also use economic incentives to promote smart growth. Developers that follow smart-growth principles can be deemed eligible for reduced fees that help offset the costs of smart-growth development, such as environmental impact fees. Conversely, developers that do not follow smart-growth principles can be subjected to higher fees.[18] Finally, governments can also invest public funds in projects and land uses that facilitate smart growth, such as mass-transit systems, recreation areas, and schools conveniently situated in neighborhoods.[2]

Objections to Smart Growth

Although smart growth appears to be a promising alternative to urban sprawl that could benefit public health and the environment, it has met with stiff resistance in some communities.[11,13,15,18,26] The following are five of the most frequently voiced objections to smart-growth philosophies and policies:

1. **Smart growth can decrease property values.**[11–13] Property values may be adversely affected when high-density housing units are built in an area where low-density housing prevails because the increase in population density may exacerbate local traffic, congestion, and crime, which reduces property values. Property values may also be negatively affected by commercial development in a residential area, because commercial development can increase traffic and crime. Crime may also increase when mass transit connects a residential area to a location where crime is more prevalent, such as the inner city.
2. **Smart growth can decrease the availability of affordable housing.**[14,15] Requiring developers to build planned communities with mixed uses, sidewalks, recreation areas, and bike paths may increase the cost of housing. Also, setting aside large undeveloped spaces can limit land available for development, which drives up the price of housing.
3. **Smart growth restricts property owners' use of their land.**[10,17,27,28] Suburbanites have complained that laws requiring residential areas to have sidewalks and bike paths deprive them of lawn space. Farmers have protested against laws that prevent development of large portions of agricultural and forest land because this interferes with their rights to sell the land.
4. **Smart growth can disrupt existing communities.**[11,12,29,30] Low-density, quiet, noncommercial living areas may become high-density, noisy, and commercial. Historically low-income minority communities may be displaced to make room for highrise, smart-growth housing complexes and upscale commercial development.
5. **Smart growth may increase sprawl instead of decreasing it.**[11,14] Some opponents of smart growth have argued that it often fails to achieve its intended effect and can actually exacerbate sprawl, traffic, congestion, pollution, and other urban problems.

Proponents of smart growth have responded to these and other objections at meetings of county planning boards and city councils, but opposition remains strong. Though smart growth has been a popular buzzword in real estate and urban development since the 1990s, some leaders of the movement worry that it has lost momentum.[13,16] One reason why smart growth has stalled is that key stakeholders involved in the debate—real estate developers, land owners, environmentalists, public health advocates, and people living in metropolitan areas affected by smart-growth projects—have divergent interests, and the political process has often been unable to resolve these conflicts.[18]

Deliberative Democracy

One approach to resolving controversial public-policy questions that may be able to help loosen the smart-growth gridlock is a procedure known as deliberative democracy. Democracy is a form of government in which citizens wield political power by directly voting on issues, as in referendums, or by electing representatives to make decisions on their behalf.[31] Deliberative democracy emphasizes public deliberation on controversial issues as an alternative to voting.[31–34] In deliberative democracy, public deliberation should meet five conditions:[31–34]

1. **Political legitimacy.** The parties to the deliberation view the democratic process as a source of political legitimacy and are willing to abide by the decision that is reached.
2. **Mutual respect.** The parties are committed to respecting each other's diverging interests, goals, and moral, political, or religious viewpoints.
3. **Inclusiveness.** All parties with an interest in the issue can participate in the deliberative process, and a special effort is made to include those parties who often lack political influence because of socioeconomic status, lack of education, or other factors.
4. **Public reason.** Parties involved in the deliberation are committed to giving publicly acceptable arguments for their positions, drawing on publicly available evidence and information.
5. **Equality.** All parties to the deliberation have equal standing to defend and criticize arguments; there is no hierarchy or presumed line of authority.

Deliberative democracy was originally proposed as a method for resolving disagreements on controversial topics for which interested parties have conflicting interests and incompatible moral or political viewpoints, such as abortion, euthanasia, and capital punishment. Proponents of deliberative democracy have argued that public deliberation about controversial topics can be more fair and effective than can traditional democratic procedures, which can be manipulated by powerful interest groups.[31–34] Critics of deliberative democracy have argued that it is an idealized theory of political decision-making whose conditions are often not met in the real world.[35] However, deliberative democracy may be worth trying when other approaches have failed to resolve controversial issues.

The debate about smart growth appears to be a good candidate for application of a deliberative approach because the parties have conflicting interests and divergent moral and political viewpoints.[10,11,18,28] Proponents of smart growth typically argue that collective action must be taken to promote common goods, such as public health, environmental integrity, or overall quality of life.[4,5,7] This type of argument is utilitarian in form because it asserts that public policies should promote the overall good of society.[36,37] Many of the property owners who oppose smart growth assume a libertarian perspective and argue that individual rights may be restricted only to prevent harm to others, not to promote the good of society.[18] According to libertarianism, the role of the state is to protect individual rights to life, liberty, or property; thus, government authority should not be used to redistribute wealth or advance social causes.[38,39] Critics who are concerned that smart growth may reduce the availability of affordable housing or adversely affect minority neighborhoods may subscribe to an egalitarian philosophy, such as Rawls's theory of justice, which holds that public policies should promote the interests of the least advantaged people in society and should not undermine equality of opportunity.[40,41] If smart growth benefits society as whole at the expense of harming its least advantaged members by reducing the availability of affordable housing or disrupting minority neighborhoods, then it would violate Rawls's egalitarian principles of justice. Thus, the debate about smart growth can be viewed as a conflict among three competing visions of social justice: utilitarianism, libertarianism, and egalitarianism.

Deliberating About Smart Growth

Smart growth is an important strategy for combating the adverse public health, environmental, and aesthetic effects of urban sprawl. Because proponents and opponents of smart growth have conflicting interests and divergent moral and political viewpoints, deliberative democracy may be a fair and effective procedure for addressing some of the controversies surrounding policy proposals designed to counteract urban sprawl. To implement a deliberative approach, governments should sponsor open community forums on issues related to sprawl and smart growth, such as focus groups, public debates, and town-hall meetings. The deliberations that occur at these public forums should supplement the discussions that take place on county planning board or city council meetings. The goal of these public forums should be to foster open debate, information sharing, constructive criticism, and mutual understanding. Forums should be well-publicized and open to all parties with an interest in the proceedings. A special effort should be made to invite participants from groups that lack political influence.[42] Many communities have already held open forums on smart growth that embody some of the principles of deliberative democracy, but many others have not.[11,18,26] Communities that have not tried the deliberative approach should attempt it; those that have already held open forums should continue deliberating.

All references for articles included in *Taking Sides: Clashing Views in Sustainability* can be found on the Web at www.mhhe.com/cls.

EXPLORING THE ISSUE

Does Sustainable Urban Development Require More Policy Innovation and Planning?

Critical Thinking and Reflection

1. What is "urban sprawl"? What are its causes and impact on the environment? Do you believe that what researchers call "urban sprawl" is an important part of the American dream?
2. What is "smart growth"? What are its basic elements and why are its policies and programs designed to minimize urban sprawl? Is smart growth an attack on American freedom of mobility?
3. Does smart growth and other forms of planning require a level of governmental intervention into the lives of Americans that many would find to be invasive?
4. What are the characteristics of "sustainable communities," and are they a feasible solution to the way that Americans presently design their communities? Would you like to live in a sustainable community? What would be the advantages and disadvantages of living in this type of community?
5. Do people who live in compact communities enjoy a healthier way of living than people who live in sprawling, automobile-oriented suburbs? Cities are more compact communities than suburbs. Can Americans be attracted back to cities?

Is There Common Ground?

It is generally accepted that the present pattern of American urbanization that emphasizes horizontal spread development causes adverse environmental impacts. This pattern can reduce water quality by increasing the amount of surface runoff, which brings sediments and other pollutants into streams and rivers. Poor water quality is associated with a variety of health problems including diseases of the gastrointestinal tract, kidney disease, and cancer. In addition to air and water pollution, adverse environmental impacts of horizontal development include loss of valuable agricultural land, deforestation, loss of natural ecosystems, and disruption of wildlife habitat. A sprawling development pattern consumes land, energy, and materials in the construction of roads and other man-made infrastructure.

Yet, Americans seem to desire the traditional suburban pattern of development. The emphasis on the single-family home sitting on a large land parcel is considered to be a main feature of the American dream. This model encourages the heavy use of automobiles to connect home, work, and play and is considered to be an unsustainable pattern of urban development. The challenge will be to convince Americans that compact living is as desirable as suburban decentralization and that other types of living styles, ones that are more sustainable, are just as desirable. The question of whether this can be done will determine whether Americans can move toward sustainable living in their communities.

The questions that can be asked are as follows: Can new communities be designed with access to reliable public transportation? Should sustainable urban development be based on resource conservation? Will it be possible for urban areas to produce a significant amount of food supply locally? Will sustainable urban development improve the social and racial inequities that urban sprawl has caused? Is the federal government willing to get involved as a leader in the revitalization of urban areas with meaningful policies and programs? All of these are challenges for urban sustainability.

Internet References

Bas, B. (2004). Is Smart Growth a Smart Adaptation Strategy? Examining Ontario's Proposed Growth Under Climate Change. *Ekistics,* vol. 426, pp. 57–62.

Bengston, D., Fletcher, J., Kirsten, K. (2004). Public Policies for Managing Urban Growth and Protecting Open Space: Policy Instruments and Lessons Learned in the United States. *Landscape and Urban Planning,* vol. 69.

Malin, C. (2010). Perspectives in Urban Sustainability. *Public Management,* vol. 92, no. 7, pp. 20–22.

Mc Elroy, R., Rich, T. (2007). The Congestion Problem. *Public Roads,* vol. 71, no. 1, pp. 2–10.

Mills, S.E. (2007). The Attrition of Urban Real-Property Rights. *The Independent Review,* vol. 12, no. 2, pp. 199–210.

O'toole, R. (2004). A Portlander's View of Smart Growth. *Review of Austrian Economics,* vol. 7, no. 2–3, pp. 203–212.

Rivers, R. (2004). The Price of Sprawl in Ontario, Canada. *Ekistics,* vol. 71, pp. 52–57.

Southworth, M. (2005). Designing the Walkable City. *Journal of Urban Planning and Development.* vol. 131, no. 4, pp. 246–255.

Taylor, R.W. & Carandang, J.S., 2010. Sustainability Planning for Philippine Cities. Manila:Yuchengco Center—De La Salle University, p. 4.

> Managing growth, reducing traffic, creating sustainable development, and making smart transportation investments, these are all challenges we face today. New urbanism is a development strategy that addresses these issues and more by creating communities that are livable, walkable, and sustainable, while raising the quality of life.
>
> **http://www.newurbanism.org/sustainability**

Vertical farms, many stories high, will be situated in the heart of the world's urban centers. If successfully implemented, they offer the promise of urban renewal, sustainable production of a safe and varied food supply (year-round crop production), and the eventual repair of ecosystems that have been sacrificed for horizontal farming.

http://www.vertical/farm.com

The Greenbelt permanently protects precious lands and supports a healthier environment for all Ontarians. By protecting 1.8 million acres of sensitive land from development, the Greenbelt protects the water we drink and the air we breathe. It also means that farmers can continue to grow the food we eat closer to home.

http://www.greenbelt.ontario.ca

The Green Globes system is a revolutionary building environmental design and management tool. It delivers an online assessment protocol, rating system, and guidance for green building design, operation, and management.

http://www.greenglobes.com

Hirschorn, Joel. *Why Is Sprawl So Hard to Curb: Time to Face Sprawl Politics and the Sprawl Lobby* (2003).

http://www.progress.org

Smart Growth Gateway. *A Smart Growth Primer* (2010).

http://www.smartgrowthgateway.org/njplan

ISSUE 12

Should Water Be Privatized?

YES: Roy Whitehead Jr. and Walter Block, from Excerpts from "Environmental Takings of Private Water Rights—The Case for Water Privatization," *Environmental Law Reporter* (October 2002), reprinted in www.eli.org

NO: David Hall and Emanuele Lobina, from *The Private Sector in Water in 2009* (Public Services International Research Unit, Business School, University of Greenwich, March 2009)

Learning Outcomes

After reading this issue, you should be able to:

- Understand the general positions of both those opposed to and in favor of water privatization.
- Have an understanding of the complexities of the water privatization debate.
- Evaluate whether there really is a yes or no answer in the water privatization debate.
- Understand the important roles that water conservation and adequate regulations play in the debate.
- Generally understand the issues pertaining to providing water service to the unserved and underserved as well as how to procure funding for that on an ongoing basis.

ISSUE SUMMARY

YES: Professor Roy Whitehead from the University of Central Arkansas and Professor Walter Block from Loyola University, New Orleans, make a case for the sanctity of private property rights and how privatization of water resources leads to economic development and a more habitable earth for people to live.

NO: David Hall is Director of Public Service International Research Unit (PSIRU) at the Business School of the University of Greenwich, London and Emanuele Lobina specializes in water research as PSIRU.

Both writers boldly state that the "experiment with water privatization has failed."

Water resources have been a principle challenge for humanity from the early days of civilization (Dolayar & Gray, 2000). They constitute a key issue for global sustainability. The United Nations estimates that over four billion people living in the developing world do not have access to clean water. They also indicate that between 30 and 50 percent of the water is diverted for irrigation purposes or is lost though leaking pipes (Morrissette & Borer, 2005). Presently, the United Nations states that 20 countries in the developing world suffer from water stress (defined as per capita available freshwater of less than 1,000 cubic meters) and projects that this will rise to 25 countries by 2050. The question of how best to provide freshwater to a growing global population could be one of the most pressing problems that face sustainability scholars and policymakers.

Many in the ongoing debate over global water supplies argue that water is not a commodity, but rather, a basic necessity of life that should be a common property and not a private good. One of the leading opponents of the commercialization and privatization of water is the environmental activist, physicist, writer, and eloquent orator, Vandana Shiva, who wrote the book *Water Wars* in 2002. Shiva contends that, "More than any other resource, water needs to remain a common good and requires community management. In fact, in most societies, private ownership of water has been prohibited" (2002, preface). The Center for Public Integrity stated the case even more forcefully when it asked, "How can anyone market something that is both vital to life and has no alternative?" (ICIJ, 2003, p. 16).

The United Nations Committee on Economic, Social and Cultural Rights stated in November 2002 that "[t]he human right to water is indispensible for leading a life of human dignity. It is a prerequisite for the realization of other human rights" (2002). On September 30, 2010, the United Nations Human Rights Council affirmed by resolution that water is a human right and that such right is legally binding in international law (United Nations, 2010). Others see water as purely a commodity, while some organizations, such as the World Bank, pragmatically believe that the resource must be properly priced to reflect true costs and to promote efficiency, conservation, and extensions of service.

Religious organizations, such as the Interfaith Center for Corporate Responsibility, have even joined the fray, arguing with biblical support that water is a human right (Interfaith Center on Corporate Responsibility). To help define the parameters on the entitlement to water, the American Association for the Advancement of Science, the Right to Water Programme of the Center for Housing Rights and Evictions, the Swiss Agency for Development and Cooperation, and UN-HABITAT teamed together to publish a "Manual on the Right to Water and Sanitation" in 2007 (UN-HABITAT). The consumer/environmental activist group Public Citizen, along with a host of other like-minded organizations, drafted a "U.S. Water Declaration" premised on the idea that "water is essential to sustain life. It is enshrined in the right to life and dignity, as set forth in the International Bill of Human Rights (Public Citizen)."

Supporters of the privatization of the water supply point out that putting a price on a resource generally leads to greater efforts at conservation. The World Bank believes that providing water at little or no cost does not provide the right incentive for consumers to use the resource wisely (World Bank, 2003, *A Priority for Responsible Growth and Poverty Reduction, An Agenda for Investment and Policy Change*). Accordingly, the Bank has advocated policies to privatize water supplies and to introduce full-cost pricing. In the context of privatization, full cost commonly includes profit margins. The Bank has stressed that water and sanitation services should be paid for by the public and private entities that use them (World Bank, 1992). Both the World Bank and the International Monetary Fund (IMF) have made water privatization a condition of financing and debt restructuring. In India, for example, pressure from the World Bank and the IMF led the government to sell off water rights to international water corporations such as Suez and Veoila (formerly Vivendi) and to industries that are heavy water users (Barlow & Clarke, 2002). To address the problematic issue of providing adequate water for the disadvantaged, the World Bank rightly suggests that water rates should be progressive—with minimal rates charged for low, efficient consumption and increasing rates charged for higher water usage. In any event, the Bank contends that that any fee schedule must collectively provide for the full recovery of costs in order to ensure the financial viability of the water purveyor (World Bank, 1992). According to the World Bank, the poor are impacted much more when water services are poorly managed. The poor in cities in the developing world pay far more for water if they are not served by formal water systems. As one example of the benefits of water privatization for the poor the Bank points to the city of Buenos Aires, which increased its service base by 1.5 million customers as a result of awarding France's Suez a concession to take over the city's water supply system. The Bank concludes that well-designed water infrastructure and market-oriented reforms can provide the poor with growth and opportunities rather than hardships as many anti-privatization advocates have contended. It maintains that sustainable growth and poverty reduction in many developing nations are inexorably linked with improved water development and management (World Bank, 2003, *Water Resources Sector Strategy: Strategic Directions for World Bank Engagement*).

The opponents of water privatization argue that corporations have an incentive to manage for profit but not for the long-term sustainability of the resource base. In England, where large-scale divestiture of public water systems originated in the 1980s, consumers' water bills reportedly doubled between 1989 and 1995 while the profit margins of the private sector water purveyors reportedly jumped almost 700 percent in the same period. In France, where privatized water has been around since the mid-1800s, cities with private water systems reportedly charge 30 percent more than those with public water systems (Barlow & Clark, 2002). The World Bank has reported on one study of water privatization in the United Kingdom that found the primary beneficiaries to be not the public, but rather, the stockholders of the water corporations that had benefitted from price increases (World Bank, 2001).

The critics of privatization point to examples where corporate efficiencies reportedly did not translate into water system improvements or savings. One cited example is a Suez subsidiary in the United Kingdom that was found by the government to have the second worst operational record in the county. The City of Potsdam, Germany, terminated its contract with Suez when the company demanded higher rates because it discovered that water consumption was much lower than it had anticipated. In Buenos Aires, where Suez's successes have been much heralded, considerable rate increases followed the company's takeover of the water system and led to profits that consistently amounted to well over 20 percent of revenues. Similar criticisms have been leveled at Veolia for its operations in Puerto Rico, Argentina, Kenya, and Germany (Barlow & Clarke, 2002). Comparable concerns about water privatization resonate from many quarters, with critics pointing out that claims by corporations to improve service and infrastructure while maintaining or reducing rates are illusory. Rather, the opponents to privatization cite a purported operational reality that includes rate hikes, service disconnects, decreased water quality, contract violations, layoffs and labor violations, billing irregularities, and a general lack of accountability (Naegele, 2004).

The YES selection, that water should be privatized, is written by professors Roy Whitehead and Walter Block. They take a more philosophical position in the debate over privatization by stating that the "cause of justice" would be enhanced if all bodies of water were transferred from the public to the private sector. The authors make a case for the sanctity of private property rights and how privatization of water resources, while a more difficult argument to make, would lead to increased economic development and make the earth a more habitable place to live.

The NO selection, that water should not be privatized, is written by two water researchers, David Hall and Emanuele Lobina. They open their piece by boldly and unequivocally stating that, "The experiment with water privatization has failed." They throw down the anti-privatization gauntlet at the outset and make it clear that privatization was only an "experiment" and that it has "failed" in every instance.

**Roy Whitehead Jr.
and Walter Block**

Environmental Takings of Private Water Rights—The Case for Water Privatization

Water Privatization

Introduction

Private property rights have benefitted every arena of human experience they have touched.[1] The economy of the Soviet Union fell apart mainly because of the absence of this system.[2] The U.S. economy is one of the foremost in the world largely due to its relatively greater reliance on this institution.[3] And yet, there are vast areas in which private property rights play no role at all: namely, oceans, seas, rivers, and other bodies of water. But why should we expect that there would be any better results from such "water socialism" than we have experienced from socialism on land? Indeed, the evidence is all around us attesting to this fact: whales are an endangered species; fish stocks are precipitously declining; oil spills are a recurring problem; droughts are becoming increasingly severe and prolonged, and not only in the underdeveloped countries of the world; rivers are polluted, some so seriously that they actually catch fire; lakes are becoming overcrowded with boaters, swimmers, fishermen, etc., and there is no market mechanism to allocate this scarce resource amongst the competing users; deep sea mining (manganese modules) is in a state of suspended animation due to unclear titles; and the legal status of offshore oil drilling rigs is unclear. Most revealing, water covers some 79% of the earth's surface,[4] but accounts for only a small percentage of world gross domestic product (GDP).[5] While no one expects an exact proportionality between surface coverage and contribution to economic welfare, such a strong disparity suggests that the economic system pursued in these two realms may not be totally unrelated to these results.

Our claim is that we have no warrant to believe that socialism, the absence of private property rights, is any more workable on land than on water. It is time—indeed, it is long past time—to explore ways in which this institution can be applied to aqueous resources.

The Case for Privatization

Privatization is the process of transferring governmental ownership, management, and control from governmental to private hands. The case for privatization, in general, is straightforward. It consists of utilitarian and deontological reasons extolling the benefits of this course of action.

What is the utilitarian case? Individual firms, owned by private persons, are better able to promote consumer sovereignty[6] than are statist agglomerations. This comes about mainly through the weeding out process.[7] Those entrepreneurs who cannot satisfy customers are forced into bankruptcy through such competition.

Similarly, the deontological case for privatization is simple and straightforward. Individuals, but not governments, can come to own land and other resources through homesteading,[8] the only method which can justify ownership on the basis of the libertarian legal code.[9] Any attempt on the part of the state to engage in this activity is fatally compromised by its essentially coercive nature. Government ownership of resources is only legitimate in the statist philosophy of coercive socialism or fascism. (We here abstract from the limited government libertarian perspective which makes an exception for what it characterizes as legitimate state functions: armies to repel foreign invaders, police to reduce invasive acts on the part of local miscreants, and courts to determine who is who in this regard.)[10]

If the case for privatization is a simple one, so too does this apply to many specific instances of this doctrine. For example, the privatization of public housing,[11] state enterprises in western countries,[12] in the Soviet Union and other former communist countries,[13] and the U.S. Post Office.[14] Privatization of roads, highways, and sidewalks are perhaps more conceptually complex, but even here there is a plethora of literature attesting to benefits and justification.[15]

Perhaps the most difficult case to make on behalf of privatization concerns water resources. Under this rubric we include anything'[16] which admits of moisture: aquifers, brooks, canals, channels, drinking water, drainage ditches, ducts, estuaries, flumes, groundwater, icebergs, irrigation ditches, lagoons, lakes, oases, oceans, ponds, puddles, reservoirs, rivers, runoff, seas, springs, streams, swamps, underground water, water basins, watercourses, waterfalls, watermains, watersheds, water tables, water traps on public golf courses, waterways, wetlands, etc.[17] In addition, privatization of air and water where appropriate is a necessary if not sufficient condition for reducing the harm done to mankind by clouds, flooding, fog, hurricanes, storms, tidal waves, tornados, torrential rain, tsunamis, typhoons, whirlpools, winds, etc.

Why is it necessary that extra care and thoroughness be taken in the attempt to build the case for privatizing water? There are several reasons.

Opposition to Water Privatization

Rare

Water privatization has rarely, if ever, been done. Apart from a few small private lakes and ponds used for fishing, boating, and swimming, there are no cases[18] of private ownership, even in countries ostensibly devoted to free enterprise but which in actuality practice various versions of "water socialism."

Unexplored

Indeed, the topic has not even been explored in the literature. There is, of course, a wealth of information available concerning water resources, but very little of it speaks in favor of full privatization.

Out of Fashion

Water privatization is simply not in keeping with current intellectual opinion. People balk at privatization for roads and other facilities mentioned above on the rare occasions the subject is acknowledged at all in what might be characterized as the "mainstream" literature. It is probably no exaggeration to predict that when and if the typical public policy wonk hears of the thesis which motivates the present enterprise, to wit, to privatize bodies of water as fully as land masses, he will dismiss it out of hand as a particularly noxious form of lunacy.

Interconnectedness

According to that old song, "the hip bone is connected to the thigh bone, is connected to. . . ." In like manner, most bodies of water are joined with most others. That is, although we call the various oceans of the world by different names, e.g., Atlantic, Pacific, Indian, they all touch upon and flow into each other at their common boundaries, Even a seemingly isolated lake is not totally separated from other bodies of water insofar as it has streams feeding into and out of it. These water avenues lead to still others and eventually to the sea, where they are linked to all others.

But the colors of the rainbow also shade into one another,[19] Yet we have no trouble distinguishing one from the other, except of course at their very boundaries. And with precision scientific instruments, we can at least mark off an agreed upon fence post. Land, too, is all interconnected, apart from where it ends at water's edge. And, for a time, our society had difficulty marking off one man's holdings from that of another. But with the advent of fencing materials, particularly barbed wire, this became easier and easier.

No, the interconnectedness of even all bodies of water constitutes no overwhelming objection to privatization. The reason we have no "fences" to place in the water is not because this is an impossible idea: it is rather due to the fact that absent aqueous property rights, there has been no financial incentive to engage in research to this end. But imagine the opposite. Suppose, that is, that property rights in bodies of water were recognized by law. It takes no great leap of imagination to suppose that scientists and engineers would soon be able to offer new technology which could distinguish between "mine and thine."

Nor need these water fences be used only to demarcate the property holdings of one firm from that of another. They can also be used to corral fish, whales, and other ocean livestock. For all too long these creatures have been free to roam the range of the oceans. It is time, it is past time, for we humans to do for them what we have done for land-based animals[20]: to tame and domesticate them,[21] and to bring them within the purview of economic rationality. [22]

Not only is water connected in the horizontal realm, so to speak, the same pertains in the vertical. That is, the three-quarters of the earth's surface on which water rests is the horizontal axis, while the vertical dimension refers to the fact that a molecule is at one time water in the ocean, and at another it evaporates and travels into the clouds, whereupon it rains down onto the surface of the earth, either on land or in the sea, but eventually comes to rest in the latter, after traveling through the river system. It is thus insufficient to ascertain only who is the owner of water in the sea; ownership must also be determined, if we are to specify a complete system, for water while it is on its way up to the sky through evaporation, while it is in a (temporary) state of "rest" in the clouds, and when it is on its way down again in the form of rain.

But these are mere technical issues. Where there is a will (and a legal system that supports it), there is a way. The reason this has not yet occurred is not entirely due to costs; a large part of the blame must rest, also, with the fact that we have not pushed the private property rights envelope far enough, yet, in terms of water.

Arrogance[23]

The idea that man should own the oceans and the seas will appear as arrogance to some people. The "tower of Babel" story in the *Bible*[24] would appear to be apropos. When man's pride and ambition got him above himself, God struck back by making it difficult for him to communicate with his fellows.

But why should land be any different than water? If it is not morally sinful to aspire to ownership of the former, why should this apply to the latter? One might with as much reason claim that walking is justified, but that driving a car, sailing a boat, or, perish the thought, flying an airplane are perversions, or somehow impious.

Legal Nightmare

Suppose a river, such as the Mississippi, changes its course, and starts moving over previously dry land. Or that any river overflows, flooding surrounding farms and neighboring houses. If the river in question were privately owned, a charge could be made that this would create a legal nightmare.

This is only legally problematic, however, because there are no precedents, and there are no precedents, in turn, because rivers are presently unowned. Their mismanagement hence now constitutes an "act of God." Instead of blaming the Deity, we would do well to attempt to address these dangers and inconveniences. Just as there should be no "fish freedom" the same should apply to rivers. Flooding and course changes[25] should be seen for the mismanagement they are. The reason there has been no private investment in taming these unruly bodies of water is that there are no economic incentives to do so. It would not pay for any single farmer located on the banks of a river to attempt to take on so gargantuan a task. Neoclassical economists would characterize this as a "market failure," since such a farmer would not be able to recoup an amount even near to his total investment. But these so-called external economies stem not from anything intrinsic to the situation; rather, they are the result of lack of ownership and responsibility.

Of course, there will be complications when this arena of the law is recognized. Absent any contract to the contrary, for example, a river owner should not be liable for all damages caused, say, by flooding due to heavy rain,[26] but only for those in excess of the amount that otherwise would have ensued. For example, if it can be shown that ordinarily, under river socialism, a storm of a certain severity would cause $100 in damages, and that in the actual event it caused only $75, then plaintiff should not be able to collect anything at all from the river owner. On the other hand, if under our assumptions $125 worth of harm was inflicted upon owners whose property abuts the river, then the defendant would be responsible for at most $25.

Equity

Another argument against private ownership is that the rich would hog it all up, and leave the short end of the stick for the poor. There is no doubt that this fear motivates, at least in part, the United Nations (U.N.) Law of the Sea Treaty, according to which "the oceans are the common heritage of all of mankind,"[27] and that therefore no individual nation, let alone private person, should be allowed to own any of it. The fear on the part of the U.N. bureaucrats who hail from the underdeveloped nations is that they do not have the requisite technology to mine manganese nodules at the bottom of the ocean, for example, and that it is "unfair" for those with this ability to be able to make us of it on their own accounts. Another way of putting this matter is that the landlocked nations would be at a disadvantage vis-à-vis those which border on the sea, and that the former are poorer than the latter, and thus it would be "inequitable" to allow a competitive race to take advantage of such watery resources.

The implication seems to be better that no one should be able to own aqueous possessions than that the rich be afforded this opportunity. One difficulty with this position is that it equates "equity" or "fairness" with "egalitarianism." But nothing could be further from the truth. If it were so, then advocates of this position would be willing to give up their own "excessive" intelligence, or "IQ" points, were this possible, to their less intellectually well-endowed brethren. That no one has even taken this position shows that even its advocates shrink in horror from the logical implications of their own system.

Insofar as is the economic well-being of the poor of the earth is concerned, it is clear that the wealth of the less fortunate would be enhanced, not worsened, by allowing economic opportunity to the rich. This is because economic development is a positive, not a zero, sum game. Under capitalism, the wherewithal enjoyed by both parties to a transaction, at least in the ex ante sense, is increased. The rich do not increase their income at the expense of the poor; rather, their income rises by *enriching* the less well to do. It is no accident that the poor in the more capitalist West enjoy a standard of living that is the envy of those at the bottom of the income distribution, and even in the middle of it, in countries infected by coercive socialism.[28]

Monopoly

There is the fear that under private ownership of seas, there could be monopolistic encroachment. For example, A owns an island which is completely surrounded by ocean,[29] and B owns the surrounding patch of water. Thus, A would be trapped on his island prison. Some property rights for A!

But a moment's reflection should convince us that this is an unlikely, if not an impossible, situation.[30] First of all, the primary and first user of the waterway surrounding A's island is likely to be A himself. According to homesteading theory. A would thus be the rightful owner of the surrounding aqueous area, not B. Second, if B first homesteaded the water, and only then, later, did A come upon the island to take up ownership over it, the latter would never have done so unless his access and egress rights were clearly specified in such a manner so as to not preclude the economic viability of ownership of the island in the first place. Third, there are airplanes and helicopters available, at least in the modern era.[31]

The thesis of this discussion is that all bodies of water should be fully and completely privatized. Consider the ocean in this regard. This would mean not merely that fishing should be limited to those who purchase rights to do so, but that the whole kit and kaboodle would be treated in much the same way as are land holdings.[32] That is, the surface of the ocean would be owned, just as railroads presently are, at least in the United States, and just as roads and highways would be, at least as contemplated by authors who advocate such a situation.[33] This is not to say that it is contemplated that all of the oceans, every single cubic foot of them, should immediately be privatized. Many of them are as presently worthless for prospective owners, for all practical purposes, as are some of the more out of the way acreage in Alaska, Antarctica, and Siberia.[34] All we are deliberating upon is the *legal status* of these places. At present, it is impossible to own them both because the law does not allow it, and, in many cases,[35] ownership is not yet economically viable. What is being advocated is a change in the *law*, such that those parts of the watery domain for which private property is now economically workable would be allowed at once to be owned, and that more and more of them could come to be owned when their economic status changed so as to make private ownership a paying proposition.

One clear benefit would be that world GDP would rise. At present, the oceans and seas, as we have seen, account for a large part of the earth's surface, but only a small percentage[36] of world GDP. It need not be the case that each and every acre of the earth's surface account for the same proportionate contribution to GDP as every other. Deserts are less productive than fertile land. But at least a large part of the vast disparity between productivity on land and in and on water must be due to the beneficial effects of private property rights on land that do not apply to water.[37]

On land, man went through the hunting and gathering stage, during which his standard of living was appropriate to the stone age. When he graduated from this precarious existence to one of farming, his standard of living exploded in an upward direction, as did sustainable population size. After that came manufacturing, and then the information age, with similar upward spurts in how well man could live, and how many of his species could be supported.

As far as the seas are concerned, however, we are still back in a "caveman" type of development, wherein hunting and gathering are . . . only avenues open to us. It was not until the institution of private [main] property took hold on the land that farming, herding, and later developments could be supported. It is a well-known fact, at least within the free market environmental community, that the cow prospered, due to private property rights which could avert the tragedy of the commons, while the bison almost perished as a species due to lack of same. Nowadays, happily, this problem has been remedied with regard to the buffalo.[38] But the whale, the porpoise, edible fish, and other sea species are dealt with, at present, in precisely the same manner that almost accounted for the disappearance of the bison.

Individual transferable quotas (ITQs), of course, are a vast improvement over nonownership, with attendant and uneconomic overfishing. But they constitute only a quasi-private property rights system, not the pure form of this institution. In order to see this, consider imposing ITQs on buffalo, or elephants. This would mean that these animals would still be free to roam as they wished, but it would be legal for only certain people to be able to hunt them. The point is, we would still be in the hunting stage of human existence with regard to such species. But if economic history has taught us anything, it is that herding is far more efficient than hunting. E.g., corralling fish in the open ocean is far more effective than fishing, or hunting, for them.

This scenario assumes, of course, that the necessary complementary technological breakthroughs occur, such as either genetic branding, or perhaps better yet, electrified fences, which can keep the denizens of the deep penned in where deep-sea fish farmers want them. Yes, this seems unlikely at present, given that under present law there would be no economic benefit to such inventions. But this is due, in turn, not to any primordial fact of nature or law. Rather, it is because the law has not yet been changed so as to recognize even the possible future scenario in which ocean privatization would be economic. The public policy recommendation stemming from this analysis is merely that the law should now be changed so as to recognize fish ownership in a given cubic area of ocean when and if such an act becomes technically viable. Then, whether or not it actually occurs is only an empirical question. It will, if and only if the complementary technology is forthcoming to make it feasible. But under this ideal state of affairs, there would be no legal impediment, as there now is, in this direction. That is, suppose that the needed innovations never occur, or are always too expensive, compared to the gains to be made by herding fish instead of hunting them. Then, of course, there can be no private property rights used in this manner in the ocean, as a matter of fact. But as a matter of *law*, things would still be different under the present proposal. There would always be the contrary to fact conditional in operation that *if* technology were such, *then* it would be legal to fence in parts of the ocean for these purposes. Under this state of affairs, there would be no legal impediments to the development of the requisite technology.[39]

Another benefit would be making the earth a more habitable place in which to live. Consider in this regard clouds, flooding, fog, hurricanes, storms, tidal waves, tornados, torrential rain, tsunamis, typhoons, whirlpools, winds, etc. At present, these are considered acts of God. If the oceans and the air, from which and in which these disasters emanate, were allowed by law to be owned by firms or individuals, at least in principle, this might well set up the first steps in mankind's long journey to quelling these "natural" disasters. How else could this ever be done, other than by employing the institution of private property rights, which is responsible for so much else we include under the category of "good works?"

Notes

1. Tom Bethell, The Noblest Triumph: Property and Prosperity Through the Ages (1998); Richard Pipes, Property and Freedom: The Story of How Through the Centuries Private Ownership Has Promoted Liberty and the Rule of Law (2000).
2. F.A. Hayek, *Socialist Calculation I, II, & III,* in Individualism and Economic Order (1948); Hans-Hermann Hoppe, A Theory of Socialism and Capitalism (1989); Ludwig Mises, Socialism (1969); Walter Block, *Socialist Psychology; Values and Motivations,* 5 Cultural Dynamics 260 (1992); Peter J. Boettke, Why Perestroika Failed: The Politics and Economics of Socialist Transformation (1993); Peter J. Boettke, The Collapse of Development Planning (1994); Peter J. Boettke & Gary Anderson, *Perestroika and Public Choice: The Economics of Autocratic Succession in a Rent Seeking Society,* Pub. Choice, Feb. 1993, at 101; Peter J. Boettke & Gary Anderson, *Soviet Venality: The USSR as a Mercantilist State,* Pub. Choice, Jan./Feb. 1997, at 93.
3. James Gwartney et al., Economic Freedom of the World, 1975–1995 (1996).
4. The New Encyclopaedia Britannica 320 (15th ed, 1998) ("The planet's total surface area is roughly 509,600,000 square km (197,000,000 square miles), of which about 29 percent, or 148,000,000 square km (57,000,000 square miles), is land. The balance of the surface is covered by the oceans and smaller seas.").
5. Mike Dowling, *Interactive Table of World Nations Sorted by Gross Domestic Product, at* http://www.mrdowling.com/800gdp.html (last updated Sept. 5, 2000).
6. *See* William H. Hutt, *The Concept of Consumers' Sovereignty,* Econ. J., Mar. 1940, at 66. For the related concept, individual sovereignty, which is even more in accord with libertarian free enterprise principles, see Murray N. Rothbard, Man, Economy, and State (1962).
7. Henry Hazlitt, Economics in One Lesson (1979).
8. Walter Block, *Earning Happiness Through Homesteading Unowned Land: A Comment on* Buying Misery With Federal Land *by Richard Stroup,* J. Soc. Pol. & Econ. Stud., Summer 1990, at 237; Hans-Hermann Hoppe, The Economics and Ethics of Private Property: Studies in Political Economy and Philosophy (1993); John Locke, *An Essay Concerning the True Origin, Extent, and End of Civil Government, in* Two Treatises of Government 17 (P. Laslett ed., 1960) (1690); John Locke, Second Treatise of Civil Government (Henry Regnery Press 1955) (1690); Murray N. Rothbard, Power

and Market: Government and the Economy (1970); Murray N. Rothbard, For a New Liberty (1978); Murray N. Rothbard, The Ethics of Liberty (New York Univ. Press 1998) (1982).

9. Terry Anderson & P.J. Hill, *An American Experiment in Anarcho-Capitalism: The* Not *so Wild, Wild West,* 3 J. Libertarian Stud. 9 (1979); Randy E. Barnett, The Structure of Liberty: Justice and the Rule of Law (1998); Bruce L. Benson, *Enforcement of Private Property Rights in Primitive Societies: Law Without Government,* J. Libertarian Stud., Winter 1989, at 1; Alfred G. Cuzán, *Do We Ever Really Get Out of Anarchy?,* J. Libertarian Stud., Summer 1979, at 151; Anthony DeJasay, Against Politics: On Government, Anarchy, and Order (1997); David Friedman, The Machinery of Freedom: Guide to a Radical Capitalism (2d ed, 1989); Hans-Hermann Hoppe, Democracy, the God That Failed: The Economics and Politics of Monarchy, Democracy, and Natural Order (2001); Jeffrey Rogers Hummel, *National Goods Versus Public Goods: Defense, Disarmament, and Free Riders,* 4 Rev. Austrian Econ. 88 (1990); Stephan Kinsella, *Estoppel: A New Justification for Individual Rights,* Reason Papers No. 17, Fall 1992, at 61; Andrew P. Morriss, *Miners, Vigilantes, and Cattlemen: Overcoming Free Rider Problems in the Private Provision of Law,* 33 Land & Water L. Rev. 581 (1998); Franz Oppenheimer, The State (1914); Aeon J. Skoble, *The Anarchism Controversy, in* Liberty for the 21st Century: Essays in Contemporary Libertarian Thought 77 (Tibor Machan & Douglas Rasmussen eds., 1995); Larry J. Sechrest, *Rand, Anarchy, and Taxes,* J. Ayn Rand Stud., Fall 1999, at 87; Lysander Spooner, No Treason (1870); Edward Stringham, *Justice Without Government,* J. Libertarian Stud., Winter 1998–1999, at 53–77; Patrick Tinsley, *With Liberty and Justice for All: A Case for Private Police,* J. Libertarian Stud., Winter 1998–1999, at 95–100; Morris Tannehill & Linda Tannehill, The Market for Liberty (1984); William C. Woolridge, Uncle Sam the Monopoly Man (1970).
10. For an articulation of the minarchist free market philosophy, see Tibor Machan, *Against Nonlibertarian Natural Rights,* J. Libertarian Stud., Fall 1998, at 233; Charles Murray, What It Means to Be a Libertarian (1997); Leonard E. Read, Anything That's Peaceful (1964).
11. Jane Jacobs, The Death and Life of Great American Cities (1989).
12. Privatization and Development (Steven H. Hanke ed., 1987); Michael A. Walker, Privatization: Tactics and Techniques (1988); T.M. Ohashi et al., Privatization Theory & Practice (1980); Madson Pirie, Privatization in Theory and Practice (1986); Bruce L. Benson, To Serve and Protect: Privatization and Community in Criminal Justice (1998); Terry L. Anderson & Peter J. Hill, *Privatizing the Commons: An Improvement,* 50 So, Econ. J. 438 (1983); The Mechanics of Privatization (Eamonn Butler ed., 1988).
13. Peter J. Boettke, *The Austrian Critique and the Demise of Socialism: The Soviet Case,* in 17 Austrian Economics; Perspectives on the Past and Prospects for the Future 181 (Richard M. Ebeling ed., 1991); David Conway, A Farewell to Marx; An Outline and Appraisal of His Theories (1987); Raimondo Cubeddu, The Philosophy of the Austrian School 109 (1993); James Dorn, *Markets True and False in Yugoslavia,* J. Libertarian Stud., Fall 1978, at 243; Richard M. Ebeling, *Economic Calculation Under Socialism: Ludwig von Mises and His Predecessors, in* The Meaning of Ludwig von Mises 56 (Jeffrey Herbener ed., 1996); Nicolai Juul Foss, *Information and*

the Market Economy: A Note on a Common Marxist Fallacy, 8 Rev. Austrian Econ. 127 (1995); David Gordon, Resurrecting Marx: The Analytical Marxists on Freedom, Exploitation, and Justice (1990); Friedrich A. Hayek, *Socialism and War: Essays, Documents, Reviews, in* 10 The Collected Works of F.A. Hayek (B. Caldwell ed., 1997); Robert Heilbroner, *Analysis and Vision in the History of Monetary Economic Thought,* J. Econ, Literature, Sept. 1990, at 1097; Hans-Hermann Hoppe, *De-Socialization in a United Germany,* 5 Rev. Austrian Econ. 77 (1991); Steven Horwitz, *Money, Money Prices, and the Socialist Calculation Debates,* 3 Advances in Austrian Econ. 59 (1996); Willem Keizer, *Schumpeter's Walrasian Stand in the Socialist Calculation Debate, in* Austrian Economics in Debate (Willem Keizer et al. eds., 1997); Peter G. Klein, *Economic Calculation and the Limits of Organization,* 9 Rev. Austrian Econ. 3 (1996); Don Lavoie, Rivalry and Central Planning: The Socialist Calculation Debate Reconsidered (1985); Peter Lewin, *The Firm, Money, and Economic Calculation,* Am. J. Econ. & Soc., Oct. 1998, at 499–519; Yuri N. Maltsev, Requiem for Marx (1993); Ludwig von Mises, Socialism (1981); Gerald P. O'Driscoll Jr., Economics as a Coordination Problem (1977); George Reisman, Capitalism: A Treatise on Economics 135–39, 267–82 (1996); Morgan O. Reynolds, *The Impossibility of Socialist Economy,* Q. J. Austrian Econ., Summer 1998, at 29; Murray N. Rothbard, *How and How Not to Desocialize,* 6 Rev. Austrian Econ. 65 (1992); Joseph T. Salerno, *Ludwig von Mises as a Social Rationalist,* 4 Rev. Austrian Econ. 26 (1990); David Ramsey Steele, *From Marx to Mises: Post-Capitalist Society and the Challenge of Economic Calculation* (1992); Karen I. Vaughn, *Economic Calculation Under Socialism: the Austrian Contribution,* Econ. Inquiry, June 1980, at 535.

14. Douglas K. Adie, *Why Marginal Reform of the U.S. Postal Service Won't Succeed,* in Free the Mail: Ending the Postal Monopoly (Peter J. Ferrara ed., 1990) [hereinafter Free the Mail]; Thomas Gale Moore, *The Federal Postal Monopoly: History, Rationale and Future,* in Free the Mail, *supra*; George Priest, *The History of the Postal Monopoly in the United States,* 18 J. Law & Econ. at 33 (1975); Stuart M. Butler, *Privatizing Bulk Mail,* 6 Mgmt. at 155 (1986); Stephen Moore, *Privatizing the U.S. Postal Service, in* Privatization (Stephen Moore & Stuart Butler eds., 1987).
15. David T. & Linda Royster Beito, *Rival Road Builders: Private Toll Roads in Nevada, 1852–1880,* Nev. Hist. Soc. Q., Summer 1998, at 71; Walter Block, *Free Market Transportation: Denationalizing the Roads,* J. Libertarian Stud., Summer 1979, at 209; Walter Block, *Congestion and Road Pricing,* J. Libertarian Stud., Fall 1980, at 299; Walter Block, *Public Goods and Externalities: The Case of Roads,* J. Libertarian Stud., Spring 1983, at 1; Walter Block, *Theories of Highway Safety,* Transp. Res. Rec. No. 912, 1983, at 7; Walter Block, *Road Socialism,* 9 Intl, J. Value-Based Mgmt. 195 (1996); Walter Block & *Matthew Block, Roads, Bridges, Sunlight, and Private Property Rights,* J. des Economistes et des Etudes Humaines, June/Sept. 1996, at 351; Walter Block, *Roads, Bridges, Sunlight, and Private Property: Reply to Gordon Tullock,* J. des Economistes et des Etudes Humaines, June/Sept. 1998, at 315; Fred Foldvary, Public Goods and Private Communities; The Market Provision of Social Services (1994); Michelle Cadin & Walter Block, *Privatize the Public Highway System,* The Freeman, Feb. 1997, at 96; Dan Klein et al., *From Trunk to Branch: Toll Roads in New York,* 1800–1860), in Essays in Economic and Business History 191 (Dan Klein ed.,

1993); Dan Klein & G.J. Fielding, *Private Toll Roads: Learning From the Nineteenth Century, Transp.* Q., July 1992, at 321; Bertrand Lemennicier, *La Privatisation des Rues,* J. des Economistes et des Etudes Humaines, June/Sept. 1996, at 363; John Semmens, *The Privatization of Highway Facilities,* Transp. Res. F., Nov. 1983, at 54–59; John Semmens, Why We Need Highway Privatization (1991); John Semmens, *Privatizing Vehicle Registrations, Driver's Licenses, and Auto Insurance,* Transp. Q., Fall 1995, at 125–35.

16. However, it must be underscored that only scarce resources are candidates for property ownership. *See* Gene Callahan, *Rethinking Patent Law,* Mises Inst., July 18, 2000, at http://www.mises.org/fullstory.asp?control=468&FS=Rethinking+Patent+Law (last visited June 4, 2002); Julio H. Cole, *Patents and Copyrights: Do the Benefits Exceed the Costs?, at* http://www.economia.ufm.edu.gt/Catedraticos/jhcole/Cole%20_MPS_.pdf (last visited June 4, 2002); Stephan N. Kinsella, *Against Intellectual Property,* J. Libertarian Stud. (forthcoming 2002); Stephan N. Kinsella, *Is IP Property or Not?*, Nat' Post, Feb. 22, 2001, at 1; Stephan N. Kinsella, *In Defense of Napster and Against the Second Homesteading Rule, at* http://www.lewrockwell.com/orig/kinsella2.html (last visited June 4, 2002); Wendy McElroy, *Intellectual Property: Copyright and Patent in* Liberty, *at* http://www.zetetics.com/mac/intprol.htm (last visited June 4, 2002); Moore D. Adam, *A Lockean Theory of Intellectual Property,* 21 Hamline L. Rev. 65 (1977).
17. The head police character in the movie the *Fugitive* demanded of his minions that they search every "house, barn, shed, palace, outhouse, doghouse, etc. . . ." Our goal is to be as exhaustive as he.
18. We here abstract from such things as backyard swimming pools, jacuzzis, bath tubs, showers, water faucets, cesspools, water foun tains, septic tanks, etc., which already fall under private control.
19. Similarly for the boundaries between radio and television stations on the electromagnetic spectrum. For the case in favor of privatization in this regard, see Ronald H. Coase, *The Federal Communications Commission,* 2 *J.* Law & Econ. 1 (1959).
20. For the argument that elephants, rhinos, and other endangered species would benefit from being barnyardized, e.g., fenced in with electrically charged wires, see Randy Simmons & Urs Kreuter, *Herd Mentality: Banning Ivory Sales Is No Way to Save the Elephant,* Pol'y Rev. Fall 1989, at 46; *Environmental Problems, Private Property Rights Solutions, in* Block, Reconciliation, *supra* note 40; Walter Block, *Environmentalism and Freedom: The Case for Private Property Rights,* J., Bus. Ethics, Dec. 1998, at 1887.
21. For the argument that this is symbiotic, e.g., beneficial to both mankind and animal and fish species, see Henry E. Heffner, *The Symbiotic Nature of Animal Research,* 43 Persp. in Biology & Med., Autumn 1999, at 128.
22. It is time, too, to jettison such socialist and profoundly anti-private property rights songs as; "Home, home on the range," and "Where the deer and antelope play."
23. We owe this objection to Marybeth Block.
24. *See also* Aristophanes' theory of love.
25. This applies only to *unwelcome* flooding and course changes. But railroads and highways sometimes change their location. If there is an economic need for this in the case of a river, and it is accomplished at minimal cost, then this constitutes an exception to the claim made in the text.

26. We assume, for the moment, that the level of technology, or of the law, is such that the clouds themselves are not owned, and that thus no one is liable for their excessive and unwarranted rain on the river. For some people, to blame rain or storm on the state is only a joke. This is not the case at present. Had the government not taken as much of the GDP as it has, to fritter it away on warfare and welfare state considerations (and for numbered bank accounts in Switzerland), there would have been just that much more available to address private needs. Some of this, undoubtedly, would have been spent in an effort to domesticate weather conditions.
27. *Available at* http://www.un.org/Depts/los/convention_agreements/texts/unclos/closindx.htm.
28. *See, e.g.,* James Gwartney et al., Economic Freedom of the World, 1975–1995 (1996).
29. This is by definition.
30. A similar objection with regard to private roads and streets has been dealt with in Walter Block, *Free Market Transportation: Denationalizing the Roads,* J. Libertarian Stud., Summer 1979, at 209.
31. This, of course, invites discussion of ownership of the relevant air travel rights, a topic we address below.
32. There are several publications whose titles indicate they are compatible with this very radical enterprise, but they are misnomers. *See, e.g.,* Terry L. Anderson & Donald R. Leal, Free Market Environmentalism (1991), The authors call their chapter nine *Homesteading the Oceans,* a policy taken seriously in the present paper, but these authors discuss only schemes to quasi-privatize fish. Similarly, the title employed by Birgir Runoflsson, *Fencing the Oceans,* Reg., Summer 1997, at 57, is misleading in that it also advocates only individual transferable quotas (ITQs) in fish, as its subtitle (*A Rights-Based Approach to Privatizing Fisheries*) makes clear. A similar analysis applies to Ross D. Eckert, The Enclosure of Ocean Resources: Economics and the Law of the Sea (1979). For a critique of tradeable emissions rights (TERs), the air analogue of ITQs in the water, see Robert W. McGee & Walter Block, *Pollution Trading Permits as a Form of Market Socialism, and the Search for a Real Market Solution to Environmental Pollution,* 6 Fordham U. L. & Envtl. J. 51 (1994). Kent Jeffreys, in *Who Should Own the Ocean?*, Competitive Enterprise Inst. Update, No. 8, Aug. 1991, at 1, perhaps comes the closest of the material cited in this footnote to our own vision of full water privatization, but even he focuses mainly on the problem of overfishing, and contemplates "permitting . . . outright ownership of limited ocean areas. For example, offshore rigs. . . ." *Id.* But why not outright ownership of *all* as opposed to "limited" ocean areas? Private ownership of offshore rigs, moreover, is already a staple of present sea law.
33. *See supra* note 155.
34. Actually, these are particularly inept examples, in that land in none of these three places is fully open for private holdings.
35. Apart from those areas of the seas which are located near population centers. There is no doubt that did the law but allow it, for example, private individuals would be willing—indeed, and more than willing—to own the Hudson River.
36. *See supra* notes 144, 145 and accompanying text.

37. See on this point the extensive "tragedy of the commons" literature. Indeed, one could expand this so as to include the literature on the failure of socialism, "water socialism" in this case.
38. And other previously endangered species also, such as the elephant, the rhinoceros, the alligator.
39. It is on this point that the "Chicago School" analysis of property rights goes wrong. In that perspective, private property rights only arise when technology, an exogenous force, makes them economically practicable. There can *be* no private property rights in the ocean unless and until electric sea fences are invented. Science is the dog, while the law is the tail that is wagged. In contrast, in the libertarian vision that underlies the present paper, technology is endogenous. It is the tail that is wagged by the legal dog. Private property rights to *anything* will *always* be recognized in law, as a matter of course, stemming from homesteading: when and if ocean owners stake claims, based on mixing their labor with this element, for which new presently nonexisting technology is available, then it will be recognized. The difference in this case is a subtle one: in the libertarian legal code, the law gives incentives for such innovations, by guaranteeing recognition of such property titles when they are achieved; in the Chicagoite tradition, the law does not. For the Chicago view of property rights, see Richard A. Posner, *Killing or Wounding to Protect a Property Interest,* 14 J.L. & Econ. 201 (1971); Richard A. Posner, Economic Analysis of Law (5th ed. 1998) [hereinafter Posner, Economic Analysis]; Ronald H. Coase, *The Problem of Social Cost,* J.L. & Econ., Oct. 1960, at 1; Harold Demsetz, *Some Aspects of Property Rights,* J.L. & Econ., Oct. 1966, at 61; Demsetz, Harold, *Toward a Theory of Property Rights,* 57 A m. Econ. Rev. 347 (1967). For the libertarian critique, see Walter Block, *O. J. 's Defense: A Reductio Ad Absurdum of the Economics of Ronald Coase and Richard Posner,* 3 Eur. J.L. & Econ. 265 (1996); Roy E. Cordato, *Subjective Value, Time Passage, and the Economics of Harmful Effects,* 12 Hamline L. Rev. 229 (1989); Roy E. Cordato, Welfare Economics and Externalities in an Open-Ended Universe: A Modern Austrian Perspective (1992); Elizabeth Krecke, *Law and the Market Order: An Austrian Critique of the Economic Analysis of Law,* J. des Economistes et des Etudes Humaines, Mar. 1996, at 19; Commentaries on Law & Economics 86 (Robert W. McGee ed., 1997); Gary North, Tools of Dominion: The Case Laws of Exodus (1990); Gary North, The Coase Theorem (1992). For the debate between Block and Demsetz on these matters, see Walter Block, *Coase and Demsetz on Private Property Rights,* J. Libertarian Stud., Spring 1977, at 111; Harold Demsetz, *Ethics and Efficiency in Property Rights Systems, in* Time, Uncertainty and Disequilibrium: Explorations of Austrian Themes (Mario Rizzo ed., 1979); Walter Block, *Ethics, Efficiency, Coasean Property Rights and Psychic Income: A Reply to Demsetz,* 8 Rev. Austrian Econ. 61 (1995); Harold Demsetz, *Block's Erroneous Interpretations,* 10 Rev. Austrian Econ. 101 (1997); Walter Block, *Private Property Rights, Erroneous Interpretations, Morality, and Economics: Reply to Demsetz,* Q. J. Austrian Econ., Spring 2000, at 63.

David Hall and
Emanuele Lobina

The Private Sector in Water in 2009

Private Limitations

The Failed Experiment

The experiment with water privatisation has failed. Since about 1990 privatisation has been actively promoted by the international institutions, donors, and private companies themselves as a way of delivering investment, efficiency and building effective water operators in developing countries. These expectations have not been delivered. . . .

Turkey and Failed Water Privatisations

Turkey has first-hand experience of the problems of failed water privatisations, including in the city of Istanbul itself, where the 5th World Water Forum is being held. These three failures all concern very large contracts; they involve each of the three largest water companies: Veolia, Suez and Thames Water; and cover each of three major forms of privatisation—a concession, a lease 'affermage contract, and a BOT.'

Istanbul: The First Failed Water Privatisation?

In 1882 the French water company **Generale des Eaux** (now **Veolia**) obtained a water supply concession in Istanbul, at that time still called Constantinople, and the capital of the Ottoman empire. The concession was awarded under an early version of IMF conditions—west European countries, principally Britain and France, took control of the public finances of the Ottoman empire in 1881, in order to repay debts to foreign investors, and this committee used its power to give concessions to western companies: "Monopolies were granted to foreign companies to control the tobacco industry, and to operate railways, tramways, ports, gas, electricity and waterworks."[1] The Turkish republic was created in the 1920s, and the foreign financial control committee was dismantled. The democratic authorities of the new state decided that the concession was failing in terms of coverage, investment and technology, and in the 1930s the concession of Generale des Eaux was terminated, and the water services of

Istanbul were taken into public ownership. This decision reflected not only the economic, social and technical failure of the contract, but also the new-found confidence of an independent and democratic state.[2]

Corruption, Excessive Pricing and the Yuvacik Dam

The Yuvacik Dam near Izmit in Turkey was built by the English water company **Thames Water** under a build-operate-transfer (BOT) concession contract. The contract, signed in 1996, stated that the water would be purchased over 15 years at a price set in the contract. This price was so high that both industrial users and neighbouring municipalities refused to buy water from the plant, but the purchase of water was guaranteed by the Turkish Government, which has thus paid far more than it should have done for water which is unaffordable for its intended customers. The Turkish Court of Accounts, the national public audit body, recommended in November 2003, that nine former ministers and the former mayor of Izmit should be investigated for corruption. In 2005 Thames Water initiated international arbitration against Turkey on the grounds that the Treasury was not paying the guaranteed amount in full. The project has been summarised in a World Bank report as: "plagued by inadequate project preparation, the inappropriate granting of Treasury guarantees, and lack of competition in bidding. It has resulted in huge payments by Treasury for water that is not used." These problems are typical of infrastructure BOT contacts signed in Turkey: a study by the Inspectors of the State Supervision Agency discovered irregularities in almost all such contracts, and estimated that the return on shareholders' capital on many projects was around 320% *per year.*[3]

Antalya: The Failed Affermage Contract

The privatisation in Antalya was imposed by the World Bank as a condition for a $100m. loan. The contract was awarded in 1996 to a **Suez** subsidiary. The privatisation was not supported by the city council but imposed by the World Bank; the Bank also specified that the private operator should only employ 50% of the existing staff, as a result of which the existing staff preferred to transfer to other municipal jobs. Non-revenue water was actually much higher at the end of the concession, at 63%, than it was at the start (46%). Suez demanded that water prices should be more than doubled, and that sewerage prices should be increased by over 10 times. When the municipality refused, the Suez subsidiary declared itself bankrupt and abandoned the contract in 2002. The service was taken over by the municipal water company, and prices decreased slightly in the subsequent years.[4]

Water Multinationals and the Economic Crisis

The water multinationals began retreating in 2003, with much encouragement from local people. They remain committed to maximising profitable market opportunities, however, and remain highly influential in policy circles. Both **Suez** and **Veolia** have suffered from the crisis and recession. **Veolia** especially has problems with net debt of €17bn, eight times its annual earnings. Rather than expanding, **Veolia** now plans to sell €3billion worth of assets in 2009, while investment plans have been cut by 45%; **Suez** has cut investment plans by 25%. Between March 2008 and March 2009 **Veolia** shares fell by 70%, compared with an overall drop of 46% for the Paris stock exchange as a whole (comparable data for **Suez** Environnement is not available).[5] **Suez** states that it plans to raise money by issuing a bond some time in 2009, but the interest rates it will have to pay are very high: in February 2009 companies were having to pay investors over 4% per annum more than government bonds.[6]

Both companies face the prospect of losing their water distribution contracts for the city of Paris, which is remunicipalising the service from the 1st January 2010, **Veolia** also faces losing some or all of its water contract in the Ile-de-France region around Paris, worth €370m. per year. Although the regional authority voted against remunicpalisation in December 2008, the contract will be re-tendered and possibly in smaller packages. **Suez**, on the other hand, supports retendering in smaller packages.[7] This is an unusual disagreement between the two groups, who have formed many joint ventures in French cities to ensure they share the business without having to compete against each other. The competition commission ruled in 2002 that these were anti-competitive, and after 7 years the two companies have responded by agreeing between themselves who will have which city—**Suez** keeps the city of Lille, for example, while **Veolia** gets Marseille.[8]

The profitability of waste management, the other sector in which both **Suez** and **Veolia** operate, has been hit by the collapse of the market for recycled 'secondary' materials, and by some major industrial companies closing factories where **Suez** have contracts. **Veolia's** transport business, Connex, is also experiencing problems, with the loss of the Stockholm metro contract.[9]

Despite attempts to privatise water worldwide, the major French water companies all now depend on the French government as their dominant shareholder. **Suez** Environnement became an independent company in July 2008 when the energy division of **Suez** was merged with the former state-owned Gaz de France to form GdF-**Suez**, But the French state now owns 36% of GdF-**Suez**, and GdF-**Suez** in turn still holds 35% of **Suez** Environnement, and so the state is indirectly much the largest shareholder, and de facto has a controlling stake. **Veolia** is 10% owned by the Caisse des Depots et Consignements (CDC), an investment agency of the French state; while the third company, **SAUR**, which now operates only in France, is 47% owned by the CDC.[10] These holdings are part of a deliberate cross-party strategy to use state ownership to prevent foreign takeovers of strategic French companies like **Suez** and **Veolia**: 'a militant state capitalism to keep the decision-making centres of the big private companies in France, starting with infrastructure, property and healthcare.'[11]

The economic crisis affects not just individual companies, but all forms of public-private partnerships (PPPs) which include investment financed by a private company. This is because banks and investors will not lend money to private companies, for fear of them defaulting, even in the richest countries. The IFC , the private sector financing arm of the World Bank, thinks the problem is even greater in developing countries, because international investors are even less willing to lend to PPPs there. The IFC estimates that $110 billion worth of proposed PPPs may be delayed or cancelled, and that $70 billion of existing PPPs are at risk because of increased costs of financing these projects for the private sector.[12]

The only internationally active British water company, **Biwater-Cascal**, has had better financial results in the last year, but faces numerous other problems. The Panamanian government has told the company that it wants to terminate its contract prematurely;[13] it has been revealed that **Biwater** paid nearly £100,000 in school fees at a prestigious English boarding school for the children of the president of Ghana, Jerry Rawlings, between 2000 and 2003[14] (in 2008 **Biwater** was awarded a €45 million contract for a water supply expansion scheme at Tamale, in northern Ghana[15]); and that a lake in Guatemala continues to suffer from severe pollution because **Biwater** abandoned a contract to build 3 wastewater treatment plants.[16]

The Myth of the Southern/Local Companies

The PPIAF has reluctantly been obliged to abandon its promotion of multinational companies as the panacea for water problems. But even in this 'mea culpa' paper, it is attempting to purvey an even less convincing myth, namely that new private sector investors and operators are springing up in developing countries (or 'emerging markets', as the PPIAF calls them). Another PPIAF paper in 2008[17] also presents 'the phenomenon of this new breed of investors'[18]; and the OECD has devoted large amounts of expensive conference time to discussing these new saviours. But the evidence is desperately weak and ludicrously exaggerated: the PPIAF. 'mea culpa' paper claims that "It would be hard to overestimate the importance of this new trend," but the paper itself proceeds to do exactly that.[19]

Their list of 'local' companies does not bear much examination:

- It includes a number of companies which no longer own or operate the water services listed
- It includes one 100% public sector company, a company which has only invested in water in the UK, and a company which is a subsidiary of Veolia
- Some companies are historical oddities which cannot be imitated
- A surprising number are owned by companies based in tax havens thousands of miles from the conctracts
- Many of the examples are residual stakes in former privatisations where the multinationals have withdrawn
- Many of the contracts continue to create problems of affordability and under performance. Local investment is weak, but local opposition is very strong.

In Chile, the PPIAF lists three different local companies as holding a total of 6 water concessions. But the Chilean group Fernandez Hurtado sold its majority stake in a water concession (Esval) in 2007, to the Ontario Teachers Pension Plan, which is a pension fund of Canadian public sector employees, not a private southern investor. A year later, the Solari group also sold its concessions (Essat, Essar, Esmag) to an infrastructure fund run by Santander, a large Spanish multinational bank.[20]

The other Chilean interest on the PPIAF list is the Essan concession, in Antofagasta, in the northern mining area of Chile. The PPIAF wrongly describes it as owned by the 'Luksic Group.' The concession is held by Antofagasta plc, one of the oldest companies listed on the London stock exchange, which originated as a 19th century British railway venture in Chile and Bolivia, and is now overwhelmingly a copper mining group, for whom the water concession in Antofagasta represents only 2.5% of their total sales; it is running it because it happens to be in an area where the company has most of its mining operations, and the company has shown no interest in operating any other water service.[21] It is 70% owned and controlled by the Luksic family—not the 'Luksic group'—which is estimated to be the 140th richest in the whole world,[22] who own these shares through trusts which are registered in the European tax haven of Leichtenstein.[23] The Daily Telegraph observed that "the family paid little heed to criticism that they fell short of corporate governance standards normally expected of FTSE companies."[24] It is not clear which aspects of this company are thought to provide a model.

Puncak Niaga is mentioned as a Malaysian company with contracts in Kuala Lumpur and Selangor state. This is not likely to last, however: its operation in Selangor state will be taken back into the public sector very soon (as of March 2009)—the national government and the state government disagree only about who should renationalise it![25] Another Malaysian company, Salcon, is a Malaysian engineering company, and the great majority of its activity is involved in traditional engineering work, including a number of treatment plant BOT contracts, rather than water distribution service operations. It has one general concession in China, at Linyi. Its latest move has been to withdraw rather than expand, by selling its stake in one of its Chinese subsidiaries, Chenggong Salcon Water.[26]

The list also includes the Moroccan company ONEP, which is not a private company, it is the 100% state-owned national water company of Morocco. The Malaysian company YTL, also on the list, is an energy company which has only one investment in water, Wessex Water in the UK; it has never expressed any interest in investing in water activities in developing countries. A similar approach is taken by Cheung Kong Infrastructure of Hong Kong (not mentioned on the PPIAF list), which, like YTL, owns an English water company—Cambridge Water—but has invested in no other water operations. It is not clear that investing Asian money in English water companies has any relevance for development of water services in the south.

Another company on the PPIAF list, Tata, is another historical oddity which is not a replicable model. The Tata group is the second largest private corporation in India and one of the largest multinationals in the world—Tata Steel ranks 315 on the Fortune top 500 companies. It covers many sectors in addition to steel, including engineering, IT, energy, and services—but water is not mentioned by the group even as one element of its services activities. Tata operates water supply services in only one place, Jamshedpur, which is a company town, or 'Tata township' as the company calls it, built by Tata at the start of the 20th century as a site for its factories and workers. Tata has effectively owned the whole town of Jamshedpur, including the water and electricity company, JUSCO, since 1907.[27] There is no effective regulation or political accountability, and slum inhabitants are reported to be charged 10,000 rupees for a household connection.[28] This very untypical water company is now seeking a private management contract in the city of Mysore, Karnataka, where it is being bitterly opposed.[29] Its only international activity has come as a result of the ADB choosing the company as a 'partner' for the National Water Supply and Drainage Board of Sri Lanka.

The South African company WSSA is also a surprising model to choose. It was a joint venture between Suez and a South African partner, which was awarded four private water contracts by the apartheid regime in the early 1990s. The track record gives no reason to treat the company as any kind of model: "The contract in Nkonkobe (formerly Fort Beaufort) was terminated by a court ruling in 2001 that the contract was invalid. . . . The contract at Stutterheim, signed in 1993 . . . allowed the company to 'cherry-pick' the profitable white and coloured areas, which already received dependable water supplies, while much of the official Stutterheim township (Mlungisi) remained unserved and the unofficial neighbouring townships (Cenyu, Kubusie, Cenyulands) were left almost entirely outside the network."[30]

The new owners of the Jakarta water companies (originally privatised, without competitive tendering, by the dictatorship of Suharto, whose family and friends were given a share of the profits by Suez and Thames Water)[31] are also unusual. One, Astra International, is a subsidiary of Jardine Matheson, the British 19th century multinational at the centre of the opium trade forced onto China, now relocated in the British tax haven of Bermuda. The company was caricatured by the British Conservative prime minister Benjamin Disraeli as "McDruggy from Canton, with a million of opium in each pocket, denouncing corruption and bellowing free trade"[32]; a more recent commentator has suggested in 2005 that Astra is now providing for Jardines "the sort of profit margins Jardine Matheson made in China's opium days."[33] The owner of the other Jakarta company is Acuatico, an investment fund set up in Singapore by two companies of obscure parentage from another tax haven, the British Virgin Isles, which was originally rejected as an unsuitable bidder for the Jakarta water.[34] Neither can be described as 'local' investors.

One of the successors to the Manila multinational concessions is similar—the owner of Maynilad Water is a joint venture between DMCI, a Philippine construction company, and MPIC, which is 96% owned by First Pacific Company Ltd (based in the tax haven of Bermuda), whose stake in DMCI-MPIC is 50% backed by Ashmore, a UK-based investment fund managing $23 billion.[35] The owners of the other, Manila Water, include the Philippine Ayala Group (the main shareholder), and also the UK company United Utilities, the Japanese multinational Mitsubishi, and the World Bank itself through the IFC—the main driver of the original privatisations in 1998.[36] These may be seen as welcome (and rare) foreign direct investments, but it is unconvincing to suggest that Manila Water represents a new breed of autonomous, southern, local investors. The company has a declared policy of expanding internationally, but so far has gained only three (or possibly four) contracts outside the Manila concession; a highly controversial and non-competitive bulk water supply contract with MWSS in Bulucan, awarded without competition; another in the Philippines for a specific tourist development, not for extending services to residents; and one in Vietnam, a 5-year consultancy and technical assistance contract on leakage.[37]

The other cases on the list are the various private companies which have collected stakes in the former multinational subsidiaries in Argentina, Brazil, Colombia. The largest of these, Latinaguas, shows the limitations of local companies: it has been warned, criticised and or fined for underinvestment or poor customer service in 2 of its 3 concessions, and supported by public subsidies in the third; and it has expanded internationally only in a joint venture in a small town in Peru. The process of restructuring the water services of Latin America in the aftermath of the multinational exits is a large one, involving a complex set of public and private interplay. But these private companies—mostly local construction companies taken on as partners by the multinationals—are not credible sources of investment finance, nor are they remotely comparable to the public sector operators of the region as a pool of expertise.[38]

Multinationals as Consumers: The World Economic Forum Report

The World Economic Forum (WEF), held at Davos, Switzerland every year, is the main platform for business interests discussions of the world economy. The WEF has created a Global Agenda Council on Water Security, chaired by Margaret Catley-Carlson, who is also the patron of the Global Water Partnership (GWP), of which she was the president for many years.[39] At the January 2009 WEF meeting a paper on water[40] was produced, in partnership with a group of people from drinks, food, chemicals, mining, and agribusiness multinationals—**Coca-Cola, PepsiCo, Nestlé, SABMiller, RioTinto, Dow Chemical, Syngenta, Hindustan Construction Company, the International Federation of Agricultural Producers.**[41] There was only one NGO representative, from WWF—an organisation which has received $23.75m. from **Coca-Cola** since 2007.[42]

These companies are all big consumers of water. Business as a whole consumes nearly three times as much water, globally, as households: **Coca-Cola** alone uses 300 billion cubic metres of water per year.[43] For these companies, water is a valuable input to their products, which they will try and get as cheaply as possible, even at the expense of other water users, such as households and farmers: **Coca-Cola** warns its shareholders that increasing demand for water means that the company "may incur increasing production costs or face capacity constraints which could adversely affect our profitability or net operating revenues in the long run."[44]

A number of these companies have also shown that they are prepared to prioritise their profits at the expense of people in developing countries.

- **Nestle** has persisted in selling milk powder for babies in developing countries, where the profitable substance may be mixed with unsafe water by the millions who still lack a public water supply. The result is lethal: "in areas with unsafe water a bottle-fed child is up to 25 times more likely to die as a result of diarrhoea."[45] Nestle continues to promote its powder "in a systematic and institutionalised way" despite repeated calls from the medical profession and worldwide campaigners to stop.[46]
- **Coca-Cola** has shown that it is prepared to compete with local people for water resources by abstracting water from aquifers in various countries, including India and El Salvador, in such quantities that local farmers have complained of falling water levels, and elected councils and courts have withdrawn their licenses.[47]
- Both **Nestle** and **Coca-Cola** have attempted to suppress trade union organisation in their subsidiaries in developing countries.[48]
- **Dow Chemical** owns the company whose toxic chemical leaks killed over 20,000 people in Bhopal, India, and continues to contaminate the groundwater. **Dow** continues to deny liability for the leak, refuses to offer compensation, and says the environmental cleanup must be paid for by the public authorities, despite continuing campaigns in India and across the world, including by Amnesty International.[49]

The WEF report also describes private investors buying water sources to re-sell at a profit as 'innovative investments'. It offers two examples of this 'innovation:'

- The oil billionaire T. Boone Pickens bought thousands of acres of land over the Ogallala Aquifer in a semi-arid part of Texas, and now (2009) expects to sell the water rights for $75m. to the public water authority in the nearby, fast-growing town of Lubbock.[50]
- A Canadian hedge fund, Sextant Capital Management, bought the rights to glaciers in Iceland. The plan is to bottle and sell the melting glacier waters. Sextant has recorded great profits and extracted large management fees, claiming that the value of the glaciers had appreciated by over 900%. Unfortunately, the Ontario Securities Commission (OSC) is now prosecuting Sextant for inflating the values of its holdings, and its assets have been frozen by court order. The OSC "found no independent valuation to substantiate the claim that these icefields are now worth close to 10 times what Sextant paid for them."[51]

Enron did this, too. In 1999, Enron's water division started developing a 'water bank' at Madera, in California, planning to sell most of the water to public and private customers under long-term contracts, but: "*the remaining 20 percent of the storage capacity will be retained by Azurix for the purpose of trading and optimisation. Trading will be maximized during dry and drought years when demand far exceeds supply.*"[52] The private venture was ferociously opposed by local people, and the project collapsed. The bank is now being revived as a public water storage project, subject to demands for local democratic oversight to prevent water from being sold to maximise profit.[53]

Conclusion

The failure of water privatisation is well-established. Turkey itself has witnessed the problems first hand. The water multinationals no longer have the wish or ability to invest in any but the most securely profitable activities in developing countries. The so-called 'local' companies are not a credible source of investment finance nor of expertise. The multinational consumers of water, or the opportunistic investors in water resources, should be seen as problems rather than solutions.

There is no rational basis for giving these organisations such a prominent role at a conference that should be focussed on developmental needs. The forum should focus on the need to support and develop public finance and public sector operations, rather than continue to focus on the interests of these companies.

All references for articles included in *Taking Sides: Clashing Views in Sustainability* can be found on the Web at www.mhhe.com/cls.

EXPLORING THE ISSUE

Should Water Be Privatized?

Critical Thinking and Reflection

1. Why will the issue of water availability be a major challenge to world population in the twenty-first century? What parts of the world will be underserved and what are the various strategies to serve these populations?
2. Given the complexities and nuances inherent in the privatization of water debate, can there be a yes or no answer, or just a "maybe"?
3. Municipal governments often provide water to their residents instead of contracting these services out to private companies. Do you know who supplies your local water? If your local supplier is a municipally owned system, does it provide good service at reasonable prices? If your local water supplier is a private contractor, does it provide good service at reasonable prices? Compare and contrast these two types of systems with examples from your region.
4. How do we procure the necessary funding on an ongoing basis to provide water service to the underserved? Is the issue of financing more important to consider than the issue of water as a human right?
5. Should as much focus be placed on ensuring sufficient regulatory schemes as is placed on the "anti" privatization debate?
6. Should water conservation issues play a bigger role in the debate? Discuss ways that you could conserve water.

Is There Common Ground?

The issue of providing clean, potable water for global populations will constitute one of the key challenges for sustainability in the twenty-first century. Already, some experts are viewing the availability of potable water as even a greater challenge to mankind than oil, and that future conflicts will be fought over water more than oil. Water is a major requirement for human survival, not only for direct intake, but also for its overwhelming use in food production for a world of seven billion people. How to get clean, potable water to people in their communities at an affordable price constitutes a major issue for sustainability.

The common ground is the recognition that many people are un-served or underserved, that water infrastructure needs funding, and that there is limited funding to provide for those needs. The following questions should be posed if a common ground is to be achieved. Given limited available funding, how do you see public funding being mobilized and increased for existing or new water infrastructure? What are your positions on public-private partnerships and public-public partnerships? Are you aware of privatization failures and why they failed? Are you aware of privatization successes and why they succeeded? What role do you see regulation playing in the debate? Is water underpriced, and if so, for which users and for which types of uses?

Internet References

Food and Water Watch provides a good overview of the privatization debate from the no perspective. As with many no sources, it does not delve deep enough into the debate to fully flesh out the nuances, but it nevertheless provides a comprehensive overview of the scope of the debate and of the no perspective. As a solution to water problems in the United States, F&WW does push for dedicated public funding for that purpose.

www.foodandwaterwatch.org/water/private-vs-public/

The World Water Council is a beneficial resource for obtaining information on the debate. It is a multi-stakeholder organization with extensive and more balanced information on the debate, although it is criticized as being a proponent of privatization or of entities that favor privatization (its membership list speaks of some diversity of perspectives). The WWC's periodic World Water Forums/Fora are also criticized by antiglobalization proponents.

www.worldwatercouncil.org/

The Pacific Institute's (PI) website is a valuable source for information on water issues. Although PI falls more in the no camp, its information and analysis has more depth and substance. The PI publishes biennially "The World's Water," which is an up-to-date compendium on water issues globally. The publication is valuable for numerous disciplines (environmental, economics, engineering, etc.) considering water issues.

www.pacinst.org/

Internet References . . .

The Nature Conservancy

This is the blog for The Nature Conservancy, an international organization that strives to protect "ecologically important lands and waters for nature and people."

http://blog.nature.org/2009/01/value-of-nature/

Food and Agriculture Organization of the United Nations

This website is sponsored by the United Nations and provides current reference material, scientific, professional, and otherwise, regarding key programs, global issues, and core activities that include those in agriculture and natural resources.

www.fao.org/

NOAA National Marine Fisheries Service

The National Marine Fisheries Service's website provides information on issues pertaining to biodiversity and habitat conservation, access to management plans developed by the eight regional councils under Magnusson-Stevens, and current information on fishery restrictions.

www.nmfs.noaa.gov

International Union for the Conservation of Nature

The website of the International Union for the Conservation of Nature, it seeks to provide pragmatic solutions to finding solutions to our most pressing environmental and development challenges.

www.iucn.org

Natural Resources Defense Council

This website is the public face for the Natural Resources Defense Council which is a grassroots organization that promotes environmental action to "safeguard the Earth: its people, its plants and animals and the natural systems on which all life depends."

www.nrdc.org

UNIT 4

Natural Resources

Sustainability by necessity deals with issues related to natural resources. How can our natural resource base be maintained? Should a national accounting system be established that debits the use of natural resources rather than count their use as an asset toward the GDP? Because natural resources are often finite, their use today means that future generations will be deprived of their use. One of the underlying principles of sustainability is the notion of transferring resources to future generations, a form of intergenerational equity.

One key issue related to natural resources is food supply? Will there be enough land to support a growing world population or will we need to get higher yields in agriculture? While organic agriculture is more sustainable than traditional chemical-based agricultural systems, can organic agriculture feed a world population that now exceeds 7 billion people? Will the demand for protein that occurs when a world becomes more affluent mean even greater use of land for animal feed?

The human impact on the natural resource base has produced significant impacts. One impact is the incursion of humans into the habitats of other species, with dire results. This has led to the extinction of species and a general human–wildlife confrontation. Certainly, biodiversity is threatened by human activities. Is biodiversity important for ecosystem preservation and human survivability? Or does it mean that we need to adjust to a different world. Will natural gas resources contained in shale formations solve our needs for a cleaner and more abundant form of fossil fuel energy?

These questions and issues are centered on debates over estimates of known quantities of natural resources, that is, known oil deposits; the best ways to use these resources; and whether we can develop sustainable practices to reduce or mitigate their use. Understanding these issues will allow us to make better decisions on their use.

ISSUE 13

Can Our Marine Resources Be Sustainably Managed?

YES: Benjamin S. Halpern, from "The Impact of Marine Reserves: Do Reserves Work and Does Reserve Size Matter?," *Ecological Applications* (February 2003)

NO: Andrew A. Rosenberg, Jill H. Swasey, and Margaret Bowman, from "Rebuilding U.S. Fisheries: Progress and Problems," *Frontiers in Ecology and the Environment* (August 2006)

Learning Outcomes

After reading this issue, you should be able to:

- Understand the threat to fisheries posed by overfishing and the various mechanisms currently in use globally to relieve overfishing pressure.
- Recognize how behavior on land affects the oceans, including issues pertaining to pollution, nutrient run-off, sediment erosion, fossil fuel, extraction, and global climate change.
- Discuss the ecological benefits of marine protected areas and other management strategies.
- Describe the positive and negative effects of fish farming and aquaculture on wild fisheries.
- Understand why the local public must be considered if management strategies are to be successful.

ISSUE SUMMARY

YES: Benjamin S. Halpern, marine biologist and project coordinator of Ecosystem-Based Management of Coastal Marine Systems for the National Center for Ecological Analysis and Synthesis (NCEAS) demonstrates how marine protected areas (MPAs) and marine reserves, tools for sustainably managing marine resources, are producing positive results based on four biological measures: density, biomass, size of organisms, and diversity.

NO: Andrew A. Rosenberg, biologist and oceanographer and presently dean of the College of Life Sciences at the University of New

Hampshire, states that the Magnuson–Stevens Fishery Conservation and Management Act has not significantly altered overfishing and the rebuilding of fish stocks in the United States due mainly to pressures from the commercial and recreational fishing communities.

The term "marine resources" comprises a myriad of issues, including maintaining sustainable fisheries (threatened by overfishing, habitat destruction, and consequences of fish farming), the maintenance of marine biodiversity (threatened by introduced species, the harvesting of tropical fish, corals, other invertebrates for the aquarium trade, etc.), and the maintenance of habitat availability and diversity (threatened by destructive fishing practices, marine and terrestrial pollution, and climate change–caused coral bleaching). To the public, although, "marine resources" generally means just one thing—sustainable commercial fisheries and sustainable seafood. There are many methods currently in use to manage fisheries for long-term sustainability, including marine protected areas (MPAs), industry-wide fishing restrictions, individual catch shares (individual transferable quotas, ITQs), fish farming to supplement the wild-fish market, and international treaty agreements. The question to address is whether any of these management approaches have the capacity to sustainably preserve marine ecosystems.

One of the most celebrated management approaches to maintain marine ecosystems are MPAs. These areas are intended to provide safe havens for species to regenerate and maintain sustainable population levels, and by doing so increase four primary biological indicators: organism size, total biomass, population density, and biodiversity. There are hundreds of studies supporting the use of MPAs. They have been shown to increase abundances and individual fitness within reserve boundaries (Halpern, 2003) as well as outside of reserve boundaries (Kellner et al., 2007) by means of the phenomenon known as "spillover," whereby populations outside of reserve boundaries are supplemented by the overabundances within the boundaries. As Kellner et al. (2007) describe, the increase in abundance is such that "fishing the line" is seen in many areas whereby fishers hug the edges of reserves to take advantage of the spillover.

Of late, some scientists have criticized the "hype" of MPAs, attempting to clarify what MPAs can and cannot accomplish. Hilborn (2006) despairs that managers have forgotten lessons learned throughout recent decades, writing "a community of belief has arisen whose credo has become 'fisheries management has failed, [that] we need to abandon the old approaches [using solely] marine protected areas and ecosystem-based management.'" Jennings (2009) makes a further case against MPAs, noting that their implementation in poor and developing nations as the sole management strategy is doomed for failure. He writes that although MPAs do generate ecotourism income, "the reliance of many poor people on an environment that cannot meet their food and income needs will inevitably lead to unsustainable fisheries." He concludes that support for the creation of alternate income sources will be required before management plans can produce positive results.

Other marine management strategies are catch shares and fishing restrictions. Both approaches serve to limit a fishery's total yearly catch, but in very different ways. Fishing restrictions limit the total industry-wide catch, often including size and sex limits (especially as regards females carrying eggs or of breeding size), bans on certain types of fishing gear, and season time limits. Catch shares, of which ITQs are one type, provide catch limits to individual fishers, in effect providing "ownership" of a fraction of a yearly or seasonal catch. In this way, a fisher is free to fish at his leisure (within season limits) up to his or her max catch allowance, or to trade away his catch license. In 2008, Costello et al. found that the implementation of ITQs can halt and even reverse a trend toward fishery collapse both on the local and on the global scale. He suggests that this is due to the fact that individual fishers no longer need to "race to fish" in order to secure their financial security. Although this data would imply that fishers can benefit economically from participating in catch share programs, other studies have shown declining fishing-related livelihoods resulting from closures and fishing restrictions (Murray et al., 2010).

Another marine management approach is to utilize international treaties to control overfishing and trafficking in illegally fished products. Many proponents of international cooperation and stewardship point to these agreements, including the International Whaling Moratorium of 1986 (www.iwcoffice.org), as success stories. Even today, however, several countries still refuse to subscribe to the whaling ban, as of this writing, Norway, Iceland, Greenland/Denmark, and Japan all continue domestically sanctioned whaling. The Convention on International Trade of Endangered Species (CITES) protects against the trade of threatened oceanic wildlife, although the effectiveness of any such treaty is debatable, especially as regards the cessation of illegal poaching. The most frequent criticism in the literature is not the scope of these laws but rather the lack of enforcement.

Our domestic laws, namely the Magnuson–Stevens Fishery Conservation and Management Act of 1976, are condemned in the literature for similar reasons. The Magnuson–Stevens Act officially phased out all foreign fishing within 200 miles of the U.S. coastline and created eight regional fishery management councils responsible for approving species- and ecosystem-specific management plans. Rosenberg et al. (2006) find fault with the enforcement of Magnuson–Stevens, noting that 72 percent of stocks managed under the Act are still overfished, and that only 3 fisheries out of the 69 covered by the Act have been successfully rebuilt to healthy population levels. He specifically cites the failure of "many plans to reduce exploitation sufficiently to end overfishing." Safina et al. (2005) find that most threatened stocks do have sufficient population levels to allow for rebuilding within a 10-year timeframe, the amount of time that Magnuson–Stevens provides for rebuilding stocks. However, many plans approved by the regional councils have been implemented with timeframes exceeding the 10-year standard, phasing in management strategies instead of immediately ceasing overfishing and fishing pressures. Safina et al. (2005) found that stocks that had management strategies phased in over time have still not recovered, while the few stocks that have seen drastic and immediate measures put into place have seen swift recoveries.

Still, another way to sustainably manage marine resources is through fish farming and aquaculture. Although farms are not exactly a strategy to directly rebuild fisheries, some experts have suggested increasing aquaculture as a way to relieve pressure on wild stocks. Benetti et al. (2006) noted in one study that fish farming does not have a detrimental impact on marine ecosystems, finding no evident eutrophication of waters upstream or downstream of offshore fish cages. Moreover, no changes to the substrate directly under the cages were found, and the authors found evidence that companies invested in fish farms had also invested in finding fish-free alternatives to fishmeal. Conversely, Naylor et al. (2000) found that fish farms increased fishing pressures on wild fisheries due to the need for wild-caught fish for fishmeal feeds. Rather than relieving pressures, they were instead shifted down the food chain to species at lower trophic levels. Naylor et al. (2000) also found that fish farms pollute their surrounding environments causing eutrophication of the waters around the farms (in direct contradiction to the Benetti et al. (2006) study), and that they further increase nonfishing pressures on wild stocks in the form of genetic drift and the increased prevalence of communicable diseases.

There is a growing consensus that fisheries management is a complex interdisciplinary problem in need of a complex interdisciplinary answer. Most scientists and managers agree that consideration of the public needs to be incorporated into future management plans. Jennings (2009) goes so far as to suggest that the creation of alternate income sources for fishermen is perhaps more important than any fisheries management strategy. These and other social concerns will likely be the subject of much fisheries management research in coming years.

The article for the YES position is presented by marine biologist Benjamin S. Halpern. He demonstrates how MPAs and marine reserves are producing positive results based on four biological measures: density, biomass, size of organisms, and diversity. He finds that the relative benefits of both large and small reserves are equal, clearly outlining evidence that size, biomass, diversity, and density all increase regardless of reserve size. He clarifies the benefits of all MPAs of any size, while making the point that when creating new MPAs, the size and location must be determined from local scientific data and not by politics and convenience.

The article for the NO position is presented by biologist and oceanographer Andrew A. Rosenberg. He states that the Magnuson–Stevens Fishery Conservation and Management Act has not significantly altered overfishing or encouraged the rebuilding of fish stocks in the United States. The reason is due mainly to poor implementation of management plans as a result of pressures coming from the commercial and recreational fishing communities. He provides examples where implementation has successfully increased fishing stocks (Atlantic sea scallops, Georges Bank haddock, and North Atlantic swordfish) and other examples where poor implementation of recovery plans has resulted in no increase in fish population (snowy grouper, South Atlantic black sea bass and cod). He concludes that the immediate cessation of overfishing and a decrease in fishing pressures, as required by Magnuson–Stevens, is essential if the United States is to maintain sustainable wild fisheries.

Benjamin S. Halpern

The Impact of Marine Reserves: Do Reserves Work and Does Reserve Size Matter?

Introduction

Marine reserves (also called marine protected areas, no-take zones, marine sanctuaries, etc.) have recently become a major focus in marine ecology, fisheries management, and conservation biology. Interest stems in part from the realization that traditional forms of fisheries stock management are inadequate, as evidenced by the historical and recent collapse of many fisheries. In addition, traditional management methods such as maximum sustainable yield estimates are inadequate for addressing the multiple types of anthropogenic impacts on marine life such as overfishing, certain fishing methods, pollution, coastal development, and other human-derived impacts. Marine reserves have been proposed as an efficient and inexpensive way to maintain and manage fisheries while simultaneously preserving biodiversity and meeting other conservation objectives as well as human needs (Plan Development Team [PDT] 1990, Ballantine 1992, Dugan and Davis 1993, Bohnsack 1996, Nowlis and Roberts 1997, Allison et al. 1998, Lauck et al. 1998).

Despite the popularity of marine reserves as a management tool, decisions on the design and location of most existing reserves have largely been the result of political or social processes (Jones et al. 1992, Agardy 1994, McNeill 1994); until very recently, little work has been done to understand or include biological considerations in reserve placement or design. A fair amount of recent work has attempted to try to understand and quantify the biological impact of marine reserves. However, these efforts have been scattered around the world and in the scientific literature, so the results are often not easily accessible to people trying to design marine reserves. Relatively little work has been done to assess the success of reserves in general (Roberts and Polunin 1991, 1993, Jones et al. 1992, Dugan and Davis 1993), and all of it has been anecdotal in nature. In an attempt to draw together all of these results, I have reviewed and synthesized the findings of marine reserve evaluations in order to assess the effectiveness of marine reserves. In particular, I evaluated how marine reserves have affected four biological measures (density, biomass, size, and diversity of organisms) within the reserves, and examined if reserve size influences the magnitude of these reserve effects. Specifically, I asked:

1. What are the impacts of marine reserves on the above four biological measures?
2. Is the magnitude of the effect of a reserve on biological measures related to the size of the reserve (i.e., does size matter)?
3. Does trophic structure change with the implementation of a reserve?
4. Does the goal of a reserve (e.g., fishery management vs. biodiversity conservation) influence how large a reserve needs to be?
5. What biases or problems exist in the current literature regarding reserve assessment and/or reserve design, and what can be done to remedy these problems?

Theoretical endeavors have produced some predictions for a few of these questions. Modeling efforts aimed at fisheries management have suggested that biomass of reproductively active fish (spawning stock biomass) should generally increase as a result of reserve protection (Polacheck 1990, DeMartini 1993, Quinn et al. 1993, Attwood and Bennett 1995, Man et al. 1995). Concomitantly, reserves are predicted to increase spillover of fishes to areas outside of the reserve, an effect that is likely to be positively correlated with higher density of fishes inside the reserve (Russ et al. 1992, Hockey and Branch 1994). Organism size and diversity are generally assumed to follow these trends as well, since reserve protection should allow for individual organisms to grow larger (i.e., not be fished out of the system once they reach a certain size) and may also provide protection for species that are normally fished to local extinction. This review will help assess the validity of these predictions.

No direct efforts have been made to evaluate how reserve size itself affects the impact of reserves on any of these biological measures, although it is usually assumed that bigger reserves will always be "better." The literature on "The Theory of Island Biogeography" (MacArthur and Wilson 1967; reviewed by Diamond and May 1976) predicts that species diversity should increase with area, and so larger reserves should contain more species. However, the theory of island biogeography does not address how reserve protection might influence species diversity at a particular location, and so few predictions can be made about how reserve size might affect the impact marine reserves have on species diversity. This review in particular addresses if reserve size affects the impact marine reserves might have on all four biological measures (density, biomass, size, and diversity).

Marine reserves are also predicted to lead to trophic cascade effects, in that protection from fishing may allow top predators to become more abundant in a reserve, which may in turn reduce the abundance of prey, releasing the subsequent trophic level from predation pressure, etc. (Steneck 1998; see also Sala et al. 1998). If this general pattern holds across reserves, then large increases in carnivore abundance and/or size should be associated with smaller differences or even reductions in prey populations.

Independent of the many predictions of the above models, most people simply assume that marine reserves provide the functions expected of them (such as increasing numbers of fish within and outside a reserve). Reserve success stories end up serving as the primary evidence for these assumptions, even though many examples exist where reserves did not provide the necessary functions. The main goal of this review is to evaluate the success of marine reserves in a quantitative way and to assess what role reserve size plays in determining the magnitude of the reserve effect. . . .

Results

General Descriptions of Reserves Studied

Reserve size varied over six orders of magnitude. Mean reserve size was 44.1 km^2, although half of the reserves were between 1 and 10 km^2 and the median reserve size was 4.0 km^2. The largest reserve (which was actually a collection of reserves) was 846 km^2; the smallest reserve was 0.002 km^2. The number of species surveyed in each study also varied widely, but the majority of studies fell into one of two categories: almost half of the measurements were of five or fewer species, and almost half were of 50 or more species.

The distribution between studies conducted in tropical climates and those conducted in temperate climates was fairly equal. Forty-one percent of studies were conducted in temperate regions and the rest were conducted in tropical areas. . . .

The Role of Reserve Size in Determining Reserve Effect

I also used the qualitative data to investigate whether reserve size influences the trends seen in the previous section. For instance, were reserves that showed the largest differences more likely to be larger reserves? In all cases but one, the mean size of reserves for each of the three trend categories (less than, no difference, and greater than) . . . w[as] statistically indistinguishable (one-way ANOVA, $P \geq 0.07$ for all cases). This result implies that the proportional effect of a reserve is independent of reserve size. The only case where reserve size appeared to have an effect was for overall biomass. In this case, reserves were never associated with lower biomass levels. . . .

Effects of Reserve Size

. . . The slopes of the regressions for all measures in all functional groups vs. reserve size are not significantly different from zero (linear regression analysis, $P > 0.12$ for all cases), indicating that reserve size has no apparent impact on proportional differences. There were only four data points for herbivore size, and so regression analysis was not possible for this case. Thus, the relative impact of reserves on all biological measures in each functional group was significantly positive, and this relative impact appears to be independent of reserve size. I discuss the implications of this in the Discussion.

Discussion

These results demonstrate that reserves are associated with higher values of density, biomass, organism size, and diversity for overall values as well as for all functional groups. This is strong support for the many claims made that marine reserves "work." The results of this study also support the predictions of many fisheries models; reserve protection should increase biomass (Polacheck 1990, DeMartini 1993, Quinn et al. 1993, Man et al. 1995) and density (which is probably correlated to the spillover of fish to nonreserve areas; Russ et al. 1992, Hockey and Branch 1994) within a reserve. This is an encouraging conclusion in that at least some of the fishery and conservation expectations for current and future marine reserves have been met and can be realized.

These results also provide some guidelines for the magnitude of change in biological measures we can expect as a result of marine reserve protection. On average, creating a reserve appears to double density, nearly triple biomass, and raises organism size and diversity by 20–30% relative to the values for unprotected areas. It is important to remember, however, that these values have considerable variance and cannot be used to predict how a specific reserve will affect particular organisms and communities.

It is also important to distinguish between how diversity is affected by reserve protection as distinct from the other three biological measures. Diversity in this review is actually species richness, which is not measured per unit area or effort, as are density and biomass. While it is quite possible for both small and large reserves to have the same initial values of density or biomass . . . , larger reserves almost always initially contain more species than smaller reserves. Therefore, finding equal proportional increases in diversity for small and large reserves actually indicates a greater absolute increase in species numbers for the larger reserve. Furthermore, a single individual of a new species has a large impact on species richness measures, whereas a single individual has little impact on overall density, biomass, or organism size. Larger reserves are more likely to contain rare species simply because they encompass a greater area. In addition, diversity values will be somewhat dependent on the effort used to measure them; a long search will more likely produce a rare species than a short search. However, effort was not standardized in any way between studies.

A surprising result of this review is that the relative magnitude of the effect of a reserve on a biological measure appears to be independent of reserve size. A small reserve can double biomass per unit area just as likely as a large reserve can. This result holds even for extremely small reserves; for example, reserves in both St. Lucia (0.026 km^2) and Chile (Las Cruces: 0.044 km^2) were associated with significantly larger values in the biomass and size of the organisms within the reserve compared to nonreserve areas (Castilla and Bustamante 1989, Roberts and Hawkins 1997). The reserve in St. Lucia is particularly noteworthy because even large, mobile fishes seemed to benefit from the small reserve, suggesting that small reserves can work even for mobile organisms. Furthermore, many of the small reserves were located haphazardly, yet still positively affected the organisms within them. If small reserves are more strategically placed, for example on spawning grounds or along migratory routes, their impact may be even greater.

When considering the results of this review it is extremely important to keep in mind the distinction between absolute and relative effects of reserve protection. Even small reserves appear to be able to increase density, biomass, size, and diversity of organisms, and small and large reserves can show the same proportional differences relative to nonreserve areas, but the absolute impacts of small and large reserves will be very different. For example, doubling fish numbers in a small reserve from 10 to 20 fish is substantially different from doubling the fish numbers in a large reserve from 1000 to 2000 fish, even though the relative change in density might be the same for both reserves. The goals of reserve and fishery managers often include some minimum benefit level from reserves (e.g., total catch outside the reserve, all species present and abundant enough to be self-sustaining, etc.), goals that may not be achieved if only proportional differences are considered.

Small reserves may also be insufficient for several other reasons. Alone, small reserves may not be able to provide significant export functions. This review does not examine the possibility that reserves serve as sources for unprotected areas (sensu Pulliam 1988), even though it is often assumed and expected that they provide this service. Models have addressed how current regimes might influence dispersal (e.g., Roughgarden et al. 1988, Roberts 1997), but only a few studies have tried to infer or measure the impact of reserves on reproductive output (Davis 1977, Davis and Dodrill 1980, Polacheck 1990, Stoner and Ray 1996, Sluka et al. 1997, Edgar and Barrett 1999; all suggest that reproductive output can be higher in reserves). An increase in numbers or size of organisms in a reserve will obviously increase reproductive output, but small reserves will only be able to increase reproductive output a small amount relative to target areas. For reserves to serve as larval sources they must be large enough to sustain themselves as well as supply the rest of the target areas.

Another potential drawback of small reserves is their susceptibility to catastrophic events. For example, if an oil tanker runs aground near a small reserve, it is likely that the entire reserve will be impacted by the spill. If the accident occurred near part of a large reserve, on the other hand, it is possible that some of the reserve would escape harm. The unaffected part of the reserve could considerably, then, aid in the recovery process of the damaged region.

It is also possible that very large reserves (e.g., >500 km^2) might provide proportionally larger values when evaluated by density, biomass, etc. If fish within a reserve use several habitats throughout their life histories, it may require a very large reserve to encompass and protect all life stages adequately. This review would most likely not be able to detect a size threshold effect such as this, since only seven of the reserves studied covered 50 km^2, and the only one 460 km^2 came from pooled data from a collection of seven smaller reserves. Furthermore, nearly three quarters of all the reserves studied covered 10 km^2. Such shortcomings in the data leave open the possibility that large reserves affect biological measures in a way not detectable here. While it would be desirable to test how such a large reserve would affect such measures, the logistics of such studies would be very difficult.

An important variable not analyzed here is the role that the length of protection plays in determining the magnitude of a reserve effect. Examples exist where the magnitude of the reserve effect increased over time (e.g., Watson et al. 1996, Russ and Alcala 1998*a*, *b*). Conan (1986) described how lobster biomass initially increased over several years but then receded to original levels. In all of these cases, results would have been different had population surveys been made at a single point in time (or over a relatively brief period of time), as they were in most of the studies I reviewed here. It is difficult to determine, therefore, if the populations had actually reached equilibrium at the time of measurement. Furthermore, the impact of a reserve is certainly not instantaneous, but little is known about how long it takes for a population to reach equilibrium, or even if it ever does. I address in depth the role that length of protection plays in determining the effect of marine reserves elsewhere (Halpern and Warner 2002).

Many other variables could also influence the impact of reserves on the biological resources contained within them. Species composition (PDT 1990, Carr and Reed 1993, Ballantine 1992, 1995, 1997, Dugan and Davis 1993, Tegner 1993, Rowley 1994), the fishing intensity around the reserve (Polacheck 1990, Russ et al. 1992, Carr and Reed 1993, Rowley 1994, Nowlis and Roberts 1997), adult mobility or home range size of fish within the reserve (Kramer and Chapman 1999), and the types and quality of habitats both inside and outside the reserve (Salm and Clark 1989, Hockey and Branch 1994, Agardy 1995, Nilsson 1998) have all been proposed as variables that could be important in determining how an organism responds to reserve protection. These sorts of observations were usually not reported in the empirical studies on marine reserves I used, and so I was unable to evaluate them here. However, these other factors should certainly be considered when setting goals and expectations for marine reserves.

Despite that many empirical studies found trophic cascade effects as a result of marine reserve protection (Kenya: McClanahan and Muthiga 1988, McClanahan and Shafir 1990, McClanahan 1994, 1995, 1997, Watson and Ormond 1994; Chile: Castilla and Duran 1985, Duran and Castilla 1989; Mediterranean: Sala et al. 1998*a*), this pattern did not emerge from my large-scale analyses. Instead, the densities of invertebrates, herbivorous fishes, planktivorous/invertebrate eating fishes, and carnivorous fishes all increased almost exactly the same amount. A possible explanation for this is that trophic cascades appear to be more likely to occur when only a small subset of a community is observed (Polis and Strong 1996). For example, in Kenya (e.g., McClanahan and Shafir 1990), the trophic cascade occurred between humans, triggerfish (Balistidae), and a few species of sea urchins, and was not evident in other families of fish and species of urchins that were studied. Similarly, in Chile (Castilla and Duran 1985, Duran and Castilla 1989) the cascade occurred between humans, a single gastropod, a single mussel, and algae. Thus, trophic cascades may be masked when entire communities are measured. In the study by McClanahan and Shafir (1990), total fish densities as well as densities for four fish families (Labridae, Balistidae, Diodontidae, and Lagocephalidae) and urchins were measured. Urchin densities were nearly 200 times higher outside the reserve, while Balistid density was nearly 10-fold greater inside the reserve, exemplifying a classic trophic cascade. When all four fish families were considered (all are planktivorous fishes/invertebrate eaters), fish densities dropped to only 28% higher inside the reserve, obscuring the trophic cascade. When family or species results are incorporated into an entire functional group, as was the case here, trophic cascade effects can often become muted.

Empirical tests of the effect of reserve size are needed to test the robustness of the results suggested here. To date, only one study (Edgar and Barrett 1999) has tried to assess empirically the potential effects of marine reserve size on biological attributes of species contained within the reserves. They studied four reserves in Tasmania, three of which were 0.6 km^2 and a fourth that was about 7 km^2. The largest reserve showed many significant differences relative to nonreserve areas, while the smaller reserves had only a few notable differences. For example, in the large reserve, overall fish size, density of large fish, abalone size, size of crayfish, mean plant cover, and species diversity of fish, invertebrates, and algae all increased significantly compared to control sites. In the other three sites, significant differences were found only for density and diversity of large fish in one reserve and density of algae in another. Although the observations from the large reserve were not replicated, these results offer some empirical evidence suggesting that large reserves can provide biological functions not possible in small reserves. This conclusion is in stark contrast to the results of this review, in which even small reserves appeared to have a positive impact on most biological measures. In order to assess adequately the role of area in reserve function, a real need exists for studies that make observations in reserves of many sizes within the same biogeographic region.

Success in the design and function of a marine reserve is closely tied to the goals of the reserve. For example, fishery reserves need to increase abundance, biomass, and organism size within the reserve in order to sustain the reserve populations as well as supply the harvested areas. Conservation reserves, on the other hand, focus more exclusively on the maintenance of diversity and abundance of organisms within the reserve itself. Fortunately, marine reserves appear to lead to higher values of all of these biological measures, implying that both goals can be met with the same reserve.

The impact of marine reserves on the organisms contained within them will never be completely predictable. Variation among reserves and a level of uncertainty will always exist when examining how marine reserves affect specific biological measures. Goals set for marine reserves should account for this variation (Walters and Holling 1990, Clark 1996, Hall 1998, Lauck et al. 1998). Ultimately, though, it is encouraging to know that reserves of any size appear to function well, in terms of producing higher densities, sizes, and diversity of organisms.

Inherent Problems and Necessary Caveats

The enormous variation in type and quality of the observations from marine reserves made it difficult to compare or analyze the results of the studies I reviewed (see also Jones et al. 1992). The primary problems include:

1. results are more likely to be reported for species that are actually affected by reserves (either positively or negatively) than for unaffected species, especially for single-species studies;
2. methodologies often differ drastically among different observations and among scientists within a study;
3. characteristics of reserves being studied (such as location, habitat type, current regimes, temperatures, etc.) are not the same;
4. observations are rarely replicated temporally or spatially (usually because there is only one reserve available for study);
5. reserves are not always adequately protected from poaching;
6. the length of protection varies among reserves;
7. numbers and types of organisms studied vary between experiments;
8. the intensity of fishing outside of the reserve may enhance or even create the perceived affect on biological measures of reserve protection.

As many have argued, the intensity of fishing occurring outside a reserve (or where a reserve is before it becomes a reserve) can have a large impact on the perceived effects of reserve protection (Polacheck 1990, Russ et al. 1992, Carr and Reed 1993, Rowley 1994, Nowlis and Roberts 1997). If an area is nearly completely fished out, the ratio of postprotection to preprotection values of abundance, biomass, etc. will be much higher than for an area that had been lightly fished (assuming all else is equal, and that new fish can be imported to the fished areas from elsewhere). It is difficult to compare fishing intensities in different parts of the world, and this can lead to inaccuracies when combining data.

Another problem many studies face is the lack of consistency in protection level for the reserves. Even fully protected reserves often suffer some poaching (e.g., Klima et al. 1986). This potential problem was rarely quantified, largely due to difficulties in monitoring a clandestine act. Because information on actual protection level is lacking, it is difficult to know exactly how long and to what degree a reserve has been protected. Reserve effects can change over time (see Russ and Alcala 1998*a*, *b* for examples of this), so knowing the length of time protection has been in place can be a critical part of analysis. To be able to make more accurate predictions of the effect of marine reserves, actual fishing effort within reserves must be measured and accounted for (Polacheck 1990, Russ et al. 1992, Carr and Reed 1993, Rowley 1994, Nowlis and Roberts 1997) and the length of complete protection identified.

The lack of temporal and spatial replication in many of the studies further complicates interpretation of the results. Snapshots in time and space can provide clues to the effects of reserves, but it is very difficult to eliminate the possibility that observed effects were not simply a result of spatial or temporal differences, especially with inside/outside reserve studies. Before/after studies offer a possible solution to these problems and should be coupled with control observations in non-reserve areas over the same time period, across several spatial scales within a biogeographic region. However, such studies are often logistically difficult to implement.

One of the largest problems with the empirical literature on marine reserve effects is that methodologies used for different studies and the characteristics of reserves and control sites (such as substrate rugosity, depth, current regime, etc.) differ dramatically. Few people make efforts to accommodate the problems mentioned above, let alone measure the same variables in the same way. For example, sample sizes in many studies were not large enough to draw statistically significant conclusions. Other studies did not report the statistical significance of their results, even though this might have been possible. Empirical work on marine reserves needs to reflect the rigorous standards of the rest of the scientific literature.

Finally, results are often only reported when a reserve actually had an effect on an organism, whether negatively or positively. This was unlikely to be a problem for studies that looked at entire communities, but was potentially a large factor influencing single-species studies. Single-species studies can often be useful, especially for fisheries management, but it is important to remember that not every species will respond to reserve protection.

Despite these potential sources of error, my analyses uncovered clear and significant positive effects of reserve establishment on the organisms dwelling within reserve boundaries.

Conclusions

The most important lesson provided by this review is that marine reserves, regardless of their size, and with few exceptions, lead to increases in density, biomass, individual size, and diversity in all functional groups. The diversity of communities and the mean size of the organisms within a reserve are between 20% and 30% higher relative to unprotected areas. The density of organisms is roughly double in reserves, while the biomass of organisms is nearly triple. These results are robust despite the many potential sources of error in the individual studies included in this review.

Equally important is that while small reserves show positive effects, we cannot and should not rely solely on small reserves to provide conservation and fishery services. Proportional increases occur at all reserve sizes, but absolute increases in numbers and diversity are often the main concern. To supply fisheries adequately and to sustain viable populations of diverse groups of organisms, it is likely that at least some large reserves will be needed.

Finally, it is paramount that we explicitly state our goals when creating marine reserves. These goals help guide the design of reserves and are critical for assessing whether or not a reserve has functioned successfully.

All references for articles included in *Taking Sides: Clashing Views in Sustainability* can be found on the Web at www.mhhe.com/cls.

Andrew A. Rosenberg, Jill H. Swasey, and Margaret Bowman

Rebuilding U.S. Fisheries: Progress and Problems

The problem of overexploitation of fisheries has been well-documented globally, showing a widespread pattern of severe resource depletion occurring over centuries, but particularly during the past 50 years (Jackson et al. 2001; Myers and Worm 2003; Pauly and Maclean 2003; Millennium Ecosystem Assessment 2005). Over the past decade, there has been a concerted effort in national and international fishery policy to end overfishing and recover overfished resources (FAO 1995). In the United States, a very strong statutory mandate to end overfishing and rebuild depleted fishery resources came into effect with the Magnuson–Stevens Fishery Conservation and Management Act of 1996 (NOAA 1996; Safina et al. 2005). The law sets out specific timelines for action to rebuild depleted fisheries, establishes requirements for the rebuilding management plans, and requires accountability for implementing plans in a timely manner.

Here we review the implementation of the rebuilding provisions of the Magnuson–Stevens Act. We consider the record of implementation based on a plain reading of the law's requirements and the public record of action by the responsible entities set out in the law, including the Secretary of Commerce (Secretary), NOAA's National Marine Fisheries Service (NMFS), and the eight Regional Fishery Management Councils created by the Act.

We address three basic questions fundamental to reviewing the program: (1) Have the effects of overfishing been reversed? In this, the 9th year since the mandate, fewer than 5% of fish stocks subject to rebuilding plans have been rebuilt and only 13% are no longer experiencing overfishing or are overfished (i.e., they are no longer depleted due to previous overfishing). However, biomass appears to be increasing in 48% of the stocks. From our review of all federal rebuilding plans, the basic premise of the theory of fishing holds that if overfishing is ended, stocks will begin to recover. (2) Why is rebuilding failing to occur for so many stocks? Nearly half of the stocks for which there are rebuilding plans are still subjected to overfishing, so that fishing pressure is still too high to allow stock recovery. In many cases, rebuilding timeframes have been extended, plans have not been adjusted even when catches are clearly too high, and there have been other delays in implementing effective controls on fisheries. (3) What are the barriers to greater success in rebuilding fisheries? Ending overfishing immediately is fundamental to rebuilding these resources. Too often, effective reductions in fishing pressure are subject to protracted political debate, while the resource continues to decline. It is essential that the fisheries are protected until an adequate rebuilding plan is in place and if a plan isn't working, adjustments must occur rapidly to prevent further depletion.

The Mandate to End Overfishing and Rebuild

In the law, Congress found: "Certain stocks of fish have declined to the point where their survival is threatened, and other stocks of fish have been so substantially reduced in number that they could become similarly threatened as a consequence of (A) increased fishing pressure, (B) the inadequacy of fishery resource conservation and management practices and controls, or (C) direct and indirect habitat losses which have resulted in a diminished capacity to support existing fishing levels" (NOAA 1996). The Magnuson–Stevens Act requires the following four steps: (1) The Secretary (through the NMFS) shall annually evaluate all fisheries to determine if they are being overfished (i.e., overexploited, overfishing is occurring) and/or the fishery is in an overfished condition (i.e., depleted due to past overfishing) based on objective and measurable criteria. Congress and the Regional Fishery Management Councils are to be notified of those stocks in need of rebuilding. (2) The relevant Regional Fishery Management Council responsible for a stock where overfishing is occurring and/or is in an overfished condition shall, within one year, prepare a plan to end overfishing and rebuild the resource. If the Council fails to do so, the Secretary must develop such a plan within 9 months. (3) The plan must end overfishing and rebuild the resources in as short a time as possible, given the biology of the resource and considering the needs of fishing communities. The rebuilding time period is not to exceed 10 years, unless the biology of the fish, environmental conditions, or international agreements dictate a longer time frame. (4) Rebuilding plans shall be based on the best science available and be reviewed by the Secretary for adequate progress at least every 2 years. If adequate progress to end overfishing and rebuild the resource is not made, then revisions shall be made.

Developing management measures for rebuilding is always contentious, because the need to reduce fishing pressure usually requires the implementation of additional restrictions on businesses and individuals engaged in fishing. However, the process is clear: identify, plan within the time frame, and regularly review progress. In fisheries science and in law, it is clear that the fundamental control variable is the exploitation rate or fishing mortality rate for a given stock—that is, the proportion of the stock removed each year due to fishing. Other variables concerning which fish are harvested (e.g., age, size, gender) are also important in relation to the overall exploitation rate. Thus, ending overfishing requires the reduction of fishing pressures to, at most, the level that would give maximum sustainable yield (FMSY).

Similarly, the goal of rebuilding should be apparent: to rebuild a given fish stock to at least the abundance (usually expressed as biomass) that can support, in the long term, maximum sustainable yield.

Have the Effects of Overfishing Been Reversed?

Under the requirements of the Magnuson–Stevens Act, 74 commercially or recreationally important fish stocks have been identified at some point by the NMFS as requiring rebuilding (NEFSC 2005; NMFS 2005). Sixty-seven stocks have been included in rebuilding plans already (Table 1), while the remaining seven stocks have only recently been identified as overfished and rebuilding plans are currently being developed. An additional four stocks have been identified as experiencing overfishing, but are not yet overfished (i.e., depleted). For these four stocks, plans are required to end overfishing, although rebuilding per se is not required.

Of the 67 stocks, three (Atlantic sea scallop, Pacific whiting, and Pacific lingcod), or less than 5%, have been rebuilt to the biomass levels that are expected to support maximum sustainable yield, the goal of rebuilding. Less

Table 1

Status of Stocks Requiring Rebuilding

Regional Fishery Management Council (FMC)	Stocks requiring recovery	Rebuilding plans	Overfishing is occurring[3]			Overfished[4]			Stocks rebuilt
			Y	*N*	*ND*[5]	*Y*	*N*	*ND*[5]	
New England FMC	19	18	10	8	1	14	5	—	1
Mid-Atlantic FMC	7	5	1	5	1	3	4	—	0
South Atlantic FMC	14	14[1]	11	3	—	11	2	1	0
Caribbean FMC	3	3	1	2	—	3	—	—	0
Gulf of Mexico FMC	8	8	4	4	—	6	2	—	0
Western Pacific FMC	1	1	—	—	1	1	—	—	0
Pacific FMC	9	9	—	9	—	5	4	—	2
North Pacific FMC	4	4	—	4	—	4	—	—	0
High migratory species	9[2]	5	7	2	—	7	2	—	0
TOTALS	**74**	**67**	**34**	**37**	**3**	**54**	**19**	**1**	**3**

[1] Ten plans are pre-SFA and have not yet been updated.
[2] Includes Large Coastal Shark Complex; within this complex 15 species are overfished. An additional two have recovered from a previously overfished condition.
[3] The fishing mortality rate in the most recent year, according to the stock assessment, is greater than the reference rate of mortality set by the Council and NMFS as defining overfishing.
[4] The current biomass level is below the threshold level set by the Council and NMFS as defining an overfished stock. By regulation, the NMFS sets a reference point of one half the biomass that would support maximum sustainable yield to determine whether a stock is depleted such that rebuilding is required. This biomass reference point is not the target of rebuilding plans but it is intended to serve as guidance on status of resources as is the exploitation rate that is expected to produce maximum sustainable yield.
[5] Current status not determined

The table includes all stocks that are reported by the NMFS to be in need of a rebuilding plan and those that are currently managed under rebuilding plans (NEFSC 2005; NMFS 2005). Statuses reported here are for all stocks requiring recovery, those currently included in a rebuilding plan and those awaiting plan development.

than 14% of the stocks are no longer experiencing overfishing or remain in an overfished condition. The majority of stocks undergoing rebuilding continue to be overexploited (45%; i.e., with fishing pressure in excess of FMSY) and/or their population biomass remains overfished (72%; i.e., depleted below reference levels).

As a consequence of the continued overexploitation of many stocks, and the often very long rebuilding plan timelines created by the Councils (see below for further explanation of rebuilding timelines), most stocks that should be rebuilt around the country are still in poor shape. Consideration of plans by Council indicates the scope of the problem. Only three stocks have been declared rebuilt, one in New England and two in the Pacific. New England and the South Atlantic lead in numbers of stocks that require rebuilding. The Mid-Atlantic and the Pacific lead in recovering stocks. Clearly, developing effective rebuilding plans is a complex process and there are numerous case-specific circumstances concerning each fishery that affect the successful development and implementation of a rebuilding plan. Here we focus on the overall results, because the success of the program ultimately depends on the actual rebuilding of exploited stocks, not in the process itself.

Why Is Rebuilding Failing for So Many Stocks?

For 45% of the stocks under rebuilding plans, overfishing is still occurring. In some cases, overfishing has persisted more than 5 years into a supposed rebuilding plan. For 7% of the stocks, biomass has continued to decrease since the rebuilding plans were implemented, but for 45% there is insufficient information to determine biomass trends under the rebuilding plans. Biomass is increasing in 48% of the stocks under rebuilding plans, reflecting real progress as a result of the program, albeit in less than half the cases. In 37% of the stocks, biomass is increasing and fishing mortality rates are decreasing. In other words, the clear principle of rebuilding holds: fishing pressure must be reduced in order to recover these resources.

These numbers indicate that the fishery rebuilding efforts have not been very successful over the past 9 years. Most stocks have the potential to be rebuilt within 10 years (Safina et al. 2005), so this lack of demonstrable progress is disappointing. Unfortunately, despite the statutory mandate to rebuild in as short a time as possible, not to exceed 10 years except under special circumstances, the actual rebuilding time frames implemented have almost invariably been a decade or longer. To make matters worse, 15 plans (mostly in New England) reset rebuilding deadlines back to year one when the plans were revised, instead of using the existing time frame. This can be a mechanism for extending rebuilding time frames well beyond the plain language of the statute. Taking into account the true time frames of the plans (not the reset times), only two plans currently in place have a rebuilding time frame of less than 10 years (9 years for bluefish; the 5-year plan for northeast spiny dogfish has expired, but the directed fishery is closed). Twenty-one plans (31%) currently have the 10-year maximum, while 36 (54%) have a time frame extending beyond a decade, due either to the exception for biological conditions or to a reset 10-year rebuilding time. An additional eight timelines are undefined.

Overcoming the Barriers to Successful Rebuilding

The US rebuilding program is impressive in concept and scope. Upholding the principle of the Magnuson–Stevens Act to ensure that the capacity to produce the maximum sustainable yield is conserved is the first major step towards sustaining marine ecosystems. Along with broader ecosystem protection from human impacts, the effort to end overfishing and rebuild fishery resources is an important component of a more effective system of management for the oceans according to the principles laid out by the US Commission on Ocean Policy (USCOP 2004). According to the analysis presented here, however, problems in implementation remain. Principal among these is the failure, in far too many cases, to end overfishing itself, an imperative for rebuilding. A related factor is the establishment of long rebuilding time frames, resulting in delays in reducing fishing pressures.

The fundamental factors leading to a failure to rebuild can be illustrated by some of the stocks currently under rebuilding plans. For many stocks, including snowy grouper, South Atlantic black sea bass, and cod, fishing mortality rates are extremely high despite recovery plan implementation. Examples of successful rebuilding are also available, and these underline the importance of implementing large decreases in fishing mortality rates quickly. Sustained rebuilding can be achieved when fishing mortality rates are reduced to below or at least close to reference levels. The examples chosen here illustrate these basic points without regard to the details and specific circumstances surrounding the development and implementation of the rebuilding plans. While the same patterns occur in many other stocks, these examples provide a few clear instances of such patterns.

In several cases, plans have even been revised one or more times, without managing to reduce overfishing to below the reference level. As a consequence, stock biomass has not rebounded. In effect, management has not been held accountable for the most basic provision of a rebuilding plan, that fishing mortality should be greatly reduced and that the plan needs to be monitored to redress any problems.

An additional cause underlying the lack of demonstrable progress in rebuilding overfished resources appears to be the absence of consistent monitoring and revision of the plans that are not showing signs of progress. The trend in fishing mortality rate is unknown for 51% of the stocks under various plans, although in some cases this is because of very recent implementation. Similarly, in 45% of stocks the trend in biomass is unknown. This may be evidence that the scientific information is unable to keep up with the management plans, and there is certainly a strong case to be made for more resources to improve and extend the ability to provide scientific advice for fishery management. However, the fact that overfishing is persisting in nearly half (45%) of the stocks under rebuilding plans indicates that even when there is advice on overfishing and stock status, management has not been held fully accountable for the lack of rebuilding success.

The continued overfishing of so many resources ostensibly under rebuilding plans indicates that the approved plans themselves are failing nearly half of the time. Those developed by the Regional Councils must be approved by the Secretary, through the NMFS. So, it is fair to ask, why is the Secretary approving plans that so often fail to address overfishing?

One reason is that the Secretary may only approve or reject a Council-developed plan; once submitted, Council plans cannot be modified by the Secretary. Rejecting a plan results in a default situation (i.e., no action to rebuild). The Secretary's decision thus rests on whether a plan is better than nothing, rather than on the best actions for the fishery over the long term. It should be noted that the Secretary does have the authority to develop a plan absent a Council recommendation, although this has been done only very rarely. Since the Act sets up a Council process it is expected that the Councils should develop management plans, rather than the Secretary. Based on the fact that Secretarial plans are rare, it is clear that most decisions are made regarding Council-developed plans that are submitted for approval. The US Commission on Ocean Policy recommended that a conservative default management plan be implemented for overfished fisheries, while a comprehensive rebuilding plan is developed (USCOP 2004). Creating a strong conservation action that will remain in place while the plan is being developed can prevent further depletion of the resource and shorten rebuilding times overall. Examples of such actions could include an immediate large reduction in fishing pressure through catch limits, closed areas, or closed seasons. Furthermore, this creates a strong incentive for those charged with developing a rebuilding plan to complete the process as quickly as possible. Clearly, the course of rebuilding will be variable, depending on environmental conditions and other factors. However, it should be clear that progress must be made at each step. While the productivity of a fish population cannot be fully controlled, we can and should have much better control over mortality rates caused by fishing. There must be a performance standard in place for both fishing mortality and for rebuilding, with rapid updates if the targets are not met.

There is often strong pressure from the fishing community to phase in reductions in fishing pressure slowly. In theory, this gives businesses time to adjust to the new restrictions. Unfortunately, it also means that the stocks are further depleted, sometimes severely, before rebuilding can begin, resulting in many more years of reduced yields. The overall economic impact is likely much greater as a result of a long continued decline and delay in rebuilding than from a short-term reduction in catch in order to rebuild populations quickly (Sumaila and Suatoni 2006).

In summary, the [United States] has a very strong fisheries law with clear requirements to end overfishing and rebuild overfished resources. The program has been broadly implemented as a result of major efforts by scientists and managers throughout the country. Rebuilding fisheries is a difficult prospect, but an important one. The depletion of these public resources has long-lasting effects on coastal communities, consumers, industry, and the nation as a whole. Our results indicate that, after nearly a decade, the outcomes of the rebuilding program are disappointing. In order to rebuild resources, overfishing must be ended quickly and kept to low levels—unfortunately, not the norm so far. There is a need for more comprehensive and timely scientific information but, most of all, the program must become results oriented, not process oriented. It is true that there are complications pertaining to each fishery, but it is the results in the water that must be the ultimate measure of the program's success, and a great deal remains to be done.

EXPLORING THE ISSUE

Can Our Marine Resources Be Sustainably Managed?

Critical Thinking and Reflection

1. Why are marine fisheries under threat, and how do the strategies currently in use attempt to alleviate the pressure?
2. In what ways are aquaculture and fish-farming operations detrimental to marine systems? How can they potentially help the overfishing crisis?
3. Why do some people resist fishery conservation attempts? What are their primary concerns, and how can these concerns be balanced with conservation priorities?
4. What evidence is there that MPAs are successfully protecting marine resources, and why are some experts questioning their effectiveness?
5. Individual fishery habitats can overlap national boundaries. What strategies are in use to coordinate conservation between multiple nations? Are these strategies successful?
6. Overfishing is not the only threat to global marine resources. What other detrimental effects do humans have on our oceans? What can we do to minimize these effects?

Is There Common Ground?

It is beneficial to all parties that marine fisheries be biologically, and therefore commercially, viable for decades and centuries to come. However, it is here that opinions and desires begin to diverge. Scientists and managers believe that fishing pressures are continuing to have a detrimental effect on fishery stocks, while many fishermen and industrial concerns maintain that current management strategies are sufficient, if not overly restrictive. The scientific evidence is admittedly thin in many cases and is not likely to increase in scope in the future unless scientific funding is increased. Political pressures from these groups are similarly often greater than that from the scientific community. However, other groups are beginning to speak up—consumers are noticing higher prices at local markets; some fishermen and recreational fishers are noticing declining catches in poorly managed stocks, while others are noticing greater annual revenues in fisheries that are well maintained. Lobster fishermen, as an example, are quite vocal in their support of managing strategies implemented in their fisheries. Moreover, residents of local areas historically overfished by foreign commercial fleets are standing up in defense of their marine resources, requesting increased protection for their reefs and oceans, both for their ecotourism potential and as a necessary protein resource for their families. Perhaps these new voices will be heard as new fisheries management strategy plans are debated.

Internet References

Benetti, D., Brand, L., Collins, J., Orhun, R., Benetti, A., O'Hanlon, B., Danylchuk, A., Alston, D., Rivera, J., Cabarcas A. (2006). Can offshore aquaculture of carnivorous fish be sustainable? Case studies from the Caribbean. *World Aquaculture,* vol. 37, no. 1, pp. 44–47.

Costello, C., Gaines, S.D., Lynham, J. (2008). Can catch shares prevent fisheries collapse? *Science,* vol. 321, pp. 1678–1680.

Jennings, S. (2009). The role of marine protected areas in environmental management. *ICES Journal of Marine Science,* vol. 66, pp. 16–21.

Kellner, J.B., Tetreault I., Gaines, S.D., Nisbet, R.N. (2007). Fishing the line near marine reserves in single and multispecies fisheries. *Ecological Applications,* vol. 17, no. 1, pp. 1039–1054

Murray, G., Johnson, T., McCay, B.J., Danko, M., St. Martin, K., Takahashi, S. (2010). Creeping enclosure, cumulative effects and the marine commons of New Jersey. *International Journal of the Commons,* vol. 4, no. 1, pp. 367–389.

Safina C., Rosenberg, A.A., Myers, R.A., Quinn, T.J., Collie, J.S. (2005). US ocean fish recovery: keepers, throwbacks, and staying the course. *Science,* vol. 309, pp. 707–708.

The website of the National Marine Fisheries Service provides information on issues pertaining to biodiversity and habitat conservation, access to management plans developed by the eight regional councils under Magnuson–Stevens, and current information on fishery restrictions. The status review page provides up-to-date status reports on endangered and threatened marine and anadromous species (the species that migrate between fresh and marine waters). These reviews are used in determining eligibility for listing under the Endangered Species Act.

http://www.nmfs.noaa.gov

This website, administered by the National Oceanic and Atmospheric Administration, provides information about the U.S. network of more than 1600 marine protected areas. Maps, data, and regulatory information are provided on individual reserves. The site also provides information on the public nomination of new MPAs.

http://www.mpa.gov

The website of the U.S. Geological Survey provides a good overview of sustainability issues as related to beaches and beach erosion. It also provides information on endangered species usage of beach and ocean resources.

http://www.usgs.gov

The website of the Marine Conservation Biology Institute provides information and data on projects currently being funded by the institute, as well as a range of topics including the effect of climate change on the oceans, ocean acidification, and destructive fishing practices.

http://www.mcbi.org

The MarineBio Conservation Society is a 501(c)(3) organization founded by David Campbell, an environmental geoscientist. Campbell has created a science-heavy organization with a board of directors consisting of experts in various areas of marine biology and oceanography. The website provides resources on a range of issues including marine conservation biology, sustainable fisheries, habitat conservation, ecotourism, and endangered species in an attempt to raise awareness of marine conservation issues.

http://www.marinebio.org

The aquarium's Seafood Watch program is a comprehensive source for sustainable seafood listings and related information. The website provides good basic overviews of sustainability issues pertaining to both wild-caught fish and aquaculture.

http://www.montereybayaquarium.org

ISSUE 14

Is Natural Gas Hydraulic Fracturing Safe?

YES: Seamus McGraw, from "Is Fracking Safe? The Top 10 Controversial Claims About Natural Gas Drilling," *Popular Mechanics*, www .poularmechaniscs.com retrieved September 18, 2012

NO: Russell McLendon, from *Big Frack Attack: Is Hydraulic Fracturing Safe?* (December 2010), www.mnn.com

Learning Outcomes

After reading this issue, you should be able to:

- Understand how natural gas production in the United States has increased.
- Understand why natural gas may be considered to be a "clean" fossil fuel.
- Understand the dynamics and characteristics of shale gas hydraulic fracturing.
- Understand the potential benefits of shale gas hydraulic fracturing.
- Understand the potential risks of shale gas hydraulic fracturing.
- Understand why shale gas hydraulic fracturing may be a "bridge technology."
- Understand whether the benefits of hydraulic fracturing outweigh the risks.

ISSUE SUMMARY

YES: Writer Seamus McGraw in *Popular Mechanics* discusses 10 controversial claims about natural gas drilling in the hope of setting the record straight on this heavily debated issue. He notes the abundance of natural gas which will fuel America's future providing both inexpensive energy and a potential resurgence in manufacturing. He also points out that hydraulic fracturing does not use as much water as other activities; that deep-injected fluids will not migrate into groundwater; and that groundwater contamination, while possible, is not probable.

NO: Journalist Russell McLendon writing in the *Mother Nature Network* points out that hydraulic fracturing poses many concerns and is skeptical about whether it is "safe." The problem, he believes, is that hydraulic fracturing has not been studied enough and that the public just does not have the answers to important issues relating to public health and environmental risk. For instance, he states that there is no study that proves conclusively that "fracking" fluids cannot end up in groundwater and even migrate into aquifers. He is also concerned over the extent of methane migration in hydraulic fracturing; its ability to cause earthquakes; and the overuse of water resources.

Across the United States, the development of natural gas reserves within shale geologic formations is growing. Hydraulic fracturing is the process by which natural gas is extracted from the shale formation. The process consists of pumping a mixture of water, sand, and chemicals (also known as hydraulic fluids) at a high pressure through a well into the formation thus creating fractures. The fractures exploit natural gas trapped in the subsurface. Horizontal wells are commonly installed to maximize the amount of gas extracted during fracturing. Several horizontal wells can be installed from a common wellbore (opening at ground surface). An average of 15 fracturing wells are present in a given wellbore. Once a well has been installed, the casing is perforated and hydraulic fluids are pumped into the formation. Hydraulic fluids contain a proppant (commonly quartz, sand, or ceramic material) and/or a gel to keep the fracture open during gas extraction. The well is plugged with cement after all the extractable natural gas has been removed. The overall objective of hydraulic fracturing is to create a conductive and highly permeable channel through which fluid and gas can flow from the subsurface to ground surface (Arthur, 2008a; Considine et al., 2009; Kargbo et al., 2010; Lee et al., 2011).

The extraction of natural gas via hydraulic fracturing has impacts on the environment, economy, and society that can be either beneficial or destructive. Proper environmental management of natural gas extraction could facilitate a hydraulic fracturing process where benefits outweigh the risks.

The Benefits of Hydraulic Fracturing

Increasing the development of domestic natural gas resources enables the United States to decrease its import dependency and high trade deficits associated with imported fossil fuels. The supply and delivery system of domestic natural gas is more reliable and less subject to interruption compared to imported energy. A large amount of energy is lost during transport due to the inefficiency of conversion as well as along transmission and delivery chains. Having a local and domestic energy hub decreases the amount of energy lost during transportation. The United States currently has a well-developed infrastructure to transport natural gas which makes it easy to use in a number of domestic applications, such as vehicles, electricity, and heating. The versatility of natural gas is significant, with major uses in transportation, electricity generation, feedstock in petrochemical industry, and underpins production of nitrogen-based fertilizers (Hughes, 2011; Kargbo et al., 2011).

Another consideration is that the United States needs to reduce its production of greenhouse gases (GHG), a contributor to global climate change. Natural gas emits approximately 30–60 percent less carbon than coal or oil. Increasing the development of natural gas can assist society in achieving lower levels of GHG. For example, natural gas can be used in conjunction with coal-fired power plants to reduce GHG. It can also be used with other renewable energy sources, such as wind and solar, to balance system load due to the intermittent nature of these energy sources. The reduction of GHG can promote carbon permit trading and support renewable energy portfolio standards. Also, natural gas is gaining in prominence in the transportation sector. In 2009, Considine et al. reported that approximately 130,000 vehicles were running primarily on compressed natural gas. The current market of natural gas in transportation is mainly for fleet vehicles, which include buses, taxis, and government vehicles. With the increased development of natural gas, its role in the transportation market can penetrate high-density urban areas (due to the delivery network required). This would also decrease overall GHG emissions.

Another consideration is that new technology has decreased the negative impacts of hydraulic fracturing. First, 3-D Fracture Simulation models are used to design the ideal fracture for a specific formation. The models take into account reservoir characteristics, mechanical properties, and natural stress profiles within the formation. The use of 3-D models to guide the placement of fractures decreases the number of fractures and wells needed, in turn decreasing the environmental footprint and overall costs (Arthur, 2008b). Second, energy companies have transitioned from installing conventional vertical wells to horizontal wells resulting in the reduction of the total number of wells necessary to extract natural gas. Several horizontal wells can be installed from one well pad. By decreasing the number of well pads necessary to extract the natural gas, the amount of access roads, pipeline routes, and production facilities required are also decreased. By downsizing infrastructure, the impact on the environment is significantly decreased. For example, the use of horizontal wells over vertical wells minimizes forest fragmentation, decreases surface disturbance (which can lessen impacts from noise, traffic, and landscape), and overall environmental footprint (Reeder, 2010). Finally, new marketing opportunities in natural gas have supported innovative technologies to reduce the impacts of hydraulic fracturing. For example, using environmentally friendly hydraulic fluids derived from natural gas processing can minimize waste water treatment and disposal issues; STW Thermal Evaporation can treat and recycle wastewater onsite, increasing efficiency and reducing demands on local systems; Dry Frac can replace the use of carrier fluid altogether by using carbon dioxide; a Rotary Steerable Closed-Loop System can increase the efficiency of drilling sites by improving directional drilling, potentially reducing the number of well sites needed (Lee et al., 2011). These innovative technologies could eventually be used in other markets to reduce impacts caused by drilling and extraction of other natural resources.

And finally, high-paying employment opportunities can be generated to perform tasks associated with developing natural gas reserves, such as exploration, legal transactions, building infrastructure, drilling, and transportation. Employment opportunities are also created in support industries, for example, suppliers of steel, machines, and equipment. These direct and indirect business activities generate additional income locally, regionally, nationally, and even globally. Considine et al. have calculated that for every $1 the Marcellus Shale industry spends locally, it generates close to $2 in total economic value output. Local economies benefit from state product, income, and tax receipts, and local landowners and municipalities are boosted by bonus payments and royalties from the development and production of natural gas.

The Risks and Negative Externalities of Hydraulic Fracturing

Technological challenges remain in the economic recovery of gas from unconventional reservoirs, as is the case for shale gas (Lee et al., 2011). Shale has low porosity, low permeability, and the matrix pore spaces are poorly connected, which creates a challenge in accessing and extracting the gas from this formation. The extracted gas is commonly in the "wet" form, meaning the gas also contains propane, ethane, butane, and other heavier gases. The "wet" components of the gas need to be removed prior to transporting and selling the "dry" natural gas to distribution companies. In addition to accessibility and extraction, the decommissioning of the well is difficult. If the annulus of the wellbore is improperly sealed natural gas, fracturing fluids and formation water containing high concentrations of dissolved solids may leak into the subsurface and potentially drinking water aquifers. Zoback et al. mention an event in 2007 when a well not properly decommissioned allowed gas to travel through the annulus into an underground aquifer. The methane built up over time and eventually created an explosion in a resident's basement (Zoback et al., 2010).

Drilling often occurs at great depths to maximize the number of fractures and increase the volume of gas-bearing pore spaces from which to extract gas. Pressure and temperature of subsurface formations increase with depth. Increased pressure and temperature accelerates the impact of saturated brines and acid gases, as well as creates difficulty in removing drill cuttings and installing well casings. Higher temperatures also affect cement setting behavior by causing poor mud displacement and lost circulation, which causes cementing the deep exploration and production wells challenging. Kargbo et al. referenced a report of natural gas observed in water supplies of residents of Dimock, Pennsylvania. Inspectors from the Department of Environmental Protection discovered casings on some of the gas wells were improperly cemented and potentially allowed contamination to occur.

Another negative externality of shale gas hydraulic fracturing is its high economic costs. It costs approximately 3–5 million dollars to complete the installation of a typical horizontal well using multistage fracturing technique. Due to the low permeability of shale, spacing between natural gas wells are two-times closer than conventional energy wells. Therefore the funds necessary to install natural gas wells are significant. Also, installing an extraction well is a risky endeavor. The proppants injected into fractures can quickly precipitate causing the fracture to close after the hydraulic fracturing process. If the fracture is closed the well cannot be used, creating major financial loss. The production efficiency of shale gas wells decreases significantly over time. Conventional gas wells decline by up to 40 percent in their first year, while shale gas wells can decline up to 85 percent. Hughes (2011) suggests that the recoverable reserves have been overstated by energy companies; therefore the profit of natural gas extraction is overstated. In addition to the costs associated with extraction wells, a significant amount of funds are needed to build the network of pipelines, rail facilities, truck terminals, and processing facilities. Research into energy-impacted communities has shown that local municipalities are often required to provide infrastructure and other services during energy development. These services are often costly and new sources of revenue are not equal to these costs (Jacquet & Stedman, 2011).

Natural gas development also affects tourism, an important source of revenue for many rural areas. Tourism in natural gas development areas includes recreational use (e.g., fishing, hiking, and hunting) and the wine industry. Natural gas development puts stress on water resources and often changes the topography. Beautiful landscapes are replaced by industrial views degrading visitor experiences. Temporary accommodations, such as hotels, are occupied by energy development workers, leaving less availability for tourists. And finally, workers place a great deal of wear and tear on local facilities, and are not subject to hotel room tax, a revenue source for the tourist industry (Rumbach, 2011).

Probably the most significant risk caused by hydraulic fracturing is environmental pollution and the stress on water resources. Up to 10 million gallons of water is used in drilling and hydraulic fracturing process per well location. The water is used during drilling to create a circulating mud that cools the drill bit and carries the rock cuttings out of the wellbore. Due to the large costs of transporting water long distances energy companies usually obtain water from local sources. Withdrawal of large amounts of water from a single location could have negative ecological impacts. This is especially true for water systems that are sensitive to drought conditions, low seasonal flow, aquatic communities that depend on clean, cool water, and locations with already stressed water supplies.

The Safe Drinking Water Act excludes the regulation of hydraulic fracturing by the United States Environmental Protection Agency. Due to this exemption, the fluid mixture used to fracture the geologic formation has been kept confidential. Hydraulic fracturing liquids contain proppants to keep fractures open, gels to increase fluid viscosity, acids to remove drilling mud, biocides to prevent microbial growth, scale inhibitors to control precipitation, and surfactants to increase recovery of injected fluids. Regulated compounds by both state and federal agencies for their carcinogenic and human health risk characteristics (e.g., pesticides, fluorocarbons, naphthalene, butanol, and formaldehyde) have been reported to be present in fluids. These additives are not 100 percent removed from the subsurface during extraction, leaving potential residual contaminants in the formation. Up to 80 percent of injected fluids may not be recovered prior to placing the well in production. The lack of information provided on the components of hydraulic fracturing liquids has also caused issues for treatment facilities during wastewater processing. These treatment facilities are not prepared to detect or handle hazardous materials. Therefore toxic compounds in the wastewater are often untreated and are discharged into waterways. The shale formation itself may also have naturally occurring high concentrations of salts, radioactive material (NORM), and other contaminants including arsenic, benzene, and mercury. The formation may also contain historically human deposited chemicals. Injecting and extracting from a contaminated formation could cause migration of existing contaminants and/or pollutes the wastewater. Other sources of pollution include emissions from processing plants, transportation vehicles, and drill rigs. The emissions contain GHG, particulate matter, and volatile organic compounds. Methane from shale gas production of natural gas often escapes to the atmosphere via vents and leaks. Methane is a GHG and is reported to be emitted significantly more during natural gas extraction/production than conventional gas (Hughes, 2011; Kargbo, 2010; Lee, 2011; Zoback, 2010).

And finally, preliminary activities that are conducted prior to drilling an extraction well include leveling the topography, constructing gravel roads, and installing pipelines. These activities could potentially create a negative impact on sensitive ecosystems or habitat for flora and fauna. Large energy development operations require moving large amounts of supplies, equipment, and vehicles to remote drill sites. This could lead to erosion and sediment overload that could threaten small watersheds. The risk of spills and leaks is also prominent during drilling and transportation activities. When drilling in a tight formation, such as shale, a risk of hitting permeable gas reservoirs is possible. Encountering a permeable gas reservoir may cause shallow gas and underground blowouts in the subsurface. Currently, regulations do not exist to mandate the installation of a blowout preventer. Incidents have been recorded where blowouts occurred from encountering an unexpected pocket of methane under higher pressure. Other geological hazards associated with hydraulic fracturing include disruption and alteration of subsurface hydrological conditions including the disturbance and destruction of aquifers, ground subsidence, and low magnitude seismic earthquakes (Kargbo, 2010; Zoback, 2011).

In the YES selection, journalist and blogger Seamus McGraw in *Popular Mechanics* discusses 10 controversial claims about natural gas drilling in the hope of setting the record straight on this heavily debated issue. He notes the abundance of natural gas which will fuel America's future providing both inexpensive energy and a potential resurgence in manufacturing. He also points out that hydraulic fracturing does not use as much water as other activities; that deep-injected fluids will not migrate into groundwater; and that groundwater contamination, while possible, is not probable.

In the NO selection, journalist Russell McLendon writing in the *Mother Nature Network* points out that hydraulic fracturing poses many concerns and is skeptical about whether it is "safe." The problem, he believes, is that hydraulic fracturing has not been studied enough and that the public just does not have the answers to important issues relating to public health and environmental risk. For instance, he states that there is no study that proves conclusively that "fracking" fluids cannot end up in groundwater and even migrate into aquifers. He is also concerned over the extent of methane migration in hydraulic fracturing; its ability to cause earthquakes; and the overuse of water resources.

Seamus McGraw

Is Fracking Safe? The Top 10 Controversial Claims About Natural Gas Drilling

Members of Congress, gas companies, news organization, drilling opponents: They've all made bold claims about hydraulic fracturing (fracking) and the U.S. supply of underground natural gas. We take on 10 controversial quotes about natural gas and set the record straight.

Claim No. 1

"WE ARE THE SAUDI ARABIA OF NATURAL GAS."

SEN. JOHN KERRY, D-MASS., MAY 2010

Less than a decade ago, industry analysts and government officials fretted that the United States was in danger of running out of gas. No more. Over the past several years, vast caches of natural gas trapped in deeply buried rock have been made accessible by advances in two key technologies: horizontal drilling, which allows vertical wells to turn and snake more than a mile sideways through the earth, and hydraulic fracturing, or fracking. Developed more than 60 years ago, fracking involves pumping millions of gallons of chemically treated water into deep shale formations at pressures of 9000 pounds per square inch or more. This fluid cracks the shale or widens existing cracks, freeing hydrocarbons to flow toward the well.

These advances have led to an eightfold increase in shale gas production over the past decade. According to the Energy Information Administration, shale gas will account for nearly half of the natural gas produced in the United States by 2035. But the bonanza is not without controversy, and nowhere, perhaps, has the dispute over fracking grown more heated than in the vicinity of the Marcellus Shale. According to Terry Engelder, a professor of geosciences at Penn State, the vast formation sprawling primarily beneath West Virginia, Pennsylvania, and New York could produce an estimated 493 trillion cubic feet of gas over its 50- to 100-year life span. That's nowhere close to Saudi Arabia's total energy reserves, but it is enough to power every natural gas—burning device in the country for more than 20 years. The debate over the

Marcellus Shale will shape national energy policy—including how fully, and at what cost, we exploit this vast resource.

Claim No. 2

"HYDRAULIC FRACTURING SQUANDERS OUR PRECIOUS WATER RESOURCES."

Green Party of Pennsylvania, April 2011

There is no question that hydraulic fracturing uses a lot of water: It can take up to 7 million gallons to frack a single well, and at least 30 percent of that water is lost forever, after being trapped deep in the shale. And while there is some evidence that fracking has contributed to the depletion of water supplies in drought-stricken Texas, a study by Carnegie Mellon University indicates the Marcellus region has plenty of water and, in most cases, an adequate system to regulate its usage. The amount of water required to drill all 2916 of the Marcellus wells permitted in Pennsylvania in the first 11 months of 2010 would equal the amount of drinking water used by just one city, Pittsburgh, during the same period, says environmental engineering professor Jeanne VanBriesen, the study's lead author. Plus, she notes, water withdrawals of this new industry are taking the place of water once used by industries, like steel manufacturing, that the state has lost. Hydrogeologist David Yoxtheimer of Penn State's Marcellus Center for Outreach and Research gives the withdrawals more context: Of the 9.5 billion gallons of water used daily in Pennsylvania, natural gas development consumes 1.9 million gallons a day (mgd); livestock use 62 mgd; mining, 96 mgd; and industry, 770 mgd.

Claim No. 3

"NATURAL GAS IS CLEANER, CHEAPER, DOMESTIC, AND IT'S VIABLE NOW."

OILMAN TURNED NATURAL-GAS CHEERLEADER T. BOONE PICKENS, SEPTEMBER 2009

Burning natural gas is cleaner than oil or gasoline, and it emits half as much carbon dioxide, less than one-third the nitrogen oxides, and 1 percent as much sulfur oxides as coal combustion. But not all shale gas makes it to the fuel tank or power plant. The methane that escapes during the drilling process, and later as the fuel is shipped via pipelines, is a significant greenhouse gas. At least one scientist, Robert Howarth at Cornell University, has calculated that methane losses could be as high as 8 percent. Industry officials concede that they could be losing anywhere between 1 and 3 percent. Some of those leaks can be prevented by aggressively sealing condensers, pipelines, and wellheads. But there's another upstream factor to consider: Drilling is an energy-intensive business. It relies on diesel engines and generators running around the clock to power rigs, and heavy trucks making hundreds of trips to drill sites before a well is completed. Those in the industry say there's a solution at hand to lower emissions—using natural gas itself to power the process. So far, however, few companies have done that.

Claim No. 4

"[THERE'S] NEVER BEEN ONE CASE—DOCUMENTED CASE—OF GROUNDWATER CONTAMINATION IN THE HISTORY OF THE THOUSANDS AND THOUSANDS OF HYDRAULIC FRACTURING [WELLS]"

SEN. JAMES INHOFE, R-OKLA., APRIL 2011

The senator is incorrect. In the past two years alone, a series of surface spills, including two blowouts at wells operated by Chesapeake Energy and EOG Resources and a spill of 8000 gallons of fracking fluid at a site in Dimock, PA, have contaminated groundwater in the Marcellus Shale region. But the idea stressed by fracking critics that deep-injected fluids will migrate into groundwater is mostly false. Basic geology prevents such contamination from starting below ground. A fracture caused by the drilling process would have to extend through the several thousand feet of rock that separate deep shale gas deposits from freshwater aquifers. According to geologist Gary Lash of the State University of New York at Fredonia, the intervening layers of rock have distinct mechanical properties that would prevent the fissures from expanding a mile or more toward the surface. It would be like stacking a dozen bricks on top of each other, he says, and expecting a crack in the bottom brick to extend all the way to the top one. What's more, the fracking fluid itself, thickened with additives, is too dense to ascend upward through such a channel. EPA officials are closely watching one place for evidence otherwise: tiny Pavillion, WY, a remote town of 160 where high levels of chemicals linked to fracking have been found in groundwater supplies. Pavillion's aquifer sits several hundred feet above the gas cache, far closer than aquifers atop other gas fields. If the investigation documents the first case of fracking fluid seeping into groundwater directly from gas wells, drillers may be forced to abandon shallow deposits—which wouldn't affect Marcellus wells.

Claim No. 5

"THE GAS ERA IS COMING, AND THE LANDSCAPE NORTH AND WEST OF [NEW YORK CITY] WILL INEVITABLY BE TRANSFORMED AS A RESULT. WHEN THE VALVES START OPENING NEXT YEAR, A LOT OF POOR FARM FOLK MAY BECOME TEXAS RICH. AND A LOT OF OTHER PEOPLE—ESPECIALLY THE ECOSENSITIVE NEW YORK CITY CROWD THAT HAS SETTLED AMONG THEM—WILL BE APOPLECTIC AS THEIR PRISTINE WEEKEND SANCTUARY IS CONVERTED INTO AN INDUSTRIAL ZONE, CRISSCROSSED WITH DRILL PADS, PIPELINES, AND ACCESS ROADS."

New York Magazine, *September 21, 2008*

Much of the political opposition to fracking has focused on the Catskill region, headwaters of the Delaware River and the source of most of New York City's drinking water. But the expected boom never happened—there's not enough gas in the watershed to make drilling worthwhile. "No one has to get excited about contaminated New York City drinking water," Penn State's Engelder told the *Times Herald-Record* of Middletown, NY, in April. The shale is so close to the surface that it's not concentrated in large enough quantities to make recovering it economically feasible. But just to the west, natural gas development is dramatically changing the landscape. Drilling rigs are running around the clock in western Pennsylvania. Though buoyed by the economic windfall, residents fear that regulators can't keep up with the pace of development. "It's going to be hard to freeze-frame and say, 'Let's slow down,'?" Sen. Robert P. Casey Jr., D-PA said last fall. "That makes it more difficult for folks like us, who say we want to create the jobs and opportunity in the new industry, but we don't want to do it at the expense of water quality and quality of life."

Claim No. 6

"NATURAL GAS IS AFFORDABLE, ABUNDANT AND AMERICAN. IT COSTS ONE-THIRD LESS TO FILL UP WITH NATURAL GAS THAN TRADITIONAL GASOLINE."

REP. JOHN LARSON, D-CONN., CO-SPONSOR OF H.R. 1380, A MEASURE THAT WOULD PROVIDE TAX INCENTIVES FOR THE DEVELOPMENT AND PURCHASE OF NATURAL GAS VEHICLES, MARCH 2011

That may be true. Plus, there's another incentive: Vehicles powered by liquefied natural gas, propane, or compressed natural gas run cleaner than cars with either gasoline or diesel in the tank. According to the Department of Energy, if the transportation sector switched to natural gas, it would cut the nation's carbon-monoxide emissions by at least 90 percent, carbon-dioxide emissions by 25, and nitrogen-oxide emissions by up to 60. But it's not realistic: Nationwide, there are only about 3500 service stations (out of 120,000) that offer natural gas–based automotive fuel, and it would cost billions of dollars and take years to develop sufficient infrastructure to make that fuel competitive with gasoline or diesel. And only Honda makes a car that can run on natural gas. That doesn't mean natural gas has no role in meeting the nation's short-term transportation needs. In fact, buses in several cities now rely on it, getting around the lack of widespread refueling opportunities by returning to a central terminal for a fill-up. The same could be done for local truck fleets. But perhaps the biggest contribution natural gas could make to America's transportation picture would be more indirect—as a fuel for electric-generation plants that will power the increasingly popular plug-in hybrid vehicles.

Claim No. 7

"DO NOT DRINK THIS WATER"

HANDWRITTEN SIGN IN THE DOCUMENTARY GASLAND*, 2010*

It's an iconic image, captured in the 2010 Academy Award—nominated documentary *GasLand*. A Colorado man holds a flame to his kitchen faucet and turns on the water. The pipes rattle and hiss, and suddenly a ball of fire erupts. It appears a damning indictment of the gas drilling nearby. But Colorado officials determined the gas wells weren't to blame; instead, the homeowner's own water well had been drilled into a naturally occurring pocket of methane. Nonetheless, up to 50 layers of natural gas can occur between the surface and deep shale formations, and methane from these shallow deposits *has* intruded on groundwater near fracking sites. In May, Pennsylvania officials fined Chesapeake Energy $1 million for contaminating the water supplies of 16 families in Bradford County. Because the company had not properly cemented its boreholes, gas migrated up along the outside of the well, between the rock and steel casing, into aquifers. The problem can be corrected by using stronger cement and processing casings to create a better bond, ensuring an impermeable seal.

Claim No. 8

"AS NEW YORK GEARS UP FOR A MASSIVE EXPANSION OF GAS DRILLING IN THE MARCELLUS SHALE, STATE OFFICIALS HAVE MADE A POTENTIALLY TROUBLING DISCOVERY ABOUT THE WASTEWATER CREATED BY THE PROCESS: IT'S RADIOACTIVE."

ProPublica, *November 2009*

Shale has a radioactive signature—from uranium isotopes such as radium-226 and radium-228—that geologists and drillers often measure to chart the vast underground formations. The higher the radiation levels, the greater the likelihood those deposits will yield significant amounts of gas. But that does not necessarily mean the radioactivity poses a public health hazard; after all, some homes in Pennsylvania and New York have been built directly on Marcellus shale. Tests conducted earlier this year in Pennsylvania waterways that had received treated water—both produced water (the fracking fluid that returns to the surface) and brine (naturally occurring water that contains radioactive elements, as well as other toxins and heavy metals from the shale)—found no evidence of elevated radiation levels. Conrad Dan Volz, former scientific director of the Center for Healthy Environments and Communities at the University of Pittsburgh, is a vocal critic of the speed with which the Marcellus is being developed—but even he says that radioactivity is probably one of the least pressing issues. "If I were to bet on this, I'd bet that it's not going to be a problem," he says.

Claim No. 9

"CLAIMING THAT THE INFORMATION IS PROPRIETARY, DRILLING COMPANIES HAVE STILL NOT COME OUT AND FULLY DISCLOSED WHAT FRAC KING FLUID IS MADE OF."

Vanity Fair, *June 2010*

Under mounting pressure, companies such as Schlumberger and Range Resources have posted the chemical compounds used in some of their wells, and in June, Texas became the first state to pass a law requiring full public disclosure. This greater transparency has revealed some oddly benign ingredients, such as instant coffee and walnut shells—but also some known and suspected carcinogens, including benzene and methanol. Even if these chemicals can be found under kitchen sinks, as industry points out, they're poured down wells in much greater volumes: about 5000 gallons of additives for every 1 million gallons of water and sand. A more pressing question is what to do with this fluid once it rises back to the surface. In Texas's Barnett Shale, wastewater can be reinjected into impermeable rock 1.5 miles below ground. This isn't feasible in the Marcellus Shale region; the underlying rocks are not porous enough. Currently, a handful of facilities in Pennsylvania are approved to treat the wastewater. More plants, purpose-built for the task, are planned. In the meantime, most companies now recycle this water to drill their next well.

Claim No. 10

"THE INCREASING ABUNDANCE OF CHEAP NATURAL GAS, COUPLED WITH RISING DEMAND FOR THE FUEL FROM CHINA AND THE FALL-OUT FROM THE FUKUSHIMA NUCLEAR DISASTER IN JAPAN, MAY HAVE SET THE STAGE FOR A 'GOLDEN AGE OF GAS'."

WALL STREET JOURNAL *SUMMARIZING AN INTERNATIONAL ENERGY AGENCY REPORT, JUNE 6, 2011*

There's little question that the United States, with 110 years' worth of natural gas (at the 2009 rate of consumption), is destined to play a major role in the fuel's development. But even its most ardent supporters, men like T. Boone Pickens, concede that it should be a bridge fuel between more polluting fossil fuels and cleaner, renewable energy. In the meantime, the United States should continue to invest in solar and wind, conserve power and implement energy-efficient technology. Whether we can effectively manage our natural gas resource while developing next-gen sources remains to be seen. Margie Tatro, director of fuel and water systems at Sandia National Laboratories, says, "I think natural gas is a transitioning fuel for the electricity sector until we can get a greater percentage of nuclear and renewables on the grid."

Russell McLendon

Big Frack Attack: Is Hydraulic Fracturing Safe?

Natural gas is cleaner than coal and oil, helping make it the hottest fossil fuel in America lately. But a controversial drilling technique known as "fracking" has some wondering if a U.S. natural gas boom is worth the risks.

In the 1953 Looney Tunes cartoon "Much Ado About Nutting," a frustrated squirrel hauls a coconut around New York City, aware it's a feast but unable to crack it open. It's reminiscent of an even trickier and more tantalizing jackpot that had, until recently, eluded the United States for nearly two centuries: shale gas, the hard-shelled dark horse of fossil fuels.

That squirrel never tasted the fruits of his labor, however, while the United States started figuring out shale gas by the late 1990s and early 2000s, after nibbling at it since the 1820s. But as shale fever sweeps the country—courtesy of a gas-drilling trick called hydraulic fracturing, aka "fracking"—some Americans have begun to wonder if, like the squirrel, we might be hurting ourselves as much as the protective husk around our prize.

Shale gas is natural gas that's embedded in ancient rocks known as shale, which are smashed by geologic pressure over millions of years into dense, impermeable slabs. This made them an unwise energy source for most of the 20th century, but gas companies never forgot that America is sitting on a gold mine—some estimates put the country's recoverable shale gas reserves as high as 616 trillion cubic feet, enough to meet current demand for 27 years. And thanks to advances in drilling technology, namely fracking, armies of gas rigs have suddenly uncorked an ample new power source just as many of the planet's known fossil fuel reserves are fading. By 2011, the Department of Energy predicts 50–60 percent of all growth in known U.S. gas reserves will come from shale.

It's not hard to see the appeal. Natural gas emits fewer greenhouse gases than other fossil fuels—about half as much carbon dioxide as coal, for example—and thus contributes less to global warming. It also has mostly avoided the bad press that plagues coal and oil, from mountaintop removal and mine explosions to recent oil spills in Alaska, Utah, Michigan, and the Gulf of Mexico. And with natural gas prices expected to rise in coming years, America's shale mania may have only scratched the surface.

Despite its potential, though, a movement has welled up lately to block the shale gas boom. Some critics say embracing natural gas so heartily will slow the rise of renewable energy, but the biggest beef with shale isn't as much about its gas—it's about how we get it out of the ground. Shale gas would likely still be a novelty fuel without modern advances in hydraulic fracturing, yet the need for fracking is also starting to seem like shale's fatal flaw. The practice has sparked major environmental and public health concerns near U.S. gas fields, from diesel fuel and unidentified chemicals in groundwater to methane seeping from sink faucets and even blowing up houses.

With gas drillers still vying for vast U.S. reservoirs like the Barnett Shale in Texas or Appalachia's sprawling Marcellus Shale, many federal and state officials across the country have begun to question their hands-off attitudes toward fracking. The EPA is in the early stages of a two-year study to assess the practice's risks, and in November it subpoenaed energy giant Halliburton for information on specific fracking chemicals it uses. It also recently ordered a Texas gas company to stop all work after methane and benzene appeared in nearby drinking-water wells. Some states and cities are also taking notice—Pittsburgh banned fracking within city limits in November, for example, and the New York Legislature followed suit with a statewide ban passed this month. Pennsylvania also outlawed fracking in its state forests, and Colorado and Wyoming have new disclosure laws on the books regarding fracking chemicals. Hollywood has even jumped into the fray, recently sending out actor Mark Ruffalo to the front lines.

But what's the big deal about fracking? What does that word even mean? And is it really risky enough to justify putting an abundant, relatively clean energy source on the back burner? Below is a brief look at how the process works, how it might affect the environment, and what its future may hold.

How Does Fracking Work?

The problem with shale gas is that it's not just stuck in some rocky reservoir like many gas deposits; it's actually embedded in the rock itself. That's because shale, a mudstone formed by the buildup and compression of sediments, often contains ancient organic debris, which can make it a "source rock" for oil and gas. It may also act as a cap for the underground caverns that collect its seeping contents, and drilling companies used to bypass it in favor of the free-flowing fossils below. But now, as Earth's shallowest and easiest energy reserves increasingly dry up, the industry has turned back to shale, using high-tech directional drilling and fracking to make the stubborn stone give up its gas.

- **Directional drilling:** One of the reasons shale was left alone for so long was its tendency to form wide but shallow layers. Drilling straight down into these doesn't produce much gas, since the drill hits too little surface area before passing through. The best way to get more gas out is to drill in sideways, which became much easier in the 1980s and 1990s as the gas industry improved its directional drilling skills. But that still wasn't enough to make shale worth the trouble—the rock is just too dense and impermeable, with lots of pores to hold natural gas, yet too few connections between them to let it flow.

- **Hydraulic fracturing:** That's where fracking comes in. Drillers pump pressurized water, sand, and chemicals down a newly drilled well, forcing them through perforations in its casing so they blast out to the surrounding shale, opening new cracks and widening old ones. Water may constitute up to 99 percent of this mixture, while the sand serves as a "propping agent" to keep the cracks open after the water is pumped out. This technology has existed for decades, but recent breakthroughs now let drillers use more water—2 to 5 million gallons per well—while new "slick-water" fracking chemicals help them slash friction. That increases the water pressure, and thus the amount of fracturing.

  "Without directional drilling and slick-water hydraulic fracturing, you can't get gas out of shale," says Tony Ingraffea, an engineering professor and fracturing expert at Cornell University. "It's been known for many decades that there's a lot of gas in the Marcellus Shale, but it just wasn't economical to get it out. . . . If you drill directionally, though, you have almost unlimited access, but you really have to break up the rock. That's what it's about: creating a lot of surface area."

Where Does Fracking Happen?

Shale is scattered generously across the United States, but each deposit has its own personality, Ingraffea points out. "Materials, pressures, gases—all those things vary among geologic regions," he says. "They even vary within a particular formation like Marcellus. That's just how nature is. No two mountains look alike, do they?"

Because of these variations, gas companies can't just take what works at one deposit and expect it to work somewhere else. That became clear after the 1990s Barnett Shale boom in Texas, when drillers who'd been capitalizing on innovations by Mitchell Energy—the drilling firm that pioneered modern fracking—tried to apply those methods elsewhere. There was a steep learning curve, especially as companies started digging into the Marcellus Shale, but they eventually picked up steam as they learned the region's geological quirks. "After three years of experimentation in Pennsylvania," Ingraffea says, "they're zeroing in on what they think will be the best way to get gas from the Marcellus while putting the least money down the well."

Barnett and Marcellus are two of the hottest shales in America lately, evolving into testing grounds for the country's fracking revolution. But they're not alone, joined by other big shales buried under Arkansas, Louisiana, New Mexico, Oklahoma, and Wyoming, to name a few. See the map on the next page for a look at all known shale gas reserves in the lower 48 states (click to enlarge).

Even with all this diversity, though, Marcellus has emerged as the king of U.S. shales; dipping under parts of seven states plus Lake Erie, it may hold as much as 516 tcf of natural gas. It was born nearly 400 million years ago after a continental collision between Africa and North America, which helped push the early Appalachian Mountains about as high as today's Himalayas. Clay and organic matter washed down their steep slopes into a shallow sea, buried over time by the up-and-coming Appalachians.

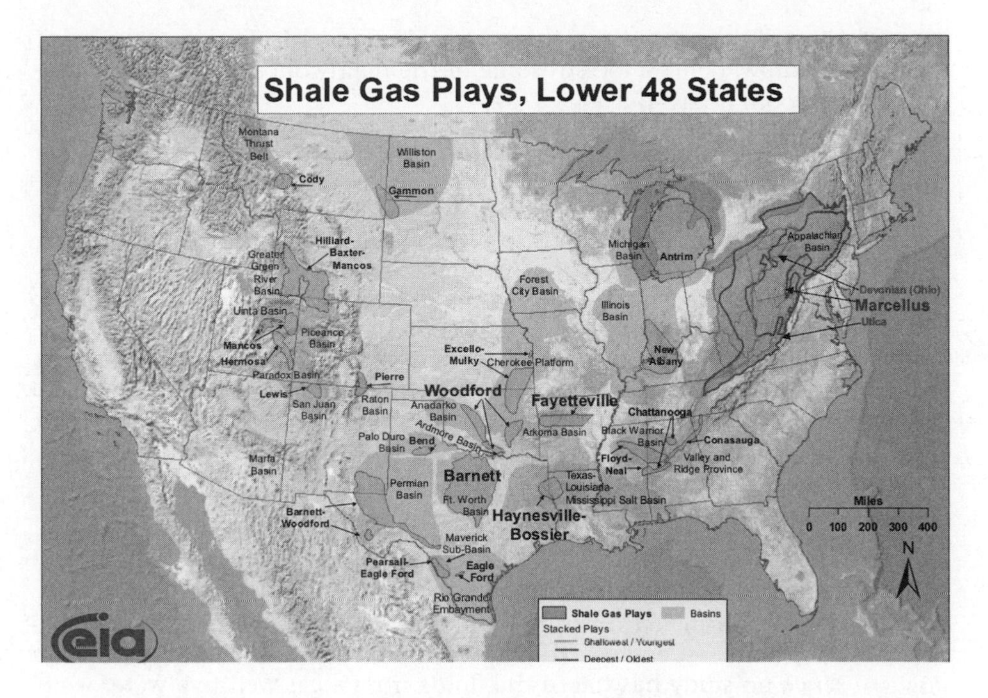

The formation of such shales is painfully slow but also heated and high-pressure—much like the political climate surrounding the Marcellus Shale today. The gas boom took Pennsylvania by storm in just a few years, stirring up ill will from residents who say fracking pollutes their groundwater, and those concerns have since spurred bans on fracking in state forests and Pittsburgh. The controversy has also spilled over into neighboring New York, where the state Legislature recently approved a temporary ban on fracking until its environmental effects are better understood.

Is Fracking Dangerous?

The EPA study follows years of pressure from environmental and public health groups, especially since Congress exempted fracking from the federal Safe Drinking Water Act in 2005. That already angered many foes of fracking, but their calls for more oversight have only grown louder since the Gulf oil spill. While BP allegedly broke federal offshore drilling laws, they point out, no such rules even exist for fracking.

The industry often counters that fracking has never been directly linked to a case of water pollution, saying it should be presumed innocent until proven guilty. Supporters also argue that halting the gas boom could hinder U.S. job growth and energy output when they're needed most. But with shale drilling poised to explode across America—especially if natural gas prices recover from the recession as expected—critics say the health risks outweigh the economic bounty, and that the burden of proof should fall on gas companies, not their customers and communities.

The burden of proof is currently on the EPA, but since its study won't yield results for at least another two years, Americans will apparently remain in the dark until then about any threats fracking presents. For an overview of what we do know, here's a look at some of the main concerns about fracking and the gas boom it has spurred:

- **Fracking fluids:** Hydraulic fracturing is a bit like using a garden hose, Ingraffea says: "You're trying to pump large volumes of fluid at high pressure through something that's six inches wide and two miles long, so a lot of energy is lost." Diesel fuel was commonly used in the past to reduce friction while fracking, but since it contains carcinogens like benzene, the EPA and major gas companies reached a "memorandum of agreement" in 2003 to stop using it. The industry then switched to a cocktail of friction-reducing chemicals that are considered trade secrets, meaning their identities aren't public knowledge. But they still sometimes reveal themselves, such as when 8000 gallons of fracking fluids spilled at a natural gas site near Dimock, PA, last year—the loose chemicals included a liquid gel called LGC-35 CBM, which is considered a "potential carcinogen" in humans. (No people were hurt in that spill, but fish were found dead and "swimming erratically" in a nearby stream.) The industry insists there's no proof such fluids get into aquifers, but the EPA estimates only 15–80 percent return to the surface, and no study has ever shown where the rest end up. That has set off an array of health alarms, but since no study has traced the fluids from a gas well to a water well, either, communities near gas fields are left to sit in legal limbo for now. "Theoretically, it's not hard to demonstrate how a high-volume, slick-water hydraulic fracturing event at some depth could cause fractures, or existing joints or faults, to receive the fracturing fluid and transport it vertically to groundwater," Ingraffea says. "What is hard is proving that such theoretical events have actually occurred."
- **Methane migration:** Methane is an explosive, asphyxiating chemical with more potent climate-changing powers than carbon dioxide, and it makes up anywhere from 70–90 percent of most natural gas. It has also begun appearing in water supplies near gas fields across the country, but—as with fracking fluids—no firm evidence has been found that implicates gas drilling. Methane occasionally enters wells through natural fractures, too, and it can be removed by venting the gas out of the water. While that's one advantage of having methane in your well instead of fracking fluids, which can't be removed, the risks from those chemicals is largely a mystery compared with the well-known dangers of methane. When it seeps into tap water, it's suspended in bubbles that later pop as the water exits a faucet or shower head. Both the methane-laden water and the air where it escapes will become flammable, eventually erupting in a fireball if exposed to a spark. So-called "methane migration" has grown increasingly common, along with gas drilling, in several Pennsylvania counties over the last six years; in one case the gas was detected in water samples spanning 15 square miles, while another in 2004 resulted in the explosion of a house that killed a couple and their 17-month-old grandson. Texas, Wyoming, and other shale gas hotspots have also seen anecdotal outbreaks of methane migration in the last few years.

Earthquakes: Blasting pressurized water so deeply into the Earth's crust has the potential to do more than just widen small cracks in the bedrock—if it hits the right underground fissure at the right angle and speed, it can actually trigger an earthquake. This is a problem gas companies share with many other subterranean industries, such as oil drillers and dam builders; even renewable, emissions-free geothermal power can be an earthquake enabler, taking blame for clusters of moderate tremors from Southern California to Switzerland. Fracking has also become a prime suspect for such "microquakes," which sometimes spike in regions where deep fracturing takes place. Earthquakes are rare in Texas, for example, but the area around Fort Worth has suffered at least 11 earthquakes in the last two years, a trend seismologists say may be linked to increased fracking at the nearby Barnett Shale. On top of all the usual problems that go along with earthquakes, gas-drilling areas are especially at risk because they tend to host gas pipelines, which transport extracted gas to market. While some pipelines are built to withstand seismic jiggling, a strong quake could nonetheless be disastrous, possibly causing a gas leak or even an explosion.

- **Water use:** Aside from allegedly adding methane and various chemicals to groundwater supplies, fracking has also come under fire for the amount of water it consumes. The 21st-century version requires around 3 million gallons of water for every well that's fracked, putting the high volume under intense pressure to break open shale formations buried a mile or more deep. According to the only estimate the EPA currently offers, somewhere between 15 and 80 percent of all fluids pumped into a well are pumped back up to the surface, where they may be placed in a containment area or may be treated and recycled. But much of the water is lost somewhere underground, adding stress to local water supplies that may already be polluted from fracking or other sources.

Following a series of public meetings in 2010 meant to inform the overall design of the EPA's fracking study, the agency is set to actually kick off the investigation in January 2011, with a time frame for initial results given only as "late 2012 [http://www.epa.gov/hfstudy/]." According to Ingraffea, who has studied hydraulic fracturing for 30 years, the EPA will likely crack down on certain fracking fluids, but gas companies will already have replacements ready. Just as some drillers continued using diesel after 2003 because it's cheaper than other friction reducers, Ingraffea says the industry has resisted switching to safer fracking chemicals because of the added cost.

"If the EPA announced tomorrow that hydraulic fracturing is now regulated, it would take 48 hours for the companies to say 'Ah! We've been working in the lab and have developed these other chemicals that are safer, so now we can start hydraulic fracturing again,'" he says. "Of course, they'd have to throw out their vast stocks [of current fracking fluids] that they've collected and are planning on using. But if you can't hydraulic fracture, you lose the industry."

EXPLORING THE ISSUE

Is Natural Gas Hydraulic Fracturing Safe?

Critical Thinking and Reflection

1. After reviewing both the pros and cons of natural gas hydraulic fracturing, do you believe that it is a safe process? Define what you mean by "safe" and give examples of how it might be "unsafe."
2. Do the benefits of natural gas hydraulic fracturing outweigh the risks and costs?
3. Does the United States need natural gas hydraulic fracturing to achieve energy independence? Should the United States even seek energy independence if it can obtain cheaper energy from outside the country?
4. Do you believe that the economic and job creation benefits of natural gas hydraulic fracturing outweigh any of the negatives?
5. Will cheaper natural gas fuel a resurgence of manufacturing in the United States? Is the energy future of the United States dependent on natural gas?
6. Is natural gas the only realistic option for America to lower its greenhouse gas emissions which contribute to global warming and climate change?

Is There Common Ground?

There is no doubt that natural gas hydraulic fracturing has the potential to provide enormous economic benefits to the United States. But many of these benefits appear to be nationwide while the negative externalities and risks are largely localized. In order for natural gas hydraulic fracturing to be safe, both stakeholder involvement and tighter regulations are necessary. Stakeholder involvement includes conversations among energy companies, regulators, municipalities, regional organizations (such as a watershed commission), and landowners. Initiating conversation among stakeholders prior to the commencement of energy development can ensure that long-term benefits outweigh the risks and costs. The stakeholders should discuss all potential risks and responsible parties for construction of infrastructure, management of water and environmental resources, management of wastewater, and remedial actions should be engaged. During preliminary activities, energy companies negotiate with landowners and municipalities concerning land leasing and royalties. These negotiations should be performed either with all stakeholders involved or with a coalition that is educated on the impacts of hydraulic fracturing and represents the local community. The party negotiating with the energy companies should implement mandates in the leasing contract. These mandates can include regulating the components of hydraulic fluids, treating and recycling wastewater, implementing environmentally friendly technologies, confirming proper sealing of the installation and decommissioning of wells, and compensation for environmental, economic, and societal impacts. A large amount of power lies with the owner of the property and therefore the owner should become the main driver in implementing beneficial items.

Tighter regulations include local, regional, and national regulators implementing standards and guidelines to monitor the components of hydraulic fluids, and the installation and decommissioning of wells. Regulators can also create incentives to influence energy companies to implement environmentally friendly technologies. Lastly, since a large amount of power lies with the property owners, regulators should educate locals on the impacts and benefits of hydraulic fracturing. In addition to regulators, environmental managers should be involved during negotiations to ensure benefits associated with impacted communities and to insure safety. Proper involvement of all stakeholders can create an energy development practice that has low impact and high benefits.

References

Pros

Arthur, J. D., Bohm, B., Jo Doughlin, B., and Layne, M. (2008a) *Hydraulic fracturing considerations for natural gas wells of the Fayetteville Shale.* All Consulting (Tulsa, OK).

Arthur et al. discusses the use of modern technology and modeling to decrease the environmental and societal impact caused by hydraulic fracturing. The article also discusses the economic benefit to local and regional communities from royalties and employment opportunities.

Arthur, J. D., Bohm, B., Jo Doughlin, B., and Layne, M. (2008b) *Hydraulic fracturing considerations for natural gas wells of the Marcellus Shale.* All Consulting (Tulsa, OK).

Arthur et al. provides a detailed description of the use of computer simulations to evaluate hydraulic fracturing designs. The authors suggest initial use of these models will assist in maximizing fracture placement and in turn decrease environmental impact and costs.

Considine, T., Watson, R., Entler, R., and Sparks, J. (2009, Jul 24) *An emerging giant: Prospects and economic impacts of developing Marcellus Shale natural gas play.* Department of Energy and Mineral Engineering, College of Earth & Mineral Sciences, The Pennsylvania State University.

Considine et al. presents several direct and indirect economic benefits from natural gas production at local (e.g., employment and local business) and national (e.g., market for natural gas) regions. The document also discusses how implementing natural gas on a large scale will assist in the reduction of greenhouse gases due to lower carbon emissions.

Jacquet, J. and Stedman, R. C. (2011) Natural gas landowner coalitions in New York State: Emerging benefits of collective natural resource management. *Journal of Rural Social Sciences, 26*(1), pp. 62–91.

Jacquet et al. discusses how landowner coalitions can take advantage of negotiations with energy companies by ensuring long-term benefits outweigh costs.

Reeder, L. C. (2010) Creating a legal framework for regulation of natural gas extraction from the Marcellus Shale formation. *William & Mary Environmental Law & Policy Review, 34*(3), pp. 999–1026.

Reeder discusses how stakeholders can create a model framework to protect the interests of municipalities and landowners, as well as the impact on the environment, while ensuring regulations are not so stringent to prohibit energy companies from development. Reeder also discusses implementing innovative technologies to decrease the overall impact to the environment and society.

Cons

Hughes, J. D. (2011, May) *Will natural gas fuel America in the 21st century*. Post Carbon Institute.

Hughes discusses challenges faced during hydraulic fracturing, including the low permeability characteristic of shale, use of a large amount of water resources, toxic constituents in hydraulic fluids, and the decline of production efficiency in a well over time. Hughes also discusses environmental issues related to hydraulic fracturing.

Kargbo, D. M., Wilhelm, R. G., and Campbell, D. J. (2010) Natural gas plays in the Marcellus Shale: Challenges and potential opportunities. *Environmental Science & Technology, 44*, pp. 5679–5684.

Kargbo et al. presents negative impacts of hydraulic fracturing in five categories: (1) exploration and drilling; (2) water resources; (3) hydraulic fracturing process; (4) wastewater management; and (5) radioactivity. Kargbo et al. suggests improvement of best management practices or implementation of alternatives to alleviate negative impacts from hydraulic fracturing.

Lee, D. S., Herman, J. D., Elsworth, D., Kim, H. T., and Lee, H. S. (2011) A critical evaluation of unconventional gas recovery from the Marcellus Shale, Northeastern United States. *KSCE Journal of Civil Engineering, 15*(4), pp. 679–687.

Lee et al. states that technological challenges remain in the economic recovery of gas from unconventional reservoirs, as is the case for shale gas. Shale has low porosity and permeability which creates a challenge in accessing and extracting the gas from this formation.

Rumbach, A. (2011) *Natural gas drilling in the Marcellus Shale: Potential Impacts on the tourism economy of the Southern Tier*. Prepared for the Southern Tier Central Regional Planning and Development Board, with support from the Appalachian Regional Commission.

Rumbach discusses the impact natural gas drilling can have on local tourism. Impacts mentioned include creating an industrial landscape, environmental pollution, stressors on wildlife and water resources, and reduction in temporary accommodation availability.

Zoback, M., Kitasei, S., and Copithorne, B. (2010, Jul) *Briefing Paper 1—Addressing the environmental risks from shale gas development.* Worldwatch Institute.

Zoback et al. discusses several environmental risks and potential best management practices that could be implemented. The environmental risks include subsurface contamination to groundwater, blowouts, seismic activity, surface water and soil contamination, air pollution, strain on water resources, and topography impacts.

ISSUE 15

Can Species Preservation Be Successfully Managed?

YES: Dale D. Goble et al., from "Conservation-Reliant Species," *BioScience* (vol. 62, pp. 869–873, October 2012), www//mc.manuscriptcentral.com/ucpress-bio

NO: Craig Hilton-Taylor et al., from *State of the World's Species* (IUCN, Gland, Switzerland, 2009)

Learning Outcomes

After reading this issue, you should be able to:

- Provide examples of international organizations and global conventions that are instrumental in species preservation.
- Understand the role of the United States Endangered Species Act (US-ESA) in protecting U.S. biodiversity by preventing species extinctions and promoting their recovery.
- Evaluate whether the ESA has been successful at protecting U.S. biodiversity.
- Understand the concept of conservation-reliant species and how it establishes management practices that can successfully reduce species extinction.
- Understand the importance of reducing species extinction and biodiversity conservation.
- Understand the complexities of protecting endangered species; where they are protected and the types of species that are protected.
- Discuss how governments are failing to stem the rapid decline of species extinction.

ISSUE SUMMARY

YES: University of Idaho law professor Dale D. Coble with John A. Wiens, PRBO Conservation Science; Michael Scott, University of Idaho, College of Natural Resources; Timothy D. Male, Defenders of Wildlife; and John A. Hall, Department of Defense, Strategic Environmental Research and Development Program/Environmental

Security have developed the concept of "conservation-reliant species" to show how species extinctions can be reduced through successful management planning.

NO: Researcher Craig Hilton-Taylor, manager of the International Union for Conservation of Nature "Red List of Threatened Species," leads a team that shows the rapid decline in biodiversity as a result of unsustainable human–wildlife confrontation.

Scientists believe that we have now entered the Anthropocene, a new geological era where humans are the major cause for the alteration of ecosystems and the main driver of planetary processes. Nowhere is this more evident than in the threats to global biodiversity. There is estimated to be close to 9 million species that inhabit the world, but only 1.2 million of them have been scientifically identified and classified (Mora et al., 2011). With an average of 15,000 species being discovered every year it will take 450 years to identify all the living organisms with which we share the world (Dirzo & Raven, 2003). Most species that become extinct are never scientifically documented. This represents a problem, since there are an estimated 7.5 million species that remain undiscovered. Undiscovered species often go extinct due to the lack of knowledge concerning ecology and anthropogenic threats affecting them. It is unimaginable how undiscovered species can survive when there is an existing threat to species already well studied.

For the past 500 years there has been an estimated 875 extinctions worldwide with 500 of these being documented in the past century. With a current average of 10 known species becoming extinct each year, half of the world's species will become extinct or threatened by 2100 (WWF). Modern extinctions have been attributed to climate change and human interactions. While species extinction have been happening for millions of years, today it is occurring at a quicker and higher rate, not allowing species to adapt to the modern world (Wilson, 2003). At present, more than 5000 species are considered endangered or threatened worldwide (IUCN), with the earth experiencing a major biodiversity crisis.

Preserving biodiversity is an important dimension in supporting ecosystem health. Species health and habitat biodiversity are important bioindicators, often used as cues to determine habitat and ecosystems health. Biodiversity is also needed to maintain ecosystem productivity such as the production of oxygen, absorption of carbon dioxide, and providing fertile soils for agriculture. Therefore biodiversity is important to maintain human welfare. Human welfare can also be achieved through careful biocentrism, a mindset that emphasizes the value of natural resources for biodiversity preservation.

Concern over the protection of endangered species started in the early 1900s. The United States has over 70 conservation and environmental laws protecting imperiled species such as the Endangered Species Act (ESA) which is considered the most powerful environmental law in the United States. Since the 1970s, there have been many successes in recovering species from the brink of extinction but there are others that describe these successes as merely winning small battles while losing the war.

Globally, the largest threat to endangered species is habitat loss followed by international trade in wildlife. Habitat loss is currently not globally regulated because land use management is a sovereign issue. Yet, the International Union of Conservation of Nature (IUCN) issues a global report of all imperiled species which provides nations, government agencies, and NGOs with this valuable information. Area-based action is done by establishing protected areas, and this approach has proven to be the most successful conservation strategy in protecting endangered species (Hoffmann et al., 2008). Key biodiversity areas (KBAs) also aid in the protection of endangered species by identifying biodiversity hotspots. KBAs are used in species conservation in over 100 countries worldwide (Eken et al., 2004i; Langhammer et al., 2007).

Examples of successful management of species conservation have been linked to the initiatives of the IUCN, the Convention of International Trade in Endangered Species (CITES), the Convention on the Conservation of Migratory Species of Wild Animals, and the US Endangered Species Act (US-ESA). The IUCN was created in 1948 with the purpose of conserving biodiversity. It provides the most current scientific data on the current status of globally threatened species which allows for the identification, conservation, and rehabilitation of many imperiled species. CITES restricts the international trade of endangered and threatened species and their products, making it the strongest trading restriction of imperiled species. The Convention on the Conservation of Migratory Species of Wild Animals aims at protecting flight and aquatic animals and is crucial to the protection of imperiled species beyond local conservation limits. And lastly, the US-ESA has listed 2000 species as threatened or endangered since its enactment. In the last 38 years, 50 species have been delisted. It requires every listed species to have a recovery plan, designated critical habitat, and prohibits the taking of any listed species.

The US-ESA has led to recovery of certain imperiled species. Bald eagle populations had plummeted to 416 mating pairs as a result of habitat loss, killings, and the use of pesticides. After the listing of the eagle under the US-ESA and the banning of the use of the pesticide DDT, bald eagles started showing signs of recovery. By 1997, with a population of almost 10,000 mating pairs, the bald eagle was delisted. The peregrine falcon is another success story. In the 1940s there were a recorded 1500 mating pairs but this number dwindled by 80–90 percent due to habitat loss and use of the pesticide DDT. With the banning of DDT and the implementation of breeding programs in 1999 the peregrine falcon has recovered to 3000 mating pairs and have been delisted.

While few species have been delisted from the US-ESA, Male and Bean (2005) reported that approximately 52 percent of all listed species are recovering or are stable, while 48 percent are declining. Species currently considered to be recovering are the whooping crane, Kirtland warbler, gray wolves, grizzly bear, Hawaiian goose, Virginia big eared bat, California sea otter, and the Florida red wolf. At first glance, the numbers provided by Male and Bean suggest that the US-ESA is not very efficient at protecting species. But the authors note that most of the declining species have been newly listed and are in the first phase of their recovery plans. They also conclude that the longer a species is listed the more likely of it recovering or being stable. After two years of being listed, 23 percent of species were improving, while after 12–13 years of being listed, 68 percent of species were considered to be stable or improving. Their study shows that US-ESA recovery plans over time can protect endangered species.

As of 2012, there are close to 2000 species listed in the US-ESA as threatened or endangered with approximately 60 percent having active recovery plans (USFW). Over the course of the Act less than 50 species have been delisted, with 22 due to signs of recovery and the remainder for other reasons. Critics of the Act point out that species recovery was due mostly to the bans on DDT and whaling and not specifically to the Act. They note that a critical flaw is that it takes too long for a species to be listed and that their populations might be too depleted to be saved (Schwartz, 2008). Also, attention is directed at species that have the best chance at survival rather than the ones that require immediate action; and that the Act aims to conserve critical habitats essential to the survival of the species rather than in protecting overall biodiversity An imperiled species can be an indicator of a poor habitat that can trigger investigations of other species to avoid further listings. This also works in reverse. If a species is considered threatened, the remaining species within the habitat can play an important role in the recovery of the target species and therefore should also be protected. Biodiversity and meta-populations are essential signs of healthy habitats (Rohlf, 1991; Campbell et al., 2002).

In the YES selection, that species conservation can be successfully managed, is provided by University of Idaho law professor Dale Goble and his associates. He introduces the concept of "conservation-reliant species" as a way to successfully manage species conservation. He states that "the threats that most species face cannot be eliminated, only managed. The scale of anthropogenic alteration of most ecosystems means that many imperiled species will require conservation management actions for the foreseeable future to maintain their targeted population levels." He views the US-ESA as an effective approach to recognizing "taxa that are on the brink of extinction and defining the steps to reverse their downward trajectory," but recognizes that society needs a full management tool-kit to conserve habitat and maintain species recovery. He concludes by saying that decision-makers must develop sensible ways of "assigning conservation priorities in which both the magnitude of management required and the potential benefits of management and conservation actions are considered."

In the NO selection, that we are not presently successfully managing our species conservation, is Craig Hilton-Taylor, wildlife researcher and manager of the International Union for Conservation of Nature's "Red List of Threatened Species." He leads a team of investigators that have produced a report that discusses how governments around the world are failing to manage the rapid decline of biodiversity. In the conflict between humans and wildlife, wildlife is losing. The report states that the world is losing almost half of its coral reef species, a third of its amphibians, and a quarter of its mammals due to the human destruction of wildlife habitat, the effects of man-induced climate change, and the inability of governments to prevent animal extinction. Unless governments undergo major changes to society, the report predicts the irreversible lose of wildlife species in what they refer to as a "wildlife crisis."

Dale D. Goble et al.

Conservation-Reliant Species

Humans have been altering the Earth's ecosystems for millennia (Diamond and Veitch 1981, Pyne 1995, Flannery 2001, Jackson et al. 2001). Since the onset of the Industrial Revolution, however, the temporal and geographic scales of these modifications have increased at an accelerating rate. The cumulative impact is such that it has been proposed that the world has entered a new geological era—the Anthropocene (Crutzen and Stoermer 2000). Regardless of the descriptor, the message is simple and damning: The accumulated effects of individual and societal actions, taken locally over centuries, have transformed the composition, structure, and function of the global environment (Janzen 1998, Sanderson et al. 2002, McKibben 2006, Kareiva et al. 2007, Wiens 2007). Ecological lows have become the new baseline (Pauly 1995). Although climates have always been dynamic, and threats have always existed, recent anthropogenic threats to the integrity, diversity, and health of biodiversity are unprecedented, not only causing additional stress to ecosystems but also challenging our ability to respond (Julius and West 2008). How do we manage species and ecosystems in a world of global threats and constant change (Botkin 1990)?

One response in the United States to the endangerment and loss of species was the enactment of the Endangered Species Act (ESA). The Act's goal is to bring species at risk of extinction "to the point at which the measures provided pursuant to this Act are no longer necessary" (ESA § 3(3)). The ESA's drafters envisioned this as a logical progression: Species at risk of extinction would be listed under the Act in a process that would identify the risks the species faced, a recovery plan to address these risks would be drafted, the management tools required to conserve the species would be identified and implemented at relevant scales, the species would respond by increasing in numbers and distribution, the recovery goals would be achieved, and the species would then be delisted as *recovered*. In the interim, it would be protected by the ESA's suite of extinction-prevention tools (e.g., prohibitions on taking listed species or adversely modifying their critical habitats; Goble 2010). With recovery and delisting, the formerly listed species would achieve the ESA's goal of planned obsolescence when the Act is no longer necessary. To the extent that management would be needed, it would be provided through existing federal and state regulatory mechanisms.

The past nearly four decades have demonstrated the naivete of this vision. The path to recovery is far more winding than had been imagined. Even species that have met their biological recovery goals often require continuing, species-specific management, because existing regulatory mechanisms are seldom sufficiently specific to provide the required ongoing management (Goble 2009). For example, few species have thrived as easily as the now-delisted Aleutian cackling goose (*Branta hutchinsii leucopareia*), whose populations recovered once foxes that preyed on breeding birds and chicks were eliminated from nesting islands and for which the Migratory Bird Treaty Act's monitoring and take restrictions are sufficient. The threats that most species face cannot be eliminated, only managed. The scale of anthropogenic alteration of most ecosystems means that many imperiled species will require conservation management actions for the foreseeable future to maintain their targeted population levels. Adequate postdelisting management (i.e., regulatory assurances), however, is seldom possible, because for most species, no sufficiently focused and powerful regulatory mechanism is available to replace the ESA (Goble 2009, Bocetti et al. 2012 [in this issue]).

This is hardly surprising. The species listed under the ESA all became imperiled despite existing state and federal management systems. The problems remain: Most states lack regulatory systems that address nongame and plant species (Goble et al. 1999); funding is often tied to hunting and fishing license fees and remains insufficient (Jacobsen et al. 2010). Although existing management systems (e.g., the Marine Mammal Protection Act) may be sufficient for species such as the gray whale (*Eschrichtius robustus;* Goble 2009), the expectation that our work would be done once recovery goals have been met turns out to have been wishful thinking. Just how wishful was suggested by Scott and colleagues (2010), who examined the management actions required by recovery plans for species listed under the ESA. Scott and colleagues (2010) found that 84 percent of the species are conservation reliant, because their recovered status can be maintained only through a variety of species-specific management actions. Even if the biological recovery goals for these species are met, continuing management of the threats will be necessary. Reed and colleagues (2012 [in this issue]) provide insight into this problem by describing the challenges to recovery and to postrecovery management for one of the world's most management-dependent communities: the endemic birds of Hawaii. These species are "conservation reliant" in the sense described by Scott and colleagues (2005).

The ESA is focused on moving species to the recovery threshold. The magnitude of conservation reliance makes it clear that attention must also be given to postrecovery management (Goble 2009, Scott et al. 2010). Furthermore, species not currently listed but at risk because of declining populations or range contractions are also likely to be conservation reliant. In this context, a range of management actions may be required to preclude the need to list the species under the ESA. Although comprehensive wildlife conservation strategies developed by states with funding from the federal government provide a blueprint for sustaining nongame species and their habitats, the available state funding for these management efforts is widely viewed as insufficient (Jacobsen 2010).

Earlier, we addressed the question of conservation-reliant species in the context of the ESA (Scott et al. 2005). We did so in part by placing species along a gradient of levels of human intervention and management. At one end were those species now known only in captivity, such as the Guam kingfisher (*Todiramphus cinnamominus cinnamominus*), or sustained in the wild only through repeated releases of individuals reared in captivity, such as the California condor (*Gymnogyps californianus*). These species require the greatest degree of human intervention to achieve the basic conservation objective: the prevention of extinction. At the other end of the gradient are species such as the peregrine falcon (*Falco peregrinus*), whose recovery, once the major threat of DDT (the insecticide dichlorodiphenyltrichloro-ethane) had been eliminated, was secured by its ability to adapt to human-dominated environments by nesting on skyscrapers and foraging in cities on pigeons (*Columba livid*) and starlings (*Sturnus vulgaris*). The falcon thus thrives under existing federal regulations that protect all birds used in falconry and no longer requires species-specific management. The species is no longer conservation reliant. Between these extremes are a variety of species that will require differing intensities and forms of management intervention to persist in the wild. The point along this gradient at which a species becomes conservation reliant is determined by the necessity of continuing, species-specific intervention, rather than the type of intervention. The need for continuing intervention is, in turn, determined by the threats that species face. In some instances, the threats can be eliminated through appropriate actions. The key to the recovery of peregrine falcons was the banning of the pesticides that contributed to eggshell thinning and reproductive failure. For the Aleutian cackling goose, it was the removal of an introduced predator on its breeding grounds. Both species now thrive under the general provisions of the Migratory Bird Treaty Act and are no longer conservation reliant. When, however, the threat cannot be eliminated but only controlled and conservation goals can be achieved only through continuing management intervention, the species will remain conservation reliant. . . .

The recognition that conservation reliance is a deeper and more widespread problem for listed and at-risk species than we (and others) initially thought has led us to a more nuanced perspective on this problem. In fact, two forms of conservation reliance affect species: population-management reliance and threat-management reliance. Although the ability of a species to persist is ultimately related to the characteristics and condition of both populations and the threats they face, conservation actions are often focused primarily either on managing populations or on managing threats. For example, species such as the northern Idaho ground squirrel (*Spermophilus brunneus*) live in isolated patches of habitat and may require some level of direct human intervention to move among those patches, even after local population sizes are stable (Garner et al. 2005). In contrast, other species may persist without direct population management if appropriate habitat is available. Given current land uses (and other pressures of the Anthropocene), however, human intervention may be required to maintain the habitat. As a result, it is not only species that are conservation reliant but entire ecosystems and the associated disturbance regimes (such as fire) and ecological succession pathways that define them. For example, the Karner blue butterfly (*Lycaeides melissa samuelis*), the red-cockaded woodpecker (*Picoides borealis*), and Kirtland's warbler (*Dendroica kirtlandii*) rely on periodic fire to maintain their habitat. The natural fire regimes that shaped the habitats and habitat associations of these species no longer occur, so prescribed burns must be used instead. Species such as these will continue to require threat management for the foreseeable future, even after the direct management of populations is no longer required. The two forms of conservation reliance are not independent of each other. For example, threats often influence what population actions are necessary: Where habitat encroachment has isolated small populations from each other, manipulation of the habitat may reduce habitat loss and fragmentation and may increase gene flow between the populations.

The conservation challenge is clear. The number of species that will require ongoing management is already large, and it will get larger as climate change, land-use change, human population growth, and other manifestations of the Anthropocene push more and more species to their limits. The ESA has been an effective approach for recognizing taxa that are on the brink of extinction and defining the steps needed to reverse their downward trajectory. The need for continuing intervention, even for "recovered" species, was not anticipated. We now face the conundrum that building on our conservation success will require long-term investments.

Paradoxically, continued listing under the ESA for many currently listed species may not be the best way to achieve long-term persistence. The legal restrictions imposed by the ESA may preclude some appropriate management actions. For example, landowners are often reluctant to manage their land in ways that might attract an endangered species because of the regulatory constraints imposed by the ESA (Wilcove 2004). Similarly, the paperwork and its concomitant costs in time and money are disincentives to the use of available conservation tools such as habitat conservation plans, candidate conservation agreements, and safe harbor agreements (Lin 1996, Burnham et al. 2006, Fox et al. 2006). However, delisting a species may open the door to an increasing array of unregulated threats that push it back into peril. For example, the delisting of gray wolves (*Canis lupus*) in the Northern Rocky Mountains resulted in unsustainable mortality from hunting and other pressures (Creel and Rotella 2010), which led to a judicial decision to relist the species (US District Court 2010) and a congressional decision to again delist the species through a budget rider (US Congress 2011).

To avoid such costly and contentious course reversals, a mechanism is needed to ensure that the appropriate management actions are implemented once the recovery goals for a species are met. Although no changes to the ESA are necessary to make this possible, we do need to acknowledge that continuing management is often needed after a species meets its biological recovery goals: We need a tool kit of management structures that will facilitate the transition from listed to delisted. Fortunately, examples are plentiful. The Robbins' cinquefoil (*Potentilla robbinsiana*) was delisted under a postdelisting management agreement under which the landowner (the US Forest Service) and a recreational group (the Appalachian Mountain Club) agreed to monitor and manage both the species' habitat and the threat (hikers) in order to maintain the recovered population (Goble 2009). Similarly, the Bureau of Land

Management acquired nearly 3000 hectares of habitat for the Columbian white-tailed deer (*Odocoileus virginianus leucurus*) and agreed to manage its habitat through prescribed burning, grazing modifications, and restoration actions. In addition, Douglas County, Oregon, adopted a series of land-use and zoning ordinances designed to maintain habitat and corridors for the species (Goble 2009). The conservation management agreement for the grizzly bear (*Ursus arctos horribilis*) in the Greater Yellowstone Area is an example of an agreement among federal, states, and tribal land- and wildlife-management agencies that can provide a structure through which postdelisting management can be assured (USFWS 2007). Such agreements operate like candidate conservation agreements that have been used to preclude the need to list at-risk species (Lin 1996).

Bocetti and her colleagues (2012) provide an example of how a biologically and legally defensible postrecovery conservation management agreement can be developed and funded. The biggest challenges lie in finding conservation partners and obtaining funding to implement the needed management actions at ecologically relevant scales. This can be complicated on an American landscape in which two-thirds of listed and other at-risk species occur on private lands outside protected areas (Groves et al. 2000). No single mechanism can meet all needs. Instead, we envision a suite of conservation tools that can be matched to the species and landscapes that meets both the conservation threats and the diverse needs of landowners with different economic and personal interests. Funding through tax rebates, real estate transfer taxes, excise taxes, general funds, and private dollars are tools that have all been used to support wildlife and their habitats (Mangun and Shaw 1984, Smith and Shogren 2001). In addition, nongovernmental groups such as the Rocky Mountain Elk Foundation, Ducks Unlimited, Trout Unlimited, and Pheasants Forever have been formed to actively manage selected species and their habitats.

Management actions undertaken to benefit conservation-reliant species offer opportunities to accelerate the removal of species from the endangered species list and to prevent other species from becoming endangered (USFWS 2001). What is required is demonstrably effective management agreements that include management and funding commitments outside the framework of the ESA. But our focus needs to shift to abating those factors that lead to endangerment, and a conservation-reliant framework may be of assistance in doing so (Averill-Murray et al. 2012 [in this issue]). Given the criticisms of the ESA and the lower potential costs of conserving species before they are listed, understanding the ongoing management requirements of a species and responding before listing is needed has the potential to be a universal societal goal regarding species conservation. The challenge will be in creating reliable alternative funding and management structures.

The barriers to conserving and eventually delisting species are nowhere more apparent than in the Hawaiian Islands. In a thoughtful examination of our recurrent failure to implement identified recovery actions, Leonard (2008) suggested several not unrelated reasons: a lack of funding (Restani and Marzluff 2001), a lack of understanding both in the islands and on the mainland of the importance and urgent need for conservation action, and social and political barriers that reflect conflicting management goals for areas in which endangered species occur (e.g., hunting mouflon sheep [*Ovis aries orientalis*] versus maintaining the integrity, diversity, and health of palila [*Loxioides bailleui*] habitat; Banko 2009).

The consequences of failing to implement needed management actions are not trivial. The refusal to remove feral ungulates from the critical habitat of the species, despite its priority in a 1977 recovery plan and several court orders, has resulted in the continuing decline of the palila (Banko 2009). On Kauai, despite a 1984 recovery plan (Sincock et al. 1984) that called for the removal of feral ungulates from the core habitat of endangered forest birds, no action was taken until 2011. In the interim, five species went extinct (Pratt 2009) and two more species have been added to the list of endangered wildlife (USFWS 2010). The failure to act on the information in the recovery plans was a consequence of social and political pressures resulting from the perceived conflict between management intervention to recover endangered species and the continued hunting of introduced ungulates. A lack of funding also contributed to the problem.

The task we face is daunting. There are nearly 1400 listed species and indications that the actual number of at-risk species is an order of magnitude or greater more (Wilcove and Master 2005). At this point, it is naive to continue to assume that funding will be available for the management needed to prevent the listing of at-risk species or to recover and manage listed species. The average expenditure for the recovery of listed species is less than a fifth of what is needed (Miller et al. 2002), and expenditures for recovery are often distributed among species for nonbiological reasons (DeShazo and Freeman 2006, Leonard 2008). Furthermore, the number of warranted but precluded decisions by the US Fish and Wildlife Service (USFWS) is increasing, and recovery has been designated a fourth-tier priority in the USFWS's guidelines for recovery planning.

Continuing business as usual, in which the majority of recovery funds are used to conserve a few iconic species while others are only monitored or simply ignored, will achieve little of lasting value. Even with increased funding, it is unlikely that we can conserve all species facing extinction, particularly as the queue gets longer. We must either develop sensible ways of assigning conservation priorities in which both the magnitude of management required and the potential benefits of management and conservation actions are considered. Information about the degree of conservation reliance of a species is central to developing sensible conservation priorities.

All references for articles included in *Taking Sides: Clashing Views in Sustainability* can be found on the Web at www.mhhe.com/cls.

Craig Hilton-Taylor et al.

State of the World's Species

A Species Rich World

The magnitude and distribution of species that exist today is a product of more than 3.5 billion years of evolution, involving speciation, radiation, extinction and, more recently, the impacts of people. Estimates of the total number of eukaryotic species in existence on Earth today vary greatly ranging from 2 million to 100 million, but most commonly falling between 5 million and 30 million (May 1992, Mace *et al.* 2005), with a best working estimate of about 8 to 9 million species (Chapman 2006). But of these, just under 1.8 million are estimated to have been described (Groombridge and Jenkins 2002, Chapman 2006) although it has been argued that the number may be closer to 2 million (Peeters *et al.* 2003).

While scientists debate how many species exist, there are growing concerns about the status of biodiversity, particularly population declines (e.g., the Living Planet Index which monitors population trends in 1,686 animal species shows an overall decline of 30 percent for the period 1970 to 2005 (Loh *et al.* 2008)) and the increasing rates of extinction of both described and undescribed species as a direct and indirect result of human activities. Although only a very small proportion (Table 1) of the world's described species have been assessed so far, The IUCN Red List provides a useful snapshot of what is happening to species around the world today and highlights the urgent need for conservation action. . . .

Highlights of the 2008 IUCN Red List

Some of the highlights of the 2008 update of The IUCN Red List include the following:

- A complete reassessment of the world's mammals showed that nearly one-quarter (22 percent) of the world's mammal species are globally threatened or Extinct and 836 (15 percent) are Data Deficient (Schipper *et al.* 2008).
- The addition of 366 new amphibian species, many listed as threatened, and the confirmed extinction of two species, which reaffirms the

From *Wildlife in a Changing World—An Analysis of the 2008 IUCN Red List of Threatened Species,* by Jean-Christophe Vie, Craig Hilton-Taylor, and Simon N. Stuart, eds., 2009, pp. 15–41.

Table 1

Numbers and Proportions of Species Assessed and Species Assessed as Threatened on the 2008 IUCN Red List by Major Taxonomic Group

	Estimated number of described species[7]	Number of species evaluated	Number of threatened species[8]	Number threatened, as % of species described[8]	Number threatened, as % of species evaluated[8,9]
Vertebrates					
Mammals[1]	5,488	5,488	1,141	21%	21%
Birds	9,990	9,990	1,222	12%	12%
Reptiles	8,734	1,385	423	5%	31%
Amphibians[2]	6,347	6,260	1,905	30%	30%
Fishes	30,700	3,481	1,275	4%	37%
Subtotal	**61,259**	**26,604**	**5,966**	**10%**	**22%**
Invertebrates					
Insects	950,000	1,259	626	0%	50%
Molluscs	81,000	2,212	978	1%	44%
Crustaceans	40,000	1,735	606	2%	35%
Corals	2,175	856	235	11%	27%
Arachnids	98,000	32	18	0%	56%
Velvet Worms	165	11	9	5%	82%
Horseshoe Crabs	4	4	0	0%	0%
Others	61,040	52	24	0%	46%
Subtotal	**1,232,384**	**6,161**	**2,496**	**0.20%**	**41%**
Plants[3]					
Mosses[4]	16,000	95	82	1%	86%
Ferns and allies[5]	12,838	211	139	1%	66%
Gymnosperms	980	910	323	33%	35%
Dicotyledons	199,350	9,624	7,122	4%	74%
Monocotyledons	59,300	1,155	782	1%	68%
Green Algae[6]	3,962	2	0	0%	0%
Red Algae[6]	6,076	58	9	0%	16%
Subtotal	**298,506**	**12,055**	**8,457**	**3%**	**70%**
Others					
Lichens	17,000	2	2	0%	100%
Mushrooms	30,000	1	1	0%	100%
Brown Algae[6]	3,040	15	6	0%	40%
Subtotal	**50,040**	**18**	**9**	**0.02%**	**50%**
TOTAL	**1,642,189**	**44,838**	**16,928**	**1%**	**38%**

Notes:

1. The number of described and evaluated mammals excludes domesticated species like sheep (*Ovis aries*), goats (*Capra hircus*), Dromedary (*Camelus dromedarius*), etc.
2. It should be noted that for certain amphibian species endemic to Brazil, it has not yet been possible to reach agreement on the Red List Categories between the Global Amphibian Assessment (GAA) Coordinating Team, and the experts on the species in Brazil. The numbers for Amphibians displayed here include those that were agreed at the GAA Brazil workshop in April 2003. However, in the subsequent consistency check conducted by the GAA Coordinating Team, many of the assessments were found to be inconsistent with the approach adopted elsewhere in the world, and a "consistent Red List Category" was also assigned to these species. The "consistent Red List Categories" are yet to be accepted by the Brazilian experts; therefore the original workshop assessments are retained here. However, in order to ensure comparability between results for amphibians with those for other taxonomic groups, the data used in various analyses (e.g., Baillie *et al.* 2004; Stuart *et al.* 2008; the Global Amphibians analysis on the Red List web site) are based on the "consistent Red List Categories." Therefore, numbers for Amphibians in Table 1 above will not completely match numbers that appear in other analyses, including the analysis later in this chapter.
3. The plant numbers do not include species from the *1997 IUCN Red List of Threatened Plants* (Walter and Gillett 1998) as those were all assessed using the pre-1994 IUCN system of threat categorization. Hence the numbers of threatened plants are very much lower when compared to the 1997 results. The results from this Red List and the 1997 Plants Red List should be combined together when reporting on threatened plants.
4. Mosses include the true mosses (Bryopsida), the hornworts (Anthocerotopsida), and liverworts (Marchantiopsida).
5. Ferns and allies include the club mosses (Lycopodiopsida), spike mosses (Sellaginellopsida), quillworts (Isoetopsida), and true ferns (Polypodiopsida).
6. Seaweeds are included in the green algae (Chlorophyta), red algae (Rhodophyta), and brown algae (Ochrophyta).
7. The sources used for the numbers of described plant and animal species are listed in Appendix 3.
8. The numbers and percentages of species threatened in each group do not mean that the remainder are all not threatened (i.e., are Least Concern). There are a number of species in many of the groups that are listed as Near Threatened or Data Deficient (see Appendices 4–8). These numbers also need to be considered in relation to the number of species evaluated as shown in column two (see note 9).
9. Apart from the mammals, birds, amphibians and gymnosperms (i.e., those groups completely or almost completely evaluated), the numbers in the last column are gross over-estimates of the percentage threatened due to biases in the assessment process toward assessing species that are thought to be threatened, species for which data are readily available, and underreporting of Least Concern species. The true value for the percentage threatened lies somewhere in the range indicated by the two right-hand columns. In most cases this represents a very broad range; the percentage of threatened insects for example, lies somewhere between 0.07 percent and 50 percent. Hence, although 38 percent of all species on The IUCN Red List are listed as threatened, this percentage needs to be treated with extreme caution given the biases described above.

extinction crisis faced by amphibians; nearly one-third (31 percent) are threatened or Extinct and 25 percent are Data Deficient.

- A complete reassessment of the world's birds indicates that more than one in eight (13.6 percent) are considered threatened or Extinct; birds are one of the best-known groups with less than 1 percent being listed as Data Deficient (BirdLife International 2008a).
- For the first time 845 species of warm water reef-building corals have been included on the Red List with more than one-quarter (27 percent) listed as threatened and 17 percent as Data Deficient (Carpenter *et al.* 2008).
- All 161 species of groupers are now assessed; over 12 percent of these highly sought after luxury live food fish species are threatened with extinction as a result of unsustainable fishing; a further 30 percent are Data Deficient.
- All 1,280 species of freshwater crabs have been assessed, 16 percent of which are listed as threatened with extinction, but a further 49 percent are Data Deficient (Cumberlidge *et al.* 2009).
- 359 freshwater fishes endemic to Europe, with 24 percent listed as threatened and only 4 percent listed as Data Deficient (Kottelat and Freyhof 2007).

BOX 1. SUMMARY OF THE 2008 IUCN RED LIST UPDATE

The 2008 update of The IUCN Red List (as released on 6th October 2008) includes conservation assessments for 44,838 species:

- There are 869 recorded extinctions, with 804 species listed as Extinct and 65 listed as Extinct in the Wild;
- The number of extinctions increases to 1,159 if the 290 Critically Endangered species tagged as "Possibly Extinct" are included;
- 16,928 species are threatened with extinction (3,246 are Critically Endangered, 4,770 are Endangered and 8,912 are Vulnerable);
- 3,796 species are listed as Near Threatened*;
- 5,570 species have insufficient information to determine their threat status and are listed as Data Deficient;
- 17,675 species are listed as Least Concern, a listing which generally indicates that these have a low probability of extinction, but the category is very broad and includes species which may be of conservation concern (e.g., they may have very restricted ranges but with no perceived threats or their populations may be declining but not fast enough to qualify for a threatened listing).

Note that The IUCN Red List is a biased sample of the world's species, and for the incompletely assessed groups, there is a a general tendency to assess species that are more likely to be threatened. It is therefore not possible to take the Red List as a whole (in which 38 percent of listed species are threatened), and say that this means that 38 percent of all species in the world are likely to be threatened.

* Includes species listed as Conservation Dependent (LR/cd), an old Red List Category which is now subsumed under the Near Threatened category.

The Status of Amphibians, Birds, Mammals, and Plants

In previous analyses of the Red List, the general analysis has looked at facts, figures, and trends across all the major taxonomic groups. However, a more thematic approach has been adopted in this review and hence because freshwater and marine groups are covered in other chapters, the main focus of the rest of this chapter is on the terrestrial groups. In particular an analysis is presented of the three comprehensively assessed vertebrate groups for which we have a relatively rich knowledge, namely the amphibians, birds, and mammals. Plants are also included, but are not analyzed to the same extent as the vertebrates because much of the supporting documentation for such an analysis is not yet available. The only invertebrate groups for which there is reasonable assessment coverage are the corals, dragonflies, and freshwater crabs, but as these are all covered in other chapters, they are not discussed any further here.

Amphibians

Current Status

The first comprehensive assessment of the conservation status of all amphibians was completed in 2004, and the results were included in the 2004 IUCN Red List. The amphibian assessment is one of several initiatives led by IUCN and its partners with the aim of rapidly expanding the geographic and taxonomic coverage of The IUCN Red List. Since 2004 there have been two updates of the amphibian data, one in 2006, and the most recent in 2008.

Ninety-nine percent of all known amphibian species (6,260 species; see Table 1) have been assessed, and of these, nearly one-third (32.4 percent) are globally threatened or Extinct, representing 2,030 species. Thirty-eight are considered to be Extinct (EX), and one Extinct in the Wild (EW). Another 2,697 species are not considered to be threatened at present, with 381 being listed as Near Threatened (NT) and 2,316 listed as Least Concern (LC), while sufficient information was not available to assess the status of an additional 1,533 species (Data Deficient (DD)). It is predicted that a significant proportion of these Data Deficient species are likely to be globally threatened.

Documenting population trends is key to assessing species status, and a special effort was made to determine which amphibians are declining, stable, or increasing. The assessment found declines to be widespread among amphibians, with 42.5 percent of species reported to be in decline. In contrast, 26.6 percent of species appear to be stable and just 0.5 percent are increasing. Because trend information is not available for 30.4 percent of species, the percentage of amphibians in decline may be considerably higher.

Extinctions are often difficult to confirm. Using the most conservative approach to documenting extinctions, just 38 amphibians are known to have become Extinct since the year 1500. Of greater concern, however, are the many amphibians that can no longer be found. Until exhaustive surveys to confirm their disappearance can be carried out, these species cannot be classified as Extinct, but rather are flagged as "Possibly Extinct" within the Critically Endangered category. Currently there are 120 such "Possibly Extinct" amphibian species.

Unfortunately, there is strong evidence that the pace of extinctions is increasing. Of the 38 known extinctions, 11 have occurred since 1980, including such species as the Golden Toad *Incilius periglenes* of Monteverde, Costa Rica. Among those amphibians regarded as "Possibly Extinct," most have disappeared and have not been seen since 1980. Fortunately, a few amphibians that previously were thought to be Extinct have been rediscovered. For example, *Atelopus cruciger* was not seen in its native Venezuela after 1986, until a tiny population was found in 2003.

BOX 2. SUMMARY OF RESULTS FOR AMPHIBIANS

- Nearly one-third (32 percent) of the world's amphibian species are known to be threatened or Extinct, 43 percent are known not to be threatened, and 25 percent have insufficient data to determine their threat status.
- As many as 159 amphibian species may already be Extinct. At least 38 amphibian species are known to be Extinct, one is Extinct in the Wild, while at least another 120 species have not been found in recent years and are "Possibly Extinct."
- At least 42 percent of all species are declining in population, indicating that the number of threatened species can be expected to rise in the future. In contrast, less than 1 percent of species show population increases.
- The largest numbers of threatened species occur in Latin American countries such as Colombia (214), Mexico (211), and Ecuador (171). The highest levels of threat, however, are in the Caribbean, where more than 80 percent of amphibians are threatened or extinct in the Dominican Republic, Cuba, and Jamaica, and a staggering 92 percent in Haiti.
- Although habitat loss clearly poses the greatest threat to amphibians, the fungal disease chytridiomycosis is seriously affecting an increasing number of species. Perhaps most disturbing, many species are declining for unknown reasons, complicating efforts to design and implement effective conservation strategies.

Birds

Current Status

Birds are probably the best-known taxonomic group. Since 1988, the BirdLife International Partnership, working with a global network of experts and organizations, including the IUCN SSC bird Specialist Groups, has conducted five comprehensive assessments of birds, with the most recent assessment of all 9,990 known species being completed in 2008. Less than 1 percent of bird species on the 2008 IUCN Red List have insufficient information available to be able to assess them beyond Data Deficient.

It is clear, however, that being well-studied does not provide immunity from decline and high extinction risk. More than one in seven bird species (13.6 percent) are globally threatened or Extinct, representing 1,360 species. Of these, 134 species are Extinct, four species no longer occur in the wild, and a further 15 are Critically Endangered species flagged as "Possibly Extinct," making a probable total of 153 bird extinctions since the year 1500.

Although 8,564 bird species (85.7 percent) currently are not considered threatened, 835 of these (8.4 percent of all known birds) are Near Threatened; the remaining 7,729 species are Least Concern.

Examining the current population trends for birds provides further confirmation that it is not just the threatened birds that are at risk as 40.3 percent of extant birds are recorded to be declining. A further 44.4 percent of bird

BOX 3. SUMMARY OF RESULTS FOR BIRDS

- Birds are the best-known group of species, with less than 1 percent having insufficient data to determine their threat status. More than one in seven (14 percent) bird species are globally threatened or Extinct, 86 percent are not threatened.
- At least 134 birds have become Extinct since the year 1500, four species have become Extinct in the Wild, and a further 15 species are "Possibly Extinct."
- The highest numbers of bird species are found in South America, with Colombia supporting 18 percent of the world's birds (1,799 species). Africa and Asia are the next most diverse regions for bird species.
- Ninety-seven percent of the world's countries hold at least one globally threatened bird species. The highest numbers of threatened birds occur in Brazil (122 threatened species) and Indonesia (115 threatened species).
- Although they are much less diverse than tropical countries on the continents, oceanic island nations hold the highest proportions of threatened and extinct species. The majority (88 percent) of known extinctions since the year 1500 have been on islands.
- Agriculture, logging, and invasive species are the most severe threats driving bird species toward extinction. The most common stress affecting bird populations is habitat loss and degradation.

species have stable populations and 6.2 percent are increasing. The population trend for 9.1 percent of birds is unknown or uncertain.

Mammals

Current Status

The mammal data on the 2008 IUCN Red List includes 5,488 species, 412 subspecies and 21 subpopulations. The primary focus of the current assessment, and hence this analysis, is at the species level. This is the second time that all mammals have been assessed, the first being in 1996 (Baillie and Groombridge 1996).

Nearly one-quarter of species (22 percent) are globally threatened or Extinct, representing 1,219 species. Seventy-six of the 1,219 species are considered to be Extinct (EX), two are Extinct in the Wild (EW), and a further 29 are flagged as "Possibly Extinct," making a total of 107 mammal extinctions since the year 1500.

Although 3,433 mammal species (63 percent) are not considered to be threatened at present, 323 of these (6 percent of all known mammals) are listed as Near Threatened (NT); the remaining 3,110 species are listed as Least Concern (LC).

Documenting population trends is a key part of assessing the status of species. Looking at current population trends in the extant mammal species, 30 percent are recorded to be decreasing. In contrast 25 percent of species are said to be stable and only 1.5 percent are increasing. Trend information is not available for 44 percent of species, hence the percentage of species in decline may be significantly higher.

BOX 4. SUMMARY OF RESULTS FOR MAMMALS

- Nearly one-quarter (22 percent) of the world's mammal species are known to be globally threatened or Extinct, 63 percent are known to not be threatened, and 15 percent have insufficient data to determine their threat status.
- There are 76 mammals which have gone Extinct since 1500, two are Extinct in the Wild and 29 are "Possibly Extinct."
- The most diverse country for mammals is Indonesia (670), followed closely by Brazil (648). China (551) and Mexico (523) are the only other two other countries with more than 500 species.
- The country with by far the most threatened species is Indonesia (184). Mexico is the only other country in triple figures with 100 threatened species. Half of the top 20 countries for numbers of threatened species are in Asia; for example, India (96), China (74) and Malaysia (70). However, the highest levels of threat are found in island nations, and in particular the top three are islands or island groups in the Indian Ocean: Mauritius (64 percent), Réunion (43 percent), and the Seychelles (39 percent).
- Habitat loss, affecting over 2,000 mammal species, is the greatest threat globally. The second greatest threat is utilization which is affecting almost 1,000 mammal species, especially those in Asia.

There was insufficient information available to assess the status of 836 species (15 percent) hence these are listed as Data Deficient (DD). While a number of these DD listings are due to taxonomic uncertaintities, in many cases they are due to inadequate information on population size, trends, distribution and/or threats. Most (80 percent) of the Data Deficient mammals occur in the tropics and 69 percent are bats and rodents which are hard to catch because of their nocturnal habits and difficult to identify.

Plants

The 2008 IUCN Red List includes assessments for 12,055 species of plants, 8,457 of which are listed as threatened. However, as only about 4 percent of the estimated 298,506 described plant species have been assessed, it is not possible to say that based on The IUCN Red List that 3 percent of the world's flora is threatened.

Since the plant and animal Red Lists were combined in the 2000 *IUCN Red List of Threatened Species*™ the number of plant assessments on the Red List has increased very slowly compared to other taxonomic groups. Of the 12,055 plants evaluated, 70 percent are listed as threatened. This partially reflects a bias amongst the botanical community to focus primarily on the threatened species, but there is also a tendency to not report on the species that have been assessed as Least Concern. The focus on threatened species is clearly illustrated by the assessments of bryophytes (mosses, liverworts, and hornworts), where the subset of 95 species was specifically chosen in order to "provide the public with general information as to which bryophytes are threatened with extinction" (Tan *et al.* 2000). The same is partly true of the assessments for ferns and fern allies (includes club mosses, spike mosses, quillworts, and true ferns); in this case, the 211 species assessed (although only 1 percent of the species) represent a widely distributed geographic sample and so might be more representative of the threats faced by this plant group, but it would be misleading to extrapolate from these results to the whole group.

A strong bias in the plant assessments in the 2000 *IUCN Red List* was toward threatened tree species because of the inclusion of the 7,388 species (includes species in all categories from Data Deficient to Extinct) listed in *The World List of Threatened Trees* (Oldfield *et al.* 1998). That bias has been reduced slightly through the inclusion of non-tree assessments. However, the trees still form 66 percent of the plants on the 2008 *IUCN Red List* (7,977 species), 5,643 of which are listed as threatened.

Many of the recent plant assessments are now introducing a geographic bias as they are single country or sub-country endemics (e.g., Cameroon, China, Ecuador, Madagascar, Mauritius, Namibia, Saint Helena, South Africa, Yemen (Soqotra), and the United States (Hawaii)).

The seemingly very large figure of 8,457 threatened plant species is proportionally very small relative to the total number of described plant species worldwide. The proportion threatened is even smaller if the higher estimate for the number of described plants is used (422,127 as opposed to 298,506 species). It is therefore premature at this stage to attempt any detailed analysis of the plants as the low numbers assessed and the strong biases toward trees and certain geographic areas misrepresents the overall picture for plants. . . .

Are Species Becoming More or Less Threatened with Extinction?

In those taxonomic groups about which we know most, species are sliding ever faster toward extinction. IUCN Red List Indices (RLIs; see description in Vié *et al.*) show that trends in extinction risk are negative for birds, mammals, amphibians, and reef-building corals. Although successful conservation interventions have improved the status of some species (Box 5), many more are moving closer toward extinction, as measured by their categories of extinction risk on The IUCN Red List.

The groups vary in their overall level of threat; for example, amphibians have a higher proportion of species threatened (i.e., lower RLI values) than mammals or birds. Groups also vary in their rate of deterioration, with the rapid declines in reef-building corals since 1996 being driven primarily by the worldwide coral-bleaching events in 1998 (Polidoro *et al.* this volume, Carpenter *et al.* 2008). Whereas the RLI for birds shows that there has been a steady and continuing deterioration in the status of the world's birds between 1988 and 2008. Over these 20 years, 225 bird species have been uplisted to a higher category of threat because of genuine changes in status, compared to just 32 species downlisted.

What Are the Geographic Patterns in Declines?

Species are deteriorating in status worldwide, but some regions have undergone steeper declines and have more threatened faunas. The Indomalayan realm showed rapid declines in birds and mammals, driven by the rapid increases in the rate of deforestation during the 1990s, particularly in the Sundaic lowlands, combined for mammals with high rates of hunting, particularly among medium- to large-bodied species. Amphibians are also highly threatened in the Indomalayan realm. Birds in the Oceania realm are substantially more threatened (with lower RLI values) than in other realms, largely owing to the impacts of invasive alien species. Amphibians are most threatened in the Neotropical realm, in particular owing to chytridiomycosis.

A downwards trend in the graph line (i.e., decreasing RLI values) means that the expected rate of species extinctions is increasing, i.e., that the rate of biodiversity loss is increasing. A horizontal graph line (i.e., unchanging RLI values) means that the expected rate of species extinctions is unchanged. An upward trend in the graph line (i.e., increasing RLI values) means that there is a decrease in expected future rate of species extinctions (i.e., a reduction in the rate of biodiversity loss).

While the RLI is not very sensitive to small-scale changes in the status of species (as reflected in population trendbased indicators), it has global scope and coverage, and hence is not biased geographically in the way that global population trend-based indicators may be.

Species Loss and Human Health

The 2008 IUCN Red List clearly shows that many species are under threat of extinction mainly as a direct or indirect result of human activities. But why should humans be concerned about this and why should we invest time and money on saving species?

For as long as humans have existed we have used the species around us for our own survival and development. Even today, with vast numbers of people living in towns and cities, seemingly far removed from nature, we still need plants and animals for our food, materials, and medicines, as well as for recreation and inspiration for everything from the sciences to the arts. In the developing countries, where wild animal and plant species can make a significant contribution to human diets and healthcare, maintaining a healthy biodiversity is of particular importance. . . .

The overwhelming message from the results presented . . . is that the world is losing species and that the rate of loss appears to be accelerating in many taxonomic groups. The number of threatened species grows with each update of the Red List. Although this growth is to a large degree the result of increased taxonomic coverage, the downward Red List Index trends calculated for those groups that have been completely assessed clearly indicate that the rate of biodiversity loss is increasing. Even a simple examination of the 223 species which changed status in 2008 for genuine reasons (i.e., become less threatened due to conservation efforts or become more threatened due to ongoing or increased threats), shows that while only 40 of these were species that became less threatened, 183 were listed in a higher category of threat.

The 40 species that showed improved conservation status in 2008 do provide a glimmer of hope. Conservation actions are being taken for many species around the world. These range from species-specific actions to broad changes in national, regional, or global policies. Measuring the efficacy of these actions in relation to individual threatened species is just beginning. But there are many case studies which show that well-focussed and concerted species-centred actions can succeed in reducing threats and improving the conservation status of species and their habitats (Box 5).

BOX 5. SUCCESS STORIES SHOW THAT CONSERVATION CAN WORK, BUT MORE IS NEEDED

Over the last 20 years, a number of species have been downlisted to lower categories of threat on the IUCN Red List as a consequence of successful conservation action that has mitigated threats, halted, or reversed declines and hence increased the population and/or range size. Examples include the following.

In Mauritius, the Pink Pigeon and Mauritius Kestrel, Fody, and Parakeet have all improved in status sufficiently to have been downlisted to lower categories of threat on the IUCN Red List during recent years. Control of alien invasive species, habitat restoration, and captive breeding and release have been important actions behind these successes, leading to reduced threats, and reducing, halting, and reversing population declines. In some cases, these interventions were only just in time: the Mauritius Kestrel was brought back from the brink of extinction when the population fell to just four individuals in 1974.

In Brazil, Lear's Macaw was until recently classified as Critically Endangered owing to its population having been reduced to little more than 200 birds by 2001 through unsustainable exploitation for the cage bird trade, and habitat loss. Successful conservation actions including control of trade, nest protection, and habitat management have now increased numbers to almost 1,000 individuals, allowing it to be downlisted to Endangered.

African Elephant has moved from Vulnerable to Near Threatened, although its status varies considerably across its range. Poaching for ivory and meat has traditionally been the major threat to the species. Across the continent, the total population is believed to have suffered a decline of approximately 25 percent between 1979 and 2007, which falls short of the 30 percent threshold required for a Vulnerable listing. It is believed that the change in status reflects recent and ongoing population increases in major populations in southern and East Africa, largely due to the implementation of highly successful conservation efforts, and which have been of sufficient magnitude to outweigh the decreases that are taking place elsewhere across their vast range, especially in West and Central Africa.

(*Continued*)

Among amphibians, conservation success stories are few, but they do exist. The Mallorcan Midwife Toad occurs on the Balearic Island of Mallorca (Spain), where it is confined to the Serra de Tramuntana. The major threat to this species has been identified as predation and competition from introduced species such as green frogs and, more significantly, the Viperine Snake, a semi-aquatic serpent that preys upon both tadpoles and adult toads. In 1985, at the invitation of the Mallorcan government the Durrell Wildlife Conservation Trust (DWCT) initiated a species recovery programme for the toad. This recovery program has proven extremely successful in reversing the decline of the toad, and the species is now listed as Vulnerable.

All references for articles included in *Taking Sides: Clashing Views in Sustainability* can be found on the Web at www.mhhe.com/cls.

EXPLORING THE ISSUE

Can Species Preservation Be Successfully Managed?

Critical Thinking and Reflection

1. Do governments in both developed and developing countries have the political will to effectively manage global species extinction and loss of biodiversity?
2. What are the problems and obstacles to species conservation that are faced in different regions of the world?
3. What are the best public policies to preserve species and maintain biodiversity?
4. Why should society have strong policies to prevent the loss of biodiversity? Is this a moral argument or an argument that is important for human survivability?
5. Do you believe that society cannot eliminate species extinction but only manage it?

Is There Common Ground?

There are a number of ways in which a common ground on this issue can be achieved. First, it is important to recognize and address the reasons why there are species extinctions. Humans have expanded their habitats and encroached upon the habitats of other species for purposes of livelihood and to meet immediate needs. We have altered global ecosystems, interfered with natural processes, and are a factor in climate change. Forests have been cut down so that humans can access wood for housing construction, cooking fuel, and agricultural and pasture land use conversion for food supply. By establishing a human need to preserve biodiversity, a key aspect of sustainability, a case can be made for the preservation of habitats and frail ecosystems. While this maintains a somewhat utilitarian view of ecosystem preservation, it may nevertheless be an effective strategy for dealing with the real threat of species extinction.

It is well known that conflict between humans and wildlife is fast becoming a serious threat to the survival of many species in the world. This conflict is global in nature and is not restricted to any particular region. Instead, it is common to all areas where wildlife and human populations collide and share limited resources. Case studies from countries all over the world have shown the severity of this conflict. They also show that these conflicts have similar causes and impacts. Thus, a better understanding of the root causes and conflict management options are fundamental for successful policy and mitigation. Accurate and detailed information, scientific research, and stakeholder commitment are all important considerations for seeking a common ground on this issue.

Governments need to usher in major changes if earth biodiversity is to be maintained and species extinction reduced. This necessitates a program of sustainability that ranges from energy and material use reduction to seeking solutions to the challenges of climate change. Species extinctions can be minimized through good management practices and through incentives designed to change behavior. Innovative approaches such as natural resource use compensation systems, community-based natural resource management schemes, and various incentive and insurance programs can provide useful models. A combination of short-term mitigation strategies and long-term preventive strategies can aim us in this loss of biodiversity.

Internet References

Dirzo, R., & Raven, P.H. (2003) Global state of biodiversity and loss. *Annual Review of Environment and Resources, 28*, pp. 137–167.

Census of Marine Life (2011, August 23) How many species on Earth? About 8.7 million, new estimate says. *ScienceDaily*. Retrieved on December 1, 2011 from www.sciencedaily.com/releases/2011/08/110823180459.htm

CITES. What is CITES? Web. Retrieved on 4 December, 2011 from www.cites.org/eng/disc/what.php

IUCN. Red List overview. Web. Retrieved on December 3, 2011 from www.iucnredlist.org/about/red-list-overview

Center for Biological Diversity, CBD. The road to recovery. 100 success stories for Endangered Species Day 2007. Web. Retrieved on December 3, 2011 from www.esasuccess.org/reports/

Lamoreux, J., Akcakaya, H. R., Bennun, L., et al. (2003) Value of the IUCN Red List. *Trends in Ecology and Evolution, 18*, pp. 214–215.

Mora, C., Tittensor, D. P., Adl, S., Simpson, A.G.B., and Worm, B. (2011) How many species are there on earth and in the ocean? *PLoS Biology, 9*(8), p. e1001127. doi:10.1371/journal.pbio.1001127

United State Department of Fish and Wildlife. All United States species information. Web. Retrieved on 3 December, 2011 from www.fws.gov/endangered/

Wilcove, D. S. and Lawrence, M. L. (2005) How many endangered species are there in the United States? *Front Ecology Environment, 3*, pp. 414–420.

World Wildlife Foundation. Tiger facts and future. Web. Retrieved on 2 December, 2011 from wwf.panda.org/what_we_do/endangered_species/tigers/

World Wildlife Foundation. Western low land gorilla. Web. Retrieved on 2 December, 2011 from wwf.panda.org/what_we_do/endangered_species/great_apes/gorillas/western_lowland_gorilla/

www.fao.org/docrep/012/i1048e/i1048e00.htm

It contains comprehensive reports about human wildlife conflict in Africa; causes, consequences, and management strategies. Sponsored by Food Agricultural Organization (FAO).

www.humanwildlifeconflict.org/

It describes the collaborations among human-wildlife conflict (HWC) practitioners who realized that sharing ideas, information, and experiences is essential to preventing and minimizing conflict in the areas where they work. Currently The Wildlife Society is the host for this website as well as serves as its fiscal agent.

www.humanwildlife.org/

This site is designed to help Virginia residents and municipal leaders identify potential sources of assistance when confronted with problematic wild animal concerns. Sponsored by the Mid Atlantic Information Node.

wwf.panda.org/about_our_earth/species/problems/human_animal_conflict/

It describes WWF strategies based on lesson learning from their field projects dealing with human wildlife conflict. Sponsored by World Wildlife Fund (WWF).

www.swmminc.com/conflict.html

It provides a service to area property owners experiencing problems with the expanding whitetail deer herd. Suburban Whitetail Management of Maryland sponsored by Howard County (Maryland) Department of Recreation and Parks.

www.iucn.org

The website of the International Union for the Conservation of Nature seeks to provide pragmatic solutions to finding solutions to our most pressing environmental and development challenges.

Abbitt, R. J. F. and Scott, J. M. (2001) Examining differences between recovered and declining endangered species. *Conservation Biology, 15*, pp. 1274–1284.

Species with lesser threats and higher numbers often recover much faster than species with numerous threats and whose populations have been immensely depleted. In the future more time and resources will need to be focused on the more endangered species in order to recover them.

Campbell, L. M., Godfrey, M. H., and Drif, O. (2002) Community based conservation via global legislation? Limitations of the Inter-American Convention for the Protection and Conservation of Sea Turtles. *Journal of International Wildlife Law and Policy, 5*, pp. 121–143.

Global laws and treaties benefit endangered species if properly implemented. Yet, there are many that do not adequately impose the regulations therefore not adequately reflecting current local conservation thinking and responsiveness.

Eken, G., Bennun, L., Brooks, T. M., Darwall, W., Fishpool, L. D. C., Foster, M., Knox, D., Langhammer, P., Matiku, P., Radford, E., Salaman, P., Sechrest, W., Smith, M. L., Spector, S., and Tordoff, A. (2004) Key biodiversity areas as site conservation targets. *BioScience, 54*, pp. 1110–1118.

Many governments and agencies heavily depend on the KBAs to identify critical habitats but the designation of such should not only be based upon the KBAs data but should also take into consideration wildlife ranges and post monitoring efficiency when delineating the sites.

Hoffmann, M., Brooks. T. M., da Fonseca, G. A. B., Gascon, C., Hawkins, A. F. A., James, R. E., Langhammer, P., Mittermeier, R. A., Pilgrim, J. D., Rodrigues, A. S. L., and Silva, J. M. C. (2008) Conservation planning and the IUCN Red List. *Endangered Species Research, 6*, pp. 113–125.

The IUCN value is beyond measure yet conservation planning should not solemnly focus upon it. In the future the development of quantitative methods and criteria for measuring threats should be considered.

Langhammer, P. F., Bakarr, M. I., Bennun, L. A., Brooks, T. M., Clay, R. P., Darwall, W., De Silva, N., Edgar, G. J., Eken, G., Fishpool, L. D. C., da Fonseca, G. A. B., Foster, M. N., Knox, D. H., Matiku, P., Radford, E. A., Rodrigues, A. S. L., Salaman, P., Sechrest, W., and Tordoff, A. W. (2007) Identification and gap analysis of key biodiversity areas: Targets for comprehensive protected area systems. IUCN (Best Practice Protected Area Guidelines Series 15), Gland, Switzerland. *BioScience, 57*(3), pp. 256–261.

The International Union for the Conservation of Nature (IUCN) Red List of Threatened Species is essential at guiding and determining the most urgent areas for conservation.

Male, T. D., and Bean, M. J. (2005) Measuring progress in U.S. endangered species conservation. *Ecology Letters, 8*, pp. 986–992.

The article is a review of the U.S. Endangered Species Act. An important conclusion is that species recovery correlates to length of species listing.

Nijman, V. and Shepher, C. R. (2007) Trade in non-native, CITES-listed, wildlife in Asia, as exemplified by the trade in freshwater turtles and tortoises in Thailand. *Contributions to Zoology, 76*, pp. 207–212.

The trade of turtles in Thailand often include CITES listed species is of immense volume becoming of serious conservation concern. CITES regulations in Thailand and other countries need to be strictly implemented in order for imperiled species to be protected.

Rohlf, D. J. (1991) Six biological reasons why the endangered species act doesn't work and what to do about it. *Conservation Biology, 5*, pp. 273–282.

The laws that protect endangered species often lack biological reasoning behind them making them inefficient, therefore biologist should contribute in the law making process.

Schalaepfer, M. A., Hoover, C., and Dodd, K. (2005) Challenges in evaluating the impact of the trade in amphibian and reptiles on wild populations. *BioScience, 55*, pp. 256–264.

The United States import and exports of amphibian and reptiles reflect an immense trade including CITES regulated species. US makes up 12 percent of the trade between the CITES bounded countries signifying the trade to be of much greater magnitude and impact upon wild populations.

Schwartz, M. W. (1999) Choosing the appropriate scale of reserves for conservation. *Annual Review of Ecology Evolution and Systematics, 30*, pp. 83–108.

Conservation plans are unique per species, habitat, geography, country among others. This article provides a wide variety of methods by which species can be conserved under different conservation scales.

Schwartz, M. W. (2008) The performance of the Endangered Species Act. *Annual Review of Ecology Evolution and Systematics, 39*, pp. 279–299.

The ESA is described as being capable of protecting species once listed. Therefore the dilemma is the ESA's inability to be fully implemented.

Wilson, E. O. (2003) *Pheidole in the New World: A Dominant, Hyperdiverse Ant Genus.* Harvard University Press: Cambdridge, MA.

ISSUE 16

Can Sustainable Agriculture Feed the World?

YES: International Fund for Agricultural Development, from "Sustainable Smallholder Agriculture: Feeding the World, Protecting the Planet" (2012), www.ifad.org

NO: Craig Meisner, from "Why Organic Food Can't Feed the World," *Cosmos Magazine* online, www.cosmosmagazine.com, retrieved October 10, 2012

Learning Outcomes

After reading this issue, you should be able to:

- Understand the challenges facing food production for a growing global population.
- Identify the characteristics of sustainable agriculture.
- Understand the differences between organic and industrial agriculture.
- Identify some of the downstream, external costs of industrial approaches to agriculture.
- Identify the positive effects on society from organic approaches to food production.
- Discuss the relationship between sustainably designed agriculture and ecosystem preservation.
- Consider coexistence of organic, industrial, and genetically modified (GM) systems.

ISSUE SUMMARY

YES: The International Fund for Agricultural Development, a global NGO, argues in its position paper that the future for world food security rests with a sustainable agriculture that protects local ecosystems and relies on smallholder farmers. They believe that smallholder farmers, when guided by coherent policies and fair incentives, can feed the world through the use of organic production methods and various green technologies and innovations.

NO: Cornell University professor Craig Meisner, while supporting many of the goals of sustainable agriculture, sees some limitations in the reliance of organic production methods for the poor in developing countries. Through his personal experience in Bangladesh he notes that the poor farmer's ability to implement organic approaches are increasingly challenged by daily survival and economic factors. For example, he notes that a key component of organic farming is the use of green manure, nitrogen fixing crops, which he sees as competing with food crops and decreasing the overall income potential of poor farmers.

Global population is now exceeding 7 billion people. How will all these people be fed while protecting and preserving the global ecosystem? Food production consumes large amounts of natural resources, that is, water, land, and minerals. Also, industrial agricultural practices based on chemical pesticides and herbicides have caused public health risks and ecosystem pollution while increasing yield. Can the world provide food for its growing population and still maintain a viable environment?

One solution is to move to more sustainable agricultural practices. Sustainable agriculture is farming which emphasizes certain goals: it is farming which conforms to basic principles of ecology; it encourages local production rather than foods that travel long distances, consuming much fossil fuel and adding carbon to the atmosphere; it stresses the use of organic farming methods instead of relying on industrial fertilizers, pesticides, and herbicides; and it favors the use of green technologies and policies that stimulate small-scale production. But can sustainable agriculture, while being an environmentally sustainable practice, be able to feed the world? Will it be able to produce the yields necessary to feed a global population projected to be close to 9 billion people by 2050?

The industrial revolution, with massive increases in fossil fuel production and use, spurred dramatic growth of human population and economies (Hall et al., 2003, LeClerc & Hall, 2007). Globalization of market forces, agricultural industrialization, migration, public policy, and cultural changes have transformed agriculture from a diverse, traditional, and smaller scale system into a agro-industrial system dependent on chemical inputs and mechanization (Angelsen & Kaimowitz, 2001; Conway & Rosset, 1996; Perfecto et al., 1996). In *The Potential for a New Generation of Biodiversity in Agroecosystems of the Future* (2007), scientist and farmer Fred Kirschenmann points out the basic assumptions for industrial agriculture. They are: production efficiency can best be achieved through specialization, simplification, and concentration; intervention is the most effective way to control undesirable events; technological innovation will always be able to overcome production challenges; control management is the most effective way to achieve production results; and cheap energy to fuel this energy-intensive system will always be available. There is no doubt that industrial agriculture has created higher yields to meet the needs of a growing global population.

But, industrial agriculture has often led to environmental degradation (MEA, 2003). Negative effects of industrial agriculture include biodiversity loss, loss of species and genetic diversity, severe degradation of health of inland and coastal waterways, high-energy use, and reduced or eliminated ecosystem resiliency. The 21st century has arrived with many believing that most of industrial agriculture's assumptions need to be replaced with a more sustainable agricultural production process.

Over the past several decades, many writers point out that the trajectory of rapid growth of the past two to three centuries, with its reliance on natural resources and energy, may reach an environmental threshold or tipping point in the future (Meadows & Randers, 2004; Odum & Odum, 2001; Wackernagel et al., 2002). Industrial agriculture worldwide is energy intensive. They also point out that industrial agriculture, conventionally accepted worldwide, has reduced soil carbon content in Midwestern U.S. soils from 20 percent carbon in the 1950s to its current 1–2 percent. This contributes greatly to increasing soil erosion, vulnerability to drought, and decreasing nutrient values. Industrial practices break down soil carbon resulting in atmospheric release of CO_2, contributing nearly 20 percent of the total atmospheric carbon dioxide emissions in the United States. Globally, these conventionally accepted agricultural practices contribute 12 percent of global greenhouse gas (GHG) emissions. Increasing population and industrial food production practices have led to excessive nitrogen buildup that eventually ends up in rivers and streams. This leads to eutrophication and episodic and persistent hypoxia in coastal waters worldwide (Nixon et al., 1996; NRC, 2000a). Synthetic production of chemical fertilizers, pesticides, fungicides, and herbicides has resulted in large-scale industrialized energy-consumptive agriculture that many contend is not compatible with ecosystem preservation.

Writer and organic farmer Wendell Berry has admonished farmers for decades to preserve the fertility and ecological health of the land. Society, he contends, must recognize this need, and learn, or relearn, to integrate their activities *with* natural ecosystems, including and especially integrating sustainable agro-ecosystems. Day et al. (2009) maintain that the functioning of natural ecosystems and the health of the human economy have been intrinsically linked throughout our evolution. They go on to say that solar-driven ecosystems powered the preindustrial world; materials such as food, fuel, and fiber, as well as ecosystem services, such as clean freshwater, fertile soils, wildlife, and assimilation of wastes through inherent regenerative and assimilative capacities, were largely dependent on solar-driven ecosystems and agro-ecosystems.

Many believe that efficient, sustainable ways to support food production through regenerative and mutualistic ecological design while requiring less energy is currently available. Studies in Mesoamerica provide scientific evidence that certain agricultural landscapes and practices contribute to biodiversity conservation while simultaneously contributing to increased food production and rural income (Daily et al., 2003; Estrada & Coates-Estrada, 2002; Mayfield & Daily, 2005; Pretty et al., 2003). Heterogeneous agricultural landscapes that retain abundant tree cover (as forest fragments, fallows, riparian areas, live fences, dispersed trees, or canopies) provide complementary habitats, resources, and landscape connectivity for a significant portion of the original biota (Harvey et al., 2006; Sekercioglu et al., 2007). Landscape configurations that connect forests, maintain a diverse array of habitats, and retain high structural and floristic complexity generally conserve species (Bennett et al., 2006; Benton et al., 2003).

Organic agricultural practices can often provide the means for building agricultural and associated ecosystem resiliency in the face of climate change. Regenerative organic agricultural practices can increase biological activity in soil organic matter. This improves carbon sequestration of soil by removing carbon from the air, while also increasing water retention and improving system resiliency. Manure-based soil systems show an increase in carbon storage over legume-based organic systems. Also, energy use and carbon dioxide emissions are substantively reduced through organic practices. In a farm study of organically grown corn/soybeans, Pimentel (2006) demonstrated that a 33 percent reduction in fossil-fuel use was possible. By adopting an organic system that used cover crops or compost instead of chemical fertilizer, GHG emissions were reduced.

Critics of sustainable agriculture often point out that reliance on organic production methods and not on newer industrial technologies can limit our capacity to produce enough food to feed the world. These writers emphasize coexistence, an agricultural practice where different primary production systems, that is, organic, industrial, and genetically modified (GM) systems, occur simultaneously or adjacent to one another while contributing mutual benefit (Altieri, 2006). One important component for consideration to coexistence is GM agriculture. GM agriculture is based on the movement of genetic DNA from one species to another, altering the capacity of that species to withstand natural predators, climatic changes, and previous tolerance levels. Proponents of GM agriculture view it as a technological innovation that can substantially increase yield while contributing much less ecosystem damage than traditional industrial agriculture. The Royal Society of London et al. (NRC, 2000) maintains that growing global population needs will require either high yield agricultural production or more conversion of natural biomes and marginal land into agricultural product. They argue that the use of trans-genes reduces the need for chemical pesticides and herbicides as biotechnology selects genetic input that strengthens predator resistance. Food output would increase if spoilage could be limited; and food shelf life could be extended genetically, particularly for high value fruits and vegetables, while placing less stress on natural ecosystems. Also, the loss of topsoil could be minimized through a no till application of seed.

Critics of GM agriculture reject the view that GM agriculture is a sustainable agricultural practice since it places natural ecosystems at risk. They point out that trans-genes cannot be contained, and will spread naturally beyond their intended destinations causing contamination with non-GM crops and promote the transfer of trans-genes from crops to other plants, and can transform wild/weedy plants into new or more invasive weeds (Marvier, 2001; Rissler & Mellon, 1996; Snow & Moran-Palma, 1997). They maintain that unless whole regions are declared GM free, the development of distinct systems of agriculture will be compromised. This, of course, would damage natural ecosystems.

The YES selection is presented by the International Fund for Agricultural Development. This report holds that global food security rests with a sustainable agriculture based on organics that protects local ecosystems and relies on smallholder farmers. They believe that smallholder farmers, who currently produce four-fifths of the developing world's food, can utilize appropriate technologies and innovations combined with relevant training to feed their growing populations while protecting local ecosystems and natural assets. By removing disincentives such as subsidies to support industrial agriculture, organics can compete on an even playing field utilizing green technologies and innovations. They conclude by saying that adequate financing and insuring the right to land will allow smallholders to feed the world.

The NO selection is presented by Cornell University professor Craig Meisner. He presents his experience in agricultural research and development, carried out for the past 25 years in Bangladesh. He does not believe that organics, a key component for sustainable agriculture, can feed the world. In his field work in Bangladesh he noticed that the poor's ability to implement organic approaches are increasingly challenged by daily survival and economic factors. For example, rather than use animal dung to develop nutrient-rich compost for crop production, they instead use this resource, gleaned from their own livestock or purchased, to heat their homes and stoke their fires in order to cook food. Therefore, they often do not have access to organics because they are less available or affordable. He agrees that although organics enrich the soil, they are not feasible alternatives to industrial fertilizers. Additionally, each hectare of land requires too much manure for adequate handling, and natural fertilizers like green manure (crops that fix nitrogen in the soil) are not acceptable alternatives since they take acreage dedicated to food crops out of production. He remains skeptical about the capacity of organics to meet the growing demand for food security in the developing world.

YES **International Fund for Agricultural Development**

Sustainable Smallholder Agriculture: Feeding the World, Protecting the Planet

Introduction

As the world belatedly turns its attention to the pressing issues of environmental degradation, resource scarcity, and climate change, the concept of sustainability takes its rightful place at centre stage in discussions about agricultural and rural development.

Farmers face two stark realities over the next four decades: They must produce 70 percent more food by 2050 to feed a growing, more urbanized population, and they must do so facing the likelihood that arable land in developing countries will increase by no more than 12 percent.[1] That monumental challenge can be met only if sustainability is the foundation of approaches to food security and poverty reduction in every country and every community. No other strategy has a hope of feeding current populations while protecting and restoring the natural resources that future generations will need to support their livelihoods.

This means that food production must be intensified even as production methods evolve. The agriculture sector will become more community-focused, establishing an appropriate local balance of crops, livestock, fisheries, and agroforestry systems to avoid overuse of pesticides and inorganic fertilizers and to protect soil fertility and ecosystem services—while increasing production and income. It will be imperative to work within ecosystems, using natural processes and a mixture of new and traditional technologies.

Fortunately this is already beginning to happen. Farmers around the world are demonstrating the benefits of preserving natural assets and working in harmony with local ecosystems:

- In Brazil, three southern states support zero-tillage and conservation agriculture.[2]
- The African Conservation Tillage Network is bringing together farmers and policymakers who are dedicated to improving agricultural productivity while using natural resources sustainably
- The Chinese government's 11th Five-Year Plan (2006–2010) emphasized the need to reduce the environmental impact of agriculture and called for organic foods, water conservation, and sustainable practices.

From *International Fund for Agriculture Development,* 35th Session of Governing Council, 2012, pp. 2-11.

- In the Philippines, the government has stopped its fertilizer subsidy program and introduced a balanced fertilization policy that promotes location-specific combinations of organic and inorganic fertilizers.
- The Indian state of Rajasthan is supporting watershed and soil management and incentives for use of biofertilizers.
- Indonesia has banned some pesticides and introduced farmer field schools to teach integrated pest management.[3]

However, many of these successes remain piecemeal and fragmented. We know that sustainable agriculture is the only way forward—but many of the necessary policies are not yet in place to scale up successful approaches and ensure widespread adoption.

Smallholder farmers, when guided by coherent policies and fair incentives, have shown they are willing and able to change how they do business. With access to appropriate technologies and innovations and relevant training, they have produced results with multiple benefits for communities, ecosystems, and natural assets as well as themselves. But without institutional support it is unrealistic to expect poor farmers to change their practices for altruistic reasons. We need to work with smallholders and support them so they can become the developers of sustainable solutions. That is the best way to boost food production and improve livelihoods in an environmentally sustainable way.

The Crucial Role of Smallholders

Four-fifths of the developing world's food is produced on about half a billion small farms.[4] Smallholder farmers live and earn their livelihoods in the world's most ecologically and climatically vulnerable landscapes—hillsides, drylands, and floodplains—and rely on weather-dependent natural resources. They are at the forefront of the world's efforts to deal with climate change, environmental degradation, poverty, and child labor.[5] These women and men, especially indigenous people and young people, make up the largest share of people living on less than $1.25 per day and represent the bulk of the world's malnourished people. Through a mixture of ingenuity and toil, they manage to feed about one-third of humanity despite the enormous difficulties they face.

AN UNPREDICTABLE GLOBAL CONTEXT

Three uncontrollable factors—climatic conditions, pressures on natural resources, and rising prices—are deviling farmers everywhere. These uncertainties pose special challenges for smallholders, whose poverty leaves them with no margin to cushion unpredictable events.

Farmers are struggling to maintain crop yields as they confront droughts, rising sea levels, and soil degradation. The growing demand for meat and dairy products among burgeoning middle classes in populous countries is raising pressures on scarce natural resources. Rising prices for energy and inputs make farming more expensive for poor smallholders. Rising food prices could benefit them—if they get access to the inputs, technology, knowledge, and markets that will expand their productivity.

"Even with the great advances of the Green Revolution, nearly one billion people are still hungry or undernourished. Now, farmers around the world experiment with integrated soil, water and plant management methods. These methods blend modern science and traditional knowledge. At Rio+20, we should aim to accelerate an 'evergreen revolution.' This revolution will meet the growing global food demand while protecting soils, water and biodiversity. This is the way of the future."

—Rio+20 Secretary-General, Mr. Sha Zukang

Smallholder producers are the backbone of rural economies and often major contributors to national food export markets. But their immense contribution to feeding the world has even more potential. To realize this growth will require investments that will give smallholders access to a range of assets and tools: green technologies, energy, land, credit, training, infrastructure, market information, and political voice.

The International Agenda

The imperative of moving toward more sustainable agriculture practices that respect local ecosystems within broader landscapes is gaining momentum in international debates. Agriculture will be a major issue at the Rio+20 Conference on Sustainable Development next June, and it has figured prominently in debates leading to the upcoming United Nations Climate Change Conference in Durban in November–December 2011.[6] This is the context in which IFAD, as a financial institution, will promote the need to increase investment in agriculture and promote a sustainable approach to farming that empowers smallholder farmers. The goal is to unleash their potential to expand economic growth and contribute to global food security.

Shifting Paradigms

Too often it is assumed that a trade-off is inevitable between maximizing agricultural production and caring for the environment. This is a false choice. We can and we must achieve both—or we will fail in both.[7] In the long run, agricultural production cannot be sustained at the cost of undermining natural assets. In most parts of the world, we are seeing the costs of unsustainable agriculture:

- Three quarters of crop diversity has been lost since 1900.[8]
- Seventy percent of fisheries are in danger, threatened by overfishing and environmental degradation.[9]
- About 5.2 million hectares of forest are lost every year.[10]

Viewing the agriculture sector as renewable rather than extractive is the only way forward. This approach embraces the idea that agriculture is an interaction with wider ecosystems, while it simultaneously improves livelihood options for those who farm the world's approximately 500 million smallholdings. In the long term, there is no trade-off between production and sustainability. In fact, the opposite is true: without sustainability, production will suffer.

In the short run, however, without the necessary support and policy environment, smallholders operating near or under the poverty line may not always have the incentives to prioritize sustainable approaches. For example, when farmers operating under subsistence conditions are offered the opportunity to boost yields by using chemical fertilizers, they are likely to do so if it is the best means available to feed their families.[11] But where the right policies and incentives are in place, smallholders have shown they will take a long-term view, prioritizing sustainable techniques. Government policies that create disincentives for smallholders to care for natural resources lead to disastrous results, threatening the very capital that rural communities need for long-term survival.

New and sustainable approaches to agriculture offer improved livelihood opportunities for smallholders. At the same time, they typify the landscape approach to agriculture,[12] which is needed to maintain and augment the planet's natural resource base. Smallholders throughout the world are already showing us that these approaches can enrich farmers and ensure the long-term survival of communities while simultaneously renewing and preserving the world's natural assets.

The Need for Locally Specific Solutions

The global challenge of sustainable agriculture requires very local solutions. From place to place there are enormous differences in natural resource endowments, population densities, social and political relations, and market opportunities—and the results of generation upon generation of experimentation, innovation, learning, and refinement. They offer different opportunities for sustainable intensification, have different requirements, and face different constraints.

Consider fertilizer. In many areas of sub-Saharan Africa, integrating sustainable practices may call for increasing use of fertilizer as a necessary adjunct to organic methods. But in many parts of Asia, integration of crop and livestock systems and improved organic-based plant nutrient management may reduce the need for fertilizer. Different agricultural systems can become more sustainable and at the same time more productive and profitable.

Access to Green Technologies and Innovations

Sustainable innovations are bringing multiple benefits in terms of yield, profit, climate resilience, and poverty reduction:

- In Malawi and Zimbabwe, planting acacia trees in maize fields has tripled yields and improved the resilience of the soil while boosting nitrogen content and water retention capacity and moderating the micro-climate.[13]
- In an IFAD-supported project in Nyange village in Ngororero, Rwanda, students at a farmer field school are increasing yields by up to 300 percent (compared to yields under traditional methods) by using integrated pest management and applying fertilizer only when there is a demonstrated need.
- In Guangxi province in rural China, IFAD and the government have supported households in building biogas plants that use waste from farm animals and household toilets to generate energy for cooking and also produce high-quality organic fertilizer. Family health has improved, an estimated 56,600 tons of firewood have been saved annually, farm yields have increased, and the average income in the village has quadrupled.

Innovations in green energy sources are another potential solution. Solar photovoltaic pumping systems,[14] windmills, solar direct desalination, solar cookers, solar refrigerators, and solar electricity are just a few of the green technologies that are already available and could be adapted to rural communities. All offer enormous potential savings in energy and money; the major challenge for adaptation is the upfront cost.[15]

However, innovative tools will only work if supported by the right policies, infrastructure, and market structures. Green energy systems may be available, but they are useless to poor rural people if they cannot access credit to buy them or training on how to use them. New approaches may increase yields, but if the producer has no way to get her produce to markets, the effect on her livelihood will be negligible. The influence of new production technologies will be minimal if smallholder producers cannot obtain price information or access networks where they can gain fair prices for their products. In other words, innovation must come about within a fundamentally improved system that allows smallholder producers to reap the rewards of their creativity and hard work.

Scaling up sustainable agriculture will also require increasing investment in agricultural science research. Support for research in developing countries is generally inadequate and is being scaled back. Reversing this trend is urgent, says the Global Conference on Agricultural Research for Development.[16] More research funding is needed if sustainable intensification is to contribute to raising agricultural productivity. And more of it must be spent on the challenges of sustainable intensification faced by smallholder farmers.

Policies and an Enabling Environment

To repeat a key point: Where a supportive enabling environment is in place, smallholders will adapt green approaches to local contexts and scale up their successes. Policymakers have access to a range of tools to unleash the potential of smallholder producers to build sustainable livelihoods while simultaneously helping the world to protect natural resources and mitigate climate change. Following are summaries of some of the key issues to be addressed.

Removing Disincentives and Creating a Level Playing Field for Green Technologies

Distorting trade policies and subsidies together with ineffective land management policies create disincentives for farmers. Government policies in both industrialized and developing countries often leave green agricultural approaches and technologies at a disadvantage. However, there are encouraging signs that this is beginning to change:

- India, Indonesia, and the Philippines have removed insecticide subsidies and reduced insecticide use by 50–75 percent, while rice production continues to increase.
- Brazil has implemented minimum-till agriculture on 60 percent of the country's cultivatable land.
- Agroforestry is practised on 12–25 percent of agricultural land worldwide.
- The Chinese Ministry of Agriculture has developed a certification framework for agricultural products and offers a range of subsidies to promote use of organic fertilizer and minimum-tillage practices.
- The Government of Moldova, with IFAD support, is promoting regeneration of large swaths of erosion-prone farmland by supporting farmers to use no-tillage farming techniques that preserve soil fertility and enhance resilience to drought-induced crop failures.

These initiatives have been shown to increase productivity while improving the supply of critical environmental services.[17] When governments support sustainable approaches to agriculture with multiple benefits, everyone wins. Questions need to be asked as to why this is not happening on a larger scale.

Providing Financing

The resources needed by smallholders to adopt sustainable agricultural intensification are significant. Adapting to new production systems and technologies while dealing with the effects of changing climatic patterns will require significant financial support. We must ensure that:

- Financing is available to help them adapt to new production systems and climatic and environmental conditions.
- Innovative financial services are set up to support them in prioritizing sustainable natural resource management (microfinance[18] is a crucial tool here, as is payment for environmental services).[19]

Ensuring Clear Rights to Land

Weak institutional environments and unjust laws and practices regarding land ownership and tenure make it difficult for smallholders to approach farming with a long-term perspective and therefore to prioritize sustainable approaches to farming. This is a particular impediment for women, indigenous people, and young people, who are hampered by discriminatory laws and inheritance rules as well as cultural norms and practices. Providing communities and individuals with clear land rights gives them the incentive to restore or maintain environmental resources, such as replanting and managing forest areas. Equally worrying are land grabs by private entities that can deprive smallholders of their farmland and create an unstable environment.

In cooperation with the World Agroforestry Centre, IFAD is running a program called Rewarding Upland Poor for Environmental Services (RUPES) in 12 sites in China, Indonesia, the Lao People's Democratic Republic, Nepal, the Philippines, and Viet Nam. Communities are given secure land rights; in return they provide environmental services such as replanting trees, managing forest areas, and applying soil protection techniques on their plots. Other activities such as watershed preservation and the shoring-up of carbon sinks are offering knock-on benefits to lowland communities. These activities are showing that when smallholder farmers have secure rights to the land they farm, they have greater incentives to adopt sustainable, green approaches to farming.

Building Resilience

Smallholder producers generally lack safety nets to catch them in the event of calamitous weather events, crop failures, economic shocks, and illness or death of family members. This severely hampers their ability to stay out of poverty. It also affects their willingness to take on risks in the form of new livelihood strategies and approaches. If smallholders are to be protagonists in bringing about sustainable agricultural intensification, they will need support in dealing with the risks they face.

In the Tarija region of Bolivia, rural areas are highly dependent on rain cycles, and farmers are vulnerable to drought, frost, hail, flood, and other weather-related adversities. A pilot crop insurance scheme, supported by the International Labour Organization, is now offering farmers triple protection: insurance against loss of food crops, life insurance in case a close family member dies, and property insurance. Plans call for the project to be rolled out nationally in 2012. Schemes such as this empower farmers to make investments, in the knowledge that if things go wrong they will not fall deeper into poverty or be forced to migrate to cities in search of work.[20]

Engaging Private Investment

The private sector's role in driving green agricultural growth that enables smallholders to move beyond subsistence farming should be maximized. Food markets are evolving rapidly, providing enormous scope for exploring ways to help smallholders link up with food value chains that operate sustainably. The interactions between private actors (smallholder producers, intermediaries, entrepreneurs and small, medium-size, and large national and international businesses) determine production, marketing, economic, and environmental outcomes. These actors need to be actively engaged in a way that encourages investment in smallholder agriculture while protecting the welfare of rural women and men and the environment.

Public–private partnerships have the potential to reduce the risks associated with investing in smallholder agriculture, facilitate networks, identify untapped opportunities, and create win–win solutions. IFAD's forthcoming Private Sector Strategy will build a framework upon which these partnerships can be fostered and the right investments made to generate and scale up new and sustainable livelihood opportunities.

In Sao Tome and Principe, cocoa's extreme price volatility caused many producers to abandon their farms. Thanks to a partnership promoted by IFAD, between French chocolate producer Kaoka and local smallholders, superior aromatic cocoa beans were produced using traditional farming methods, fetching higher and more stable prices than common cocoa.[21]

Achieving Cooperation at Every Level

Translating sustainable agricultural practices into large-scale, coordinated plans of action will require cooperation between actors globally, nationally, and locally. It will also require effective communication between governmental ministries and all subsectors of agriculture, from production to processing and marketing. Equally importantly, it will require that farmers' voices are heard.

Fragmentation between and within local, national, and international policies frequently undermines efforts to address and combat the world's environmental and climate change challenges. A large-scale, global movement to intensify sustainable agriculture will require all parties to act together, with consistency, and to complement each other's actions. The necessary changes cannot take place if short-term political or business interests are given precedence in key decisions at any level by any actor.

Opportunities for Smallholder Farmers to Acquire New Skills and Knowledge

Practising sustainable agriculture is not easy. It is a knowledge-intensive, skilled activity based on constantly changing environmental, social, and institutional conditions that are specific to communities. Scaling up is most likely to succeed if the women and men responsible for implementing sustainable practices on their farms receive adequate training and support. Nurturing an institutional enabling environment for smallholders—with a focus on access to knowledge, inputs, credit, and markets—must form a key part of the movement toward sustainable approaches.

Basic Education

Access and quality of education in rural areas are in need of urgent attention:

- Rural–urban gaps remain wide in education enrollment and attainment rates.
- Basic education is frequently biased against agriculture.[22]

- Basic education generally fails to teach young people about agriculture in the context of sustainable development or to appreciate how it is linked to communities' development aspirations.

Vocational Training

Technical and vocational education and training (TVET) has the potential to upgrade the skills of smallholder farmers, introduce them to sustainable techniques and technologies, and enable them to improve their livelihood opportunities. However, this sector has suffered from a lack of focus in recent years. Outcomes from existing courses have not always reached expectations. This sector needs reinvigoration, with emphasis on:

- targeting young people
- reducing the gender gap in access
- facilitating the involvement of the private sector in courses
- adapting sustainable farming practices and green technologies to local realities
- ensuring participatory, inclusive approaches with curricula grounded in local conditions.

A successful example of TVET is farmer field schools. This approach brings together concepts from agroecology, experiential learning, and community development. Farmers carry out activities that help them understand the ecology of their farmlands. The knowledge they gain allows them to make crop management decisions relevant to their situation. More than 2 million farmers have participated in this type of learning since it was introduced by FAO in 1989.[23]

Inclusion of Diverse Actors

To adopt environmentally sustainable approaches to smallholder agriculture, we will need to capture and use the diverse skills of all the actors operating in the sector. But certain groups have traditionally been constrained from contributing to advancing rural agriculture: Loss of land rights by indigenous peoples has precluded many from applying their knowledge to preservation of biodiversity and renewal of the natural asset base on which agriculture depends; women's unequal access to assets such as credit and training has severely limited their productivity; and ambitious young people in search of decent work have been discouraged from farming due to its low status, low income, and uncertain prospects. We must be certain that all of these groups receive the opportunity to contribute their abilities and knowledge.

Indigenous People

Indigenous farming traditions and knowledge represent an untapped resource in efforts to protect the world's natural asset base, mitigate climate change, and improve livelihood opportunities for smallholder producers:

- Indigenous women and men possess unique, in-depth, and locally rooted knowledge of the natural world.

- Traditional indigenous land and territories contain 80 percent of the world's biodiversity, so indigenous people have the potential to play a leading role in managing natural resources.

Women

Sustainable, holistic approaches to agriculture must be based on equity. Any plan that does not use the skills of half the world's farmers and take their needs into account is doomed to failure.

- Women make up a significant portion of the agricultural workforce in developing countries.
- Large gender gaps exist in access to extension services, credit and land tenure, and these are at odds with women's contributions to agriculture.[24]
- Making full use of the skills of both women and men is urgent, particularly given the scope of the challenges ahead.

Young People

Sustainable agricultural intensification puts a premium on knowledge and innovation, and this makes it particularly well-suited to young farmers. For young people to be attracted to agriculture, however, it needs to be grounded in a new narrative that highlights the modern and innovative character of the proposed agenda and the potential of agriculture itself as a profitable activity in today's natural and market environment.

The Future

With policy support and adequate funding, sustainable agriculture could be intensified over large production areas in a relatively short period of time. The challenge facing policymakers is to find effective ways to scale up sustainable approaches so that hundreds of millions of people today and tomorrow can benefit. Some of the key needs are:

- Large-scale investment in agricultural research to find out what works, where it works, and how to adapt it to local contexts.
- Assessment of the harm caused by current practices on agroecological systems.
- Decisions at national level about which production systems are unsustainable and which sustainable approaches are suitable for scaling up.
- Work with farmers to validate and adapt approaches to local ecosystems.
- Preparation of plans for investment in appropriate policies and institutions, including farmers' organizations.
- Monitoring, evaluation, and review of progress, making adjustments where appropriate.[25]

Throughout the world, it is clear that sustainable agriculture is the answer. Farmers have produced countless examples of the advances that result when people work in harmony with nature. In the long term, this is the only way we can achieve sustainable solutions to hunger and poverty.

Notes

1. The current world population of 7 billion is projected to reach 9.3 billion by 2050 and 10.1 billion by 2100. See *2010 Revision of World Population Prospects*. United Nations Department of Economic and Social Affairs (2010). Estimate of arable land comes from Bruinsma (2009), as cited in IFAD's *Rural Poverty Report 2011*.
2. Conservation agriculture aims to achieve sustainable and profitable agriculture by promoting three principles: minimum soil disturbance, permanent soil cover, and crop rotation.
3. IFAD, *Rural Poverty Report 2011* (Rome, 2011). Integrated pest management takes into account the life-cycles of pests and their interaction with the environment, reducing the need for pesticides.
4. FAO, *Save and Grow* (Rome, 2011), chapter 1, available at www.fao.org/ag/save-and-grow/en/1/index.html. Around 97 percent of agricultural holdings in developing countries are below 10 hectares (FAO Agricultural World Census).
5. Sixty percent of all children involved in child labor are working in agriculture (ILO, 2010).
6. The 17th Conference of the Parties to the Convention on Climate Change (Durban Conference, November 28–December 9, 2011) will focus on implementation of the Convention and the Kyoto Protocol, as well as the Bali Action Plan and Cancun Agreements.
7. The Alliance for a Green Revolution in Africa estimates that in Africa environmental degradation is responsible for losses of between 4 and 12 percent of GDP.
8. FAO, *The State of the World's Plant Genetic Resources for Food and Agriculture*, Second Report (Rome, 2010).
9. See:www.iucn.org/about/work/programmes/pa/pa_what/?4646/Marine-Protected-Areas—Whyhave-them
10. FAO, *Global Forest Resources Assessment 2010*, Forestry Paper 163 (Rome, 2010).
11. With agricultural technologies and inputs, as with all else, prices influence demand. Subsidies on agrochemicals, inorganic fertilizers, or on agricultural water all encourage their use. In some regions, phasing out those subsidies makes much sense, combined with introducing subsidies for biofertilizers.
12. Landscape approaches integrate plans for food production and other land uses into broader plans for environmental preservation, clean water, clean air, and preservation of biodiversity for the long term.
13. Seehttp://www.africanagricultureblog.com/2010/11/fertilizer-tree-triples-malawi-zambia.html for more information.
14. An environmentally sustainable way of improving water access for rural households, reducing soil salinity and erosion, allowing production during dry seasons, and increasing agricultural production by up to 30 percent. A photovoltaic panel receives sun rays, which in turn produce electricity to generate a pump that is submerged in a borehole. Water is then pumped through an outlet pipe into a water tank for collection.

15. Green energy sources tend to have higher upfront costs than traditional devices. But taking into account the extremely low running and maintenance costs, the savings over the natural lifetime of such technologies are in thousands of dollars.
16. IFAD, *Rural Poverty Report 2011* (Rome, 2011), chapter 5.
17. IFAD, *Rural Poverty Report 2011* (Rome, 2011), chapter 5.
18. Innovative loan products are needed to finance rural water and renewable energy products. A survey in West Africa of prospective microfinance clients found that over 80 percent had a project related to water that would help increase farm productivity, but only 1 in 10 clients were able to acquire funding to implement their projects. In terms of energy, microfinance can allow rural households to turn away from traditional energy sources (wood, diesel, and kerosene), which are expensive and damage the environment, and toward renewable sources such as solar systems, wind energy, and biogas.
19. Payment for environmental services is another way of giving smallholders access to the resources they need as well as ensuring they receive just rewards for the important work they do. In Foz do Iguacu in Brazil, the operators of Itaipu Dam, which provides 25 percent of Brazil's energy, pay municipalities along the reservoir lake to provide the environmental service of implementing no-till agriculture to reduce siltation at the dam. In Morocco and Kenya, the Green Water Credit initiative provides regular payments to downstream water users in recognition of their important role in managing land and water resources. This enables farmers to invest time and resources in green water management, while diversifying their income and helping them to stay out of poverty (see: www.greenwatercredits.org/).
20. For more information, see: www.guardian.co.uk/global-development/poverty-matters/2011/feb/21/ micro-insurance-protect-poor.
21. IFAD launched a three-year pilot project involving 500 farmers in 11 communities, with Kaoka agreeing to supervise the project and to purchase as much certified organic cocoa as the farmers could produce. By the end of the pilot project, the farmers had produced 100 ton of certified organic cocoa that sold for two and a half times the price of common cocoa. The farmers subsequently formed a cooperative and signed a five-year contract directly with Kaoka, guaranteeing them a stable price. IFAD has scaled up organic, aromatic cocoa farming to another 12 communities in Sao Tome within its ongoing Participatory Smallholder Agriculture and Artisanal Fisheries Development Programme.
22. A recurring theme at IFAD's 2011 Governing Council High-Level Panel and Side Events was the difficulty in motivating talented young people to use their skills in agriculture; the negative image of the sector that comes out of schools is believed to be a significant contributing factor. (Proceedings available at: https://webapps.ifad.org/members/sessions/88184-34th-session-of-the-governing-council/documents/199/get/english)
23. Another impressive example of participatory community-based TVET is Songhai, established in Benin and since replicated in numerous other countries in sub-Saharan Africa. This agricultural entrepreneurship

training center focuses on providing an entrepreneurial platform for young African farmers to develop skills to practise sustainable, profitable agriculture. Activities at the centers focus on green technologies and renewable energy as well as business and life skills training. Songhai values the principle of working in harmony with local ecosystems, and centers are adapted to the realities of the environments in which the trainee farmers operate. In 2008 Songhai was promoted as a Centre of Excellence for Africa by the United Nations.

24. It is estimated that if women had the same access to productive assets as men, they could increase their yields by 20–30 percent. This would imply a 2.5–4 percent increase in agricultural output in developing countries, which in turn could reduce the number of hungry people in the world by 12–17 percent. (FAO, *The State of Food and Agriculture,* [Rome, 2010])
25. See FAO, *Save and Grow* (Rome, 2011) for more information on the challenges of sustainable agricultural intensification and how it could shape the future.

Craig Meisner

Why Organic Food Can't Feed the World

Recent studies have revisited the idea that organic methods of agriculture would be sufficient to feed the world—but they are flawed because of naïveté about agriculture in developing nations.

Can organic food feed the world? A recent study, published in the journal *Renewable Agriculture and Food Systems* provides new data that suggests it can. However, I have some grave reservations about this prospect that are based on my experience as a scientist and my time living and working with real farmers in developing nations. The authors of this study assume the major stumbling blocks to organic farming feeding the world are low crop yields and insufficient quantities of approved organic fertilizers. However, I have lived and worked in Bangladesh—as a professor of Cornell University, covering agricultural research and development—for the last 25 years, and I believe that even if these problems could be surmounted, using organic farming to feed the developing world remains a pipe dream.

Green Revolution

Bangladesh is the size of England and Wales together, but with a larger population of about 140 million people. It has achieved remarkable progress in its food productivity, even achieving self-sufficiency in flood-free years (currently we are experiencing a particularly devastating flood). The basis of the Green Revolution that saved South Asia was not organics, but the use of a dwarfing gene to stop rice and wheat from collapsing when they flourished, coupled with chemical fertilizers and irrigation systems.

Despite the burgeoning population, the Green Revolution of the 1960s is continuing today in South Asia with an increase in the use of hybrid rice and maize, conservation agriculture, deep placement of nitrogen in rice paddies, and many exciting technologies.

Heavy Burden

So, why won't the use of pure organics work in developing countries like Bangladesh?

Most supporters of the idea that organic farming can feed the world, assume that organic manures are cheap and available to all—even the poor. But this isn't often the case. I see cow dung in Bangladesh and all of South Asia as a valuable commodity. During my walks in the villages I see it collected largely by women and children and used as fuel. It's found in nearly every house, dried and formed into patties, to be sold or burned for cooking.

Straw is another organic source of nutrients, but that's not always available either. Rice and wheat straw is collected from the fields, and used for cattle feed or thatching for roofs. Even the stubble is used, which the poorest come and cut for fuel.

The authors of the study mentioned above—led by researchers at the University of Michigan in Ann Arbor, U.S.—have rightly assumed that organics can supply sufficient nutrients for plant growth. However, the quantities of organics required to sustain such productive growth makes it very difficult for the poor to handle. Organics, whether farmyard manure, compost, or cow dung, contain moisture and are heavy and difficult to carry from the homestead to the fields by the growers.

For example, to raise a 6-ton rice crop in the peak season requires 100 kg of nitrogen. Because of monsoons and the fact that several meters of rainfall drains through the soil every three months, the amount of nitrogen it carries is low. Assuming we used good quality manure, there would be about 0.6 percent nitrogen in the material, thus, requiring 17 tons per hectare to produce a 6-ton rice yield.

Can you imagine carrying 17 tons of manure, in repeated 50 Kg loads, in a basket on your head? The lack of machinery to carry that material and the labor required to apply it, compounds the challenge.

Plus, there is simply not enough manure, or even plant biomass, available to apply 17 tons per hectare, for even a single annual rice crop across the whole of Bangladesh. That's enough of a problem, but when you consider there are actually two rice crops a year, the full scale of the problem becomes apparent!

Green Manure

In answer to some of these problems, the new study proposes the use of a leguminous "green manure" crop. These pulse crops fix nitrogen into the soil from the air through a symbiotic relationship with bacteria in their roots. They provide enough nitrogen for their own growth and more, and when ploughed under provide nitrogen for a subsequent crop too.

However for such a crop to be used in Bangladesh, it would have to take the place of a food crop, effectively halving the amount of food the land can provide. The cropping intensity in many developed countries is well over two crops per year, but I have seen as many as four to five crops per year in places that are elevated and flood-free.

Besides substituting for a food crop, green manure crops would also require cutting and ploughing under the soil. While ploughing technology has increased dramatically in the last decade in many developed countries, it is mostly the two-wheel tractors or roto-tiller types, thus making it a significant challenge to plough down any high biomass green manure or crop residues into the soil.

Some propose a greater use of leguminous food crops to supply nitrogen for the proceeding cereal crop and where possible, growers would love to expand pulses. However, in South Asia, while the national pulse yields appear stable, switching to more of these crops is quite risky for individual farmers due to unseasonable rainfall, diseases, and poor growing environments.

Faced with a Choice

So, to make compost effectively, one has to have surplus plant biomass and cow dung. For the poor who have limited land and animals, this is quite difficult.

Surveys I have conducted in Bangladesh clearly show that growers that do have the ability to add organics to their land are those who are richer and have larger land holdings and animals. The poor have to rely on purchased fertilizers, whether organic or chemical. When faced with a choice based on labor and expense, the poor choose the nonorganic fertilizers.

Another recent study, published in *Nature,* revealed clearly what plant scientists have known for years—that plants take up some 20+ elements from the soil—whether it is from decomposing organics or chemical fertilizers. That study showed there was absolutely no difference in the biochemical make up of the plants grown in pure organics compared to fertilizers.

When I have asked growers in Bangladesh, most would love to be able to use more organics in their farming production. But due to the lack of availability and costs, organics are actually being used less each year. Can organic agriculture feed the world? No, but most growers understand that it benefits the soil, and as such its use is advocated as much as is possible. Unfortunately, for Bangladesh, and many developing countries, those possibilities are diminishing yearly as organics become less and less available and affordable.

EXPLORING THE ISSUE

Can Sustainable Agriculture Feed the World?

Critical Thinking and Reflection

1. Discuss the positive and negative effects of industrial agriculture on society and the environment.
2. What are the benefits in moving to sustainable agricultural practices? Are there any costs and problems associated with such a transition?
3. Given current knowledge about sustainable agriculture and the evidence of how these approaches have succeeded throughout the world, why hasn't there been a concerted effort from governments to create the policy environment for scaling up these approaches globally?
4. Can green approaches—such as conservation agriculture, sustainable forest management, and integrated pest management—be applied universally? What are the barriers to their implementation?
5. What pros and cons have been identified for consideration regarding transgenic agriculture? Should GM agriculture be considered as a sustainable practice?
6. Provide some of your own ideas on fostering ecological food production.

Is There Common Ground?

How can we feed the world while maintaining global ecosystem integrity? This is a key issue in the debate over whether sustainable agriculture can feed the world. There is abundant scientific evidence to show that sustainable agriculture based on organic production methods is environmentally sound. There is also abundant scientific work which shows that industrial agriculture places greater stress on natural systems and can degrade ecosystems. The question that is crucial to the debate is whether organic agriculture can be scaled up to meet the needs of 9 billion people on earth by 2050. Also, can we consider new industrial technologies such as genetically modified (GM) agriculture as sustainable since it increases yield while decreasing the use of industrial fertilizers, tilling, pesticides, and herbicides, or are its potentially negative, irreversible affects to all phases of natural systems of significant concern?

To produce any common ground between the YES and the NO selections, the two specific approaches to food production, the industrial and the sustainable, need to be examined from two criteria: first, does the approach preserve ecosystems and have a smaller ecological footprint; and second, does it have sufficient scale to feed the world. Any production approach that compromises ecosystem integrity should be phased out. But if sustainable agriculture cannot scale itself up to feed the world, then it is at risk of not being universally accepted. Policies need to be developed that can integrate sustainable agriculture into global political systems, and the best aspects of industrial agriculture should be integrated with organic methods to ensure larger yields that do not compromise ecosystem integrity.

To achieve sustainable food production, agricultural approaches must be compatible with ecosystem preservation. We must not only retain current high yield production, but we must produce more for an exponentially increasing global population. Approaches of industrial agriculture that are compatible with ecosystem preservation can be identified and integrated with multifaceted, holistic approaches that have been proven to diversify production, reduce consumptive energy costs, while increasing yields, particularly in times of drought and other effects of climate change.

Continued research support, education, and adoption of policies that allow sustainable agriculture to play on an even field with industrial practices could lead to a major paradigm shift in the bottom-up training of our agricultural land managers and a top-down shift in state, national, and global support of farmers who transition to responsible stewardship and husbandry of agricultural lands. And who knows, maybe there is even a way to grow genetically engineered crops in open-air environments that does not risk wild and classically bred species from cross-contamination and potential elimination; meanwhile, the precautionary principle might best guide those experiments behind closed, sealed, controlled environments, as we have much genetic diversity at risk.

References

Altieri, M. A. (2006) The myth of coexistence: Why transgenic crops are not compatible with agroecology based systems of production. *Bulletin of Science, Technology and Society, 25,* pp. 361–371.

Angelsen, A. and Kaimowitz, D., eds. (2001) *Agricultural technologies and tropical deforestation.* CAB International, Wallingford, UK.

Bennett, A. F., Radford, J. Q., and Haslem, A. (2006) Properties of land mosaics: Implications for nature conservation in agricultural environments. *Biological Conservation, 133,* pp. 250–264.

Benton, T. G., Vickery, J. A., and Wilson, J. D. (2003) Farmland biodiversity: Is habitat heterogeneity the key? *Trends in Ecology & Evolution, 18,* pp. 182–188.

Berry, W. (1990) *What are people for?* North Point Press, San Francisco, CA. ISBN:9780865474376.

Conway, M. and Rosset, P. (1996) *A cautionary tale: Failed US development policy in Central America.* Reinner Publishers, London.

Daily, G. C., Ceballos, G., Pacheco, J., Suzan, G., and Sanchez-Azofeifa, A. (2003) Countryside biogeography of neotropical mammals: Conservation opportunities in agricultural landscapes of Costa Rica. *Conservation Biology, 17,* pp. 1814–1826.

Day, J. W., Hall, C. A., Yanez-Arancibia, A., Pimentel, D., Marti, C. I., and Mitsch, W. J. (2009) Ecology in times of scarcity. *BioScience, 59*, pp. 321–331.

Estrada, A., and Coates-Estrada, R. (2002) Dung beetles in continuous forest, forest fragments and in an agricultural mosaic habitat island at Los Tuxtlas, Mexico. *Biodiversity and Conservation, 11*, pp. 1903–1918.

Hall, C. A. S., Tharakan, P. J., Hallock, I., Cleveland, C., and Jefferson, M. (2003) Hydrocarbons and the evolution of human culture. *Nature, 423*, pp. 18–322.

Harvey, C. A., Komar, O., Chazdon, R., Ferguson, B. G., Finegan, B., Griffith, D. M., Martinez-Ramos, M., Morales, H., Nigh, R., Soto-Pinto, L., Van Breugel, M., and Wishnie, M. (2008) Integrating agricultural landscapes with biodiversity conservation in the Mesoamerican hotspot. *Conservation Biology, 22*, pp. 9–21.

Harvey, C. A., Medina, A., Sanchez Merlo, D., Vilchez, S., Hernandez, B., Saenz, J., Maes, J., Casanovas, F., and Sinclair, F. L. (2006). Patterns of animal diversity associated with different forms of tree cover retained in agricultural landscapes. *Ecological Applications, 16*, pp. 1986–1999.

Kirschenmann, F. L. (2007) Potential for a new generation of biodiversity in agroecosystems of the future. *Agronomy Journal, 99*, pp. 373–376.

LeClerc, G. and Hall, C. A. S., eds. (2007) *Making world development work: Scientific alternatives to neoclassical economic theory*. University of New Mexico Press, Albuquerque, NM.

Marvier, M. (2001) Ecology of transgenic crops. *American Scientist, 89*, pp. 160–167.

Mayfield, M. M. and Daily, G. C. (2005) Countryside biogeography of neotropical herbaceous and shrubby plants. *Ecological Applications, 15*, pp. 423–439.

Meadows, D. and Randers, J. (2004) *Limits to growth: The 30-year update*. Chelsea Green, White River Junction, VT.

Millennium Ecosystem Assessment (MEA). (2003) *Ecosystems and human well-being: A framework for assessment.* World Resources Institute, Washington, DC.

National Research Council (NRC). (2000a) *Clean coastal waters: Understanding and reducing the effects of nutrient pollution.* National Academy Press, Washington, DC.

National Research Council (NRC). (2000b) *Transgenic plants and world agriculture*. The National Academies Press, Washington, DC.

Nixon, S. W., Ammerman, J. W., Atkinson, L. P., Berounsky, V. M., and Billen, G. (1996) The fate of nitrogen and phosphorus at the land-sea margin of the North Atlantic Ocean. *Biogeochemistry, 35*, pp. 141–Í80.

Odum, H. C., and Odum, E. C. (2001) *A prosperous way down: Principles and policies*. University Press of Colorado, Boulder, CO.

Perfecto, I., Rice, R. A., Greenberg, R., and Van Der Voort, M. E. (1996) Shade coffee: A disappearing refuge for biodiversity. *BioScience, 46*, pp. 598–608.

Pimentel, D. (2006) Impacts of organic farming on efficiency and energy use in agriculture (40pp). www.organicvalley.coop/fileadmin/pdf/ENERGY_SSR.pdf

Pretty, J. R., Morison, E., and Hine, R. (2003) Reducing food poverty by increasing agricultural sustainability in developing countries. *Agriculture, Ecosystems and Environment, 95*, pp. 217–234.

Rissler, J. and Mellon, M. (1996) *The ecological risks of engineered crops*. MIT Press, Cambridge, MA.

Rodale, J. I. (1948) *The organic front*. Rodale Press, Emmaus, PA.

Sekercioglu, C. H., Loarie, S. R., Oviedo Brenes, F., Ehrlich, P. R., and Daily, G. C. (2007) Persistence of forest birds in the Costa Rican agricultural countryside. *Conservation Biology, 21*, pp. 482–494.

Snow, A. A., Moran-Palma, P. (1997) *Commercialization of transgenic plants: Potential ecological risks. Bioscience, 47*, pp. 87–96.

Wackernagel, M., et al. (2002) Tracking the ecological overshoot of the human economy. *Proceedings of the National Academy of Sciences, 99*, pp. 9266–9271.

Internet References

American Society of Agronomy (ASA)

This website is sponsored by the ASA, a prominent, international scientific society headquartered in Madison, WI and provides a wealth of information to glean from including: scientific journals (*Agronomy Journal*, *Journal of Environmental Quality*, and *Journal of Natural Resources and Life Sciences Education*); insights to current agricultural news and events; and political actions regarding science policy at the Federal level. Because of their common interests, ASA, the Crop Science Society of America (CSSA), and the Soil Science Society of America (SSSA) share a close working relationship, the same headquarters office staff, and, by mission statement, are dedicated to the conservation and wise use of natural resources to produce food, feed, and fiber crops while maintaining and improving the environment.

www.agronomy.org

Food and Agriculture Organization of the United Nations (FAO)

This website is sponsored by the United Nations (UN) and provides current reference material, scientific, professional, and otherwise, regarding: key programs; global issues; and core activities that include those in agriculture, economic and social, natural resources, and details of recent and upcoming meetings. The FAO works with the UN, Academic and Research Institutions, Civil Society Organizations (CSOs), including NGOs, and the private sector in an international effort to defeat hunger.

www.fao.org

Holmgren Design Services

This website is sponsored by a private sector, free-enterprise business, Holmgren Design Services. David Holmgren, author, consultant, and educator is best known as the co-originator, with Bill Mollison, of the permaculture concept of sustainability principles. This website is a good source because it provides extensive, free background and educational material on permaculture; practical how-to information regarding permaculture; and ongoing permaculture design projects that are extraordinarily transdisciplinary in both their design and the extant of real-world sustainability problems that those designs address.

www.holmgren.com.au

Permaculture Research Institute (PRI) of the United States

This website is provided through the non-profit organization (NPO) known as The Permaculture Research Institute, whose locations and website affiliations worldwide are sponsored through private donations and monies raised through certification education in permaculture, which it also provides at reduced rates or free of charge to developing countries, inner-cities, or other areas of need where a demonstration project is held, not only for their paying customer permaculture students, but for the socio-economic and ecological benefit of local peoples. This website provides a thorough definition of permaculture in the "about" section; a world map with descriptions of ongoing permaculture design projects throughout the world in the "project" section; and other permaculture-related resources.

www.permacultureusa.org

Rodale Institute

This website is sponsored by the Rodale Institute, pioneers since 1947 of organic farming practices through scientific research and outreach that is based in Kutztown, PA. This website provides decades, in fact the longest running farm trials in the United States, of side-by-side conventional versus organic methods scientific research that repeatedly and convincingly conclude the agricultural and ecological benefits of organic agricultural practices; educational material is free and readily available online or in print; and exhaustively covers a broad range of socio-economic and ecological facets related to agriculture, local, regional, national, and worldwide.

www.rodaleinstitute.org

Internet References . . .

Nuclear Energy and Global Climate Change

This website provides information about nuclear power and global climate change. It also has articles based on issues related to nuclear energy.

http://timeforchange.org/nuclear-energy

Nuclear Energy

This website provides comprehensive information, publications, and news on nuclear power.

http://www.world-nuclear.org/

Green Business

The website is devoted to information on the best practices of corporations that follow sustainability practices and policy.

http://www.greenbiz.com

Triple Bottom Line Sustainability

This is the corporate website of John Elkington, the business consultant and author credited with originating the concept of the "Triple Bottom Line."

http://www.sustainability.com/

Cradle-to-Cradle Environmental Design

This is the website of green design consultants and authors William McDonough and Michael Braungart, better known for Cradle-to-Cradle Design.

http://www.mbdc.com/

Sustainable Communities

This is the website of Resourceful Communities, a group dedicated to pursuing conservation and social justice particularly within the communities of NC.

http://www.resourcefulcommunities.org/triple_bottom_line

World Business Council for Sustainable Development

This is the website provides case studies of sustainable businesses worldwide.

http://www.wbcsd.org

International Association of Local Governments (ICLEI)

This organization has one of the most comprehensive programs to promote sustainable cities worldwide.

http://www.iclei.org

Urbanism and Sustainability

This website discusses smart growth policy and New Urbanism.

http://www.newurbanism.org/sustainability.html

UNIT 5

Energy, Business, and Society

The three "E's" of sustainability—Environment, Economy, and Equity—deal with the capacity of society to balance each of these three needs. Two areas where sustainability practices, goals, and objectives need to be implemented are in business and urban development. It is in these two sectors that most human activity occurs. Business, for instance, places great stress on the natural environment and deals directly with issues of social justice (through community and stakeholder involvement) and economic development. Corporations have made steady advancement over the past 10 years at incorporating sustainability into their business plans and operations. But, how do businesses become more sustainable? And why should businesses become more sustainable? When the goals of business are to make profits and return value to their shareholders or owners, doesn't this conflict with the goals of sustainability? And, if there is an intrinsic distinction between the profit goals of business and sustainability goals, is it reasonable to expect business to trump sustainability over immediate profits? Can a business case be made for corporate sustainability? Is it more effective to publicly regulate business to achieve the desired goals or should greater reliance be placed on voluntary methods? These issues are highly charged and need to be addressed if society can implement sustainability goals and practices.

Another sector that is subject to great changes over the next 20 years is urbanization. Cities and urban settlements are the places where people concentrate and where most global resources and energy are consumed. This is also the sector that is undergoing the greatest change as people in developing countries such as China and India migrate to cities. Can sustainability practices become central to the way people adapt to these cities?

Understanding these issues and deciding the best course of action are central dilemmas for the twenty-first century. The scale, timing, and spatial implications for society are key challenges for sustainability.

- Can Nuclear Energy Be Green?
- Is Corporate Sustainability More Public Relations Than Real?
- Are Social Concerns Taken Seriously in the "Triple Bottom Line" of Sustainability?
- Can Cities Be Made Sustainable?

ISSUE 17

Can Nuclear Energy Be Green?

YES: A. Adamantiades and I. Kessides, from "Nuclear Power for Sustainable Development: Current Status and Future Prospects," *Energy Policy* (December 2009)

NO: Milton H. Saier and Jack T. Trevors, from "Is Nuclear Energy the Solution?" *Water, Air, & Soil Pollution* (May 2010)

Learning Outcomes

After reading this issue, you should be able to:

- Identify advantages and disadvantages of nuclear energy as a replacement for fossil fuel energy systems.
- Understand why nuclear energy is supported by some as a "green" energy source.
- Understand new developments and new technologies that are driving the resurgence of nuclear energy among policy makers.
- Fully comprehend the risks of nuclear energy for society.
- Understand why nuclear energy up until the present has not been considered as a feasible alternative to fossil fuel–based energy.

ISSUE SUMMARY

YES: Engineer and energy consultant Achilles Adamantiades and economist and writer I. Kessides discuss how burgeoning population, growing demands for energy, dependence on foreign fossil fuels, and rising concern about global climate are major reasons for the growing interest in nuclear power.

NO: Biologist Milton H. Saier and environmental scientist Jack T. Trevors argue that nuclear power is not cost-competitive compared with other green energy sources such as solar and wind, which can be installed much faster. They also discuss its inability to deal with the issue of energy security since oil is mostly used for transportation and nuclear energy is not used for this key activity.

With the release of radioactivity from the Fukushima Daiichi nuclear power plant in Japan demonstrating to the world the hazards of nuclear power, it seems unlikely that nuclear energy could be viewed as sustainable. But when viewed from the perspective of the need for large-scale energy for global economic development, the threat of global climate change due to rising carbon dioxide emissions from fossil fuels, a growing world population, and a comparison with alternative-scale power sources, nuclear power can look more attractive. Countries such as France, which receives 75 percent of its electricity from nuclear fuel, India, and China are not scaling back their nuclear energy ambitions. They believe that there is no other feasible source of large-scale energy. In the United States, which receives only 20 percent of its electricity from nuclear fuel, the natural gas–fueled utilities will dominate the future electricity landscape. But some believe that the environmental hazards of shale gas extraction will make this energy source unsustainable. Also, can alternative energy sources such as solar and wind really provide the large amounts of energy that will be required to maintain economic growth? It is with these questions in mind that the issue of nuclear energy as sustainable is broached.

What are the necessary elements for an energy source to be considered as sustainable? First and foremost it must be renewable. Fossil fuels such as oil, coal, and gas, the primary sources of global energy today, clearly are not renewable but finite resources. Eventually they will be used up and are not replaceable in their present form. A second major requirement is that the energy source must pose no threat to the global ecosystem or to human health. Renewable energy sources such as large-scale hydropower have often been criticized for altering river ecosystems and impacting fish populations. And lastly, with today's emphasis on the reduction of carbon dioxide due to its relationship to global warming and climate change, a sustainable energy source must emit little or no carbon dioxide. How does nuclear energy measure up as sustainable energy? First, it relies on uranium as its fuel source, which is, of course, a finite resource. Second, it can pose a threat to both natural ecosystems and to human health. If a nuclear accident occurs, a Chernobyl-type event, a major release of radioactivity could prove devastating to both nature and humans. But the one positive feature of nuclear energy, and the one that perhaps might outweigh its negatives, is that it releases little or no carbon dioxide directly into the atmosphere. Does that one feature entitle nuclear energy to be considered as sustainable?

Global warming is now a well-known phenomenon in the world. According to the World Nuclear Association, for thousands of years greenhouse gases were responsible for the melting of ice and the creation of a moderate climate suitable for life to evolve. But with population growth and its reliance on fossil fuel energy sources to power economic development, greenhouse gases have proliferated causing a radical warming of the globe. It is predicted that in the next 50 years, human energy consumption will grow exponentially. Most of the energy produced today comes from burning fossil fuels such as coal, oil, and natural gas. This energy is used to generate electricity, power vehicles, heat and cool homes, run factories, and so forth. These nonrenewable resources are being consumed so rapidly that it is believed that major shortages will occur later in the twenty-first century.

A major problem with fossil fuel–based energy is its emission of carbon dioxide, a major greenhouse gas, directly into the atmosphere. It is estimated that every year about 25 billion tons of carbon dioxide is emitted into the air by burning fossil fuels. This equals about 70 million tons per day or 800 tons per second. The daily emission of carbon dioxide into the atmosphere by Americans is 120 pounds per person, by Europeans and the Japanese about 50 pounds per person, and by the Chinese about 13 pounds per person (with a population of 1.4 billion) (World Nuclear Association). With increasing world population, the demands for food, electricity, water, and energy are growing. Meeting the large-scale demand for energy, some energy consultants believe, will require a combination of both renewable and nonrenewable energy sources. Renewable energy, these consultants believe, cannot meet the scale of energy demand. There will be the need for a large-scale source of energy such as nuclear to both supply world demand while fitting into global initiatives and reduce carbon dioxide emissions by 50 percent.

Renewable energy sources such as wind, biomass, geothermal, and solar are clearly better choices for sustainability. But these resources can also have their problems. Wind power, which is the fastest growing alternative energy source, is viewed by some as unreliable and inefficient. There have also been examples, particularly in the United States, of popular opposition to wind turbine farms. The International Energy Agency of the Organization for Economic Cooperation and Development (OECD) projects that alternative energies will be able to provide only 6 percent of the world's energy by 2030. If this projection is correct, how will energy demand be met while reducing global carbon dioxide? Supporters of nuclear energy see it as the answer to this dilemma. They see nuclear energy as the only alternative today that has the capacity to produce clean electricity to fulfill the growing demands on a global scale.

It has been more than 50 years since the emergence of the nuclear energy generation. It is estimated that around two-thirds of the world's population lives in countries where nuclear power plants are already generating electricity or are planning to build nuclear reactors. There are presently 440 nuclear reactors around the globe that are generating electricity. Fifteen countries today depend on nuclear reactors to supply at least 25 percent of their electricity. Europe and Japan generate more than 30 percent of their electricity from nuclear power, while the United States presently receives 20 percent of its electricity from nuclear power. France is a world leader in nuclear electricity generating 75 percent of its electricity from nuclear power and is the world's largest net exporter of electricity.

Besides producing electricity, nuclear technologies are used in medical diagnosis and cancer therapy. Today several nuclear technologies are being used not only to cure human illnesses but also to maintain livestock health, preserve food, improve agricultural productivity, and eradicate virulent pests. Clean nuclear energy can also be used to distill salt water on a scale that would provide fresh water for millions of people in drought-prone regions and in regions where water is lacking for agricultural production.

The demand for energy security, the ability of a country to be less reliant on imported energy, is another factor that might drive nuclear energy. Countries that are not fossil fuel rich can look to nuclear fuel as a way to provide some energy security to its people while meeting global demands for carbon dioxide reductions. It is believed that France decided to expand its use of nuclear energy in 1974 as a result of the turmoil in the Middle East and as a way to support its economic growth. According to the European Green Paper on the security of the energy supply in 2000, the three pillars of energy security for the future are coal, nuclear fuel, and renewable fuels. Two of these pillars—nuclear fuel and renewable fuels—can meet global demands for carbon dioxide reduction.

The persisting issue is providing a continuous and reliable supply of electricity. The question is how to generate this electricity. Many scientists feel that nuclear energy is a "green" alternative to burning fossil fuel (Totty, 2008). Nuclear energy seems to be the answer to reduce the threat of global warming and satisfy the growing demand for electricity. The United States and other countries are trying to include nuclear power into the energy mix in order to have a less carbon-intense world for present and future generations. For some countries, embracing nuclear power means independence of energy, while others see nuclear technology as an export business. Countries like China, India, Japan, Pakistan, and so forth are actively developing nuclear power for energy security and as a way to reduce their greenhouse emissions (*The New York Times,* 2010).

The article for the YES position, that nuclear energy can be considered as sustainable, is presented by engineer and energy consultant Achilles Adamantiades and economist and writer I. Kessides. They discuss the current status and future plans for expanding nuclear power. They describe all the necessary reasons for moving to nuclear energy as a safe and reliable energy source. They point out its benefits, such as its capacity to provide energy security, reduce air pollution, reduce greenhouse gas emissions, and avoid fossil fuel cost volatility. They also note that there is a growing acceptance of nuclear energy as a potential solution to global energy demand. The authors not only present the major developments and improvements in nuclear technology but also address safety concerns, accidents, and waste disposal.

The article for the NO position, that nuclear energy is not sustainable, is presented by biologist Milton H. Saier and environmental scientist Jack T. Trevors. They point out the necessary considerations that should be reviewed before policymakers in the United States decide for a nuclear energy future. Some of these considerations are its high capital costs for construction, the risk of it being targeted by terrorists, and whether it will bring any major changes in our dependence on imported oil (oil is used for transportation energy and not for electricity unless there is a major commitment to electric vehicles). Presently, the costs of future nuclear energy are not competitive with fossil fuels such as coal or natural gas, and are not even competitive with alternative sources such as wind, geothermal, or solar, which are clearly "more green." And lastly, what to do with the radioactive waste is still a challenging question for the viability of nuclear energy.

A. Adamantiades and
I. Kessides[1]

Nuclear Power for Sustainable Development: Current Status and Future Prospects

Introduction

In recent years there has been a resurgence of interest in developing nuclear power in both developed and developing countries. The United States, where construction had ceased for decades, has now formally certified new reactor designs. In Europe (with the notable exception of France), where nuclear power development has been in a holding pattern for almost two decades, nuclear energy has been the subject of continuous political debate and is now a key element in the European Union's climate-change policy. After an intense debate, Finland's parliament voted in 2002 to approve building a fifth nuclear power plant—the first such decision to build a new nuclear plant in Western Europe for over a decade. A new White Paper on Nuclear Power put nuclear energy at the core of the UK government's energy policy, and the Government's support for new nuclear build was confirmed in January 2008. In May 2008, two decades after a public referendum resoundingly banned nuclear power and deactivated the country's reactors, Italy announced plans to resume building nuclear plants within five years. And in February 2009, Sweden announced plans to overturn a near 30-year ban on new nuclear plant construction. Debates on new nuclear build are underway in Germany, Belgium, the Netherlands and Hungary.

More than 40 developing countries, ranging from the Gulf to Latin America, have recently approached United Nations officials to express interest in starting nuclear power programs (Reuters, 2008). In contrast to North America and most of Western Europe, nuclear power capacity in Asia has been growing significantly. A number of countries in East and South Asia are planning and building new reactors—21 are under construction and there are plans to add 150 more. China, Japan, South Korea and India are expected to experience the strongest growth in the region (WNA, 2008). Indonesia, Vietnam, Thailand, the Philippines and Malaysia are also expressing strong interest in nuclear power (Symon, 2008). In late 2007, Egypt announced that it would build several nuclear power plants to meet rising energy demands (Fleishman, 2007). In June 2008, the South African Cabinet approved an ambitious nuclear energy policy contemplating the installation of 20 GW of nuclear power. There are also ambitious plans to expand nuclear power in Latin America. In September 2008, Brazil's top energy official announced the country's intention to set up 50–60 nuclear power plants in the coming half century (Associated Press, 2008). Argentina is planning to double its existing nuclear capacity and Mexico may add eight more reactors by 2025. Chile, Venezuela and Uruguay are expressing strong interest in nuclear energy (Squassoni, 2009).

This paper describes the current status and future plans for expansion of nuclear power, the advances in nuclear reactor technology, and their impacts on the associated risks and performance of nuclear power. Developments in the United States are given some prominence because nuclear technology originated there and has expanded, in absolute terms, more than in any other country. In addition, the United States has developed a nuclear regulation and supervision system which is arguably the world's most elaborate and demanding. Thus, whatever happens in the United States is a bellwether of developments elsewhere. This is not to minimize accomplishments and future potential in other countries, some of which have built and operated nuclear plants and are at the forefront of industry developments. But the United States, merely by its size, is bound to have an overwhelming influence on future developments. Similarly, US policy initiatives are closely watched and often used as a springboard for action in other parts of the world.

A Renaissance in Nuclear Power—Why Now?

Several factors seem to be driving the resurgence of interest in nuclear power:

- A global desire to diversify fuel sources, reduce dependence on fossil fuel imports, and develop immunity to power disruptions;
- A desire to mitigate volatile fuel costs, given the low dependence of the price of nuclear-produced kilowatt-hours on the price of uranium;
- The need to mitigate climate change by reducing greenhouse gas emissions—specifically, carbon dioxide;
- A desire to decrease air pollution, by taking advantage of the virtual absence of air pollutants from nuclear plants; and
- A way to prepare the transition towards a hydrogen economy.

Energy Security

Dependence on energy imports carries a large risk of disrupted power supplies. Whether such disruption is caused by political events such as the oil embargo of 1973, physical events such as severe weather phenomena, or commercial events such as price disputes, the importing country will have to rely on its fuel reserves to avoid large negative economic impacts.[2] For this reason the International Energy Agency (IEA), the energy arm of the Organization for Economic Co-operation and Development (OECD), has developed plans for coping with such disruptions and requires its members to maintain minimum fuel reserves. In June 2009, the agency estimated that its 30 members had an average of 63 days of oil stocks—the four-to-five-year average cover being 57 days (Reuters, 2009).

For many countries, a large percentage of the fuel needed for their economies may be at sea (or in pipelines traversing politically unstable regions) at any given time, with all the vulnerabilities that this entails. Nuclear fuel may also have to be imported and transported. However, because of the high energy density of nuclear fuel, it is possible for countries to stockpile sufficient imported uranium to operate their nuclear supply systems for many years on the once-through fuel cycle and thus weather any realistic supply interruption. This is a major reason why France and Japan, for example, have tenaciously pursued nuclear power. Other energy resources, such as coal, could also be stockpiled, but uranium has significant advantages: the cost is low (about one-tenth that of coal for equivalent energy); storage is easy (more than four orders of magnitude less mass than the mass of coal for equivalent energy); and uranium, unlike coal, will not degrade in storage (Lidsky and Miller, 1998).

Avoidance of Fossil Fuel Price Volatility

The costs of electricity generation plants consist of three major components: capital or construction costs (those incurred during the planning, preparation and construction of a new power station); operations and maintenance (relating to the management and upkeep of a power station—labor, insurance, security, spares, planned maintenance, and corporate overhead costs); and fuel costs (reflecting the cost of fuel for the power station). Nuclear power also includes a fourth major component: back-end costs—(those related to the decommissioning of the plant at the end of its operating life and the long-term management and disposal of radioactive waste). While annual capital charges are fixed (assuming a fixed interest rate), and operation and maintenance costs should vary little unless major improvements are needed, fuel costs can create major electricity cost volatility. For nuclear power, construction accounts for most of the costs, whereas for gas-fired generation fuel is the largest component.[3] Because of the small weight of fuel cost in the overall cost of nuclear generation, nuclear plants are much more immune to fuel cost volatility relative to gas-fired stations. A doubling in the price of uranium would cause only a 5–6 percent increase in the total cost of generation; while a similar increase in the price of natural gas would lead to a 65 percent increase in gas-fired costs.[4]

Global Climate Change

In 1990 a major environmental concern emerged—the potential for climate change due to rising greenhouse gas (GHG) emissions that trap heat from the sun. That same year, the United Nations Framework Convention for Climate Change was convened and signed in Rio de Janeiro. To implement the convention, the Kyoto Protocol was then negotiated, signed, and ratified by many countries. Although the United States—until recently, the world's largest emitter of greenhouse gases (superceded by China in 2008)—withdrew its signature in 2001, the protocol was eventually ratified by the required number of countries and went into effect in 2005.

The estimated changes in the global climate have led to dramatic predictions of impacts such as accelerated melting of polar ice caps, rising sea levels, reduced availability of freshwater, redistributed agricultural patterns, more extreme weather conditions, and more rapid spread of disease and loss of biodiversity. Stern (2007) estimates that the economic impacts of global warming could reduce global GDP by as much as 25 percent, while greenhouse gas mitigation would cost about 1 percent of global GDP. Obviously, such predictions involve considerable uncertainty.

Like renewable energy sources (hydro, wind, solar, biomass, and geothermal), nuclear power is a low-GHG emitting technology. Indeed, GHG emissions from nuclear and renewable technologies are between one and two orders of magnitude below corresponding emissions from fossil fuel energy chains. Lignite exhibits the worst performance, closely followed by coal and natural gas. Among the renewable technologies, hydro and wind perform better than solar PV and wood cogeneration. Nuclear power compares favorably to most renewables with only hydro having lower life cycle GHG emissions. The operation of nuclear plants worldwide make a significant contribution to the mitigation of GHG emissions—nuclear power plants currently save some 10 percent of total CO_2 emissions from world energy use. This represents an immense saving of GHG emissions that would otherwise be contributing to global warming. If the world were not using nuclear power plants, emissions of CO_2 would be some 2.5 billion tonnes higher per year. According to the International Atomic Energy Agency (IAEA), the complete nuclear power chain, from uranium mining to waste disposal, and including reactor and facility construction, emits 30g of carbon dioxide per kilowatt-hour—compared with over 950g per kilowatt-hour emitted, on average, by coal-burning plants and just under 450g per kilowatt-hour by gas-fired plants. Direct emissions from nuclear plants are about the same as those from wind and solar energy plants, while indirect emissions from nuclear plants are estimated to be lower.

Reducing Air Pollution

Nuclear power has significant environmental benefits compared to fossil fuel generation. Under normal operations, nuclear power plants produce almost no airborne pollutants. Small quantities of radioactive gases are regularly emitted under controlled conditions imposed and supervised by regulatory authorities and pose no significant threat to plant workers or surrounding populations. By contrast, emissions from fossil fuel plants pose significant threats to human health and the environment. The main emissions from fossil fuel plants are particulate matter, sulfur dioxide, nitrogen oxides, and a variety of heavy metals—mercury being the most prominent. Thus, nuclear power almost entirely avoids the environmental effects of fossil fuel pollutants.

Changing Public Attitudes

Public opposition to nuclear power started in earnest in the early 1970s. The 1979 accident at Three Mile Island gave strong impetus to the anti-nuclear movement; after that accident, there were no new orders for nuclear power reactors in the United States, and many orders already placed were cancelled.

Similar responses occurred elsewhere in the world, and intensified after the Chernobyl accident in 1986. In the early 1980s, Sweden passed a law requiring it to retire all its nuclear units by 2010.[5]

Italy retired all of its nuclear power plants, the last one in 1990. Spain, where a substantial portion of power is nuclear (about 20 percent), passed a moratorium in 1983 on further construction of nuclear power units—though the issue is now back on the table.[6] In Germany, a country with strong technological strengths and a substantial nuclear power development program, nuclear power has become highly politicized and opposition to it is one of the main planks of the Green Party platform. The Greens, a coalition partner of the government for several years, made their participation in government conditional on retiring all nuclear power plants by 2021. Such plants currently supply about 31 percent of Germany's electricity. The issue may be brought up for fresh debate by the current government. Although opposition is still strong and expected to continue, some environmental advocates have raised their voices in favor of nuclear power to reduce greenhouse gas emissions and avert global warming (Lovelock, 2006).[7]

Public support is considered essential for a new campaign to build nuclear power plants. As the memories of the Three Mile and Chernobyl accidents fade and the security of energy supplies and need to cut greenhouse gas emissions come to the fore of public concerns and debates, attitudes toward nuclear power have gradually changed. Figure 1 shows the diverging curves of "favoring" (growing) and "opposing" (declining) opinion toward nuclear power between 1983 and 2008 in the United States. While the percentage of those favoring nuclear power declined from just under 70 percent in March 2006 to 63 percent in April 2008, a survey conducted in September 2008 indicated that a record-high 74 percent of Americans favor nuclear energy, with only 24 percent opposed. Those who strongly favor nuclear power outnumbered those strongly opposed by almost four to one. The favorability mark in September was 11 percentage points higher, and the unfavorability level 9 percentage points lower than was the case just five months before (NEI, 2008).

Recent international polls also show strong support for nuclear power. A survey of more than 10,000 people in 20 countries found that more than two-thirds of the respondents believed that their countries should begin using or increase the use of nuclear energy (NEI, 2009). Support for energy production by nuclear power stations has also grown in the European Union. Since 2005, the percentage of Europeans favoring nuclear power has increased from 37 to 44 percent while the share of those opposed to it declined from 55 to 45 percent (EU, 2008). However, these aggregate values mask significant differences in attitudes towards nuclear power across countries. For example, less than 40 percent of the Spaniards polled support nuclear power, over 48 percent still oppose it, and 75 percent would not consent to a nuclear power station being built in their municipality. Thus, it would be premature to assume that opposition to nuclear power has disappeared in Europe and it should be noted that the decision to build new nuclear plants is a country-level decision. Although there is a general public perception that energy policy must address the three objectives of secure supply, economic efficiency, and environmental protection

Figure 1

Public Attitudes Toward Nuclear Power in the United States, 1983–2008

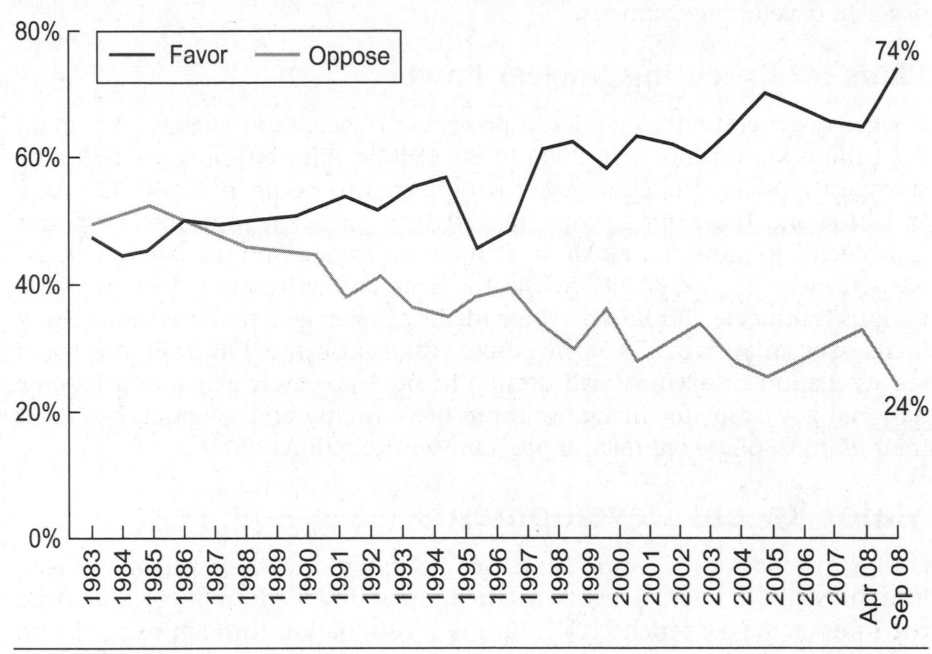

Source: NEI (2008b).

(mostly against pollution from fossil fuels), the public is often reluctant to accept the necessary trade-offs.

Current Status and Expansion Plans for Nuclear Power

Nuclear power's share of worldwide electricity production rose from less than 1 percent in 1960 to 16 percent in 1986, and that percentage has held essentially constant for almost two decades from 1986 to 2005.[8] In 2006, nuclear power's share dropped to 15 percent and in 2007 it dropped another percentage point to 14 percent (IAEA, 2008a). In June 2009, the nuclear power industry comprised (WNA, 2009a):

- 436 plants in operation in 30 countries with a total net installed capacity of 372 GW of electrical output.[9]
- 45 power reactors were under construction in 14 countries which will increase current capacity by 10.7 percent; and 131 were on order or planned, equivalent to 38 percent of present capacity. . . .

Nuclear electricity generation is concentrated in developed countries. More than half of the world's reactors are in North America and Western

Europe, and less than 10 percent in developing countries. For example, the share of nuclear in total power is 3 percent in Brazil, 2 percent in China, and 2 percent in India. But this century's greatest growth in energy demand will occur in developing countries.

Plans for Executing Nuclear Power

Electricity generation from nuclear power is projected to increase from about 2.7 trillion kilowatthours in 2006 to 3.8 trillion kilowatthours in 2030. The strongest growth in nuclear power is projected to occur in non-OECD Asia. In China and India, for example, electricity generation from nuclear power is projected to grow, from 2006 to 2030, at an average annual rate of 8.9 and 9.9 percent respectively. Outside Asia, the largest growth among the non-OECD nations is projected for Russia, where nuclear power generation is projected to increase by an average of 3.5 percent per year. In contrast, OECD Europe could see stagnation or even a small decline in nuclear power generation if some national governments, including those of Germany and Belgium, carry out their plans to phase out nuclear programs entirely (EIA, 2009).

Major Recent Developments

During the past two decades, nuclear plants around the world have realized substantial improvements in their operating performance. Moreover, the industry has experienced significant consolidation through mergers and acquisitions. And new streamlined licensing processes promise to reduce regulatory uncertainty and make it easier to build nuclear plants.

Improvements in Nuclear Plant Operating Performance

During the period of lull in new plant construction, many owners of nuclear plants have focused instead on improving their safety, technical, and financial performance. The 1986 accident at Chernobyl catalyzed the creation of the World Association of Nuclear Operators and radically changed the International Atomic Energy Agency's (IAEA) approach to plant safety. Both organizations helped create networks for conducting peer reviews, continuous and results-oriented reviews of safety practices, and exchanges of vital operating information to improve safety. . . .

Plant Ownership and Operations Consolidation

In the early days of nuclear power, the rush towards this form of electricity generation spurred many US utilities, large and small, to adopt the technology. Many utilities owned and operated one or two units, stretching their capabilities and leading to difficulties in maintaining qualified staff; regulatory oversight was also stretched. With the minor and major accidents that occurred and the ensuing public reaction, the owners got a hard-learned lesson—namely, that running a nuclear power plant safely and economically requires highly skilled staff, a variety of resources, and management dedication to maintaining a high level of capability.

As a result, the industry has seen massive consolidation in recent years through mergers and acquisitions. Today, only a handful of companies own and operate nuclear plants in the United States.[10] These companies have substantially strengthened their staff's ability to perform required duties, introduced needed improvements, responded to regulatory requirements, and, in general, maximized the performance of their plants.[11] Consider Exelon, the leader in such consolidation and the largest US nuclear utility. It owns and operates 17 reactors at 10 nuclear power plants in three states—about 20 percent of the US nuclear industry's power capacity. These plants produced a record of 132.3 million net megawatt-hours of electricity in 2007. During the same year, the fleet achieved an average capacity factor of 94.5 percent, the seventh year in a row the capacity factor was more than 92 percent.

Regulatory Process Improvement in the United States

A number of actions by the US Nuclear Regulatory Commission (USNRC) have been taken, resulting in significant process improvements. The Commission: (i) has streamlined the review and licensing processes—licensing has been simplified, standardized, and made more predictable, to the point that it should no longer be considered an impediment to a revival of nuclear power[12]; (ii) has introduced a review of standard designs rather than reviewing the design of each license submission individually and separately; (iii) reviews prospective nuclear plant sites and issues an "early site permit" even before construction plans have been made, removing another large uncertainty factor; and (iv) accepts applications for combined construction and operating licenses, where the proposed plant design and site are submitted together.[13]

Extending the Licenses—and Lives—of Nuclear Power Plants

License renewal is another important regulatory development in the United States. This renewal and the life extension it provides to nuclear plants have become more significant in the past 20 years, adding considerable periods of time for utilities to earn revenues on original investments and so markedly improving the financial performance of nuclear power plants. Originally, based on the Atomic Energy Act, the USNRC issued 40-year licenses for commercial power reactors but allowed the licenses to be renewed for up to 20 more years. The 40-year term was based on economic and antitrust considerations rather than on technical limitations.[14] Of 104 US nuclear reactors originally licensed to operate for 40 years, about 52 have already received a license renewal from the USNRC (Nuclear News, 2009). Potential environmental impacts of extending plant operations for 20 years are examined, as are plans for managing these impacts. In the process of license renewal, as with original licensing, public participation is a priority—with several opportunities for the public and stakeholders to obtain relevant information and provide inputs.

Extent, Availability, and Cost of Nuclear Fuel Resources

To assess the sustainability of nuclear power, required and available fuel resources must be examined relative to the demand for them. Uranium ore is processed near mines, producing a uranium oxide (U_3O_8) known as yellowcake. By weight, yellowcake is about 85 percent uranium metal. According to the 2007 "Red Book", the total identified amount of conventional uranium stock that can be mined at less than $80 per kilogram is approximately 4.5 million metric tonnes.[15] At the current rate of use of just over 65,000 metric tonnes a year, these resources would last about 70 years. If one considers a higher recovery cost, say $130 per kilogram, the resources grow to 5.5 million metric tonnes, not a dramatic increase, but affording a longer period (85 years) of lifetime fueling of nuclear plants (OECD/NEA-IAEA, 2008).

In recent years, growing demand and higher prices have spurred increased investment in uranium exploration and development activities—there has been an eight-fold rise in world-wide annual expenditures in these activities since 2001. Increased investment in exploration should ultimately lead to further additions to the uranium resource base. . . .

Approaches to Enriching Nuclear Fuel

The enrichment technology is highly contentious because of its potential use for the production of highly enriched uranium which can be used in making explosive devices. For most commercial nuclear power plants—i.e., light-water reactors (pressurized or boiling water)—nuclear fuel has to be enriched by 3–5 percent, depending on reactor design. Natural uranium, produced in the form of yellowcake (U_3O_8) at the mine, is first converted into uranium hexafluoride (UF_6), then enriched to the desired level in an enrichment plant. The capacity of enrichment plants is measured in terms of separative work units—a function of the amount of uranium processed and the degree to which it is enriched (i.e., the increase in the concentration of the uranium-235 isotope relative to the remainder) and the level of depletion of the remainder.[16] . . .

New Nuclear Reactor Technologies

Because of its perceived advantages, nuclear power has attracted renewed interest from policymakers, energy planners, utilities, and investors. Reflecting this interest, improved designs for nuclear reactors have emerged which are addressing many of the public health and safety risks that plagued the industry since 1979. These improvements should make nuclear reactors less vulnerable to accidents—whether due to equipment malfunctions or human error—and they include:

- Installing instruments to facilitate detection of coolant leaks before they lead to catastrophic breaks.
- Improving relief valves to make them much less likely to remain open after letting off excessive steam pressure.

- Providing excess water for steam generators to prevent them from drying out quickly after a leak of the secondary fluid.
- Introducing measures to control the potential production of hydrogen in the containment vessel after core meltdown, to prevent hydrogen explosions like the one at Three Mile Island.
- Tightening operating rules to reduce the likelihood of human error. . . .

Generation III and III+ Reactor Designs

The US Department of Energy, working with the nuclear industry, has launched a program called NuStart to stimulate and support the development of evolutionary and revolutionary reactor designs. NuStart is evaluating Generation III and III+ reactor designs, and covers both pressurized water reactors and boiling water reactors.[17] . . .

Generation IV Reactor Designs

Looking towards the more remote future, beyond 2030, governments have been analyzing innovative reactor concepts to ease the transition from fossil fuels. Complementing national initiatives are two major international projects to promote innovation: the Generation IV International Forum and the International Atomic Energy Agency's International Project on Innovative Nuclear Reactors and Fuel Cycles (USDOE, 2002; IEA/NEA/IAEA, 2002).[18] . . .

Lingering Concerns

The perils of nuclear power revolve around several basic concerns, including potential problems in the nuclear fuel cycle, fuel irradiation in the reactor core, and the generation of nuclear waste. Accordingly, the main issues that brought nuclear power development to a standstill in the United States and Western Europe and still impede nuclear expansion are concerns stemming from: (i) the possibility of a nuclear accident, which, although an event of low probability, could have large consequences; (ii) lack of comprehensive, publicly accepted plans for permanent and safe disposal of nuclear waste, especially in developing countries; and (iii) anxiety about civilian nuclear power leading or facilitating nuclear weapons proliferation.

Safety

A unique hazard of nuclear plants is the potential for exposing the public to radiation. All aspects of the nuclear cycle present safety issues, but the most critical ones occur in the operation of power plants. A major health hazard would result, for example, if a significant portion of the radioactive inventory in the core of a nuclear reactor were released to the atmosphere. Such releases are obviously unacceptable, and steps are taken to ensure that they never occur. Various techniques are used, including conservative designs, safety equipment, and physical barriers to radiation releases if all other measures fail. Physical barriers include the containment structure, a large steel and concrete structure enveloping the nuclear reactor and many of its systems.[19] . . .

Major Reactor Accidents

Two accidents have indelibly marked the history of nuclear power, leaving impressions in the public mind that, many years later, still color public reactions to this form of energy.

The Three Mile Island accident (1979): a combination of malfunctioning valves in the demineralizer and blocked valves in a backup safety system stopped the flow of feedwater to the steam generator. The turbine tripped automatically and the reactor scrammed shortly thereafter. Through a sequence of events, the core was severely damaged. The seriousness of the accident was not recognized until late the next day, when a pressure spike on the monitor printout from the previous day showed that hydrogen had burned inside the containment building. Severe overheating of the core had caused the cladding to react with steam and produce large amounts of hydrogen—some of which escaped to the containment building through the open relief valve.

The accident resulted in the release of a small part (one in 10 million) of the radioactive iodine in the core as well as larger quantities of the noble gases krypton and xenon (2 percent of core inventory). The release of the noble gases was quickly dissipated. Radiation levels outside the reactor site were quite low, mostly below 1 millirem per hour. There was no failure of the containment building during the accident, so there were no direct releases of radioactivity from the building. The releases that did occur were secondary leaks from auxiliary systems. Except for the xenon and krypton, most of the radioactivity was retained by the water in the reactor's containment building. A large government-sponsored study, lasting more than a year, concluded that no fatalities could be attributed to the accident and health effects from radiation exposure were minimal. The accident's most severe impacts were on mental health (from the resulting anxiety) and on property values near the plant. But the financial impacts on the plant were large. In addition to the complete loss of the unit's investment, the costs of handling contaminated water, dismantling the damaged core, and transporting and disposing of the debris at the Idaho National Laboratory site were in billions of dollars.

The Chernobyl accident (Ukraine, 1986): the nuclear reactor accident at Chernobyl (in Ukraine) occurred in 1986 in unit 4 of the power station. As operators were experimenting with unit 4 behavior outside prescribed limits, the reactor experienced an uncontrolled excursion of power—leading to fuel overheating and an explosion of the large reactor core. The high temperatures caused the graphite to interact with the water coolant, producing large amounts of hydrogen that burned, and carbon dioxide that was released and traveled long distances. Immediately after the accident, authorities sent in crews to quench the fire, exposing the workers to lethal doses of radiation and causing acute radiation sickness and death in a matter of days, affecting 47 people. In addition, large amounts of radioactive materials were released to the environment because the reactor lacked a containment structure. The extensive radioactive releases mainly affected Ukraine, Belarus, and Russia (all of which were members of the Soviet Union at the time), but the radioactive clouds traveled as far north as Sweden, as far west as England, and as far south as Greece and Italy, causing severe consternation among the affected populations. Multiple studies of the accident have concluded that it was the result of faulty design (of control rods and coolant channels), lack of a containment structure (stemming from the Soviet safety philosophy that severe accidents "cannot happen"), and lack of a disciplined safety culture, allowing operators to violate established norms in the operation of the reactor.

Although more than 20 years have passed since the accident, there remain many conflicting reports and rumors about its consequences. For that reason the Chernobyl Forum was created in 2003 by the International Atomic Energy Agency in cooperation with several other organizations, including the United Nations Food and Agriculture Organization, United Nations Development Programme, United Nations Environment Programme, United Nations Office for the Coordination of Humanitarian Affairs, United Nations Scientific Committee on the Effects of Atomic Radiation, World Health Organization, and World Bank, as well as authorities from Belarus, Russia, and Ukraine. The first public announcement of the forum's findings was made in late 2005, and its report was officially published in April 2006 (IAEA, 2006). The Chernobyl Forum's findings, arrived at by more than 100 experts from 12 countries, were that more than 600,000 people suffered high levels of radiation exposure, including reactor staff, emergency and recovery personnel, and local residents.[20] The forum's report predicted that the long-term death toll could rise to 4000. This includes 47 emergency workers who died of acute radiation syndrome in 1986 and from other causes in later years, 9 children who died from thyroid cancer, and 3940 people who could die from cancer as a result of radiation exposure. The report found no convincing evidence that there has been a rise in other cancers due to the accident and that "confusion over the incident's impact has arisen because many emergency and recovery workers have died since 1986 from natural causes which cannot be attributed to radiation exposure." Furthermore, the report stated that "widespread expectations of ill health and a tendency to attribute all health problems to exposure to radiation have led local residents to assume that Chernobyl-related fatalities were much higher."

Although the Chernobyl accident was large-scale and severe, it was not out of line relative to other serious industrial accidents—such as the one in Bhopal, India, in 1984, when a large leak of methyl isocyanide occurred.[21] The Chernobyl Forum's report was the product of consensus among many specialists and illustrated the wide chasm between objective assessments of effects and public perceptions. Still, political decision makers cannot afford to ignore public perceptions. Thus, the Chernobyl accident had a dramatic, decisive effect on the fate of nuclear power worldwide.

Safety Risks and Perceptions

Although objective estimates of safety levels can be made using physical laws and probability theory, setting safety criteria is often difficult and subjective. Such standards must take into account the public's view of how safe is safe enough. In this regard, Murphy's Law is often quoted: "whatever can go wrong, will go wrong." As a result, absolute guarantees of safety are sought. But a policy based on that premise overlooks the fact that every human activity entails some risk. . . .

An important factor in setting acceptable risk levels is the need to consider both the costs and the benefits of a technology relative to those of alternatives. Thus, in the case at hand, two questions are pertinent. Is nuclear energy safer than alternative sources of energy, such as coal?[22] And does a nuclear plant's benefit to society exceed its risk? Decisions on risk are made through a process of social debate in which the voices of all stakeholders must be taken into account (McGill et al., 2006).

Disposal of Nuclear Waste

Nuclear waste disposal has been one of the more recalcitrant problems facing the nuclear industry—a decisive impediment to its expansion. It could be a significant consideration in decisions to expand nuclear plants. The most visible case is the United States, because:

- It has accumulated large amounts of waste from its nuclear weapons program that need to be handled safely over the long term, regardless of waste from power production.
- It has developed the world's largest nuclear energy production system, with more than 100 reactors operating. Accordingly, it has the greatest current and projected accumulation of spent fuel.
- It is a highly litigious society, creating many opportunities for intervention and making the nuclear waste problem a societal issue.

The nuclear fuel cycle produces a variety of radioactive waste, including low- and intermediate-level waste, transuranic waste,[23] and spent fuel and high-level waste. Spent fuel and high-level waste create by far the most serious problems and so dominate the debate.

It is important to keep the issue of radioactive waste management in perspective. Although such waste is dangerous, its volume—about 12,000 metric tonnes a year from the world's nuclear power plants—is small relative to waste produced by fossil fuel plants. These latter release, every year, enormous amounts of ash, diverse air pollutants, and a large portion of a total of about 8.5 billion metric tonnes of carbon directly into the atmosphere.[24] Nuclear waste can be put in glass or ceramic containers, further encased in corrosion-resistant containers, and isolated geologically. Moreover, research is under way that would use accelerator-driven systems to reduce the volume and radioactive toxicity of nuclear waste. Because of the potential future uses of useful materials in spent nuclear fuel, disposal programs strive to ensure the retrievability of such waste, and research and development programs are aimed at achieving this goal. . . .

The Risk of Proliferation of Nuclear Weapons

. . . Worry about proliferation is hardly new. In 1963 US President John F. Kennedy said, "I am haunted by the feeling that by 1970, unless we are successful, there may be 10 nuclear powers instead of 4, and by 1975, 15 or 20." That timetable proved inaccurate. But in recent years there has been a sense around the globe that President Kennedy's prediction will soon come true. Nuclear proliferation is a complex and controversial issue, with many views expressed on its various aspects. . . .

Are the horses out of the barn? Questions about the proliferation of dangerous nuclear technologies are not subject to simple answers, because it is not clear how much sensitive information has been released, or when and to whom. Thus, risks need to be identified, assessed, and tackled accordingly. Risks posed by potential terrorist groups differ from those of troublesome countries (though a nexus may exist between the two). But given the various avenues that can lead to nuclear weapon proliferation, nuclear power technology must strive to raise the barriers to proliferation to at least the same level as that of an independent, secret route, which seems to be preferred by such countries. Technology needs to address all phases of the nuclear fuel cycle—with an emphasis on enrichment, which poses a threat entirely independent from civilian applications.

On the other hand, risks must be kept in perspective. Existing barriers are considerable. Producing weapons requires specialized skills and resources largely different from those needed to operate plants. In the broader context of technological and institutional arrangements, a complex system of additional barriers to acquiring weapons capability should also be considered. These include technical difficulties in acquiring, handling, and processing the needed materials, the wide variety of skills and resources required to design explosive nuclear devices and accessories (including conventional explosives and electronic timing devices), the alternate means available to achieving the intended goal, the potential advantages of the proposed alternatives, and the disincentives promulgated by a network of international institutions.

As of June 2009, fewer than 10 countries have developed nuclear weapons. But the number of aspiring countries may be growing—mainly due to regional competition that, as many fear, could lead to regional weapon races. Still, the connection between nuclear power and nuclear weapons is tenuous at best. Most countries that have developed nuclear weapons obtained the needed material from dedicated military facilities where the production could be optimized and kept secret. India, however, detonated an underground explosive device using plutonium from a small spent fuel reprocessing plant that got its fuel from an experimental heavy-water reactor. A similar case is that of North Korea.

Institutional arrangements for current and future nuclear power facilities can greatly ease suspicions and fears about the development of nuclear weapons, and eliminate incentives in this direction. Such measures must be supplemented by assurances of fuel supply and transfer of technology to meet the legitimate needs of non-nuclear weapon countries. The availability of enrichment and reprocessing services could go a long way toward eliminating the need for aspiring countries to develop independent national programs for such sensitive facilities, given also the uncompetitive nature of facilities for small numbers of reactors relative to the large commercial enterprises that exploit economies of scale.

Technology can provide considerable assistance, although its capacity for solutions is limited. The proliferation problem is fundamentally political and so requires political solutions. In a world where nuclear power is already a reality and many countries view it as at least a partial solution to their energy and environment problems—not least of which is a sizeable contribution to mitigating greenhouse gas emissions—a combination of technical measures and legal and institutional instruments seems the only approach with any chance of success.

Summary and Conclusion

This paper is motivated by the revived worldwide interest in nuclear power—a result of rapidly rising and volatile fossil fuel prices and concerns about the security of energy supplies and global climate change. Because nuclear power does not generate greenhouse gas emissions, it will likely play a growing role in supplying global demands for electricity. But nuclear power is perhaps the most contentious of all means of electricity generation, arousing strong passions for and against.

Although nuclear power is a well-established technology for generating electricity, it has long been considered unattractive by many environmental groups and ordinary citizens. These unfavorable attitudes emanate from concerns about the potential hazards of reactor meltdowns (and their catastrophic ecological and social impacts), unresolved issues related to nuclear waste disposal, and potential problems with diversion and proliferation of fissile material. Indeed, the word "nuclear" strikes fear in the hearts of many people.

Public opposition to nuclear power facilities and their association with nuclear weapons are not entirely justified. Still, for nuclear power to gain greater public acceptance, become a significant option for mitigating greenhouse gas emissions, and meet growing needs for electricity supply, four critical problems must be overcome: safety, waste, proliferation, and costs.

To assess the potential of nuclear power, its role as a relatively secure, largely carbon-free alternative to fossil fuels must be weighed against its technical risks. This paper describes the current status and future plans for expansion of nuclear power, the advances in nuclear reactor technology, and their impacts on the associated risks and performance of nuclear power. Advanced nuclear reactors have been designed to be simpler and safer, and have lower cost than currently operating reactors. By addressing many of the public health and safety risks that plagued the industry since 1979, these reactors offer some promise in breaking the political deadlock over nuclear power. Moreover, the lowering of capital costs via simplification and modification of plant design could substantially enhance the competitive position of nuclear power—provided these lower costs are proven by the experience from the construction of the first units to be completed under the "nuclear renaissance." Adding to these factors favoring nuclear power is the significant contribution it could make to stabilizing greenhouse gas emissions and mitigating climate change.

Notes

1. Lead Economist, Development Research Group, The World Bank. The findings, interpretations, and conclusions are the authors' own and should not be attributed to the World Bank, its Executive Board of Directors, or any of its member countries.
2. The publicly articulated rationale for the substantial nuclear power build announced by Brazil was that the political turmoil going on in Bolivia had necessitated measures to end Brazil's dependence on that country for the supply of natural gas <www.india-server.com/news/brazil-to-build-60-nuclear-powerplants-3712.html>.
3. The total fuel costs of a nuclear power plant are typically about a third of those for a coal-fired plant and between a quarter and a fifth of those for a gas combined-cycle plant.
4. We refer here to the cost of U3O8 produced at the uranium mine from ore, which is commonly called "yellowcake." The total cost of nuclear fuel is about 2.5 times the cost of uranium yellowcake.
5. However, as it was noted above, in February 2009 the Swedish government announced plans to reverse its long-standing policy of nuclear phase-out.
6. In October 2008, former socialist Prime Minister Felipe Gonzalez urged Spain's socialist government to reconsider its moratorium on building more nuclear plants (Guardian, 9 October 2008). Gonzalez was a staunch defender of Spain's moratorium on new nuclear capacity when he was in power from 1982 to 1996.
7. Patrick Moore, a co-founder of Greenpeace, is engaged in grassroots advocacy for nuclear power with the Clean and Safe Energy Coalition. His basic message is that global energy demands keep growing and that we need to diversify our energy sources to meet these needs—nuclear energy should be an important part of this diversification plan, especially since its production generates no air pollutants or greenhouse gases. More recently, four prominent environmentalists in the United Kingdom (Stephen Tindale, head of Greenpeace UK; Chris Goodall, Green Party activist; Lord Chris Smith, chairman of the Environment Agency; and the author Mark Lynas) have reversed their long-time opposition to nuclear energy (Nuclear Energy Insight, May 2009).
8. Total world net electricity production rose from 9.68 trillion kilowatthours in 1986 to 17.36 trillion kilowatthours in 2005 <http://www.eia.doe.gov/emeu/international/RecentTotalElectricGeneration.xls>. Thus, while during that period nuclear power's market share remained constant, the amount of electricity generated by nuclear plants increased substantially.
9. All measures of nuclear energy in this report—such as megawatts, gigawatts, and terawatts—are in terms of electrical output (as opposed to thermal output).
10. The acute fragmentation of the nuclear power industry seen earlier in the US was not observed nearly as much overseas.
11. There are about half a dozen nuclear companies in the United States with six or more nuclear plants: Exelon (14), Entergy (10), Dominion Generation Company (7), Duke Power (7), Nuclear Management

Company (7), Southern Nuclear Operating Company (6), and Tennessee Valley Authority, (6). Additional companies have five or fewer plants.

12. Title 10 of the Code of Federal Regulations, Part 52, contains requirements for early site permits, standard design certifications, and combined construction and operating licenses for nuclear plants.
13. The process for obtaining a combined construction and operating license takes about 30 months if a preexisting design certification is involved—not including the 12 months for the hearing process.
14. Nuclear Regulatory Commission Website: www.nrc.gov/reading-rm/doc-collections/fact-sheets/licenserenewal-bg.html.
15. Identified resources refer to uranium in known mineral deposits (reasonably assures resources) or in extensions of well-explored deposits with established geological continuity (inferred resources). At a cost of $80/kg yellowcake for recovery, the countries with the most uranium resources are Australia (with 27 percent of the global total), Kazakhstan (17 percent), Canada (10 percent), and South Africa (8 percent)—(NEA, 2008; IAEA, 2008a, b).
16. The unit should be strictly called "kilogram separative work unit," and it measures the quantity of separative work performed to enrich a given amount of uranium to a certain level. It is thus indicative of energy used in enrichment when feed and product quantities are expressed in kilograms. The unit "metric tonne separative work unit" is also used. For instance, to produce one kilogram of uranium enriched to 3 percent uranium-235 requires 3.8 separative work units if the plant is operated at a tails assay 0.25 percent, or 5.0 separative work units if the tails assay is 0.15 percent (thereby requiring only 5.1 kg instead of 6.0 kg of natural uranium feed).
17. Worldwide, the ratio of pressurized water reactors to boiling water reactors is about three to one.
18. A third effort, the Michelangelo Initiative Network, is part of the Framework Programme of the European Commission and shares the same goal—though it focuses on coordinating and supporting the research and development efforts of EU members rather than bringing integrated reactor concepts to the market.
19. Such a containment structure is not part of some early nuclear reactors of Soviet design, such as the VVER-440, and RBMK (used at Chernobyl), leading the international community to call for their early closure. All RBMK reactors at Chernobyl are now closed. One is still in operation at Ignalina, Lithuania, and 11 units are in operation in Russia.
20. The following figures from the IAEA (2006) Chernobyl Forum Report are given to place the exposure in perspective. Normal exposure to ionizing radiation averages 2.4 millisieverts (mSv) a year. Typical lifetime exposure is 100–700 mSv. Average Chernobyl-related doses between 1986 and 2005 have been 10–20 mSv a year, though some people have been exposed to hundreds of mSv annually. People living in areas of Brazil, China, India, and Iran with high levels of radioactive materials in soils are exposed to about 25 mSv a year with no apparent ill health effects.
21. In a 2004 report by Amnesty International (on the 20th anniversary of the accident), short-term deaths were estimated at 7000 and long-term

deaths at 15,000. Injuries were estimated in the many tens of thousands (Amnesty International, 2005).

22. Among other impacts, about 7500 deaths related to coal mining occur around the world every year—80 percent of them in China. In the United States over the past two decades, 30–50 fatal accidents have occurred every year, though the number has been falling—until 1985, the average was about 150 deaths a year. Each year about $600 million is paid in compensation to US miners suffering from pneumonoconiasis (black lung disease).
23. These consist of the artificial elements that are produced in the irradiation of uranium and have atomic numbers higher that that of uranium (92); hence their name as "transuranic." They include plutonium isotopes as well as americium, curium, einsteinium, and others.
24. The latest figures were published on 26 September 2008 by the Australia-based Global Carbon Project. It reveals that emissions are outpacing worst-case scenarios forecast by the Intergovernmental Panel on Climate Change.

All references for articles included in *Taking Sides: Clashing Views in Sustainability* can be found on the Web at www.mhhe.com/cls.

Milton H. Saier and
Jack T. Trevors

Is Nuclear Energy the Solution?

On October 29th, 2009, one of us, (MHS) attended a lecture on "Nuclear Responsibility" on the University of California, San Diego campus. The speaker was Rochelle Becker, Executive Director of the Alliance for Nuclear Responsibility. The information presented was both revealing and upsetting and is documented extensively in a pamphlet entitled: "Why a Future for the Nuclear Industry Is Risky." These sources emphasized the problems that exist, especially in the USA. This editorial is based on these sources.

We should all know that: first, investments in nuclear power are risky as indicated by the fact that Wall Street has chosen to stay clear; second, nuclear power plants are stated terrorist targets and carry serious risks of their own; third, nuclear power will not reduce our dependencies on foreign energy as is sometimes claimed; fourth, nuclear-generated electricity does not compare favorably with electricity derived from either the combustion of fossil fuels or renewable sources such as wind, solar, geothermal, wave, and tide, and finally, there is currently no good means of nuclear waste disposal, hence more environmental pollution.

Is Nuclear Power a Good Investment?

Promises of improved safety and performance are coupled with billions of dollars of subsidies. Nevertheless, claims that nuclear power is a necessary energy source for displacing greenhouse gases has not convinced investors. Wall Street is flat out not investing in new nuclear power plants because they do not believe that they will be safe profitable investments. In fact, as things stand, new nuclear power plants will not be cost competitive with other electricity-generating alternatives. For example, wind power and other renewable technologies, combined with energy efficiency and conservation can be more cost effective and can be deployed much sooner than new nuclear power plants. Building expensive nuclear plants will divert private and public investment from the cheaper and readily available renewable and energy efficiency options needed to protect our climate and humanity.

In competitive markets, new nuclear power plants will be bad investments. By contrast, worldwide private equity and venture capital investments in clean energy continue to grow. Worldwide annual investments in renewable energy capacity are now over 50 billion US dollars, and renewable energy markets continue to grow.

Why do investors have little or no incentives to back the construction of new nuclear power plants? The answers are multifaceted and complex.

- Nuclear construction cost estimates in the USA have been far less than final costs, by roughly 3-fold.
- Standard & Poor's has stated that "given that construction would entail using new designs and technology, cost overruns are highly probable."
- The Department of Energy's (DOE's) Energy Information Administration has clearly and concisely stated that "new plants are not expected to be economical."
- A 2003 study by the Massachusetts Institute of Technology forecasted that costs would be substantially in excess of traditional means.
- Nuclear utilities have acknowledged that there are large economic risks associated with the operation of nuclear power plants.

Are Nuclear Power Plants Potential Targets of Terrorism?

In testimony before the Select Committee on Intelligence in the US Senate in February 2005, FBI director Robert S. Mueller stated that, "Another area we consider vulnerable and target rich is the energy sector, particularly nuclear power plants. Al-Qa'ida planner Khalid Sheikh Mohammed had nuclear power plants as part of his target set and we have no reason to believe that Al-Qa'ida has reconsidered."

Over 53,000 metric tons of highly radioactive spent nuclear fuel is stored at commercial reactors in the US. Nearly 90% of this fuel is stored in cooling pools without adequate protection. Does this sound like a pollution problem? According to a study by the National Academy of Sciences, a terrorist attack on a spent fuel pool could lead to the release of large quantities of radioactive materials to the environment. Such an event could result in thousands of cancer deaths and economic damages in the range of hundreds of billions of dollars.

In the event of a major radioactive release from a nuclear power plant, public opinion would likely react strongly against nuclear power (as occurred after the Chernobyl and Three Mile Island accidents), resulting in the halting of construction of any new planned reactors.

Will Nuclear Power Relieve Our Dependency on Foreign Energy?

The USA is importing more oil each year—most of it from the world's most unstable regions—increasing our country's economical and political vulnerability and making oil dependency among the largest threats to our economy and national security.

Nuclear power's only substantial contribution to oil displacement in the USA comes in regions in which natural gas, displaced by nuclear power, can penetrate further into oil's share of the markets, such as space heating in New England.

Indeed, transportation is the sector that accounts for most of USA oil consumption. About two thirds of the country's oil consumption is used by vehicles, which corresponds to roughly 13 million barrels per day. Thus, nuclear power development would not have an appreciable influence on these statistics.

To make matters worse, almost all of the uranium produced on the planet comes from foreign sources. Moreover, like oil, there is a limited supply, suggesting that if demand increases, so will the price. Money will continue to flow out of the USA without diminution.

How Does Nuclear Power Compare with Traditional and Renewable Energy Sources?

Climate change is one of the most pressing threats of our time and it is imperative that we take swift and decisive action to avert its most severe impacts. However, building more nuclear power plants does not appear to provide an answer.

The claim that "we need all energy options" to face growing energy needs is irresponsible. In fact, we cannot afford all energy options. Further investment in nuclear power is likely to squander the limited financial resources that are available to implement meaningful climate change mitigation policies. Moreover, nuclear power plants are not CO_2 free. Enriched uranium is the required fuel for all US nuclear power plants, and the uranium enrichment process emits about 15% of the CO_2 emitted from a coal-fired power plant per unit of electrical energy.

Wind power and other renewables, such as solar and bio-energy, coupled with energy efficiency, and conservation are proving to be much more cost effective. Moreover, they can be deployed much faster. Building new nuclear power plants will divert private and public investment from the cheaper, readily available options needed to protect our global climate. Each dollar invested in electric efficiency in the USA displaces nearly seven times as much carbon dioxide as a dollar invested in nuclear power, and nuclear power saves as little as half as much carbon per dollar as wind power.

Recent studies analyzing the potential of nuclear power to combat global warming have concluded that between 1,000 and 2,000 new nuclear reactors would have to be built around the globe to achieve a meaningful impact on carbon dioxide emissions. These projections point to an infeasible schedule as new reactors would have to be completed every few weeks.

How Are We to Dispose of Nuclear Waste?

One of the riskiest elements of building new nuclear plants is that the long-term disposal of radioactive waste is far from resolved. The planned Yucca Mountain repository in Nevada is almost 20 years behind schedule and may never open. The projected opening date for this permanent spent fuel repository has been delayed countless times and, according to the DOE, the current target date of 2017 is a "best-achievable schedule."

A plan proposed by the Bush administration, the Global Nuclear Energy Partnership (GNEP) that would have allowed the reprocessing of spent nuclear fuel will face technical, legal, and political challenges and cannot be counted on as a realistic solution. Reprocessing leaves large amounts of waste still needing disposal, and much of the technology is unproven or undeveloped. Indeed, similar attempts to reprocess spent fuel in the past have been unsuccessful, and the DOE does not have a lifecycle cost analysis for the program.

Reprocessing would be a dangerous shift in the US global nonproliferation policy and would increase the likelihood that a terrorist could obtain material to build a nuclear bomb. It would increase the number of nuclear waste streams to be managed and secured and is the most polluting part of the nuclear fuel cycle. It would not alleviate the problem of used (spent) fuel storage on reactor sites or the need for a permanent waste repository.

US taxpayers are still paying billions of dollars each year to clean up contamination from reprocessing programs in the 1960s and 1970s for nuclear weapons at the Hanford Site (WA) and the Savannah River Site (SC), as well as the reprocessing of naval irradiated fuel at the Idaho National Laboratory (ID) and commercial reprocessing at West Valley (NY).

Conclusions

The genesis of nuclear power was the "Atoms for Peace Program" which was intended to make the public more comfortable with the horrifying destruction of the nuclear bomb. Originally, the promise was that technology would provide energy that would be "too cheap to meter." However, in the last 50 years, nuclear energy subsidies have totaled nearly 150 billion US dollars, amounting to more taxpayer dollars for R&D than for all other energy sectors combined. In fact, nuclear power is the energy that is "too expensive to matter."

A nuclear revival is financially risky. The likelihood of large numbers of new nuclear units being built on the basis of favorable economics is unlikely. Nuclear power is not competitive today, and for nuclear power to succeed, it must achieve major cost cuts, avoid serious accidents, resolve the nuclear waste storage and disposal issues, and achieve the status of a lower carbon-emitting power source. All of these goals must be reached before nuclear power can be considered as a rational solution to our energy needs. There are risks galore, but no guarantees.

EXPLORING THE ISSUE

Can Nuclear Energy Be Green?

Critical Thinking and Reflection

1. Is it worth taking the "nuclear power" risk? Outline the various risks to society of a nuclear energy option.
2. The world will require large amounts of energy to propel economic growth and meet the demands of 7 billion people on the earth. If renewable energy cannot meet this demand, is nuclear energy a better option than fossil fuels?
3. Nuclear energy has been touted as a "green" energy source because it constitutes a major alternative to fossil fuel–based energy. Is the threat of global climate change a greater risk than nuclear energy? Does nuclear power really generate clean energy that means zero carbon dioxide emissions?
4. Is there enough supply of uranium, a finite resource, for nuclear energy to be considered as a green energy resource?
5. Will nuclear technology be able to solve the energy "urgency" we are facing today?
6. Can we change people's perception about nuclear power? Has the meltdown of nuclear reactors in Japan as a result of a massive earthquake and the resultant tsunami doomed the nuclear energy industry in the United States?

Is There Common Ground?

Nuclear power will only succeed if it can be cost-effective, if nuclear plants are safe from accidents and terrorist attacks, and if a technology is developed to treat nuclear waste, tackle storage issues, and lessen the discharge of carbon dioxide. It will be important to fulfill these criteria before considering nuclear power as an option to sustainably meet our growing energy demands. But even if we decide to build a nuclear plant today, it will take at least 25–30 years to set it up and generate energy from it. Nuclear power will definitely create job opportunities. Public perceptions can be changed by educating people about the new designs of nuclear reactors, which are safe, and new technologies, which help take care of the radioactive waste by recycling. Hydro, wind, and solar energies seem to be competitive and better options for now. Reaching a common ground on this issue will involve taking risk and it likely will be too big of a risk to take. Our decisions today are going to affect the future generation, so one must make a conscious choice.

Internet References

European Commission Green Paper (2000). *Towards a European strategy for the security of energy supply,* COM/2000/0769 final (29 November 2000).

Hodgson P. and Marignac Y., (2001). Is nuclear power viable? *The Ecologist,* vol. 31, no. 7.

Lynas, M., (2008). Why greens must learn to love nuclear power. *New Statesman* September 22, 2008.

Mian, Z., & Glaser, A., (2008). A frightening nuclear legacy. *Bulletin of the Atomic Scientist* vol. 64, no. 4, pp. 42–47.

Totty, M., (2008). The case for and against nuclear power. The Wall Street Journal, June 30, 2008.

OECD/NEA and IAEA, Uranium, 2005.

The New York Times. (2010). *Editorial—A reasonable bet on nuclear power.* February 17, 2010.

This website provides information about nuclear power and global climate change. It also has articles based on issues related to nuclear energy.

http://timeforchange.org/nuclear-energy

This website gives information about nuclear power through publications and public information service. It also has a reactor database. The World Nuclear University is a global partnership committed to enhancing international education and leadership in the peaceful applications of nuclear science and technology.

http://www.world-nuclear.org/

This website provides all the latest news about nuclear power in the world. It provides information about environment and energy, regulation and safety, nuclear policies, and so forth.

http://www.world-nuclear-news.org/

This Web page provides an overview about why nuclear power should be considered as a sustainable source of energy.

http://www.worldnuclear.org/info/inf09.html

The article on this Web page provides an opinion about why it is important for the United States to opt for nuclear power since its other clean energy resources fail to meet the needs of a growing economy.

http://online.wsj.com/article/ SB10001424052748704224004574489702243465472.html?KEYWORDS= sustainable+nuclear+energyKEYWORDS%3Dsustainable+nuclear+ energyKEYWORDS%3Dsustainable+nuclear+energy

ISSUE 18

Is Corporate Sustainability More Public Relations Than Real?

YES: Richard Dahl, from "Greenwashing: Do You Know What You're Buying?," *Environmental Health Perspectives* (June 2010)

NO: Cristiano Busco et al., from "Cleaning Up," *Strategic Finance* (July 2010)

Learning Outcomes

After reading this issue, you should be able to:

- Understand the characteristics of corporate social responsibility (CSR).
- Identify basic characteristics of corporate sustainability.
- Understand why corporations seek to develop sustainable practices and policy.
- Know the market advantages to companies that seek a sustainability path.
- Identify the characteristics of "greenwashing."

ISSUE SUMMARY

YES: Boston freelance environmental health issues writer Richard Dahl argues that there is increasing competition between companies to portray themselves as "green" and warns that if false green claims are not controlled, then people's skepticism will grow and an important tool for sustainability will be lost.

NO: Busco et al. describe how General Electric and Procter & Gamble have operationalized corporate sustainability initiatives using management control and management accounting systems.

Corporate sustainability refers to a new way of thinking about the role of business in society. Promoted by international organizations such as the Business Council for Sustainable Development, it holds that business needs to look at the triple bottom line, how corporations affect economic growth, social equity, and environmental quality. Those who do not include these objectives into their corporate missions will lose shareholder value and fail, while those who do incorporate the objectives into their business strategy will maintain competitive advantage over their rivals (Taylor, 2008). The issue of whether corporate sustainability is real, meaning that corporations actually implement a systematic program to reduce their carbon and ecological footprints, or it is mostly illusionary is open to discussion. Some argue that corporations focus on sustainability solely as a marketing concept to enhance their public image and to reduce their regulatory risks. This approach is referred to as corporate "greenwashing," or as the campaign of companies to appear to be sustainable without really making sufficient changes in their operations that would make sustainability a reality.

Sustainability is often linked to corporate social responsibility (CSR), an approach that focuses on the function of business in society. It is the belief that corporations have an ethical responsibility to consider and deal with the needs of society as a whole, and not just to perform exclusively for the interests of shareholders (or their self-interests). Organizations such as the World Bank and the World Business Council on Sustainable Development foster the goals of CSR and define it as "the commitment of business to contribute to sustainable economic development—working with employees, their families, the local community, and society at large to improve their quality of life, in ways that are both good for business and good for development" (SIDA, 2005). CSR has been expanded recently into corporate sustainability, where social, economic, and environmental concerns motivate a company's mission and its transparency with its stakeholders. Wilson (2003) views corporate sustainability as a new and constantly changing corporate management paradigm. He believes that "corporate sustainability is an alternative to the traditional growth and profit-maximization model." He notes that there are a growing number of companies pursuing a path toward sustainability and expects this practice to continue.

Yet, some companies advocate sustainability primarily as a public relations (PR) function to enhance profits while not truly implementing sustainability. Because business is increasingly competitive, some corporations see an opportunity to capture market share by representing themselves as environmentally concerned. Today, it is well known that consumer demand for greener products is rising, and companies are competing for the attention of these "green" consumers and eventually for the profits it brings. Research shows that from 2009 to 2010, the number of "green" products has increased by more than 70 percent. Also, 95 percent of these products were found to be committing at least one of the "sins of greenwashing"—a hidden trade-off, no proof, vagueness, irrelevance, lesser of two evils, and "fibbing" (Terrachoice, 2010).

A current activity of some companies is to engage in sustainability reporting (SR), a voluntary action by companies to report to the public their level of sustainability. Borkowski et al. (2010) point out that corporations use SR for both ethical and economic reasons. The authors believe that corporations implement sustainable practices for either moral reasons—to participate in socially responsible activity, or to generate better corporate reputation that helps to increase profits and escalate shareholders' wealth. Case studies of companies such as Ben & Jerry's and Johnson & Johnson show that both reasons might be at play (Freese, 2007). Companies that adopt sustainable practices can achieve constructive results, such as saving money with cost reductions and attracting more loyal consumers. Also, sustainable companies can attract extraordinary employees and clearly differentiate themselves from their competition. In another study, Dwyer (2009) investigates the importance and ways of implementing sustainable business practices. He addresses the need for companies to implement corrective systems to minimize business losses due to environmental degradation.

Boiral (2009) discusses the importance of organizational citizenship behaviors (OCB) in improving corporate sustainability and environmental management. He points out how these behaviors can be applied to the environmental practices of organizations and be helpful to encourage eco-efficiency. Companies that are willing to operationalize corporate sustainability can benefit from environmental OCB, which directly influences their environmental performance. Taylor (2008) believes that if a business case for sustainability can be made, it must be based on traditional business principles—will the business make a profit and sustain itself financially. He quotes the Stern report that economic costs related to climate change, a major sustainability challenge, could amount to 20 percent of global gross domestic product. He believes that a "business case for sustainability" can be made by understanding and quantifying the value of ecosystems and by using a longer time-frame for decision making. And lastly, in a review of the 2009 State of Corporate Citizenship Survey, Veleva (2010) shows that support for sustainability comes from companies of all sizes. He finds that even in a time of economic disturbance, companies see benefits in corporate citizenship strategies.

There are a number of studies that support the notion that corporate sustainability is more public relations than real. Mohr (2005) believes that "greenwashing" is one of the main threats to the advancement of corporate sustainability. He suggests that businesses engage in this practice because of the present "hype" for "green" as a way to differentiate products for successful sales and as a way to enhance the company's public reputation. Gillespie (2008) warns that by engaging in "greenwashing," companies risk losing public support and causing a long-term "negative impact on public engagement with wider environmental issues." He stresses that intensified misleading environmental claims can cause consumers to undermine the environmental issues in the future. And finally, Engel (2006) analyzes why some companies seek to address environmental issues in spite of higher costs, while other companies pursue a "greenwashing" approach. She argues that, in most cases, the quality of a product, that is, the environmental, social, health, or safety impacts of a product, cannot be adequately assessed by the public, but if the product's cost is reasonable, the company's monitoring is successful, and the public's trust in the company is high, the product will be successful. Typically, when these conditions are not met, she concludes, we can expect greenwashing (Engel, 2006).

Alves (2009) states that "green marketing and advertising are ever popular strategies to reconcile business interests with ecological interests, and more precisely, with the increased concern for sustainability issues." He believes that not all companies are using green marketing legitimately, and that some use green marketing to lessen pressures to meet intensified standards. He suggests that as long as companies are not punished severely for contemptible practices, consequential progress toward sustainability will not be attained. He advocates the importance of private, public, and independently monitored regulatory systems. Similarly, Gibson (2009) discusses the importance of the Federal Trade Commission (FTC) and the Environmental Protection Agency (EPA) in setting regulations and rules to limit the illegitimate activities of corporations. However, he also adds that in spite of the great significance of the mentioned authorities' rules and limitations (named as Green Guides), these rules do not effectively control the increasing problem of "greenwashing" by themselves. He supports the idea of collaborative work to successfully deal with the problem, such as bringing the FTC and the EPA together. He concludes by stating: "With the environmental expertise of the EPA and the consumer knowledge of the FTC, the two agencies could ably protect consumers from deception in advertising and marketing, provide businesses with workable guidelines within which they must work, and ultimately protect the environment by enabling consumers to choose truly green products and services" (Gibson, 2009).

In the article for the YES position, that corporate sustainability is more public relations than real, freelance environmental health issues writer Richard Dahl argues that "greenwashing" has escalated because of constantly increasing competition between companies as well as a declining U.S. economy. He mentions that companies across different sectors are seeing the benefit of promoting themselves as "green." He gives an example from the makers of indoor cleaning products, which are believed to be among the top "green washers." Additionally, he discusses how some companies might be practicing "greenwashing" as a means of avoiding regulations. He also notes that with new FTC rules and regulations, companies are feeling more pressure to present themselves as more ecofriendly. Finally, the article argues that excessive "greenwashing" can make consumers skeptical of all green claims, thereby causing supporters of sustainability to lose a powerful tool for environmental improvements.

In the article for the NO position, that corporate sustainability is real, university researchers Busco et al. discuss that some companies operationalize corporate sustainability using management control and accounting systems. The authors use both General Electric and Procter & Gamble as examples of how corporations are integrating sustainability practices into their daily operations. They discuss how the pressures from regulatory bodies and stakeholders have influenced these decisions and point out how employees like working for companies that are environmentally and socially responsible. They also note how the financial community is paying close attention to socially responsible investments and that sustainability can lead to corporate financial gains.

Richard Dahl

Greenwashing: Do You Know What You're Buying?

In a United States where climate change legislation, concerns about foreign oil dependence, and mandatory curbside recycling are becoming the "new normal," companies across a variety of sectors are seeing the benefit of promoting their "greenness" in advertisements. Many lay vague and dubious claims to environmental stewardship. Others are more specific but still raise questions about what their claims really mean. The term for ads and labels that promise more environmental benefit than they deliver is "greenwashing."[1] Today, some critics are asking whether the impact of greenwashing can go beyond a breach of marketing ethics—can greenwashing actually harm health?

Greenwash: Growing (Almost) Unchecked

Greenwashing is not a recent phenomenon; since the mid-1980s the term has gained broad recognition and acceptance to describe the practice of making unwarranted or overblown claims of sustainability or environmental friendliness in an attempt to gain market share.

Although greenwashing has been around for many years, its use has escalated sharply in recent years as companies have strived to meet escalating consumer demand for greener products and services, according to advertising consultancy TerraChoice Environmental Marketing. Last year TerraChoice issued its second report[2] on the subject, identifying 2,219 products making green claims—an increase of 79% over the company's first report two years earlier.[3] TerraChoice also concluded that 98% of those products were guilty of greenwashing. Furthermore, according to TerraChoice vice president Scot Case, the problem is escalating.

TerraChoice also measured green advertising in major magazines and found that between 2006 and 2009, the number mushroomed from about 3.5% of all ads to just over 10%; today, Case says, the number is probably higher still. Case says researchers are currently working on another update that will be released later this year, and he predicts the number of products making dubious green claims will double.

Compounding the problem is the fact that environmental advertising—in the United States, at least—is not tightly regulated. The Federal Trade Commission (FTC), the agency responsible for protecting the public from unsubstantiated or unscrupulous advertising, does have a set of environmental marketing guidelines known as the Green Guides. Published under Title 16 of the *Code of Federal Regulations,*[4] the Green Guides were created in 1992 and most recently updated in 1998. According to Laura DeMartino, assistant director of the FTC Division of Enforcement, the proliferation of green claims in the marketplace includes claims that are not currently addressed in the Green Guides, and updated guidance currently is being developed.

The FTC originally planned to begin a review of the Green Guides in 2009, but the commission moved the schedule up, according to DeMartino, in response to a changing landscape in environmental marketing. "The reason, at least anecdotally, was an increase in environmental marketing claims in many different sectors of the economy and newer claims that were not common, and therefore not addressed, in the existing Guides," she says. "These are things like carbon offsets or carbon-neutrality claims, terms like 'sustainable' or 'made with renewable materials.'"

The FTC held a series of workshops in 2008, holding separate events for each of three areas: carbon-offset and renewable-energy claims, green packaging, and buildings and textiles. In association with each workshop, the FTC asked for comments to help shed light on consumer perception of green advertising, but DeMartino says the commission received very few. The FTC responded to this gap by commissioning a research firm, Harris Interactive, to provide that information. DeMartino says that research has been completed, and a report on it will accompany the revision announcement, which is expected soon.

How Updated Guidance Might Look

In aspiring to revise its environmental marketing guidelines, the FTC is following a trend that has been evident in other nations. In 2008, the Canadian Competition Bureau (a government agency similar in function to the FTC) updated its environmental marketing guidelines to reduce green misinformation,[5] and the Australian Competition and Consumer Commission took a similar step.[6] In March 2010, the U.K. Committee of Advertising Practice and Broadcast Committee of Advertising Practice announced an update to their codes of practice designed to curtail greenwashing.[7]

All three updates are "remarkably similar," Case says, but he suggests the Canadian revisions might provide the best "sneak peak" at what the FTC might do because the two agencies have a long history of working together on cross-border consumer matters. Attorney Randi W. Singer, a litigation partner at New York's Weil Gotschal who has defended companies accused of false advertising, agrees the moves made in Canada but also in Australia and the United Kingdom may provide a good look at what is to come in the United States. Those changes, coupled with her own analysis of the FTC workshop discussions, provide the basis for her to make several predictions about what the new U.S. regulatory scheme might look like.

Singer predicts the revisions will probably contain new definitional language for terms such as "carbon neutral" and "sustainable." She also expects the FTC will address the issue of third-party certifications—that is, the plethora of green labels consumers see on their products. She says the workshop discussions included "a lot of talk about the need for standardization of certifications, a need to have a process for certifications so it's not just people registering themselves, a need to standardize the iconography and the testing."

According to Case, there are now more than 500 green labels in the United States, and some are "significantly more meaningful" than others. "I testified before Congress last summer[8] and I pointed out a certain lawyer in Florida who set up a website and is 'certifying' products. He doesn't need to see the product, he doesn't need test results. He just needs to see your credit card number," Case says.

Meanwhile, the FTC has begun to step up enforcement regarding claims that it considers clear violations of the existing Green Guides, last year charging three companies with false and unsupportable claims that a variety of paper plates, wipes, and towels were biodegradable.[9] "When consumers see a 'biodegradable' claim they think that product will degrade completely in a reasonably short period of time after it has been customarily disposed," DeMartino says. But for about 91% of the waste in the United States, the FTC wrote in its 2009 decisions,[10] customary disposal means disposal in a landfill, where conditions prevent even a theoretically biodegradable item from degrading quickly.

In another instance, the FTC charged four sellers of clothing and other textiles with deceptively advertising and labeling various textile items as biodegradable bamboo that had been grown in a more sustainable fashion than conventional cotton, when, in fact, the items were rayon, a heavily processed fiber.[11] In January 2010, the FTC sent letters to 78 additional sellers of clothing and textiles warning them they may be breaking the law by advertising and labeling textile products as bamboo.[12]

The Health Impact of Greenwash

One major result of greenwashing, say Case and others, is public confusion. But can greenwashing also pose a threat to the environment and even to public health? Critics say greenwashing is indeed harmful, and they cite examples.

In 2008, the Malaysia Palm Oil Council produced a TV commercial touting itself in very general terms as eco-friendly; a voice-over stated "Malaysia Palm Oil. Its trees give life and help our planet breathe, and give home to hundreds of species of flora and fauna. Malaysia Palm Oil. A gift from nature, a gift for life." But according to Friends of the Earth and other critics of the ad, palm oil plantations are linked to rainforest species extinction, habitat loss, pollution from burning to clear the land, destruction of flood buffer zones along rivers, and other adverse effects. The U.K. Advertising Standards Authority agreed, declaring the ad in violation of its advertising standards; contrary to the message of the ad, the authority ruled, "there was not a consensus that there was a net benefit to the environment from Malaysia's palm oil plantations."[13]

THE SEVEN SINS OF GREENWASHING

In the course of assessing thousands of products in the United States and Canada, TerraChoice Environmental Marketing categorized marketing claims into the following "seven sins of greenwashing":

1. **Sin of the hidden trade-off:** committed by suggesting a product is "green" based on an unreasonably narrow set of attributes without attention to other important environmental issues (e.g., paper produced from a sustainably harvested forest may still yield significant energy and pollution costs).
2. **Sin of no proof:** committed by an environmental claim that cannot be substantiated by easily accessible supporting information or by a reliable third-party certification (e.g., paper products that claim various percentages of postconsumer recycled content without providing any evidence).
3. **Sin of vagueness:** committed by every claim that is so poorly defined or broad that its real meaning is likely to be misunderstood by the consumer (e.g., "all-natural").
4. **Sin of irrelevance:** committed by making an environmental claim that may be truthful but is unimportant or unhelpful for consumers seeking environmentally preferable products (e.g., "CFC-free" is meaningless given that chlorofluorocarbons are already banned by law).
5. **Sin of lesser of two evils:** committed by claims that may be true within the product category, but that risk distracting the consumer from the greater health or environmental impacts of the category as a whole (e.g., organic cigarettes).
6. **Sin of fibbing:** committed by making environmental claims that are simply false (e.g., products falsely claiming to be Energy Star certified).
7. **Sin of false labels:** committed by exploiting consumers' demand for third-party certification with fake labels or claims of third-party endorsement (e.g., certification-like images with green jargon such as "eco-preferred").

Adapted from: *The Seven Sins of Greenwashing: Environmental Claims in Consumer Markets*[2] © 2009 TerraChoice Environmental Marketing http://sinsofgreenwashing.org/findings/greenwashing-report-2009/

In 2008, the authority rebuked Dutch energy giant Shell for misleading the public about the environmental effects of its oil sands development project in Canada in the course of advertising its efforts to "secure a profitable and sustainable future."[14] While acknowledging the term "sustainable" is "used and understood in a variety of ways by governmental and non-governmental organisations, researchers, public and corporate bodies and members of the public," the authority also noted that Shell provided no evidence backing up the "sustainability" of the oil sands project,[14] which has been criticized widely for its environmental impact.[15]

Case contends that makers of indoor cleaning products are among the worst greenwash offenders. "People are attempting to buy cleaning chemicals that have reduced environmental and health impacts, but [manufacturers] are using greenwashing to either confuse or mislead them," he says. "People aren't really well-equipped to navigate the eco-babble, and so they end up buying products that don't have the environmental or human-health performances that they expect."

TerraChoice's 2009 report concluded that of 397 cleaners and paper cleaning products assessed, only 3 made no unsubstantiated or unverifiable green claims.[2] The report noted that cleaners, along with cosmetics and children's products, are particularly prone to greenwashing—a worrisome state, given that these items are "among the most common of products in most households."[2]

While companies see consumers' growing demands for green products as an opportunity to increase sales by making perhaps dubious environmental claims, they may also be doing so in an attempt to avoid regulation, says Bruno. In addition to the FTC's promises to tightening up its rules on environmental advertising, broader governmental pressures increasingly place greater burdens on producers to ensure their products are environmentally sound.

"A single ad or ad campaign may be an attempt to sway a customer. But the preponderance of green image ads, many of which are not even attempting to sell a product, combined with lobbying efforts to avoid regulation, add up to a political project that I call 'deep greenwash,'" Bruno says. "Deep greenwash is the campaign to assuage the concerns of the public, deflect blame away from polluting corporations, and promote voluntary measures over bona fide regulation."

However, several corporate and marketing professionals warn that growing consumer cynicism about these kinds of general campaigns make them risky ventures for companies who engage in them. Keith Miller, manager of environmental initiatives and sustainability at 3M, last year addressed a seminar of The Conference Board, a business-management organization, about what his company does to avoid greenwashing allegations. Summarizing his presentation for the business blog CSR Perspective, Miller said that, based on 3M's experiences, he encouraged companies to avoid making "broad environmental claims" and that any claims made should be specific to products and backed up by "compelling" data.[16]

Ogilvy & Mather advertising agency, recently released a handbook designed to guide managers in how to avoid greenwashing charges and called upon them to adopt a policy of "radical transparency" in green advertising campaigns.[17] Business for Social Responsibility, a consulting and research organization, has also published a handbook, *Understanding and Preventing Greenwash: A Business Guide,* which also emphasized the need for transparency as well as for bolstering any environmental claims with independent verification.[18]

Reining in Greenwash

In the absence of a strong regulatory scheme, consumer and environmental groups have stepped into the vacuum to keep an eye on corporate use of greenwashing. Greenpeace was one of the first groups to do so, creating a separate anti-greenwash group, stopgreenwash.org, which monitors alleged greenwash ads and provides other information on identifying and combating greenwash. The University of Oregon School of Journalism and Communication and Enviro-Media Social Marketing operate greenwashingindex.com, where people may post suspected greenwash print or electronic ads and rank them on a scale of 1 to 5 (1 is "authentic," 5 is "bogus").

Claudette Juska, a research specialist at Greenpeace, also points to numerous anti-greenwash blogs that have emerged. The result, she says, is that "there's been a lot of analysis of greenwashing, and the public has caught on to it. I think in general people have become skeptical of any environmental claims. They don't know what's valid and what isn't, so they disregard most of them."

Thomas P. Lyon, a business professor at the University of Michigan who has written and spoken extensively about greenwashing, agrees. He says companies are aware they may be criticized or mocked for making even valid claims, so they're starting to grow skittish about making green claims of any kind.[19] "That's why companies, I think, want to see the FTC act—to give them some certainty," he says.

David Mallen, associate director of the National Advertising Division of the Council of Better Business Bureaus, the advertising industry's self-regulatory body, says companies are growing increasingly aware of the dangers of greenwashing. Although some of the matters his office handles are initiated by consumers, the large majority are prompted by companies disputing competitors' claims.

"We're definitely seeing a rise in challenges about the truth and accuracy of green marketing and environmental marketing," he says. "It's certainly taking up a greater percentage of the kinds of advertising cases that we look at. Because green advertising is so ubiquitous now, there's so much greater potential for confusion, misunderstanding, and uncertainty about what messages mean and how to substantiate them."

Typically, he says, a company will be attacked for making a broad or general claim about a product being environmentally friendly "based only on a single attribute, which might not even be a meaningful one." But he says many other cases focus on a competitor's use of a word such as "biodegradable" or "renewable." He adds, "We're also seeing these aggressive, competitive green advertisements where a company will say 'Not only are we green, not only are we making significant efforts toward sustainability, but our competitors aren't.'"

Lyon says he's found the companies that are most likely to engage in greenwashing are the dirtiest ones, because dirty companies know they have a bad reputation, so little is lost in making a green claim if the opportunity arises. At the same time, he and coauthor John W. Maxwell wrote in 2006, "[P]ublic outrage over corporate greenwash is more likely to induce a firm to become more open and transparent if the firm operates in an industry that is

MAKING SENSE OF ALL THOSE LABELS

Although the sheer number of green labels can make it hard to tell legitimate claims from bogus ones, not all ecolabeling is greenwash. Many certifications and labels offer useful guidance for selecting products and services that really are produced in more sustainable fashion. Now there are resources to help consumers judge the labels they encounter.

Visitors to http://ecolabelling.org/ can search more than 300 labels by any of 10 categories (buildings, carbon, electronics, energy, food, forest products, retail goods, textiles, tourism, and other) or by world region. The site explains what products the label is used for and the steps producers and manufacturers must follow to obtain certification. Visitors can also rate and discuss each label.

Consumer Reports' Eco-Labels Center at http://www.greenerchoices.org/eco-labels/ lets users search for information by label, product category, or certifying organization or program. Each label receives a "report card" and an extensive evaluation describing the certification requirements for the label, the type and extent of input solicited in crafting those requirements, how certification is verified and by whom, and the funding and structure of the certifying body. The evaluations also tell how meaningful the label is for each product type (for instance, the USDA Organic label is deemed highly meaningful for foods but not for cosmetics). The site also describes the elements of a "good" label and offers a glossary of terms used on labels.

likely to have socially or environmentally damaging impacts, and if the firm is relatively well informed about its environmental social impacts."[19]

"It's somewhat counterintuitive, but the clean guy is likely to shut up altogether," Lyon says. "The rationale is: if you're clean and people already think you're a green company, you don't need to bother touting it so much—and if touting it puts you at risk of being attacked, just shut up and let people think you're clean."

Making Green Claims Work

But when a clean company pulls in its horns over the risks of backlash from a cynical public, Lyon believes an opportunity has been lost. He suggests clean companies can be effective green marketers if they take certain steps. First, he says, they might incorporate a full-blown environmental management system (EMS), which would detail its full environmental program in a comprehensive manner. "When a company has an EMS in place, you have a greater expectation that they actually do know what their environmental results are," Lyon explains. EMSs themselves are supposed to meet an international standard called ISO 14001 developed by the nongovernmental International

Organization for Standardization in Geneva, which sets out a variety of voluntary environmental standards.

Another step is to take part in the Global Reporting Initiative (GRI), an international organization that has pioneered the world's most widely used corporate sustainability reporting framework. The GRI was launched in 1997 by a nonprofit U.S. group called Ceres—a network of investors, environmental organizations and other public interest groups—in partnership with the United Nations Environment Programme. Lyon says the GRI can provide good green credibility at the company level.

The FTC's attention, however, is directed at products—not companies. And Lyon is one of several experts questioning how effective the looming changes to the Green Guides will be in modifying greenwashing. "Honestly, I don't think the FTC Green Guides are going to block much activity," he says. "All the FTC can do is force companies not to provide materially false information. They could potentially go into the domain of what's misleading as well, but that's very tricky. But they could . . . require companies to give you a more complete story."

To Lyon, the ideal system for regulating green marketing claims would entail comprehensive labeling and certification requirements. "You could picture a system that would be a little like the nutrition labeling that we get for food," he explains. "But whether or not that would be helpful is really unclear to me. From what I understand, there's not a lot of evidence that those nutrition labels have changed America's eating habits."

Among hundreds of green labels available today, a few are broadly recognized as highly reliable. One of them is Green Seal, which awards its seal to companies that meet standards that examine a product's environmental impact along every step of the production process, including its supply of raw materials. "It's a differentiator," says Linda Chipperfield, vice president of marketing and outreach at Green Seal. "If you're really walking the walk, you should be able to tell your customers about it."

Other labels are attained via self-certification—that is, if a company wants the label, they can buy it—and aren't so reliable. The Government Accountability Office (GAO) recently proved that in an investigation[20] of Energy Star, a joint program of the U.S. Environmental Protection Agency (EPA) and Department of Energy.

Energy Star provides labels to companies who submit data about products and seek the stamp of approval to place on their packages. "Currently, in a majority of categories [Energy Star] is a self-certification by the manufacturer, which leaves it vulnerable to fraud and abuse by unscrupulous companies," says Jonathan Meyer, an assistant director in the GAO's Dallas office. Indeed, over a nine-month period, GAO investigators gained Energy Star labels for 15 bogus products, including a gas-powered alarm clock the size of a portable generator. In addition, two of the bogus firms that GAO created as "manufacturers" of the products received phone calls from real companies that wanted to purchase products because the fake companies were listed as Energy Star partners.[20]

The EPA and DOE subsequently issued a joint statement pledging to strengthen the program.[21] The GAO report has also prompted responses from consumers and industry alike that a strong and reliable federal certification program is needed. In a story on the investigation *The New York Times* quoted the director of customer energy efficiency at Southern California Edison as saying industries affected by Energy Star hope the report will be "a wake-up call to whip [the program] into shape."[22]

Case believes an improved regulatory scheme does require some kind of certification and labeling. "I think there is room for some kind of unifying green label," he says. "But I'm not sure if the government wants to get into the business of putting 'approved' stickers on good products." He proposes that the function of providing environmental labels be handled by a new office of the EPA. Under this plan, the EPA would combine several existing environmental labels (such as Energy Star and Green Seal) under a single brand to make it easier for consumers to identify more environmentally preferable goods and services. He points to the U.S. Department of Agriculture's affirming label on organic foods as a model.

Toward a Unified Approach

The growing demands of society for greener products and corporate America's desires to meet it and make a profit make for "a fascinating interaction with cultural change," says Lyon. "The norms have really started to shift. I think that's our hope for information and labeling—that it will create a new floor that keeps rising. I don't think we're anywhere close to that yet, but I think it's starting to happen."

Case says he is "somewhat hopeful" that all involved are moving toward a unified approach to solving the challenges posed by greenwashing. "The huge danger of greenwashing is if consumers get so skeptical that they don't believe any green claims," he says. "Then we've lost an incredibly powerful tool for generating environmental improvements. So we don't want consumers to get too skeptical."

All references for articles included in *Taking Sides: Clashing Views in Sustainability* can be found on the Web at www.mhhe.com/cls.

Cristiano Busco et al.

Cleaning Up

Balancing financial performance and corporate sustainability is a challenge, especially in today's economic environment. Based on the idea that there's a trade-off between what's "good for the business" and what's "good for the environment and society," companies sometimes perceive corporate sustainability and corporate social responsibility (CSR) as an add-on cost, or they may perceive sustainability as an opportunity for "green PR." By doing so, they may miss significant opportunities for business growth, innovation, and organizational change. In this article, we describe how General Electric (GE) and Procter & Gamble (P&G) have operationalized corporate sustainability initiatives using management control and management accounting systems.

Current forces of change are driving a general rethinking of business in a more sustainable direction. Pressures come from national and international regulatory bodies as well as from business partners, stakeholders, and activists. For instance, many companies require their suppliers to comply with standards for environmental management systems certification, such as the ISO 14000 series or the European Union's Eco-Management and Audit Scheme. Moreover, thanks to new technologies, customers are increasingly informed, empowered, and active, and their demand for products that are clearly identified as sustainable is growing. Employees are also looking for companies that operate in a socially responsible way, which can impact the ability of an organization to recruit and retain talent. In addition, the financial community is paying closer attention to socially responsible investments (SRI) and investment rating systems such as the Dow Jones Sustainability Index.

These internal and external stakeholder views shouldn't be neglected by companies aiming to achieve superior corporate sustainability performance. As suggested in the May 2009 *Strategic Finance* article "Co-Creating Strategic Risk-Return Management" by Mark Frigo and Venkat Ramaswamy, "*Sustainable wealth creation* requires *balanced risk taking* by focusing on *co-creation opportunities* that can generate *superior returns* while simultaneously reducing risks for companies and their stakeholders." Since shareholder value creation is driven from creating value for others (customers, employees, suppliers, and other stakeholders), organizations can think in terms of how these stakeholders can be engaged to help define and achieve sustainability performance. This process can help identify new sustainability opportunities that will create mutual value with internal and external stakeholders. The current drivers of change and the economic downturn provide an opportunity for more integrated, strategic, and value-creating sustainability efforts.

Operationalizing Sustainability

Although the meaning of sustainability (the integration of social, economic, and ecological values) is widely accepted, the main problem is how a company interprets such a concept and, subsequently, operationalizes it. To do this, organizations need to establish well-defined sustainability strategies that identify achievable and measurable goals and then communicate them within. As suggested by the Corporate Sustainability Model (see "Implementing Corporate Sustainability: Measuring and Managing Social and Environmental Impacts" by Marc Epstein in the January 2008 issue of *Strategic Finance*), the alignment of strategy, structure, management systems, and performance measures is fundamental for organizations to coordinate activities and motivate employees toward implementing a sustainability strategy.

Sustainability principles can affect companies in many ways, so we distinguish three main levels of their impact:

1. **Strategic Thinking.** The corporate vision, mission, and strategic goals should integrate sustainability. Through innovative and environmentally friendly products, companies can pursue these new market opportunities and positively affect sales, profits, and return on invested capital. As Michael E. Porter and Mark R. Kramer suggested in "Strategy and Society: The Link between Competitive Advantage and Corporate Social Responsibility" in the December 2006 issue of *Harvard Business Review,* organizations can make the strategic connection between social responsibility and business opportunities.

2. **Managerial Processes.** To succeed, all organizational processes such as research and development (R&D), purchasing, and marketing should integrate sustainability.

3. **Operations.** Production processes and operations should align with sustainability principles to obtain continuous improvement, reduce resource consumption, eliminate waste, and recycle. A catalyst to reduce costs and achieve more efficient processes, sustainability aims to increase or maintain output value with reduced resources and costs.

Management control systems (MCS) and management accounting systems can play a vital role at each level. Once a company defines sustainable strategies and goals, MCS and, more specifically, performance measurement systems break down an organization's sustainability targets and objectives that are meaningful for local entities, business units, functions, and individuals. A performance measurement system enables the cascading down of sustainable strategic thinking into sustainable managerial processes and sustainable operations. Thanks to the cascading approach, all employees know how they can help achieve sustainable goals.

Figure 1

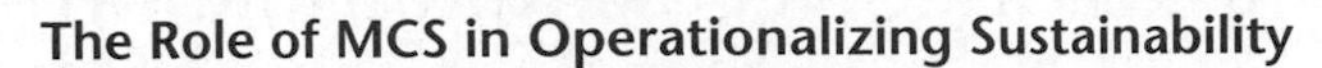

The Role of MCS in Operationalizing Sustainability

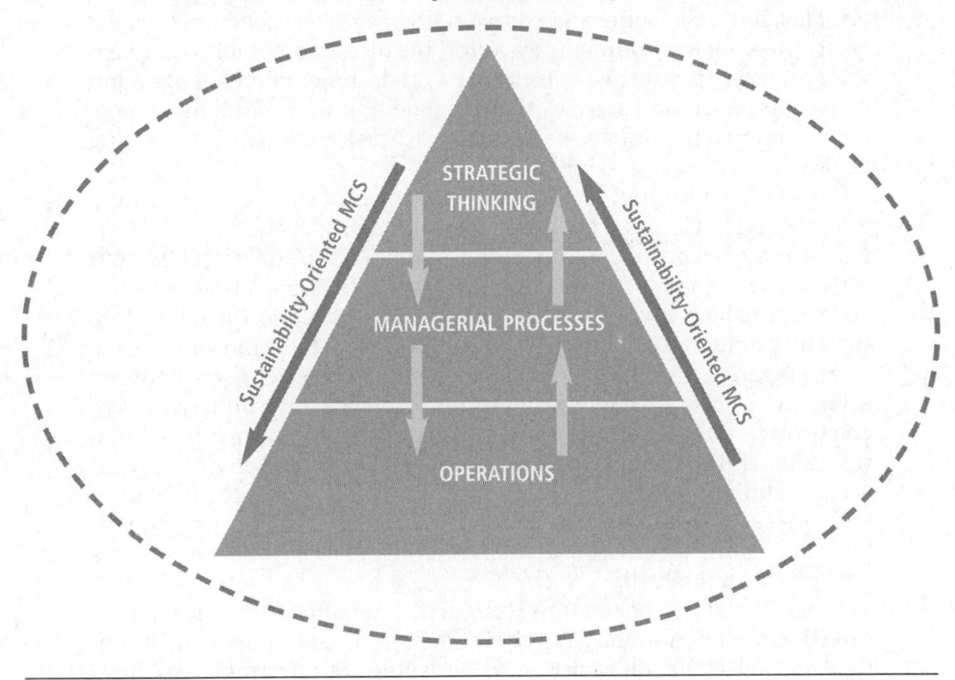

Stimulating Organizational Change

MCS have great potential to introduce new ways of thinking and operating. They support sustainability as a corporate priority by evaluating, measuring, and rewarding employees based on sustainable goals. They promote proactive behavior. Moreover, MCS introduce social and environmental issues into the organizational language and include them in the set of shared values and principles. Finally, embedding sustainability into corporate culture strengthens "enlightened" strategic thinking (see Figure 1). Leading organizations, such as General Electric, addressed sustainability issues at these levels (see "General Electric's Ecomagination Initiative").

Toward Sustainability-Oriented Management Control Systems

Social and environmental accounting (SEA) has focused on collecting information for external users, thus allowing stakeholders to know and evaluate the way in which a business operates. Many scholars and research teams have defined guidelines to integrate social and environmental aspects in external reporting. Although external reporting is a preeminent aspect of sustainability

GENERAL ELECTRIC'S ECOMAGINATION INITIATIVE

In May 2005, GE launched Ecomagination with the belief that "green is green"—that environmentally sound business is profitable. Ecomagination creates innovative solutions to environmental challenges and delivers products and services that provide significant and measurable environmental performance advantages to customers while generating profits.

Strategic Thinking: Ecomagination engages sustainability initiatives strongly tied to the business. Through an independent and externally conducted process, Ecomagination relies on the development and certification of products that might improve GE's operating performance and reduce environmental impact. The portfolio includes more than 80 products that obtain $17 billion of related revenues.

Managerial Processes: Aiming to reconcile the tensions between economic, social, and environmental performance, such initiatives have taken place through a worldwide business strategy, an organizational structure. This has helped infuse principles of social and environmental accountability into the processes through which product innovation takes place and customer satisfaction is pursued. For instance, investments in "clean" or environmentally friendly R&D are made to achieve a specific goal for 2010.

Operations: Ecomagination deals with operations since it engages in reducing greenhouse gas (GHG) emissions and water consumption as well as in improving energy savings. Many initiatives have taken place. For instance, the company installed new energy-efficient GE lamps within manufacturing plants and warehouses worldwide, obtaining huge greenhouse gas reductions and big savings. Moreover, the "eCO_2 Site Certification Program" rewards GE sites and facilities that have best embraced Ecomagination goals on energy efficiency and consumption reduction. Finally, the "Energy Treasure Hunts" initiative engages employees through a treasure-hunt process aimed at identifying energy savings and GHG reduction opportunities.

management, it isn't enough without internal mechanisms and practices to implement sustainability.

Because social and environmental issues are important dimensions of the overall corporate performance, traditional performance measurement and monitoring systems should be broadened to embrace these dimensions. Corporate goals can include not only financial terms, but social and environmental ones as well. They can be integrated with the whole planning and monitoring systems to ensure that the organization acts in a sustainable way to satisfy stakeholders' and shareholders' interests.

MCS could play a key role guiding organizations to achieve sustainability goals so that SEA could be a means to build a stronger business image and an integrated management tool. Management control systems enable managers to monitor whether the business is operating in accordance with financial, social, and environmental principles. In the meantime, managers can feed back information on sustainability performance at all organizational levels, which the company can use to set or modify sustainability strategies.

In order to broaden MCS in a sustainable direction, the first effort is to break down company-wide sustainability strategies into targets and objectives that are meaningful for local entities, business units, functions, teams, and individuals. After incorporating these goals, the MCS role is to guide management actions to achieve the desired outcomes.

To ensure that MCS would achieve social and environmental goals, formal and informal elements need to work together. An MCS, the formal element, links performance to strategies by providing a base for monitoring and assessment.

But formal elements aren't enough. Their effectiveness may be influenced by informal elements, such as leadership commitment, that are fundamental levers in promoting awareness of social and environmental strategies and encouraging the right actions. Without integrating sustainability with the set of values and principles shared within an organization, only a symbolic and formal approach will be adopted in dealing with social and environmental issues.

Although formal and informal elements of control should work together to motivate decision makers to operate in a sustainable way, some tensions between them often emerge. Social and environmental principles are embedded into informal controls that shape organizational culture, but formal tools are often focused exclusively on financial issues, probably because of the difficulties in defining appropriate quantitative and objective measures for social and environmental performances since they are often very judgmental. Adding to the challenge, the principles are usually linked to long-term time horizons, a high level of uncertainty, and impacts that are often difficult to quantify. Nevertheless, measurement and internal collection of social and environmental data represent important elements for more informed decision-making processes, increasing the chances of a convergence between the organization's actions and stakeholders' needs.

Sustainability at Procter & Gamble

Headquartered in Cincinnati, Ohio, Procter & Gamble, a large multinational company, produces a wide range of consumer goods. Since 2002, P&G's organizational structure has featured three global business units (GBUs) and market development organizations (MDOs) that have different structures and, therefore, different functions. The three GBUs—Beauty & Grooming, Health and Well-Being, and Household Care—operate at a global level dealing with business strategies, innovation, brand design, and new business development. Organized into seven geographic regions, the MDOs are local structures responsible for profits and interfacing with local markets. They deal with regional marketing, sales, and external relations and are mainly accountable for net operating sales (NOS).

Each MDO is organized into a market operation team (MOT) and a customer business development (CBD) team. Each team works on a unique project and is responsible for a unique goal. Thanks to the matrix design of P&G's structure, MOT and CBD are involved in many vertical and horizontal relationships. Moreover, they have to pursue interrelated goals by aligning their activities through some formal meetings and informal mechanisms such as trust and sense of ownership.

When Strategy Meets Sustainability

Although P&G has realized the importance of social and environmental issues since the 1960s, it didn't set up sustainable development as a strategic objective until 1999. The corporate mission became "to provide branded products and services of superior quality and value that improve the lives of the world's consumers, now and for generations to come."

P&G's mission shows the company's desire to present itself as an organization that takes care of its stakeholders and, more in general, of social and environmental issues. For more than 10 years, P&G has accounted for its social and environmental strategy in its Sustainability Report. The company communicates the importance of sustainability externally through formal claims and informally to employees through, for instance, leadership commitment.

In 1999, P&G also set up a centralized organizational structure, the Global Sustainability Department, composed of experts who provide advice to business units on sustainability matters. Deciding on centralization or decentralization in defining sustainability strategies is a very complex task for multinational companies because decentralization often leads to some advantages such as greater flexibility, increased responsiveness, and greater innovation. Centralization, however, is often the most successful choice in terms of greater control and coordination among local entities.

P&G has preserved the choice of centralization. This department currently deals with P&G's overall sustainability policy, identifies emerging sustainability issues, manages corporate sustainability reporting, builds external relations, and helps business units incorporate sustainable development into their businesses. In the meantime, the presence of sustainability ambassadors within local organizations allows the balance between local and global instances.

The P&G Planning and Control System

To better understand if and how sustainability strategies and goals are embedded in the broader planning and monitoring process, P&G's Objectives-Goals-Strategies-Measures (OGSM) system provides perspective (see Figure 2). Starting from the highest organizational levels, this OGSM cascade mechanism involves the business units and all employees. In fact, each employee is accountable for specific objectives and must contribute to achieving the overall goals. Here's how it works.

Objectives: The planning process starts with an objective that overlaps the corporate mission. This element is quite stable and doesn't have a definite duration.

Figure 2

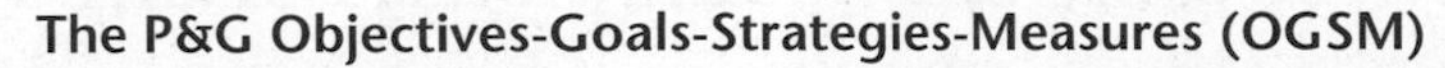

The P&G Objectives-Goals-Strategies-Measures (OGSM)

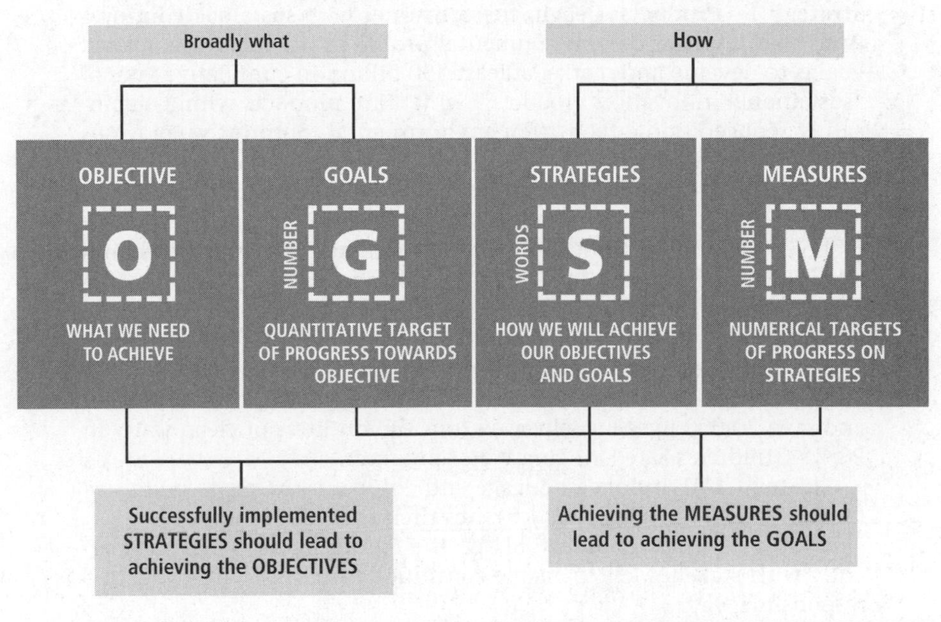

Goals: The goals, which come from the objective, remain stable in content but are reviewed annually. Communicated to shareholders, the goals are substantially related to the increase in net sales, earnings per share, and total shareholder return.

Strategies: The goals are translated into strategies. The brand and country plans reflect the strategies, which include two main areas:

- "Where to play" refers to decisions about geographical areas, markets, and businesses where P&G competes.
- "How to win" refers to decisions that create competitive advantages, create leaders, and, more in general, achieve good financial results. As an example, the slogan "consumer is boss" summarizes one strategy. This slogan means that consumers' needs and desires are the starting point for developing every organizational activity. Sustainability is perceived as a request from consumers since achieving progress in sustainability management is included in this strategy. Other strategies are related to two main levers: cost management and organization. The former relies on driving out all costs that don't deliver or build value for consumers or shareholders.

Measures: Specific measures are defined for each strategy. These measures are both quantitative and qualitative, not focused merely on financial performance since many aspects of organizational activity are recognized and considered.

PROCTER & GAMBLE SUSTAINABILITY STRATEGIES

Strategy 1—Products: Delight the consumer with sustainable innovations that improve the environmental profile of products. The related goal is to develop and market at least $50 billion in cumulative sales of "sustainable innovation products," which are products with a significantly reduced (more than 10%) environmental footprint vs. previous or alternative products.

Strategy 2—Operations: Improve the environmental profile of P&G's operations. The related goal is to deliver an additional 20% reduction (per unit production) in CO_2 emissions, energy consumption, water consumption, and disposed waste from P&G plants, leading to a total reduction over the decade of at least 50%.

Strategy 3—Social Responsibility: Improve lives through P&G's social responsibility programs. The related goal is to enable 300 million children to live, learn, and thrive; prevent 160 million days of disease; and save 20,000 lives by delivering four billion liters of clean water in P&G's Children's Safe Drinking Water program.

Strategy 4—Employees: Engage and equip all P&Gers to build sustainability thinking and practices into their everyday work.

Strategy 5—Stakeholders: Shape the future by working transparently with stakeholders to enable continued freedom to innovate in a responsible way.

The global OGSM is translated into more specific plans and programs for the various organizational structures. Starting from the overall corporate OGSM, the cascade mechanism involves all organizational levels. The global OGSM translates into specific OGSM for each global business unit, which can achieve the planned goals only if the marketing development organization's activity is aligned with them. As a consequence, action plans for regional and local markets have to be drawn.

To coordinate employee behavior within GBUs and MDOs, a continuous dialogue between them takes place since their actions should align and be consistent with the overall OGSM. Then, action plans for each market operation team and customer business development team are defined. Some specific objectives are assigned to each team that identify programs, measures, and goals consistent with the plans of the related MDO and GBU and the overall OGSM system.

Once the plan for the multifunctional team is defined, each employee plans activities and identifies goals. In fact, under the review of a supervisor, each employee has to draw up two main documents annually, such as the work development plan (WDP) and the action plan, according to the organization's expectations. The WDP contains a qualitative analysis of each employee's main strengths and weaknesses, and the action plan contains specific, measurable, achievable, and consistent goals.

The comparison between actual and target allows everyone to assess and evaluate activity. Supervisors evaluate their employees quarterly, and progress in achieving overall corporate goals is assessed quarterly. Then, through a formal document each quarter, the chief executive officer communicates the progress made to employees and investors.

Driving Sustainability into the P&G Planning and Control Systems

Sustainability issues are widely present across various stages of the OGSM cascade mechanism. The reference to sustainability is evident in the mission, namely the objective. The goals are built on financial measures, but the reference to sustainability is indirect. A sales manager attempted to justify the absence of a clear reference to sustainability in this section of the framework:

THE DEVELOPMENT OF "COOL WATER CLEANING" DETERGENT

An action aiming to implement the first sustainability strategy concerns the development of a concentrated detergent able to clean at low water temperatures, expending less energy, water, packaging, and waste. Developing this product relied on a three-fold assessment to evaluate the financial, social, and environmental profiles of the new product.

Life-cycle assessment (LCA) measured the overall environmental footprint of the new product from raw materials to disposal. The LCA of laundry detergent outlined that heating water for washing consumed far more energy than any other step in the process. As a consequence, developing a detergent that cleans in cool water clearly offered the greatest opportunity for energy reduction.

To evaluate the product's profitability, the company used traditional capital budgeting criteria, such as net present value (NPV). Since financial analysis showed high return deriving from it, the "cool water cleaning" detergent looked like a profitable investment.

Finally, the company did a "social" assessment of the product in terms of benefits that stakeholders and, more specifically, consumers would derive from it. The detergent provided relevant benefits in terms of reducing energy and water consumption for users without compromises in terms of product performance. In order to promote the importance of the new product and stimulate responsible consumer behavior, the company launched a campaign termed "Turn to 30°C." Along with key product benefits, including energy savings and brilliant cleaning, the campaign emphasized another motivation: supporting the environment in a more sustainable way. P&G further managed this innovation through partnerships with NGOs (nongovernmental organizations) and energy-focused third parties.

"We are a public company so we have to create value for our shareholders . . . but we cannot create this value without reference to other dimensions of sustainability. . . . I cannot contribute to the achievement of our first goal ("increasing net sales") without increasing sales coming from sustainable products."

Thus sustainable strategies are integrated with the traditional planning system since they are viewed as something highly related to and aligned with the business. This integration may be a key element of successful implementation of sustainability-oriented strategies because it clearly communicates to employees how to behave in order to translate the abstract sustainability principle into action.

In 2007, P&G set up five-year sustainability strategies for five broad areas: products, operations, social responsibility, employees, and stakeholders (see "Procter & Gamble Sustainability Strategies"). In terms of this framework, the first and second strategies refer to strategic thinking and operations levels, respectively.

Driving Greater Sustainability

The first sustainable strategy—products—shows that a traditional goal such as sales could be rethought in light of sustainability. Procter & Gamble focuses on how customers use its products. Some examples of sustainable products are the packaging redesign of some "beauty" products in order to save plastic and the development of a "cool water cleaning" detergent (see "The Development of 'Cool Water Cleaning' Detergent").

The second strategy—regarding operations—implies that pursuing both financial and environmental benefits leads to eco-efficiency. For example, one project will increase rail transportation in Western Europe from 10% to 30% by 2015, reduce CO_2 emissions, reduce plants' environmental footprint by installing solar photovoltaic systems, and use recycled materials and rainwater recycling systems.

Finally, sustainability is built into the business rather than treated as an additional activity. In this sense, sustainability is perceived as an opportunity to meet new customer needs and reduce costs. Moreover, P&G defines and continuously monitors specific, measurable, achievable, and consistent goals for these two strategies.

Sustainable development tracks progress in achieving sustainable goals by extrapolating socially and environmentally sensitive data from the overall information system. Data analysis and assessment follows data collection. The evidence from this process is used for internal and external reporting. Internal reporting aims to stimulate suggestions to improve future sustainable strategies and initiatives.

Moreover, internal communication and tracking of sustainability targets also facilitate the process of integrating sustainability principles within P&G's organizational culture. Since April 2009, the company has been sending a quarterly newsletter about sustainability goals to all employees in Western Europe via e-mail. The newsletter updates employees about what the company and region are doing to advance their environmental and social responsibility and engages employees "to bring sustainability into their life."

The responsibility to achieve sustainability goals belongs mainly to GBUs since they deal with product development and production, which are the most critical phases for P&G's environmental sustainability. The MDOs have an important role, too, since they have to reconcile global corporate policies and strategies with local stakeholders' interests and demands. This is a very critical task because multinational companies have to face the great heterogeneity of the various contexts in which they operate. To increase awareness of the corporate commitment to sustainability, the company organized initiatives such as Earth Day to celebrate a more sustainable way of operating. A sustainability ambassador at each site is responsible for illustrating and promoting five sustainability strategies.

Moving Forward with Corporate Sustainability

To integrate social and environmental issues within day-to-day activities, a company should adopt a strategic approach. Management control systems can play a vital role in driving sustainability throughout the organization. Here are five steps companies can consider:

1. Identify a few focused, clear, and shared "sustainability goals," and integrate them into the corporate planning and control systems. Traditional planning and control systems often work well.
2. Set measures that are able to capture social and environmental performance at the corporate level as well as throughout the organization.
3. Collect social and environmental data and information for internal users, and integrate the data in the decision-making process through supporting tools.
4. Combine formal elements (performance management systems, reward systems, internal codes) and informal ones (leadership commitment, initiatives, and practices aiming to integrate sustainability with corporate culture).
5. Manage the tension between coordination and control on one hand and flexibility on the other hand in choosing between centralization and decentralization of sustainability strategies.

Companies can use this approach to make corporate sustainability a value creator for the business and the stakeholders.

EXPLORING THE ISSUE

Is Corporate Sustainability More Public Relations Than Real?

Critical Thinking and Reflection

1. Do the corporations focus on the sustainability issues to represent themselves as environmentally cautious exclusively to increase their sales margins?
2. Can adaptation of corporate sustainability be a way of increasing the company's financial health? Does corporate sustainability bring an extra burden to the companies financially?
3. Can corporate sustainability and corporate social responsibility be considered as the same attribute?
4. Today, there are increasing numbers of businesses leaning toward "greenwashing." Do you think this will eventually lead to a pressure to adopt better practices in terms of corporate sustainability?
5. Is there a greater (bigger scale) harmful effect of "greenwashing" on the society?

Is There Common Ground?

The argument whether corporate sustainability is a genuine practice or just PR for the companies has been discussed in detail with different aspects. It seems as if there is a "common ground" on the argument such that corporations can focus on the sustainability issues (genuinely or as PR) and further increase their sales margins immensely. The green attributes of the companies are very important for today's consumers. Mostly, they are willing to support sustainable businesses and their products or services over unsustainable ones. So the common ground of the given argument could be: "corporate sustainability is an essential initiative for the companies and the society."

Additionally, in this argument, the question "does corporate sustainability become a way of increasing the company's financial health" can produce a "common ground" between the yes and no positions. A company may well be taking sustainability seriously with enthusiasm and may be adopting it strictly. As some of the articles reviewed here support, this approach can result in corporate financial growth. Therefore, the more we prove the association between this corporate financial growth and corporate sustainability implementation, the more the vast majority of the companies would be willing to adopt similar practices. This can connect the two different arguments with demonstrating the adaptability and conformity of the common ground (i.e., corporate sustainability is an essential initiative). However, we need to exhibit this with more "real life" examples and exemplify that conformity can provide increased earnings to the companies.

Internet References

Alves, I. M. (2009). Green spin everywhere: how greenwashing reveals the limits of the CSR paradigm. *Journal of Global Change and Governance.* Volume 2, Number 1. Winter/Spring 2009. Division of Global Affairs, Rutgers University-Newark. Retrieved from <http://andromeda.rutgers.edu/~gdga/JGCG/archive/Winter-Spring2009/Alves.pdf>

Boiral, O. (2009). Greening the Corporation through Organizational Citizenship Behaviors. *Journal of Business Ethics.* Volume 87:221–236 Springer 2008, pp. 221–234. DOI: 10.1007/s10551-008-9881-2.

Borkowski, S. C., Welsh, M. J. and Wentzel, K. (2010). JOHNSON & JOHNSON: A Model for Sustainability Reporting. *Strategic Finance.* Montvale: September 2010. Volume 92. Issue 3, pp. 29–37.

Dwyer, R. J. (2009). "Keen to be green" organizations: a focused rules approach to accountability. *Management Decision.* Vol. 47 No. 7, 2009. pp. 1200–1216. Emerald Group Publishing. DOI 10.1108/00251740910978377.

Engel, S. (2006). Overcompliance, labeling, and lobbying: The case of credence goods. *Environmental Modeling and Assessment,* Volume 11: 115–130. Springer 2006. DOI: 10.1007/s10666-005-9030-6.

Elkington, J. (1998). Cannibals with Forks. The Triple Bottom Line of the 21st Century. *Environmental Quality Management.* Fall 1998. Volume 8, Issue 1, pp. 37–51. DOI: 10.1002/tqem.3310080106.

Freese, W. (2007). The Business Case for Sustainability. *New Directions for Institutional Research.* no. 134, Summer 2007. Wiley Periodicals, Inc. Published online in Wiley InterScience <www.interscience.wiley.com> pp. 27–35. DOI: 10.1002/ir.210.

Gibson, D. (2009). Awash in Green: A Critical Perspective on Environmental Advertising. *Tulane Environmental Law Journal.* Volume 22, LexisNexis. pp. 423–440.

Modrak, V. and Dima, I. C. (2010). Conceptual Framework for Corporate Sustainability Planning. *International Business Management.* Medwell Journals. 4(3): 139–144, 2010 ISSN: 1993-5250.

Taylor, R. (2008). "Making the Business Case for Corporate Sustainable Development," *i-Manager's Journal on Management,* Vol. 3, No. 1, June–August, 11–18.

Veleva, V. (2010). The State of Corporate Citizenship 2009: The Recession Test. Corporate Finance Review. Jan/Feb 2010. 14, 4. ABI/INFORM Global. pp. 17–25.

Wilson, M. (2003). Corporate sustainability: What is it and where does it come from? *Ivey Business Journal Online.* Volume 1. Retrieved November 28, 2010, from ABI/INFORM Global. Document ID: 532772771.

The report on this Web page discusses the increasing demand for greener products and how companies compete for the attention of these "green" consumers.

http://sinsofgreenwashing.org/findings/greenwashing-report-2010/

This Web page has a discussion paper on greenwash and corporate sustainability.

http://docs.google.com/viewer?a=v&q=cache:GzG4LAPTaHYJ:www.tec.org.au/greencapital/

This Web page discusses the "bottom line on corporate sustainability" as a way for businesses to achieve real financial rewards.

http://green.blogs.nytimes.com/2009/09/02/the-bottom-line-on-corporate-sustainability/

This website is devoted to information on the best practices of corporations that follow sustainability practices and policy.

www.greenbiz.com

In this Internet source, examples are given of how corporations can achieve a competitive advantage and increase shareholder value by implementing sustainable practices.

http://www.iisd.org/sd/

ISSUE 19

Are Social Concerns Taken Seriously in the "Triple Bottom Line" of Sustainability?

YES: Michael Laff, from "Triple Bottom Line: Creating Corporate Social Responsibility That Makes Sense," *T + D* (February 2004)

NO: Frank Vanclay, from "Impact Assessment and the Triple Bottom Line: Competing Pathways to Sustainability?" *Sustainability and Social Science: Round Table Proceedings* (July 2004)

Learning Outcomes

After reading this issue, you should be able to:

- Define and explain the concept of the "triple bottom line" of sustainability.
- Understand how companies are building their concept of sustainability around the triple bottom line.
- Understand how companies are beginning to measure business performance on the triple bottom line.
- Know the difficulty of defining the meaning of the "social pillar" in the triple bottom line.
- Understand how companies are using the triple bottom line to improve community relations.

ISSUE SUMMARY

YES: Internet training and development blogger Michael Laff details how corporations are utilizing triple bottom line (TBL) to develop innovative approaches to improve their relationship with the local community and reduce their impact on the environment.

NO: Frank Vanclay, a professor of cultural geography at the University of Groningen in the Netherlands, discusses the inability of triple bottom line (TBL) to provide an adequate framework for organizations to assess their progress toward social equity or justice in their management functions.

Social concerns have always been difficult to measure and implement in sustainability. The "triple bottom line" (TBL) refers to the belief that sustainability must be based equally on environmental, economic, and social criteria. A sustainable practice or action must be measured on its capacity to meet these "three pillars" of sustainability. When discussing this concept relative to business, Elkington (2004) pointed out that companies must focus on both environmental and social issues equally in assessing their value. Recently, the International Institute for Sustainable Development (2010) has noted that the TBL is rapidly gaining recognition as a framework for measuring business performance and captures the spectrum of values that organizations must embrace—economic, environmental, and social equity (often called the "three E's" of sustainability). But, although this constitutes a noble mission for corporations and business, is it attainable?

Although corporations are beginning to adopt the "three pillars" as a business mission, can they be integrated into business operations, organizational structure, and corporate policies? The difficulty lies in creating practical strategies that link and equally weigh the three pillars, as opposed to allowing for trade-offs or the dominance of one pillar over the rest. Generally, economic considerations have borne the greatest weight in the decision-making process. It is an easy metric to measure and quantify, that is, profits and losses for business and U.S. dollars per capita for countries, and has traditionally been used as a guide for societal development (Norman and MacDonald, 2003). The next easiest metric to measure and quantify is environmental impact or performance. Measures of air and water pollution, the environmental impacts of corporations on society, are not only quantifiable but also are regulated by government. Until recently, these governmental requirements have been traditionally dealt with as "costs" to business. Today, sustainable business strategy has sought to integrate them into broader business strategy. But, the third pillar of sustainability, the social component based on goals of social equity and justice, has received less attention in corporate sustainability. These have not advanced because of the inability of corporations to adequately assess its social impact and the lack of quantifiable measures of success (Vanclay, 2004).

There are three different definitions for the social pillar of sustainability. They include noneconomic designations such as "social, social development, or social progress"; human development designations such as "human development or human well-being"; and finally, issues of justice and equity such as "social justice, equity, and poverty alleviation" (Kates, Parris, and Leiserowitz, 2005). Because of these wide variants in the understanding of the social equity pillar, corporations are finding it more difficult to integrate these concerns into their organizational strategies.

The main way in which corporations have addressed the three pillars and the social component of sustainability is through the TBL (Elkington, 2004; Gedro, 2010; Laff, 2010). Included in corporate TBL approaches are the design of new corporate models that go beyond mere donations or one-time community service projects to broader community strategies. These include educational grants for local schools and the responsible sourcing of products from foreign suppliers. Many publicly traded corporations do not wish to compromise their public images through unjust social practices. Some seek to integrate social equity and justice into their daily operations. This often includes an attention to employee satisfaction, involvement, and participation, and the corporation's positive involvement with its local community and society at large. For example, employees are evaluated not only for recruiting purposes, but for their contribution to sustainability initiatives (Gedro, 2010). This helps ensure that employees actively participate in corporate sustainable development strategies and are permitted to creatively and innovatively contribute to the process. This usually results in several "greening" strategies such as cutting back on paper waste, participating in recycling and reuse schemes, and increasing a workplace sense of community.

Despite the efforts of some corporations to utilize the TBL approach to genuinely incorporate social equity into business strategy, critics argue that this approach is limited and is tantamount to nothing more than "greenwashing" (Gibson, 2006; Norman and MacDonald, 2003; Vanclay, 2004). According to experts in the field of social impact assessment, there is no established method by which the social consequences of a corporation can be analyzed, monitored, and managed. Hence, there is no approach that can measure whether a corporation is truly making progress toward social equity. Instead, the TBL approach is better at providing a philosophical overview for how corporations can embrace sustainability through novel corporate social responsibility. Essentially, critics argue that TBL does not provide the guidance necessary with which to appropriately determine how to consider and incorporate social indicators into sustainability (Gibson, 2006; Norman and MacDonald, 2003; Vanclay, 2004). For corporations to be truly transparent in their sustainable development approaches, they recommend that they adopt social impact assessment methodologies to gauge whether or not their efforts are successful. Social impact assessment has an established set of values and principles that can guide corporations and help them measure their progress more efficiently under an all-encompassing and interdisciplinary approach that is more likely to yield measurable social equity indicators (Vanclay, 2004).

However, it is important to note that social impact assessment is limited since measures of social equity are ill-defined under international sustainability guidelines. At present, there exists no truly complete or complex matrix to fully incorporate social equity into sustainability (Norman and MacDonald, 2003; Vanclay, 2004).

In the article for the YES position, that social concerns are taken seriously in the triple bottom line of sustainability, Internet training and development blogger Michael Laff details how corporations are utilizing the TBL approach to develop innovative approaches to improve their relationship with the local community, thereby reducing their impact on the environment while promoting economic growth. He recounts the story of several corporations, such as Interface Carpet, Men's Warehouse, Hewlett-Packard, and Delloite, that changed business practices so as to contribute to the social betterment of their surrounding community. These corporations recognized that sustainability means going beyond donating money to community organizations, and to develop an employee-centered social responsibility model for business. This new model goes beyond mere greenwashing to actions that positively impact the surrounding community. Employees are encouraged to participate in long-term community service practices that are integrated directly into corporate performance.

In the article for the NO position, that social concerns are not taken seriously in the triple bottom line of sustainability, researcher and university professor Frank Vanclay points out the problems with using TBL as a framework for assessing social equity or justice in corporate operations and management processes. He asserts that TBL is a poorly developed and immature concept that amounts to nothing more than greenwashing. He contends that TBL, which was originally intended to provide a philosophy for sustainability that could encourage the development of corporate social responsibility, has digressed into an accounting and reporting system that lacks a complete understanding of the concept. He argues that there is an underdevelopment of the social aspect of the sustainability triad, leading to further confusion and misapplication of TBL in corporate and organizational planning, management, and assessment. TBL is, therefore, limited in scope and likely to ignore critical social indicators when utilized as an evaluative tool. He argues that TBL is neither novel nor an improvement over currently utilized and better established assessment tools. As such, he recommends that instead of using TBL to gauge social equity goals in corporations and organizations, those seeking to assess their performance utilize the social impact assessment (SIA) method that provides an all-encompassing, established, and transdisciplinary approach to measuring social equity and justice. Although he acknowledges the limitations of SIA to account for all social equity and justice issues, he believes that as a result of its established value system and more widespread use, SIA is better able to assess impacts. Ultimately, TBL should remain a philosophy for sustainability, whereas SIA should be used as a tool to measure progress of the social equity and justice pillar.

Michael Laff

Triple Bottom Line: Creating Corporate Social Responsibility That Makes Sense

On a Saturday afternoon, workers at the California plant of Interface Carpet don jumpsuits, head to a local landfill, dive into the refuse, and retrieve garbage.

Employees aren't searching for lost currency, just discarded items that could be recycled. It's a routine called "the dumpster dive." After cleaning up, staffers head back to the plant, and think about how they can preserve items for recycling that are normally discarded.

The image may sound messy, but it fits perfectly with the vision of the company's founder, Ray Anderson, who decided to transform his company's practices, in large part because of a book he read.

Besides green initiatives, Interface Carpet adopted a service mission and altered its dealings with customers, the surrounding community, and the environment. Instead of selling carpet and simply moving on to the next customer, the company implemented a service model whereby customers lease the carpet and Interface maintains the product.

Sustainable initiatives underwritten by the company range from education grants for schools studying the effects of climate change, to a "fair works" initiative whereby hand woven panels made in India are purchased at market price and sold in the United States.

Convinced of the necessity to change the business model, Anderson toured every plant to share his awakening with all employees in the company. He created an "Eco Dream Team" (that included Paul Hawken, the author whose book, *Ecology of Commerce,* shaped his vision) to advise him on initial steps.

"Donating products is the old model," says Erin Meezan, assistant vice president for sustainability at Interface.

"It's not about writing a bigger check or being smarter about who you donate products to," Meezan states. "Corporate social responsibility starts with developing and engaging your own people and integrating it into your business model."

New Model

Corporate social responsibility is moving beyond purchasing recycling bins and turning off lights at the end of the day. It encompasses how an organization conducts itself in the marketplace and the image it projects outside company walls. Employees might contribute legal or financial services to a charity in kind, or send workers to clean up after a natural disaster.

With the exception of deciding to pay a fair price to suppliers in developing countries, no single element within the chain is new, but the sophistication of responsible corporate behavior, linking disparate practices into a whole strategy, is taking different shapes.

Whether such practices will survive in the coming decades is unknown, but many organizations are conscious of what they call the triple bottom line—environment, community, and finance.

It is an acknowledgement by organizations already engaged in charitable works and greening efforts that what's good for the surrounding community can benefit the organization as well. And the best efforts do not require new purchases or carving out additional time.

"[To the question of] whether this is a fad that will fizzle out, I think the answer is no," says Peter Heslin, assistant professor of management and organizations at Southern Methodist University. "I don't see companies going in reverse."

According to a survey on the issue conducted by the Society for Human Resource Management, most organizations continue to sponsor community service activities several times each year—a sign that such efforts are more like singular events rather than strategic, long-term initiatives.

Yet, a crucial calling card for the generations entering the workforce is a company's commitment to responsible practices. MBA students report being committed to working for a socially responsible employer and are willing to take a pay cut to do so.

"This generation wants to work for a corporation that is socially responsible," says Barry Anderson, interim CEO for Gifts in Kind. "When I entered the workforce, that never came up."

What separates sustainable volunteer or charitable efforts from feel-good efforts is the amount of creativity behind them. Individuals charged with managing corporate responsibility emphasize the need to use the organization's existing talents, whether in the form of skills or supplies donated, in a way that aids recipients. The community service component should be integrated with an organization's daily functions so that it becomes more than just a one-time event.

"It shouldn't be, 'Today we're going to a homeless shelter, but tomorrow we're not doing anything,'" says Susan Sarfati, an organizational strategy consultant.

Tom Dekar of Deloitte traces the origins of the current movement to the Global Compact of 1999, when the United Nations outlined 10 principles for organizations to follow regarding fair practices such as uniform labor laws.

Expectations to meet such ideals are higher, even among for-profit enterprises. When submitting proposals for consulting contracts, Deloitte asks itself and is often asked by potential partners, to identify its corporate citizenship initiatives in detail.

"We tell our clients to look at their supply chain critically to find out if they do anything that might embarrass you," he says. "Ultimately it reflects on your business."

Giving Mood

Oftentimes, a successful campaign can draw upon the willingness of employees to chip in. At Hewlett-Packard, employees can choose a product they wish to purchase for donation, select a charity, and contribute 25 percent toward the product cost, and the company will pick up the remaining 75 percent.

Groups of employees can join to purchase a server for a charity or other recipient. The allocated budget is often used up in a matter of days as the company opens the donation through an email announcement.

Men's Warehouse provides suits for a welfare-to-work program in Greensboro, North Carolina. The clothier keeps the pipeline going with little fanfare, as the tags are removed to limit advertising or the reappearance of clothes in secondary markets, according to Anderson.

One distinguishing trait of responsible corporate behavior is its relatively low visibility. Many organizations are content to inform their staff of ongoing initiatives, but stop short of launching a media campaign, perhaps fearing a backlash if efforts fail or being criticized for acting solely out of self-interest.

While it is fair to ask whether corporate outreach efforts are done to project a favorable image above other considerations, the real test of its value is whether an individual or institution benefits from the actions of an organization.

"Critics will say that if an initiative is self-interested, and the organization benefits from it, then it should not be classified as corporate citizenship," says Marc Orlitzky, a professor of management at Penn State University at Altoona.

"I disagree. They should be judged based on how it benefits society. You can't judge an organization's motives, but you can measure outcomes."

Normally a frequent target of critics because of its labor practices, Wal-Mart made great strides in building sustainable initiatives. Orlitzky believes the retailer could become a leader in sustainability efforts within three years. Solar panels are being installed in all of the retailer's super centers. Unused energy is sold back into the energy grid, representing a cost-saving move for the community.

Wal-Mart also sells organic cotton products, and instead of passing on the cost to consumers, sells it at the same price as regular cotton products. Fish products are purchased from sustainable sources such as fish farms, and not from natural sources. To verify their contribution, a council reviews Wal-Mart's efforts.

Many of Wal-Mart's initiatives remain unseen or unknown to consumers because the company chooses not to promote its efforts. Ortlitzky says large retailers often seek to avoid any hint of exaggerating its contribution, also mindful of the attendant criticism from interest groups that the first steps are adequate, but not enough.

Not all of Wal-Mart's greening efforts succeeded. The retailer attempted an electronic recycling program, allowing customers to return laptops or monitors at the end of the product cycle. Low participation rates forced the store to scale back efforts. Orlitzky says one reason was the lack of awareness among customers.

Reuse, Rebuild

Donating unused inventory is a natural way to give back for large retailers. Home Depot launched a program whereby unused or unsold household supplies are donated to organizations charged with building affordable housing.

Called "Framing Hope," every imaginable item is set aside, including lumber, hand tools, flooring, sinks, or tiles, with the exception of potentially hazardous materials. Tom Wroblewski, senior operations analyst for Home Depot says the company will donate about $10 million worth of supplies this year. Most of the items are unsold or are being replaced by a new version.

The initiative started sluggishly because associates had to handle products for donation manually. Eventually, the store created a software suite that is integrated with existing inventory programs to earmark items for potential donation. The system helps associates make decisions and includes enough checks and balances to prevent mishandling of items, mistakes, or theft.

Store officials did not want the donation initiative to consume too much of associates' time. Nor did they want items lying in storage awaiting pickup.

"We had to design something that was simple and foolproof with support from the corporate office so that associates are not afraid to make decisions," Wroblewski says.

Another ongoing charitable partnership continues with KaBOOM!, a not-for-profit organization dedicated to building playgrounds in underserved neighborhoods such as ones damaged by Hurricanes Katrina and Rita. Home Depot donates materials and labor toward rebuilding efforts.

Various charitable initiatives are managed throughout the organization. Each store designates an individual Team Depot captain who coordinates charitable efforts within a store. At the district level, which encompasses 10 to 12 area stores, another captain oversees larger initiatives.

What dooms many organizations' efforts, Orlitzky believes, is a lack of performance evaluation methods. Managers and employees should be graded for their efforts to maintain responsible corporate practices and be given support to correct any mistakes.

"If a company takes the triple bottom line seriously, then they need a balanced scorecard with equal weight given to each category," Orlitzky says. "It has to be judged for sustainability and not just recruiting purposes otherwise it is set up for failure."

Greening initiatives are one of the more fundamental aspects of responsible corporate behavior. In contrast to donations, greening often requires little additional funding, just planning. The trade show industry is the second most wasteful industry in the country, trailing only construction.

Trade show organizers could take small steps such as cutting down on 400-page program guides, offering a newspaper opt-out program and eliminating bottled water, according to Scott Lindley, vice president of development for the National Convene Green Alliance.

"We're in step 1 of a 100-step process," he says. "Not every association can be totally green. You need to find a balance. Sponsors want visibility, but there may be a way to do that without paper. People often determine whether it was a good meeting based on how much 'stuff' they receive so it's important to explain why policies are different."

Starbucks held a recent convention in New Orleans and partnered with United Way to donate all of the furniture used during the event to individuals affected by Hurricane Katrina, according to Anderson.

"There's no extra inventory, so they have to plan this," Anderson says.

Alex DeNoble, a professor of management at San Diego State University believes that self-regulation regarding the environment is a way for businesses to stave off eventual requirements. Government mandates are always costly to businesses, so if they adopt them gradually over time, they will be in compliance when laws change.

Taking a leadership role before being required to do so demonstrates good business sense, he says. The connection between responsible practices need not be tied to quarterly reports or reaching new customers.

"Joan Kroc exhibited a social conscience without thinking about how it translates into greater profits," DeNoble says. "When the length to good business practices isn't clear, that shows a business is being proactive. There's not a clear link between running the Ronald McDonald House and the hamburger business."

DeNoble acknowledges that McDonald's menu is an ongoing source of controversy and one that is hard to reconcile with even the most aggressive charitable campaigns.

Helping Hand

Responsible practices do not always refer to aiding the community at large. They can be directed inward, providing a lift to staff.

Borders decided to focus on the well-being of its own employees when it established an "internal United Way," according to Adam Grant, professor of management at the University of North Carolina, who studies corporate social practices.

Employees who experience a qualifying event such as a death in the family, extended illness, or bankruptcy can apply for a grant and receive a reply within 48 hours, with a check if approved. Educational scholarships were initially available on a need basis, but the initiative has expanded to include merit-based awards.

All along, the goal was to keep the outreach at grassroots levels, so any employee donation was matched up to 50 percent by the bookseller. Grant found a 60 percent participation rate in charitable giving among employees. While the names of individual beneficiaries were kept confidential, Borders made sure to publicize acts of charity in its newsletter, during staff meetings, and on the intranet.

"They institutionalized the giving process," Grant says. "It showed how important a single donation could be for the life of one employee. People saw a social problem that they could help to solve."

Surveys conducted to determine the reasons for such high participation showed surprising results.

"The company thought that helping an employee would create a sense of loyalty, but the act of giving was a much more powerful source of commitment to the company," Grant says.

Even when the best intentions are philanthropic, initiatives do carry pitfalls. Especially inside companies beset by turmoil or low morale, cynical employees will question whether initiatives are undertaken to burnish the company's image instead of advancing a service mission.

Another area to avoid is charitable programs to which employees feel obligated or pressured to give, which can undermine the whole effort, according to Grant. Such initiatives should be flexible enough to allow people to contribute only in ways with which they are comfortable.

Heslin emphasizes the need to establish an identifiable purpose for the initiative other than simply raising morale that can be documented before and after the program is completed. He believes that sharing the stories of people who benefitted from donations or technical assistance is the most powerful way that employees learn the value of their contributions. Some organizations invite beneficiaries to discuss how they were affected.

"You have to clarify the links between what individuals in the company are doing and outcomes," Heslin says. "That's at the core of keeping it going. You can't get the environment to speak, but you can get people who are affected to tell their stories. It can be very powerful."

Workplace analysts agree that corporate social responsibility presents a way to think strategically about efforts within an organization more than presenting a set of ideas. After all, conservation, charitable giving, and donations are not new. But the pressure to be active throughout an organization exists more now than at anytime in the past.

"Companies could have been doing all of this 20 to 30 years ago," DeNoble says. "Now, if a company is not proactive, it could have a negative effect."

Frank Vanclay

Impact Assessment and the Triple Bottom Line: Competing Pathways to Sustainability?

Introduction

The Triple Bottom Line (TBL) is a concept that has received official imprimatur as a framework for encouraging institutional concern about sustainability. But is it achieving its goal? Although initially intended as a philosophy or way of thinking about sustainability, akin to the concept of corporate social responsibility, it has become simply a mechanism for accounting and reporting. TBL is inherently limited in what it has to offer, and is promulgated by proponents who are largely ignorant of other approaches. Although TBL is meant to add social and environment to the equation, it is often championed by people who have little understanding of what the social entails.

This paper argues that the concept of TBL is not fundamentally different to the well-established field of impact assessment, but that impact assessment and, in particular, the field of social impact assessment (SIA), have much more to offer in terms of accumulated experience and understanding, and a professional and theoretical base. The paper, therefore, is critical of TBL, not because the author is opposed to sustainability or the need to think about social and environmental, as well as economic, criteria, or the need for corporate social responsibility—far from it—but rather because the originators of TBL and its current advocates seem to be ignorant of the field of impact assessment. It is argued that impact assessment, and specifically social impact assessment, offers far more to those concerned about social justice and human welfare than does TBL.

What Is the Triple Bottom Line?

The 'triple bottom line' is variously described as:

- social, environmental and economic performance;
- sustainable development, sustainable environment, sustainable communities;

From *Sustainability and Social Science: Round Table Proceedings,* by Helen Cheney, Evie Katz, and Fiona Solomon, eds. (Institute for Sustainable Futures, July 2004), pp. 27–39. For more elaborate discussion, see: Vanclay, F., 2004. "The Triple Bottom Line and Impact Assessment: How do TBL, EIA, SIA, SEA and EMS relate to each other?", *Journal of Environmental Assessment Policy & Management* 6(3): 265–288.

- impact on society, the environment and economic sustainability;
- economic, environmental and social sustainability;
- economic prosperity, environmental quality and social justice;
- economic growth, ecological balance and social progress;
- economic growth, social progress and environmental health;
- economy, environment, equity;
- profit, people, planet (or planet, people, profit).

The term 'triple bottom line' was allegedly coined by John Elkington in 1995 (Sarre & Treuren 2001) although it did not become popularised until the widespread take-up of his 1997 book, *Cannibals with Forks: The Triple Bottom Line of 21st Century Business*. The title of this book comes from a question posed by the Polish poet Stanislaw Lec, "Is it progress . . . if a cannibal uses a fork?" (Elkington 1997: vii). Elkington applies the question to contemporary capitalism. While the wording of the question/title would suggest that Elkington was sceptical, in a confusing way he states that he believes that "it can be" (ibid.) in a logic that is not altogether apparent. Surely progress and genuine corporate social responsibility requires more than the adoption of some greenwashing to makeover big business and/or government.

Elkington's consultancy company, SustainAbility, which he founded in 1987, gives a big picture description of TBL, as well as an accounting concept.

> The triple bottom line (TBL) focuses corporations not just on the economic value they add, but also on the environmental and social value they add—and destroy. At its narrowest, the term 'triple bottom line' is used as a framework for measuring and reporting corporate performance against economic, social and environmental parameters.
>
> At its broadest, the term is used to capture the whole set of values, issues and processes that companies must address in order to minimise any harm resulting from their activities and to create economic, social and environmental value. This involves being clear about the company's purpose and taking into consideration the needs of all the company's stakeholders—shareholders, customers, employees, business partners, governments, local communities and the public. (SustainAbility 2003: website)

TBL is meant to be a way of thinking about corporate social responsibility, not a method of accounting. On this point, Elkington (1997: 70) is unambiguous, likening TBL to a Trojan Horse which is wheeled in by corporations. In the beginning they succumb to an accounting procedure, but ultimately they are meant to embrace a wider vision of sustainability. Unfortunately, too many agencies and companies have not appreciated the philosophy behind TBL, and are responding only to the reporting requirements. An Australian scoping study by TBL Victoria (Vandenberg 2002) revealed that there was considerable confusion about the definition and philosophy behind TBL in the 32 organisations surveyed.

TBL has achieved considerable imprimatur because corporations such as Shell and BP have adopted it (see BP Australia's *Triple Bottom Line Report*). The World Business Council for Sustainable Development, a coalition of 160 international companies, has also given strong endorsement to the concept (see Holliday et al. 2002). Various government agencies at all levels have been required to implement TBL, and have struggled because the social (and in some cases environmental) indicators have not been determined. But they have lost sight of the intention. TBL should be a philosophy, not a set of accounts.

Although ostensibly coined by John Elkington in 1995, and publicly articulated in 1997, the underlying concept has connections to many earlier ideas, and is totally consistent with ecologically sustainable development (ESD) thinking that was espoused in the Brundtland Report (WCED 1987) and in the 1992 Rio Declaration and Agenda 21. In Chapter 4 of his book, Elkington (1997: 70) makes this clear, saying that "none of this was new."

The essence of ESD thinking was popularised as a three-legged stool that has become so ubiquitous that establishing the originator of this concept is impossible. The three-legged stool, now somewhat contested, was a powerful metaphor because with three legs, each was equally important, and all necessary for support. If a fourth leg is added (for example, culture, see Hawkes 2001), the allegory of support no longer works. The terms describing each leg were meant to be understood broadly. Sustainability represented the intersection of all three domains. The diagram was a useful heuristic and the legs were never intended to be quantified or operationalised.

One consequence of the 1992 Earth Summit, which was reinforced by the United Nations Environment Program and the Organisation for Economic Cooperation and Development, was that many countries agreed to undertake State of the Environment (SoE) reporting. With a broad holistic definition of the environment, and with the noble sentiments of the Rio Declaration and Agenda 21, it would be expected that the social environment would be included in the SoE reports. While some writers (for instance, Goodland & Daly 1995) articulated social sustainability in general terms, the SoE reports required quantitative indicators to measure performance and change over time. Therefore, the difficult process of defining and operationalising social constructs began. As the social scientist engaged to consider the role of social science data in the National Land and Water Resources Audit, my report indicated how difficult this was (Vanclay 1998).

What Is Impact Assessment?

Impact Assessment

"Impact assessment can be broadly defined as the prediction or estimation of the consequences of a current or proposed action (project, policy, technology)" (Vanclay & Bronstein 1995: xi). Or, as on the website of the International Association for Impact Assessment (IAIA 2004), "Impact assessment, simply defined, is the process of identifying the future consequences of a current or proposed action." Impact assessment is a generic term that can mean either an integrated approach or the composite/totality of all forms of impact assessment such as environmental impact assessment (EIA), social impact assessment (SIA), health impact assessment (HIA), etc.

Environmental Impact Assessment

EIA "is a process of identifying and predicting the potential environmental impacts . . . of proposed actions, policies, programmes and projects, and communicating this information to decision makers before they make their decisions on the proposed actions" (Harvey 1998: 2). EIA emerged as a discipline in the early 1970s with the introduction of the United States National Environment Policy Act of 1969 (NEPA) (Ortolano & Shepherd 1995).

The major dilemma across the world is what does 'the environment' in EIA mean? For most writers, and Harvey (1998: 2) is typical, environmental impacts means "bio-geophysical, socio-economic and cultural." In other words, EIA is a triple bottom line phenomenon. This holistic notion of EIA is present in some regulatory contexts. In Australia, the Federal Government's Environment Protection and Biodiversity Conservation Act of 1999 (EPBC) (Section 528), defines the environment as including:

(a) ecosystems and their constituent parts, including people and communities; and
(b) natural and physical resources; and
(c) the qualities and characteristics of locations, places and areas; and
(d) the social, economic and cultural aspects of a thing mentioned in paragraph (a), (b) or (c).

Other Australian legislation, such as Local Government (Planning and Environment) Act 1990 of the State of Queensland, use a similar definition but with sub-sections (c) and (d) slightly reworded to emphasise the social.

(c) the qualities and characteristics of locations, places and areas, however large or small, that contribute to their biological diversity and integrity, intrinsic or attributed scientific value or interest, amenity, harmony, and sense of community; and
(d) the social, economic, aesthetic, and cultural conditions that effect, or are affected by, things mentioned in paragraphs (a) to (c).

In Australia, therefore, at least until recently, the 'environment' was broadly understood. Unfortunately, in some subsequent legislation, for example the Gene Technology Act 2000 (Section 10) this definition has been watered down. The Explanatory Memorandum relating to the Gene Technology Act states very bluntly, "It is intended that the definition of environment include all animals (including insects, fish and mammals), plants, soils and ecosystems (both aquatic and terrestrial)." Around the world, definitions of 'the environment' vary enormously with some countries having very limited definitions, while others reflect an holistic understanding (see Donnelly, Dalal-Clayton & Hughes 1998).

Social Impact Assessment

In short form, "Social impact assessment is analysing, monitoring and managing the social consequences of development" (Vanclay 2003: 6). SIA emerged in the early 1970s along with EIA and as a consequence of NEPA (Burdge & Vanclay 1995). To some extent, SIA is a component of EIA, especially when 'the environment' is understood broadly. However, SIA is more than a technique or step, and is more than a component of EIA, it is philosophy about development and democracy (Vanclay 2002a). Ideally, SIA considers:

- pathologies of development (i.e. harmful impacts),
- goals of development (clarifying what is appropriate development, improving quality of life), and
- processes of development (e.g. participation, building social capital).

SIA has become such a major field of activity that three levels of SIA can be conceived. This creates a confusion in developing a definition (Vanclay 2002a). At the lowest level, SIA is a technique or method within (i.e. subordinate to) EIA (the prediction of social impacts in an environmental impact statement). More commonly, SIA is considered as a methodology in its own right (the process of managing the social issues of a planned intervention) and equal in standing, i.e. comparable, and compatible with EIA. Finally, SIA can be considered as a body corporate, a group of scholars and practitioners, field of research and practice, and/or a paradigm or sub-discipline of applied social science understanding.

What Sets Impact Assessment Apart from TBL?

There are several reasons why impact assessment, and SIA in particular, is a preferable approach to TBL to consider to social implications.

SIA Is an Established Discipline

With over 30 years of existence, the SIA discipline has much to contribute, and has achieved profound learning. Unfortunately, many of the advocates of TBL are ignorant of the field, and many are ignorant of the other forerunners of TBL in the history of sustainability. It is true that in its 30 year history, SIA has had various ups and downs. However, over time, considerable progress has been made. Two significant documents have assisted in codification of the discipline. The first was the 1994 report of the Interorganizational Committee for Guidelines and Principles, *Guidelines and Principles for Social Impact Assessment,* which was developed for the USA/NEPA context. The second was the 2003 *International Principles for Social Impact Assessment* (IAIA 2003).

The variables/issues to be considered in any particular case have been documented (Vanclay 2002b), although they need to be substantiated in each case through a local scoping process. There are several textbooks (Barrow 2000; Becker 1997; Becker & Vanclay 2003; Burdge 1998; 1999; Lane et al. 2001; Taylor et al. 1995). SIA practitioners have a professional body in the International Association for Impact Assessment (www.iaia.org), an organisation which, founded in 1980, now has over 1,000 members across more than 100 countries.

The mention of SIA as a research code in the Australian Research Council's listing of Research Fields, Courses and Disciplines (RFCD 370105 Applied Sociology, Program Evaluation and Social Impact Assessment) is substantiation of its standing as a legitimate field of intellectual endeavour. Several universities teach courses in social impact assessment. By contrast, TBL has no standing and no legitimacy.

The SIA Discipline Has Articulated Its Core Values and Guiding Principles

An important feature of SIA is the professional value system that its practitioners uphold. While all professionals should have a commitment to sustainability as well as to professional integrity, SIA practitioners also uphold an ethic that advocates openness and accountability, fairness and equity, and defends human rights. The role of SIA goes far beyond the ex-ante prediction of adverse impacts and the determination of who wins and who loses. SIA also encompasses: empowerment of local people; enhancement of the position of women, minority groups and other disadvantaged or marginalised members of society; capacity building; alleviation of all forms of dependency; increase in equity; and a focus on poverty reduction. The International Association for Impact Assessment (IAIA) has articulated the *International Principles for Social Impact Assessment* (see Box 1). Such a declaration is testament to the field's maturity as a discipline as well as a statement of its commitment to social justice and human welfare.

BOX 1: EXTRACT FROM THE *INTERNATIONAL PRINCIPLES FOR SOCIAL IMPACT ASSESSMENT* (IAIA 2003)

I. The Core Values of SIA

The SIA community of practice believes that:

1. There are fundamental human rights that are shared equally across cultures, and by males and females alike.
2. There is a right to have those fundamental human rights protected by the rule of law, with justice applied equally and fairly to all, and available to all.
3. People have a right to live and work in an environment which is conducive to good health and to a good quality of life and which enables the development of human and social potential.
4. Social dimensions of the environment—specifically but not exclusively peace, the quality of social relationships, freedom from fear, and belongingness—are important aspects of people's health and quality of life.
5. People have a right to be involved in the decision making about the planned interventions that will affect their lives.

6. Local knowledge and experience are valuable and can be used to enhance planned interventions.

II(a). Fundamental Principles for Development

The SIA community of practice considers that:

1. Respect for human rights should underpin all actions.
2. Promoting equity and democratisation should be the major driver of development planning, and impacts on the worst-off members of society should be a major consideration in all assessment.
3. The existence of diversity between cultures, within cultures, and the diversity of stakeholder interests need to be recognised and valued.
4. Decision making should be just, fair and transparent, and decision makers should be accountable for their decisions.
5. Development projects should be broadly acceptable to the members of those communities likely to benefit from, or be affected by, the planned intervention.
6. The opinions and views of experts should not be the sole consideration in decisions about planned interventions.
7. The primary focus of all development should be positive outcomes, such as capacity building, empowerment, and the realisation of human and social potential.
8. The term, 'the environment,' should be defined broadly to include social and human dimensions, and in such inclusion, care must be taken to ensure that adequate attention is given to the realm of the social.

II(b). Principles Specific to SIA Practice

1. Equity considerations should be a fundamental element of impact assessment and of development planning.
2. Many of the social impacts of planned interventions can be predicted.
3. Planned interventions can be modified to reduce their negative social impacts and enhance their positive impacts.
4. SIA should be an integral part of the development process, involved in all stages from inception to follow-up audit.
5. There should be a focus on socially sustainable development, with SIA contributing to the determination of best development alternative(s)—SIA (and EIA) have more to offer than just being an arbiter between economic benefit and social cost.
6. In all planned interventions and their assessments, avenues should be developed to build the social and human capital of local communities and to strengthen democratic processes.
7. In all planned interventions, but especially where there are unavoidable impacts, ways to turn impacted peoples into beneficiaries should be investigated.

(*Continued*)

8. The SIA must give due consideration to the alternatives of any planned intervention, but especially in cases when there are likely to be unavoidable impacts.
9. Full consideration should be given to the potential mitigation measures of social and environmental impacts, even where impacted communities may approve the planned intervention and where they may be regarded as beneficiaries.
10. Local knowledge and experience and acknowledgment of different local cultural values should be incorporated in any assessment.
11. There should be no use of violence, harassment, intimidation or undue force in connection with the assessment or implementation of a planned intervention.
12. Developmental processes that infringe the human rights of any section of society should not be accepted.

The process th[at] led to the articulation of the SIA discipline's values and principles also resulted in a formal declaration of the definition of the field. "Social Impact Assessment includes the processes of analysing, monitoring and managing the intended and unintended social consequences, both positive and negative, of planned interventions (policies, programs, plans, projects) and any social change processes invoked by those interventions. Its primary purpose is to bring about a more sustainable and equitable biophysical and human environment" (IAIA 2003: 2). Such a definition reflects the deliberations of a committee in trying to address all the issues that need consideration.

The SIA Discipline Has a Broad Concept of What Is 'Social'

While there has been consensus on the need to consider the social issues more, there has been little agreement on what the social issues actually comprise. Consultants who undertake impact assessments or TBL accounting are limited to what is specified in their Terms of Reference, but they also need to be mindful of what constitutes acceptable professional practice, duty of care considerations, and to some extent professional culture. How 'social impacts' are interpreted is therefore central to what is actually considered.

Unfortunately, the case history of impact assessment presents a poor record reflecting inadequate consideration of social issues. There are many reasons for this, including the asocietal mentality[1] that exists, and the lack of SIA expertise (see Burdge & Vanclay 1995; Vanclay 1999; Lockie 2001). Too often the only impacts that are considered are economic impacts and demographic changes. A further problem is that there are groups with narrow sectoral interests who have advocated new fields of impact assessment. A limited view of what is 'social' creates demarcation problems about what impacts are to be identified by SIA, versus what is considered by fields such as health impact assessment, cultural impact assessment, heritage impact assessment, aesthetic impact assessment, or gender impact assessment. The SIA community of practitioners considers all issues affecting people, directly or indirectly, are pertinent to SIA.

SIA is thus best understood as an umbrella or overarching framework that embodies the evaluation of all impacts on humans and on the ways in which people and communities interact with their socio-cultural, economic and biophysical surroundings. SIA thus encompasses a wide range of specialist sub-fields involved in the assessment of areas such as: aesthetic impacts (landscape analysis), archaeological and cultural heritage impacts (both tangible and non-tangible), community impacts, cultural impacts, demographic impacts, development impacts, economic and fiscal impacts, gender impacts, health and mental health impacts, impacts on indigenous rights, infrastructural impacts, institutional impacts, leisure and tourism impacts, political impacts (human rights, governance, democratisation etc.), poverty, psychological and psychosocial impacts, resource issues (access and ownership of resources), impacts on social and human capital, and other impacts on societies. As such, comprehensive SIA cannot be undertaken by a single person; it requires a team approach.

Elsewhere, I have identified over 80 social impact concepts that should be considered in SIA (Vanclay 2002b). A convenient way of conceptualising social impacts is as changes to one or more of the following (Vanclay 2002a: 389; 2003: 7):

- people's way of life: that is, how they live, work, play and interact with one another on a day-to-day basis;
- their culture: that is, their shared beliefs, customs, values and language or dialect;
- their community: its cohesion, stability, character, services and facilities;
- their political systems: the extent to which people are able to participate in decisions that affect their lives, the level of democratisation that is taking place, and the resources provided for this purpose;
- their environment: the quality of the air and water people use; the availability and quality of the food they eat; the level of hazard or risk, dust and noise they are exposed to; the adequacy of sanitation, their physical safety, and their access to and control over resources;
- their health and wellbeing: where 'health' is understood in a manner similar to the World Health Organisation definition: 'a state of complete physical, mental, and social [and spiritual] wellbeing and not merely the absence of disease or infirmity';
- their personal and property rights: particularly whether people are economically affected, or experience personal disadvantage which may include a violation of their civil liberties;
- their fears and aspirations: their perceptions about their safety, their fears about the future of their community, and their aspirations for their future and the future of their children.

The field of SIA has considered these issues at length and have a far broader consideration of social than those who utilise a TBL framework.

Conclusion

The triple bottom line provides nothing original other than a quaint turn of phrase. Its originators and progenitors are largely ignorant of broader sustainability discourses and of other developments in the social sciences that could make a greater contribution. The emphasis in the TBL framework on empirical indicators is likely to reduce consideration of important social issues rather than to increase it.

Social Impact Assessment, by contrast, is a mature discipline, which has a broad understanding of social issues, a well-developed statement of values and guiding principles, and is far better placed to facilitate the path to sustainability than TBL.

Note: This essay has been edited for space and discussion points of this issue. The elaborated discussion is available in the following journal article: Vanclay, F. 2004 "The Triple Bottom Line and Impact Assessment: How do TBL, EIA, SIA, SEA and EMS relate to each other?" *Journal of Environmental Assessment Policy & Management* 6(3): 265–288.

Note

1. Burdge & Vanclay (1995: 46) consider that there is "a prevailing 'asocietal mentality'—an attitude that humans don't count—amongst the management of regulatory agencies and corporations. . . . This mentality also extends to politicians at all levels of government, public officials, physical scientists, engineers, and even economists and some planners. Persons with this mindset do not understand—and are often antithetical to—the social processes and social scientific theories and methodologies which are very different in form from those in the physical sciences in which these people are often trained."

All references for articles included in *Taking Sides: Clashing Views in Sustainability* can be found on the Web at www.mhhe.com/cls.

EXPLORING THE ISSUE

Are Social Concerns Taken Seriously in the "Triple Bottom Line" of Sustainability?

Critical Thinking and Reflection

1. Much time and work has been dedicated to drawing awareness about the growing need for sustainability and sustainable development in order to address global issues. Yet, very little work has been done to concretize how to implement sustainability and sustainable development measures into real-world strategies. Is it possible to create a methodology with which to make sustainability and sustainable development a reality?
2. Corporations are attempting to implement TBL to improve their social performance. Generally, they are using this largely philosophical approach to measure their progress. Although corporate CEOs and consultants claim that TBL is a sound approach, many disagree. Is TBL enough to measure and ensure social equity and justice?
3. Sustainability and sustainable development are largely unregulated and optional considerations at the moment. Is it necessary for government agencies to set specific regulations that corporations and organizations must meet in their daily operations? Essentially, is command and control necessary when juxtaposed to the current global ecological, economic, and social crisis that human society is facing?
4. Current sustainability and sustainable development policies tend to focus on Western values and ideologies. Although past global summits and conventions have included both developed and developing nations in the political discourse and development of sustainability and sustainable development language, is the current approach truly egalitarian and inclusive of non-Western nations?
5. The triple bottom line or three pillars of sustainability focus on integrating environmental, economic, and social issues to address global problems. Is this triad enough? Or, does true sustainability require a greater paradigm shift?

Is There Common Ground?

A "common ground" issue that continues resonating in regard to integrating social equity and justice into the "triple bottom line" of sustainability is the pressing need to address the issue from an interdisciplinary perspective. As it currently stands, multiple disciplines attempt to adopt the triple bottom line or three pillars of sustainability piecemeal. Although their efforts are laudable, the discipline-specific perspectives and distinct approaches tend to stymie true integration of the social aspect into sustainability, particularly in reference to performance measurements and operational and managerial capabilities.

Traditionally, sustainability has lacked a full consideration of its social aspects. Environmental and economic considerations have tended to dominate as these perspectives are more capable of being quantified. Environmental impacts can be measured through indicators such as loss of wetlands, air and water pollution, and the increase of carbon dioxide in the atmosphere. Economic impacts can easily be measured in consumption per capita, materials and energy increases, and gross domestic product. Social concerns, usually the purview of sociologists, have tended to be blurred as they relate to indicators that are integrated into sustainability. Usually, this has been done through indicators of human development, that is, health, education, mortality statistics, and so forth.

The most productive way for sustainability to become a viable solution to global problems is for a variety of stakeholders, that is, scholars, government officials, business leaders, and scientists, to come together with open minds, willingness to work together to solve implementation problems, and a desire to engage in an active exchange of idea sharing, and solution-driven discourse. At present, there are ambiguities around the triple bottom line concept and its ability to integrate social justice and equity into sustainability. As a consequence, it has been neglected as an idea that can be an integrating concept for sustainability.

Internet References

Coffman, M., & Umemoto, K. (2010). The Triple-Bottom-Line: Framing of Trade-offs in Sustainability Planning Practice. *Environment, Development, and Sustainability,* 12(5), pp. 597–610.

Elkington, J. (2004). Enter the Triple Bottom Line. Retrieved from http://www.johnelkington.com/TBL-elkington-chapter.pdf.

Gedro, J. (2010). HRD and the Triple Bottom Line: Creating and Sustaining Equitable Practices. 11th International Conference on Human Resource Development Research and Practice Across Europe. University of Pecs, Hungary. Jun. 2010. Retrieved from http://works.bepress.com/julie_gedro/30/.

Gibson, R.B. (2006). Beyond the Pillars: Sustainability Assessment as a Framework for Effective Integration of Social, Economic, and Ecological Considerations in Significant Decision-making. *Journal of Environmental Assessment Policy and Management,* 3(3), pp. 259–280.

Kates, R.W., Parris, T.M., & Leiserowitz, A.A. (2005). What is Sustainable Development? Goals, Indicators, Values, and Practice. *Environment: Science and Policy for Sustainable Development,* 47 (3), pp. 8–21.

Mehta, R. (2009). Sustainable Development—How Far is it Sustainable? *Proceedings of International Conference on Energy and Environment,* March, pp. 754–757.

Norman, W., & MacDonald, C. (2003). Getting to the Bottom of "Triple Bottom Line." *Business Ethics Quarterly.* Retrieved from http://www.businessethics.ca/3bl/triple-bottom-line.pdf.

Sustainability is the corporate Web site of John Elkington, the business consultant and author credited with originating the concept of the "triple bottom line." This Web site is supported by several leading conservation, environmental, sustainability, industry, and governmental organizations including The Confederation of Indian Industry, E-Square, the Ethos Institute, the International Union for Conservation of Nature, the International Finance Corporation, the United Nations Environment Program, and WWF UK, to name a few. This is a good resource because it is the originator's Web site and provides his definition of the triple bottom line and sustainable development, and provides several case studies that illustrate how corporations and institutions have adopted the approach in their business practices.

http://www.sustainability.com/

This Web site belongs to professors Norman and MacDonald who wrote the first academic critique of the triple bottom line approach to accounting for social equity and justice issues. The Web site includes a basic "general public" article on their critique and a direct link to the full academic journal article. It also provides a list of links for those interested in learning more about TBL. This is a good resource because it is an honest examination of the limitations of TBL from individuals who appear to have no hidden agenda and who are well versed in business ethics and philosophical discourse. This Web site is not sponsored or supported by any advertising or outside agencies/organizations.

http://www.businessethics.ca/3bl/index.html

This Web site is the home of Triple Pundit—a media company for the business community dedicated to disseminating information about the triple bottom line and sustainability. It is dedicated to the promotion of corporate social responsibility and are sponsored by the following organizations: EOS Climate, Dominican University of California, and Saybrook University. It is a good resource because it provides articles and op-ed pieces on TBL, advances in business sustainability/sustainable development, and green business culture.

http://www.triplepundit.com/tag/triple-bottom-line/

This is the Web site of green-design consultants and authors William McDonough and Michael Braungart, better known for Cradle to Cradle Design. Although this Web site is not specifically about the triple bottom line, it provides what is perhaps a better approach to business and design. McDonough and Braungart have contributed to the analysis of TBL and have expanded on it by exploring the Triple Top Line. The excerpted article can be accessed from the following link: http://www.mbdc.com/images/Beyond_Triple_Bottom_Line.pdf. McDonough and Braungart are proponents of "ecologically intelligent design" that eliminates waste rather than creating a waste management problem. They have a long list of clients, but do not list any sponsors.

http://www.mbdc.com/

This is the Web site of Resourceful Communities, a group dedicated to pursuing conservation and social justice particularly within the communities of North Carolina. They have utilized a TBL approach to bring together varied corporations and organizations to create economic gains, social improvements, and environmental stewardship in several disenfranchised communities throughout the state. This is a good resource because it gives real-world examples of TBL in action and demonstrates how these principles of sustainability can positively impact the lives of those who are most at risk from irresponsible corporate or governmental practices. The group is sponsored by several organizations, the most well-known of which are Environmental Defense, Heifer International Appalachia-Southeastern Program, Hollister REACH, North Carolina A&T State University, North Carolina Environmental Justice Network, and others.

http://www.resourcefulcommunities.org/triple_bottom_line

ISSUE 20

Can Cities Be Made Sustainable?

YES: Stephen M. Wheeler, from "Planning for Sustainability," in Gary Hack et al., eds., *Local Planning: Contemporary Principles and Practice* (International City-County Management Association, 2009)

NO: Giok Ling Ooi, from "Challenges of Sustainability for Asian Urbanisation," *Current Opinion in Environmental Sustainability* (December 2009)

Learning Outcomes

After reading this issue, you should be able to:

- Describe how cities can indeed be sustainable through better urban planning and designs.
- Understand why city consumption of goods and services needs to be re-evaluated.
- Understand depths of waste management issues within cities.
- Describe the vulnerability of coastal cities under extreme natural conditions.
- Discuss effect of financial constraints and budget cuts on city sustainability efforts.
- Give ideas on how cities of the South can overcome impeding shortcomings that derail sustainability efforts.

ISSUE SUMMARY

YES: Community planner Stephen M. Wheeler delineates how cities can move to sustainability by emphasizing compact urban designs, preservation of open space, adopting transport alternatives, and implementing building codes that emphasize energy conservation and efficiency.

NO: Urban geographer Giok Ling Ooi of Nanyang Technological University shows how the challenges of rapid urbanization in emerging Asian economies are making it difficult for these cities to meet the basics of sanitation, water supply, housing, and so on not to mention the most lofty goals of sustainability.

Today, nearly 50 percent of the world's population lives in cities. By 2030, this will increase to 60 percent, and cities of the developing world are expected to absorb 95 percent of this growth as a result of rural to urban migration, transformation of rural settlements into urban ones, and natural population increase. Although comprising only 3 percent of the earth's land area, cities consume 75 percent of global energy, create 80 percent of global greenhouse gas emissions, and intensely concentrate industry, people, materials, and energy (Taylor & Carandang, 2011). The question of whether cities are sustainable largely relates to the issue of whether they can be made sustainable. According to Newman and Jennings (2008), over time the urban form of cities has gone from small, compact, walking cities built on local hinterlands to large, sprawling megalopolises. In the process, cities have moved from "localness," utilizing materials and energy from their local catchment area, to global with a planetary resource base. Ecosystems that are most resilient and sustainable tend to be autotrophic, that is, they capture energy sufficient for their needs on a local or a bioregional level. Cities generally can be considered as heterotrophic, that is, they do not produce sufficient energy to meet their needs, so they must acquire their needs from adjacent or global ecosystems. The process by which cities can become more sustainable then relies on their capacity to access more energy and materials locally or regionally. It requires them to reduce their demands for global goods and instead use the resources of their local or regional communities for meeting the needs of energy demand, decomposing waste, and food production. At a time when trade is based on the global movement of goods and services, can cities really become sustainable?

There is a general consensus that cities today are not sustainable but that they can be made more sustainable through corrective policies. McDonough and Braungart (2002) suggest that cities can become more sustainable if they are "redesigned" around sustainable principles. Taylor (2011) suggests that cities can evolve toward sustainability through a sustainability planning process that emphasizes collaboration. He states that "sustainability planning is a relatively new approach to local government management that seeks to integrate urban planning with the principles and practices of sustainability." It is based on the unique characteristics of the local community; capacity building through stakeholder engagement; sustainability issue mapping (SIM) as a tool to assess the major environmental, social, and economic risks to the community; practical and cost-effective programs for sustainability; examples of "best practices" to replicate; and providing a financial implementation strategy.

Kent (2005) contends that at least 41 cities in the United States have begun sustainable city programs—programs believed to effectively improve their livability. He highlights programs such as smart growth, increased bicycle ridership, integrated pest management, urban gardens, composting, local energy generation, and recycling and waste reuse. These programs are designed to move a city toward sustainability. Cities such as Seattle, Washington, and Portland, Oregon, have been recognized as leaders in urban sustainability. These cities, like all cities that encourage sustainability, have a common theme. It is the role of public participation at the grassroots level and recognition of broader civil society, particularly non-government organizations (NGOs) in helping to shape and implement city programs. Fundamental program elements vary from city to city, but successful programs incorporate benchmarks that measure progress toward sustainability over time.

Bugliarello (2006) discusses the virtues of cities as places to implement sustainability. He notes that cities concentrate human population, resource and material use, and economic activity. They exhibit certain advantages: their compactness, creativity, and diversity of design can promote equitable and just distribution of amenities and resources; the degree and ease of contact and mobility can contribute to a more livable habitat; and integrated mixed use communities and high-density urban living can shrink per capita ecological footprints by reducing energy and material needs. Further, he describes the paradigm of the city as biological, social, and machine, complex in nature and involving three basic components and their interaction with the environment. It is biological in sense that it encompasses humans and other species that together strive to balance while sharing the same resources, exposed to heightened exposure to microbiological threats in dense urban environments. Organizations, businesses, the city government, ethnic and informal social groups, and families form the social component. At same time, the presence of structures, vehicles, and other artifacts can be said to represent the machine component of cities.

Wood (2007) is representative of writers who believe that not all cities can become sustainable. He notes that although cities in the developing world face great challenges—climate change, loss of biodiversity, and land degradation—they still continue to emphasize traditional unsustainable designs. He believes that developing countries lack the political will and are unable to consider sustainable solutions. Sometimes, sustainable issues are perceived as secondary to more "urgent" needs such as HIV/AIDS and drought and food insecurity (not seen as related to sustainability). Also, developing countries often heavily prioritize economic growth ahead of other concerns, that is, environmental considerations.

Urban sustainability is often an idea or concept neither well understood nor supported by majority populations in both developed and developing countries. There needs to be a discourse change to one that encourages greater reciprocal opportunity and perceived mutual advantage for all "eco aware" citizens. Synergistic urban lifestyles that are desirable, attainable, maintainable, and reproducible—better known as "meta design planning"—are needed by today's high-dense cities. Eco cities can be developed through a strong consensus that is inclusive of the business community, consumers, politicians, educators, bankers, and developers. Through collaboration, we can thus envision future designs, coherent and multilayered in nature, between wider groups of professionals working with urban planners and designers to create future sustainable cities.

In the article for the YES position, that cities can be made more sustainable, university professor and community planner Stephen Wheeler discusses how cities can move to sustainability by emphasizing compact urban design, preserving open space, adopting alternatives to automobile use, and implementing building codes that emphasize energy conservation and efficiency. He indicates that sustainability is indeed achievable in cities. One way is through environmental planning where cities create management plans for watersheds to protect and restore ecosystems. Another way is by reducing consumption levels and waste generation, and by adopting eco-friendly designs that reduce packaging and emphasize local production. Also, cities that reduce energy consumption and invest in alternative sources are more likely to achieve sustainability.

Wheeler notes four ways that cities can move toward sustainability. First, as mentioned above, to engage in environmental planning that emphasizes watershed management. Second, to create a land use policy that preserves agricultural land on urban peripheries, maintains open space and greenways within cities, and advocates compact mixed use development that reduces vehicular use and promotes a sense of place and community. Third, to design transportation plans aimed at reducing vehicle dependency by improved transport alternatives, for example, pedestrian and bike paths, new public transit options, car share programs, and revised street designs. And fourth, to increase both affordable housing and energy-efficient buildings through codes that emphasize solar energy use and energy conservation.

In the article for the NO position, that cities are not sustainable, urban geographer Giok Ling Ooi of Nanyang Technological University shows that the challenges of rapid urbanization in emerging Asian economies is making it difficult for these cities to meet the basics of sanitation, water supply, and housing, not to mention the more lofty goals of sustainability advocates. Professor Ooi points out the many challenges of cities in developing countries including sustainability into their development agendas. She discusses how rampant urbanization has created resource depletion, air and water pollution, traffic congestion, poor sanitation, and lack of housing. Extensive migration of rural people to urban areas in search of jobs has caused the mushrooming of informal settlements in cities with poor sanitation and health issues. Also, the lack of integrative or multisector policy at all levels of government compounds the challenges that these cities face.

Ooi points out that governance in developing countries often does not seek wide-level stakeholder input in decision making and often ignores local community knowledge. Also, the lack of transparency and credibility of officials entrusted with planning and management responsibilities can hinder the vision to pursue sustainability. And finally, limited access to information and corruption can limit the capacity for sustainability.

YES  Stephen M. Wheeler

Planning for Sustainability

The term *sustainable development* came into existence in the early 1970s, and was first used in print in two books published in 1972: *The Limits to Growth,* by Donella Meadows and a group of colleagues at the Massachusetts Institute of Technology (MIT), and *Blueprint for Survival,* by Edward Goldsmith and other staff members of *The Ecologist* magazine in London.[1] *The Limits to Growth* was especially influential and controversial. The MIT team used newly available computer technology to model global population, resource use, pollution, and economic growth. Every scenario that the group fed into its model showed the human system crashing midway through the twenty-first century, subsequently stabilizing only at lower levels of population and consumption. When the original team revisited its model in 2002, armed with thirty years of additional data, the basic predictions remained accurate.[2] Moreover, the researchers concluded that humanity had entered into a period of "overshoot," in which its needs were substantially greater than the planet could support.

Since the early 1970s, the sustainability concept has also built on a number of other events and issues. The first United Nations (UN) Conference on Environment and Development, held in Stockholm in 1972, helped catalyze concerns about the global environment, as did the second "Earth Summit" held in Rio in 1992. Public attention to energy crises in the 1970s and to climate change in more recent years has fueled calls for sustainability. In 1987, the UN-sponsored World Commission on Environment and Development produced the most widely used definition of sustainable development: "development that meets the needs of the present without compromising the ability of future generations to meet their own needs."[3]

As support for sustainable development has gathered momentum, several streams of thought have emerged. The first stream is represented by environmentalists, who focus on threats to the Earth's ecosystems. The environmental camp encompasses multiple viewpoints, ranging from mainstream environmental science, which emphasizes pragmatic development of greener industrial and development practices, to deep ecology, which argues in a more philosophical vein that the rights of other species are equal to those of humans, that humans should be seen as part of larger global systems, and that human impacts on the planet need to be very greatly reduced.[4]

The second stream of thought is represented by sustainability advocates who emphasize economics and apply such tools as cost-benefit analysis to development problems. These efforts range from attempts to assign monetary value to clean air or wilderness,[5] to a fundamental questioning of the desirability of endless growth in material production and consumption.[6] Ecological economist Herman Daly, for example, has proposed a steady-state model of the economy that emphasizes growth in quality of life rather than in material production.[7]

Advocates for social justice, many of whom are in developing countries, provide the third main perspective on sustainable development: a focus on equity issues. From this perspective, overconsumption in the developed world and maldistribution of resources are principal obstacles to sustainability. Adherents of this perspective point out, for example, that it is unfair for the United States, with about 4 percent of the world's population, to consume about 25 percent of its resources and generate about a quarter of its pollution and greenhouse gas emissions.

The fourth stream of thought on sustainable development emphasizes the ethical, cognitive, and spiritual dimensions of development debates, and stresses that industrial society must move beyond economic value as the main measure of worth. This perspective has roots in many of the world's religious traditions, but it has also found fertile ground within environmental philosophy. In particular, Aldo Leopold's 1949 formulation of a land ethic is often cited: "A thing is right when it tends to preserve the integrity, stability, and beauty of the biotic community. It is wrong when it tends otherwise."[8]

Key Themes for Planners

Although debates on sustainable development often originate in very different perspectives, several common themes have implications for planners, managers, political leaders, and community activists.

First, sustainability depends on a long-term approach to decision making. Implicit in the word *sustain* is the desire for human societies to remain healthy far into the future—far beyond the typical 10- to 20-year horizon of planning documents, the next-election focus of the political system, or the next-year or next-quarter time horizon of much corporate decision making. A consideration of the impacts of current trends 50, 100, or 200 years into the future needs to become standard planning practice.

Second, sustainability requires a holistic, interdisciplinary approach to planning that meshes traditionally separate specialties. For example, transportation planning must be coordinated with land use, housing, air quality, and social equity concerns. Equally essential is the integration of actions across different scales: the building, the site, the neighborhood, the city, the region, the nation, and the planet. Recent movements such as smart growth and new urbanism seek such integration.

Third, sustainability planning emphasizes place and context. Although some past planning theorists embraced the notion of a "non-place urban realm" in which people are so mobile as to be unattached to place,[9] both human and natural systems are always rooted in specific contexts—including, at the largest scale, the planet itself. The limits and characteristics of these settings are vitally important to planning. Emphasis on the sense of place and on place-based identity can help develop more effective planning strategies; unite constituencies around shared historical, cultural, social, or environmental resources; and promote greater stewardship for local places.

Finally, planners, managers, and community leaders must advocate for sustainability in development debates. Professionals need to assert the importance of the future, based on real threats to the health of human and ecological communities. They need to present alternatives to the public, insert underrepresented points of view into debates, assist underrepresented communities in getting organized, and call public attention to the need for long-term thinking.

Promoting Sustainability

Promoting sustainability at the local level requires attention to many subject areas, such as environmental planning, land use, transportation, housing, economic development, and social justice. Municipalities sometimes create stand-alone sustainability plans, setting forth new initiatives in such fields to promote sustainability, or sometimes integrate the theme of sustainability across all elements of their comprehensive or general plans.

Environmental Planning

In every community, no matter how urban, much can be done to protect and restore ecosystems. Strategies include restoring streams, shorelines, and wetlands; recreating wildlife habitat; landscaping streets and parking lots; reducing the use of asphalt; constructing green roofs; and landscaping with native and climate-appropriate plants. Municipalities can create overall plans for watersheds or green spaces to coordinate such actions. In addition to improving ecosystem function, these initiatives can help improve water quality, lessen runoff from impervious surfaces, reduce the urban heat island effect, provide green spaces for the public, and educate residents about the environment.

Resource use is of great importance for sustainability. "Reduce, reuse, and recycle" has long been an environmental mantra, but local governments to date have emphasized recycling, while the reduction and reuse of material goods offer greater long-term savings. Local governments can, for example, charge steeply rising rates to collect from trash containers of different sizes to encourage residents to reduce waste. They can require companies to reuse or recycle wooden shipping pallets, and can require builders to reuse or recycle construction debris. Some materials might be eliminated altogether: San Francisco has prohibited the use of nonbiodegradable plastic bags; Oakland, Portland (Oregon), and about a hundred other cities have banned the use of polystyrene foam.

Energy is yet another main concern of environmental planning, particularly because of the need to reduce greenhouse gas emissions. Some municipalities

LOCAL GOVERNMENTS AND GLOBAL WARMING

As concern mounts about climate change, local governments face the challenge of reducing greenhouse gas (GHG) emissions and adapting to a changed climate.

GHG emissions in the United States come from heating, cooling, and electricity use in buildings (about 30 percent of the total); the transportation sector (about 27 percent of the total); industry (about 20 percent of the total); and a large variety of other sources, including agricultural fertilizers, livestock, and landfills.[1] Local action can affect most of these sources.

Local governments can adopt more energy-efficient building codes that emphasize passive solar design as well as improved insulation and appliances. (The "passivhaus" movement in Germany provides interesting examples of super-efficient homes that need no heating or cooling systems.) Municipalities can also establish renewable portfolio standards, requiring utilities that sell electricity to them to generate a certain percentage of the power from renewable sources. They can reduce motor vehicle use through a three-part strategy: improving transportation alternatives, changing land use patterns, and revising economic incentives. They can lead by example by buying fuel-efficient or alternative technology vehicles. They can help industrial polluters reduce their emissions by offering technical assistance or financial incentives for change. And they can change their economic development strategies to emphasize clean or green businesses rather than those that generate GHG emissions or other types of pollution.[2]

1. World Resources Institute, "U.S. GHG Emissions Flow Chart," at cait.wri.org/figures.php?page=/US-FlowChart (accessed May 6, 2008).
2. The Cities for Climate Protection campaign of the International Council for Local Environmental Initiatives (ICLEI) offers further examples at iclei.org.

own their utilities, in which case they can develop renewable sources of power. But others can at least purchase electricity for public facilities from green sources, convert municipal vehicle fleets and buses to alternative fuels or technologies, and provide incentives and assistance to homeowners and businesses for improved energy efficiency. Changing building codes to require much greater energy efficiency and promote passive solar design is also crucial, as is discussed further on.

Land Use

Smarter and more ecologically appropriate land use, vitally important for sustainable development, includes preserving agricultural land and open space near cities; creating park and greenway systems for ecological and recreational purposes; and designing development to reduce driving and resource use and to promote social vitality, public health, and a sense of community.

For sustainability advocates, the compact city is a principal goal. Exactly how dense cities should be to achieve sustainability is a matter of some debate,[10] but certainly most North American communities could use land far more efficiently. In addition to being compact, development should be contiguous to other urban areas (to reduce driving and promote social integration), well-connected to other urban areas (to facilitate travel by many different modes), and fine-grained in terms of land use mix (to provide residents with many local destinations and enhance community vitality). Compactness and density do not require high-rise development, although places such as Vancouver, British Columbia, provide examples of how slender, high-rise towers set above street-oriented development can create an attractive living environment. A mix of low- to mid-rise housing types, in three-to-five-story buildings, can yield neighborhood densities of twenty units or more per acre while also including greenery and open space.

To limit the unending spread of cities and towns, urban areas in North America will need aggressive policies to promote infill rather than greenfield development. The amount of infill development is increasing in the United States as old shopping malls, office parks, industrial sites, train yards, and vacant lots are redeveloped, but even in jurisdictions known for growth management, such as the Portland Metro region, redevelopment and infill development account for only around 25 percent of residential housing units.[11] In the United Kingdom, in contrast, the Blair government set 60 percent as the infill target,[12] despite the fact that British cities are already far more intensively built than American ones.

Since most new development still occurs at the urban edge, strategies to regulate development there are particularly important for sustainability. Although some new urbanist planners have proposed gradually decreasing densities from a city or neighborhood center to the surrounding countryside—a model that they call the "transect"[13]-sharp edges between built and unbuilt environments make it clear where development may or may not occur, and prevent the kind of low-density, urban fringe development that consumes agricultural land or wilderness, fragments wildlife habitat, and requires residents to drive long distances. Zoning codes need to sharply divide urban densities of at least eight to twelve units per net acre from agricultural densities of one unit per ten to one hundred acres, ruling out low-density subdivisions, ranchettes, and McMansions.

Transportation

Ending the constant growth in motor vehicle use, one of the most important sustainability objectives, requires three interrelated strategies: providing a greater choice of travel modes, changing land use patterns, and revising pricing incentives. Some progress is being made on all three fronts. Many U.S. cities and towns are devoting greater attention to alternative modes of transportation by creating pedestrian and bicycle plans; revising design standards for streets; promoting car-share programs; and exploring new public transit options, including bus rapid transit, light rail, and commuter rail.

Land use regulation is beginning to change as well, in part because of the smart growth and new urbanist movements. And many municipalities have instituted higher parking charges and other incentives not to drive. In 2003, in one of the most dramatic initiatives in recent years, London introduced a congestion charge of five pounds (later raised to eight pounds) for every vehicle entering an eight-square-mile area of the central city. The program has been highly successful, cutting traffic in the central area by 20 percent and generating significant resources for public transit.[14] New York is the first American city to propose a similar scheme, and while it was rejected by the New York State legislature in April 2008, it is likely to resurface again in the future.

Housing

Many American communities lack affordable, well-located, and energy-efficient housing, which affects both the environmental and the social dimensions of sustainability. A range of strategies can help address this situation. Local governments can support nonprofit housing developers; adopt inclusionary zoning policies, which require all projects above a certain size to include a percentage of affordable units; and promote a greater range of housing types within neighborhoods. Developers of large office parks and malls can be required to provide housing for workers. To improve energy efficiency, municipalities can revise building codes to require better insulation, passive solar design, and energy-efficient appliances. Many U.S. municipalities already require public buildings to be certified by the LEED (Leadership in Energy and Environmental Design) rating system; the private sector can more fully embrace such standards as well.

Economic Development

As an alternative to growth at all costs, Paul Hawken, Amory Lovins, and Hunter Lovins have proposed a "natural capitalism" in which entrepreneurial energies are turned toward protecting and restoring the environment.[15] Michael Shuman and David Morris, among others, have called for locally based economies.[16] Jane Jacobs argued for decades for regionally based economies, in which regional products reduce the need for imports.[17] Peter Barnes has called for economic incentives to protect and enhance the commons—environmental and community assets that are not owned by any private interest.[18]

Local governments can help implement such philosophies by resisting big-box commercial development, or by refusing to grant subsidies to large industrial employers that make no long-term commitments to a community. They can encourage small, locally owned, and eco-friendly businesses through a number of means, including loans, public provision of infrastructure, development of small business incubators, workforce training, and preferential allocation of municipal contracts.

Complete community self-sufficiency, however, is unlikely and may not even be desirable from a sustainability point of view, since international trade does produce many efficiencies and benefits. Thus, a better balance must be found between a globalized economy and more place-based ones. This process will require rethinking the many unacknowledged subsidies that are currently provided to large-scale capitalism, and taking steps to make all participants in the economy bear the true costs of their activity.

Social Justice

Sustainable development is sometimes described as nurturing the "three E's"—environment, economy, and equity. Of these, equity is by far the least emphasized in American communities, in part because there is little organized constituency for it. In a society that has grown steadily more unequal for decades, the need to reemphasize equity seems more pressing than ever.

Inequities undermine sustainability. Poverty often leads directly to environmental damage, as impoverished people deforest landscapes, hunt wildlife, or seek ecologically harmful livelihoods because those are the only jobs available. Communities without adequate income cannot afford to construct energy-efficient homes, adequately process wastes, or purchase environmentally friendly products. On the other end of the spectrum, extreme wealth encourages overconsumption, setting a nonsustainable model for the rest of the population.

Local governments can improve equity by ensuring adequate and affordable housing; pursuing economic development that provides decent-paying, meaningful jobs; adopting "living wage" policies; providing education and social services for the least well-off; and instituting progressive tax structures that emphasize property and income taxes rather than sales taxes, which hit the poor hardest. Community development corporations, such as the Sawmill Community Land Trust in Albuquerque offer one way to meet the social justice dimensions of sustainable development.

Process

Public participation in local decision making is important to tap local knowledge, to allow local constituencies to shape their own future, and to foster a sense of stewardship and interdependence. But current methods of participation do not always lead to more sustainable decisions. They may empower affluent, well-organized groups at the expense of others, and they often play into the hands of those who oppose any proposed course of action.

The challenge is to develop community planning processes that are constructive, proactive, and far-sighted. This may mean avoiding excessive numbers of workshops, which can burn out community members (except those with the strongest vested interests or the greatest tolerance for group process), in favor of a few well-organized meetings over a shorter period, coupled with surveys or focus groups to obtain input from a larger range of constituencies. (For further information on public participation options, see "Civic Engagement" in Chapter 5.) During the process, planners need to frame alternatives that address long-term needs, not just for the community but for regional and global contexts as well.

Consistency between planning, zoning, and regulations at different levels is an essential prerequisite for sustainability. It does little good to adopt ambitious sustainable development goals at the state, regional, or municipal level when there is no legal requirement that day-to-day decision making reflect these goals and little systematic evaluation to indicate whether the goals are being met. Other desirable process changes include greater scrutiny of the conflicts of interest that plague U.S. land use planning, improved transparency of decision making, and reduction of the role of campaign contributions in local elections.

Local governments can use a variety of indicators to evaluate progress toward sustainable development: these are usually created through public participation and reflect the particular values of a place. The Sustainable Seattle indicators were developed by a citizen coalition; Vancouver, London, and other cities have developed their own systems.

A Long-term Task

Sustainable development seeks to ensure long-term human and ecological well-being. It reflects a worldview that emphasizes future implications, cross-disciplinary linkages, renewed attention to local place and context, and more active engagement by professionals in addressing the needs of multiple, overlapping communities at different scales ranging from the local neighborhood to the planet as a whole. Planning for sustainability is a long-term task—and not an easy one. But it can be a richly rewarding and meaningful objective for local government planners, managers, elected officials, and community leaders.

All references for articles included in *Taking Sides: Clashing Views in Sustainability* can be found on the Web at www.mhhe.com/cls.

Giok Ling Ooi

Challenges of Sustainability for Asian Urbanisation

In rapidly urbanising regions such as Asia, clearly economic growth has not been paralleled by the effort seen in placing sustainability more firmly in development agendas particularly in the cities. The so-called mega cities of Asia that characterise urbanisation in the region are challenged with major environmental problems that observers say have become legendary from traffic congestion to slum formation, air and water pollution as well as resource depletion. Many scholars have questioned the sustainability of Asian urbanisation with more consensus concerning the challenges it faces than how sustainability as it applies to cities would be defined. This paper highlights the challenges of sustainability for Asian urbanisation, arguing that Asian city governments' failure to address sustainability is a failure to identify and implement an urban development policy framework that will link the effort at more sustainable development in a variety of sectors. These policy frameworks essentially link economic with social policies in the city. Housing should provide one such framework for the provision not only of important environmental infrastructure including modern sanitation and sewerage treatment but also supplies of water and energy as well as the planning of urban transport. Another urban development framework would comprise the planning of industrial sites with transport and environmental infrastructure to reduce air and water pollution. This multi-sectoral approach however requires political and economic support that city governments might not necessarily be in a position to secure particularly with other pressing national development needs.

Sustainability and Urban Growth Trends

Sustainability remains an abstract idea and highly contested concept in spite of its frequent usage in everyday speech and even the development initiatives of governments around the world including Asia [1,2]. For cities, sustainability has been defined as development pattern that (1) satisfies the requirement for equity, social justice, and human rights; (2) meets basic human needs; (3) allows social and ethnic self-determination; (4) promotes environmental awareness and integrity; and (5) promotes awareness of interlinkages between various living beings across both space and time [3]. This definition may be

From *Current Opinion in Environmental Sustainability,* December 2009, pp. 187–191.

considered by some to be impractical. In addition, such definitions do not help their translation into programmes that are likely to promote sustainability with rapid urban growth in many of the countries in Asia and the developing world. Indeed, the sustainability of urban growth in much of the developing world has, as suggested by Glassman and Sneddon [4] been the focus of scholarly research [5–8]. More recent effort to pin down sustainability has been in the work that has been done in industrial ecology on quantitative guidelines for urban sustainability [9] and metrics to establish the standards of urban sustainability [10] does provide city governments with the desirable outcomes or goalposts towards which they should be working in the bid to achieve greater sustainability.

While there are scholars who believe that generally cities are inherently unsustainable [11,12] because of their reliance on materials and energy sources located elsewhere, there are governments in Asia who have introduced sustainable cities initiatives with the surrounding rural areas included [37] and there are networks of cities focused on working to achieve sustainability [13]. Other scholars might not necessarily agree with such a view of sustainability applied to cities but these tend to suggest that cities can focus to a greater extent on how they can be organised and managed far more sustainably than they are at present [14–17]. Briefly therefore, cities should be able to contribute a lot more to sustainability than they have done in the past.

There are scholars who argue that the effort at being sustainable will either progress in an evolutionary manner as cities gain in affluence or evolve with further development. An urban environmental transition [18] has been proposed that argues that low-income urban centres face local environmental problems that do not contribute substantively to global environmental changes, whereas affluent cities have either displaced or reduced serious local environmental hazards. These cities however can contribute to global environmental problems because of lifestyles and consumption choices. This stage-by-stage transition model highlights a unilinear process that however only suggests that cities are inherently unsustainable. They progress from local to global environmental degradation.

Another urban environmental evolution model [19,20] argues for transitions by cities at a higher level of economic development. The evolutionary model however, suggests that the trajectory followed by cities in environmental performance can be non-linear. There is recognition that cities being complex systems, they can be vulnerable to disasters and setbacks, as well as the acknowledgment of the underlying mechanisms behind the changing patterns. This model tries to understand the underlying processes to which change can be attributed including policy directions and development agendas.

There is hence, a growing recognition that the translation of sustainability into development agendas with their implementation can be highly political [21]. This is especially the case when cities particularly in rapidly growing regions like Asia are vying keenly for investment capital and business. Indeed such competitions among cities are contributing in a great way towards the rate of urbanisation and scale of urban growth seen in Asia.

Table 1

Urban and Rural Population Growth Rates, 1950–2030

	Urban population (%)				Average annual rate of change (%)			
					Urban		Rural	
	1950	1975	2000	2025	2000–2025	2025–2030	2000–2005	2025–2030
Asia	17.4	24.7	36.7	50.6	1.31	1.10	−0.80	−1.19
Eastern Asia	18.0	25.2	38.5	51.8	1.20	1.06	−0.79	−1.20
South-central Asia	16.6	22.2	30.6	44.7	1.36	1.37	−0.63	−1.18
South-eastern Asia	14.8	22.3	37.2	53.2	1.85	1.00	−1.18	−1.20
Western Asia	26.7	48.5	70.2	77.0	0.71	0.26	−1.77	−0.97
World total	29.7	37.9	47.0	58.0	0.83	0.77	−0.86	−1.12

United Nations [39], pp. 28–31; 72–73; 76–77.

Urbanisation and Development

Urbanisation in Asia is proceeding at a rate much higher than the world average (see Table 1). In particularly the burgeoning capital cities of many Asian countries that are opening up to global business and investments, rural–urban migration has contributed to the rapid growth of urban populations. The Southeast Asian region alone has 3 large capital cities—Manila, Jakarta and Bangkok—which have populations above five million. Manila, the capital city of the Philippines, has double this population.

The growth of mega cities or cities with some 10 million in population is characteristic of urbanisation in Asia and it is this among other development trends that has led scholars to question their sustainability. Certainly, the rapid urban growth seen in Asia poses strong challenges to governments in terms of the infrastructure and services that need to be developed to meet the growing needs of the population as well as the concentration of businesses and people in cities.

Infrastructure spending for cities appropriates the giant share in most Asian countries including China. Yet such spending might not mean the provision of basic needs such as, homes, clean drinking water supply, modern sanitation, sewerage treatment, public transport among others. More probably, the infrastructure being developed has been meant to serve international businesses and investors from modern airports to hotels and telecommunications services. This has serious implications for sustainability as far as urbanisation is concerned.

Implications of Rapid Urbanisation

The cities of the Asian region are what has been argues as an experience of several layered environmental conditions and burdens associated with varied and complex development forces [19]. When compared to the industrialisation and urban development seen in Europe and most of the developed world, in many countries in Asia, there has been compression, and telescoping of different development trends in space and time. This means a complexity of sustainability issues faced by cities in the region that combine several trends at one time. With the polarisation in incomes and material wealth that globalisation, there are sharp divisions in Asia's urban societies with contrasting outcomes for sustainability. The poor and low-income people who are living in the city slums and squatter settlements might be contributing to local environmental problems but the city's rich people with their different consumption patterns are certainly accountable for global environmental problems.

Air pollution is a major problem in the large cities of developing countries, such as Mexico City, Beijing, Calcutta, Bangkok, and Jakarta [22]. The reports are that 'in most of the mega cities of the South, air pollution is worsening because of increased industry, vehicles, and population. According to studies, the air pollution levels can sometimes exceed the air quality standards of the World Health Organisation (WHO) by a factor of three or more' [23]. Current trends point to urban transport in Asia as a key factor, often responsible for 70–80% of local air pollution. In fast industrialising countries such as, China, which depend on coal as a major source of energy, industry is a major source of air pollution [24].

In Asia, observers argue that the environmental performance in most countries has not matched the economic or human development progress that has been achieved in large part because of urbanisation and rapid urban growth. Indeed, pollution is one of the greatest challenges for the region. In general, Asian environmental quality has deteriorated and the environmental conditions in many countries are severely degraded. The air in Asia's cities is among the dirtiest in the world. So, the levels of ambient particulate matter, smoke particles and dust, are generally twice the world average and more than five times as high as that of industrial countries [25]. The management of the future urban growth and development in major cities and their surrounding communities will remain as one of the most serious challenges in facing this problem of air pollution that is not only an environmental or ecological problem but a health and medical issue as well.

Urban transport presents another major challenge to Asian cities. Private mode of transport, the private car, appears to have become the main mode of transport with the associated problems of congested roads and high air pollution problems. This can best be illustrated by comparing the traffic conditions in Singapore and Kuala Lumpur. While currently, only approximately one-third of resident households own a motor car (Singapore Department of Statistics, 2001), in the neighbouring middle income city of Kuala Lumpur, much higher car and motorcycle ownership rates have prevailed leading to traffic jams, more highway construction and urban sprawl [26]. One of the reasons that lower car ownership is possible in Singapore can be attributed to the mass public transport's share of daily trips of 48% in Singapore in 2004 [27], which is much higher than in most other rich cities, except for a few others in eastern Asia [28].

Similarly, 'there now seems to be widespread agreement that in low-income and middle-income countries the state alone will be unable to meet the internationally agreed targets for reducing the number of people in cities with no access to clean water and adequate sanitation' ([29], p. 333). High population density and rapid urban growth have meant degradation of water sources not only from the lack of modern sanitation and sewerage treatments works but also industry in countries such as, China [30].

Solid waste is another serious problem in cities like Manila and Beijing, and the growth in quantity and the environmental consequence of solid waste management can be both local, in terms of combustion and scavengers, and global, as a result of greenhouse gas emission from solid waste treatment. Air pollution, urban transport, clean water supply, adequate sanitation provision and solid waste management are some of the examples of problem associated with rapid urbanisation. Other examples encountered in Asia can be found in Roberts and Kanaley [31].

The city's environmental problems are likely to have the most serious outcomes for the poor. Slums and squatter settlements house at least one-third of the urban population in Southeast Asia [32,33]. Many of such settlements are located in ecologically fragile areas such as riverbanks or hill slopes. The scenario remains virtually similar in the rest of Asia. Such settlements are usually not connected to clean drinking water supply, modern sanitation, sewerage treatment or solid waste removal services. In short, the residents of such settlements make do with their own ways of securing such services, if at all.

Sustainability in Urban Development Agendas

Rizhao is a city of three million people in Shangdong province of China. The effort to install solar water heaters began in 1992. As of 2007, 99% of households in the central districts are using solar energy to provide energy, heating and lighting, and many traffic signals are powered with photovoltaic solar power. Shandong provincial government provided subsidies and funded the research and development of the solar water heater industry. The cost of a solar water heater was brought down to the same level as an electric one. Panels are simply attached to the exterior of a building. Rizhao City mandated that all new buildings incorporate solar panels and oversaw the construction process to ensure proper installations. The city has been designated as the Environmental Protection Model City by China State Environmental Protection Administration and is consistently listed in the top 10 cities for air quality in China. Rizhao city has succeeded in popularising solar energy through regulations for new development and other means, with significant improvement in air quality and reduction of CO_2 emission.

Most cities in Asia lack a policy framework that is multisectoral in nature that can address sustainability issues while implementing urban development plans. In cities such as, Singapore, a large-scale or nation-wide public housing programme provided for modern housing and environmental infrastructure including clean drinking water supply, modern sanitation and sewerage treatment. The public housing policy also enabled transport planning and the development of a comprehensive public transport network comprising bus and rail for daily commuters that have pre-empted some of the worse of the traffic congestion issues discussed above.

Policy frameworks that provide for housing and environmental infrastructure also address many of the health issues that plague Asian cities including water-borne illnesses from polluted water. Indeed, sound policies providing for such infrastructure helps address water pollution problems as would urban transport planning emphasising public transport modes.

The tendency has been for most city governments to address sustainability one issue at a time such as air pollution or the cleaning up of polluted rivers. While these are laudable in themselves, they lack the capacity to challenge the complex sustainability issues facing Asian cities today.

The governments of many Asian cities unlike their European counterparts have remained ambivalent about sustainable solutions to urban traffic congestion with relatively tepid approaches towards the development of public transport infrastructure [34]. If more Asian cities have been placing sustainable development in their agendas, there are equally as many cities in Asia that do not appear to have gone beyond the rhetoric of becoming environmentally more sustainable. While many cities have given priority to water and energy management issues for both economic as well as ecological reasons, air and water pollution issues remain largely unaddressed in Asian cities. Added to these more basic issues are those concerning poverty, poor infrastructure, financing and climate change. It is worthwhile to point out that Asian cities, including the mega cities are sites of incredible differences in terms of governance, political economy, culture, and so on. It is therefore not likely to have 'one size fits all' approach to urban sustainability in Asia.

The failure of city governments in Asia to assume a more multi-sectoral approach to urban sustainability is a result of the failure to muster the necessary political and economic support. The challenge remains in the integration of environmental thinking into mainstream economic and development decisions. Ministries of Environment and related agencies remain relatively weak and at the best most of them operate on the margins of significant policy decisions. In most Asian countries, traditional economic models have yet to incorporate the costs of environmental decline continue to drive most decisions on development.

To a large extent, the models of urban governance in the Asian region imply tensions that constrain many city governments from effectively placing sustainable development in their agendas. These tensions see differences between central and city governments in policy directions as well as intervention by the former in financial and related matters that determine the ability of city governments to act in response to urban developmental issues and problems. The tensions are examples of the issues that Asian cities and their national governments need to rise up to in order to face the challenges posed by sustainable urban development. The future of urban sustainability can only succeed when there is integration of environmental thinking into mainstream economic and development decisions.

All references for articles included in *Taking Sides: Clashing Views in Sustainability* can be found on the Web at www.mhhe.com/cls.

EXPLORING THE ISSUE

Can Cities Be Made Sustainable?

Critical Thinking and Reflection

1. How will a decrease in city vulnerability to natural forces affect a sustainable city agenda? Increasing sea level rise poses a threat to coastal regions and cities; is the present evidence of seasonal increases in temperatures and elevated consumption of electricity same in all cities?
2. How are cities responding to the current economic meltdown and prevailing budget cuts? Are sustainable projects within such cities likely to be affected?
3. Is the design and life cycle of products viewed as a major factor that could limit sustainable development of cities?
4. How are contemporary cities coping with an ever-increasing waste stream and limited technology and capacity to handle such waste?
5. Does the current urban city design promote varying cultures and social backgrounds that incorporate other valuable designs that appeal to large groups of people?

Is There Common Ground?

City sustainability is an issue that about 80 percent of the authors agree upon as worth pursuing irrespective of geographical locations, that is, developed or developing economies. The common theme is that it is necessary to start acting, given available resources, applying a polycentric approach such that individuals, organizations, institutions, NGOs, and everybody in the chain make an effort to do something to improve our environment. There is an expressed concern that urban centers be planned with people in mind and not just cars and that city or local governments are best positioned to implement sustainability practices, programs, and projects. Improving quality of cities promotes their livability while acting to encourage more sustainable behavior and land stewardship through programs like recycling, urban gardening, rain barrel usage, etc.

A good design affordable to middle group earners is desirable since they are more vulnerable to increased prices and inequality in resources and wealth distribution. Indicators to measure city sustainability are necessary tools to monitor progress toward achieving desired goals. However, city-derived benchmarks and strategies to measure sustainable progress are hugely town dependent and might not apply across other regions. What one town finds pressing and urgent might not be viewed as pressing or urgent in another. It, thus, becomes a duty for city officials and public representative committees to decide on what needs to be addressed given a strong backup and political will besides public participation.

Internet References

Bugliarello, G. (2006). "*Urban sustainability: Dilemmas, challenges and Paradigms*" Technology in Society 28 (2006), pp. 19–26.

Glaeser, E. et al. (1992). "Growth in Cities," *The Journal of Political Economy* Vol. 100, No. 6, Centennial Issue (Dec., 1992), pp. 1126–1152.

Kent, J. (2005) Civic engagement and Sustainable Cities in the United States Public Administration Review Vol. 65, No. 5.

McDonough, W. and Braungart, M. (2002). Cradle to Cradle: Remaking the Way we Make Things, North Point Press: New York.

Owen, D. (2009). *Green Metropolis: Why Living Smaller, Living Closer, and Driving Less are the Keys to Sustainability.* Publisher River Head books.

Taylor, R. & Carandang, J. (2011). "Sustainability and Redevelopment in the City of Manila, Philippines," in *Human Ecology-Journal of the Commonwealth Human Ecology Council,* Issue No. 23, January, pp. 14–21.

Wolch, J. (2007) "Green Urban Worlds" *Association of American Geographers,* 97(2), 2007, pp. 373–384.

Wood, J. (2007). Synergy city; planning for a high density, super-symbiotic society Landscape and Urban Planning Vol. 83, pp. 77–83.

David Owen site and blogs: It is a site by Owen who discusses various topics on sustainability and ways of greening cities and the globe. The site has articles and blogs on various issues touching on the environment as well. Retrieved on February 25, 2011 from http://www.davidowen.net/.

City of Fort Worth, Texas. Covers sustainability topics and projects run by the city. Retrieved on February 24, 2011 from http://www.fortworthgov.org/sustainability/.

Kent Portney: It is a site that generally covers sustainability topics, projects, and initiatives by cities in the United States. Retrieved on February 25, 2011 from http://ourgreencities.typepad.com/our_green_cities/sustainability_in_boston/.

Mathis Wackernagel: Global Footprint Network Web site, an international Web site that focuses on advancing the science of sustainability through use of the Ecological Footprint Network Web. It discusses ecological footprint basics and the science behind it. Retrieved on February 25, 2011 from http://www.footprintnetwork.org.

Web site with articles, book reviews and blogs on sustainability issues from different authors around the globe. Retrieved on February 25, 2011 from http://sspp.proquest.com/.

Contributors to This Volume

EDITOR

ROBERT W. TAYLOR is a professor of urban and environmental studies in the Department of Earth & Environmental Studies at Montclair State University of New Jersey, USA. He is responsible for teaching both graduate and undergraduate courses in the following specialties: sustainability science; best practices for sustainable management of cities and climate change adaptation; and communication strategies for environmental management and sustainability science. His administrative responsibilities include advising and directing doctoral studies in the PhD program in environmental management. He completed a Fulbright Scholar Award to the Philippines in 2009 where he was a visiting professor in the Department of Biology at De La Salle University. While there, he led a research team that produced a sustainability management report for the city of Manila, and in 2010 co-authored a book with Jose Santos Carandang, *Sustainability Planning for Philippine Cities,* published by the Yuchengco Center of De La Salle University. He was previously a Fulbright scholar/professor at the University of Jos in Nigeria where he authored *Urban Development in Nigeria—Planning, Housing and Land Policy*, published by Avebury in 1993. Taylor has published widely in the areas of urban planning and development, and is currently specializing in urban sustainability as it relates to emerging economies. His technical specialties are in brownfield redevelopment and zoning and master plans for sustainable cities. He has administered grants in both photovoltaic energy systems and energy conservation for the New Jersey Board of Public Utilities and directed investigations of potential brownfield sites for inclusion into a web-based information system for the New Jersey Department of Community Affairs, Office of Smart Growth. He has delivered over 100 international presentations in the following countries: Nigeria; United Kingdom; Sweden: United States; China; South Korea; Japan; Canada; Poland: Israel; Russia; Malaysia; Hong Kong; and the Philippines. Conference sponsors range from the Association of American Geographers; American Planning Association; International Planning History; Regional Science Association; Urban Affairs Association; and Fulbright/CULCOM, etc. He has recently been awarded a grant to teach a course, "Technologies for Climate Change Adaptation for Cities," at the Ho Chi Minh City University of Natural Resources and Environment in 2013.

CONTRIBUTING RESEARCHERS

ROBERT W. TAYLOR co-authored the "Introductions" to individual issues with the following researchers. These contributors are either doctoral students in the PhD program in Environmental Management or advanced graduate studies in the Masters in Environmental Studies at Montclair State University. They are as follows: Issue 1—Sieglinde Mueller; Issue 2—Ben Ondimu; Issue 3—Sushant Singh; Issue 4—Margaret Vyff; Issue 5—Virinder Pal Singh Sidhu; Issue 6—Sagarika Roy; Issue 7—Anila Rambal; Issue 8—Wendy Neill; Issue 9—Paola Dolcemascolo; Issue 10—Michael Pawlish; Issue 11—Elizabeth Stagg and Martha Duque; Issue 12—Christopher Stevenson; Issue 13—Rebecca Shell; Issue 14—Melissa Koberle; Issue 15—Natalie Sherwood and Aslan Aslan; Issue 16—Catherine Alexander; Issue 17—Rocio R.D. Onoro; Issue 18—Pooja Kulkami; Issue 19—Naz Onel; Issue 20—Alejandra Bozzolasco; Issue 21—Faith K. Justus.

AUTHORS

ERIK ASSADOURIAN is a senior researcher at the Worldwatch Institute and project director of the State of the World 2010.

SAAMAH ABDALLAH graduated in experimental psychology from the University of Cambridge and received an MSc in democracy and democratization at University College London. He has worked on the National Accounts of Well-being and Regional Indices of Sustainable Economic Welfare and Caerphilly Sustainability Index.

ACHILLES ADAMANTIADES is an independent consultant of energy and environment. Before his retirement from the World Bank Group in May 1996, he was energy engineer with director's responsibilities in the Industry and Energy Department for international energy issues with emphasis on developing countries.

JOHN AMBLER is senior vice president of Oxfam America, responsible for overall supervision of the relief and development operations of Oxfam America worldwide. When he wrote his article he was director, East Asia Program Development, Social Science Research Council, New York.

M. ASIF completed his PhD in the area of Applied Energy Engineering in 2002 from Edinburgh Napier University. He presently is a lecturer in the School of Built and Natural Environment at Glasgow Caledonian.

SHARON BLOYD-PESHKIN is an associate professor in the Journalism Department of Columbia College Chicago and a freelance writer with a special interest in kayaking, health, travel, the sea, and the environment.

SHARON BEGLEY wrote the "Science Journal" in *The Wall Street Journal* for 5 years and is currently writing a biweekly column, essays, and cover stories for *Newsweek*.

WALTER BLOCK is the Harold E. Wirth Eminent Scholar Endowed Chair in Economics at Loyola University. He is also an Adjunct Scholar at the Mises Institute and the Hoover Institute.

ALAN COLIN BRENT received his PhD in engineering management from the University of Pretoria and is the leader of the sustainable energy futures (SEF) research group in the resource-based sustainable development (RBSD) area of the Council for Scientific and Industrial Research (CSIR) program on Natural Resources and the Environment.

MUKADASI BUYINZA is a senior lecturer in the Department of Community Forestry and Extension, Makerere University, Kampla, Uganda. He works on resource economics and institutional change, and his current research interests include the impacts of conservation policy on smallholder farming systems and livelihoods.

CRISTIANO BUSCO is an associate professor of management accounting at the University of Siena in Italy.

J. ANTHONY CASSILS is a population researcher and writer for the nonprofit organization Population Institute of Canada, located in Ottawa, Ontario. He is a specialist in population issues and his ideas are spread by the Population Institute.

RICHARD DAHL is a Boston freelance writer who has contributed to *Environmental Health Perspectives*. He also writes periodically for the Massachusetts Institute of Technology.

LEIGH K. FLETCHER is a shareholder in the law firm of Stearns Weaver Miller Weisser Alhadeff and Sitterson located in Tampa, Florida. She obtained her law degree from Florida State University College of Law in 1996 and is currently LEED certified.

MARK L. FRIGO is director of the Center for Strategy, Execution, and Valuation in the Kellstadt Graduate School of Business and Ledger and Quill Alumni Foundation Distinguished Professor of Strategy and Leadership at DePaul University in Chicago.

THOMAS M. GEHRING is currently a professor in biology in the Department of Central Michigan University. He received his PhD degree from Purdue University with specialty in wildlife ecology.

DAVID HALL is the director of Public Service International Research Unit (PSIRU) at the Business School, University of Greenwich, London, United Kingdom.

BENJAMIN S. HALPERN received his PhD in ecology, evolution, and marine biology from the University of California, Santa Barbara. He has served as the project coordinator of ecosystem-based management of coastal marine systems for the National Center for Ecological Analysis and Synthesis (NCEAS) since 2005.

STEVEN D. HEINZ has 35 years of professional experience in energy accounting and engineering. He is the chief executive officer and founder of Good Steward Software.

CRAIG HILTON-TAYLOR is manager of the Red List Unit at the International Union of Conservation of Nature (IUCN), a nonprofit organization. He is presently located in Cambridge, United Kingdom, and has been a researcher at the National Botanical Institute Education University of KwaZulu-Natal.

INTERNATIONAL FUND FOR AGRICULTURAL DEVELOPMENT (IFAD), a specialized agency of the United Nations, was established as an international financial institution in 1977 as one of the major outcomes of the 1974 World Food Conference.

ROBERT W. KATES is an independent scholar in Trenton, Maine, and university professor Emeritus at Brown University. In 2008, he was appointed the inaugural presidential professor of Sustainability Science at the University of Maine, Orono.

BRUCE KATZ is a vice president at the Brookings Institution and founding director of the Brookings Metropolitan Policy Program. He regularly advises national, state, regional, and municipal leaders on policy reforms and in 2006 received the prestigious Heinz Award in Public Policy for his contributions to urban and metropolitan America.

I. KESSIDES is affiliated with the Economic Research Group at the World Bank in Washington, District of Columbia.

PAUL KRUGMAN is a professor of economic and international affairs at Princeton University and an economics writer for *The New York Times*.

JO KWONG is vice-president of Institute Relations, Atlas Economic Research Foundation in Fairfax, Virginia.

MICHAEL LAFF is the senior associate editor and blog writer for T + D, an organization that is dedicated to leadership in the profession of training and development.

JEAN-MARC LANDRY is widely known by farmers in Germany and France as a consultant expert in dog protection. He used to work as a consultant expert on guard dogs for a company mandated by the Department of Environmental in Switzerland from 2004 to 2009. At the same time he also holds the position as trainer of guards dogs and ethology, as well as being a behavioral expert on protection dogs.

EMILIA L. LEONE is a PhD candidate at the University of Siena, Italy.

JOANNA I. LEWIS received her PhD from the University of California, Berkeley, and is an assistant professor of science, technology, and international affairs (STIA) at Georgetown University's Edmund A. Walsh School of Foreign Service.

EMANUELE LOBINA is a researcher specializing in water at the Public Service International Research Unit (PSIRU) at the Business School, University of Greenwich, London, United Kingdom.

SEAMUS MCGRAW is an American journalist and author. He has received the Freedom of Information Award from the *Associated Press* Managing Editors.

BILL MCKIBBEN is an American environmentalist and writer who frequently writes about global warming and alternative energy and advocates for more localized economies. He is a frequent contributor to *The New York Times, The Atlantic Monthly, Harper's, Orion Magazine, Mother Jones, Granta,* and *Rolling Stone.*

RUSSELL MCLENDON is a science editor for Mother Nature Network.

CRAIG MEISNER holds a PhD in Agronomy from the University of Georgia and currently teaches at Cornell University.

T. MUNEER is a professor of energy engineering at Napier University, Edinburgh, United Kingdom.

GIOK LING OOI is an urban geographer who was on the faculty in Humanities and Social Sciences Education Department at the National Institute of Education, Nanyang Technological University. She has been president of the Southeast Asian Geography Association and is currently writing on how to make cities sustainable.

STEPHEN POLASKY is Fesler-Lampert professor of ecological/environmental economics at the University of Minnesota and research fellow at the Beijer Institute of Ecological Economics.

DAVID B. RESNIK is a bioethicist and vice-chair of the institutional review board for human subjects research at the National Institute for Environmental Health Science, National Institutes of Health. He received his PhD from the University of North Carolina at Chapel Hill and was a professor of medical humanities at the Brody School of Medicine at East Carolina University.

ANGELO RICCABONI, PhD, is a professor of management accounting and dean of the faculty of economics at the University of Siena. You can reach him at riccaboni@unisi.it.

ANDREW A. ROSENBERG is an oceanographer who is currently dean of the College of Life Sciences and Agriculture and professor of natural resources at the University of New Hampshire (UNH), Durham. Before he joined UNH, he held many positions at the National Marine Fisheries Service, both in research and administration.

MILTON H. SAIER is a professor of biology at Division of Biological Sciences, University of California at San Diego, La Jolla, California. He received his PhD from UC Berkeley and was a postdoctoral fellow at Johns Hopkins University.

CLIVE L. SPASH is an economist who is a professor of public policy and governance in the Department of Socio-Economics at WU Vienna University of Economics and Business. He is editor-in-chief of *Environmental Values* published by While Horse Press.

FREDRIK SEGERFELDT is author of the book, *Water for Sale: How Business and the Market Can Resolve the World's Water Crisis* (2005), and holds a masters degree in European politics. He is on the advisory council of the European Enterprise Institution and has experience with numerous think tanks and business organizations.

HERMANN SHEER is the founder of the nonprofit European Renewable Energy Association EUROSOLAR in 1998 and the World Council for Renewable Energy, serving as president and chairman of both organizations. He supported the establishment of the International Renewable Energy Agency (IRENA), which promotes the transition toward a sustainable use of renewable energy.

HUUB SPIERTZ is a professor of crop ecology, with emphasis on nutrient and carbon flows, at the Wageningen, the Netherlands. He has served as past-president and member of the Continuing Committee of the International Crop Science Congress (ICSC), past-president of the European Society for Agronomy (ESA), and member of the editorial board of the *European Journal of Agronomy* (EJA).

SAM THOMPSON has a PhD in psychology from the University of London and teaches psychology at the University of East London. He has been working with the New Economics Foundation since 2006.

JACK T. TREVORS is affiliated with the School of Environmental Sciences at the University of Guelph, Guelph, Ontario. He is a specialist in environmental biology/microbiology and is presently chair of environmental microbiology in the Canadian College of Microbiologists.

WAYNE C. TURNER has broad experience in energy management and has authored five textbooks and numerous articles. He is Regents Professor Emeritus of Industrial Engineering and Management at Oklahoma State University. He has conducted and supervised more than 1000 energy audits for industrial and commercial facilities.

FRANK VANCLAY is a professor of cultural geography at the University of Groningen, the Netherlands. His prior academic appointment includes a faculty position at Tasmanian Institute of Agricultural Research where he was a professor of rural and environmental sociology.

KURT C. VERCAUTEREEN is a project leader in wildlife disease at the USDA APHIS Wildlife Services, National Wildlife Research Center. He received his MS and PhD degree from University of Nebraska. He conducts research to create novel means of managing the diseases and damage of deer and elk to protect American agriculture. His current research involves addressing scientific questions that will lead to a better understanding of diseases of ungulates and to develop means to manage and control them. Many of his efforts are focused at the interface between free ranging cervids and livestock.

STEPHEN M. WHEELER is an associate professor at University of California, Davis, teaching community and regional planning, urban design, and sustainable development. He is the author of two books on sustainability—*Planning for Sustainability: Towards Livable, Equitable, and Ecological Communities* and *The Sustainable Urban Development Reader.*

ROY WHITEHEAD is a professor of Business Law at the University of Central Arkansas.

WILL WILKINSON is a writer and public intellectual based in Iowa City, Iowa. He graduated from the University of Northern Iowa, received a masters from Northern Illinois University and is working on his doctorate at the University of Maryland. He writes blogs on American politics for *The Economist* and other digital media and has published in *The Atlantic, Forbes, Policy,* and other publications. He has also been a research fellow at the Cato Institute.

ERIC A. WOODROOF has been a board member of the Certified Energy Manager (CEM) Program since 1999 and more than 15 years of experience helping clients "to make money and simultaneously help the environment." He is also a certified carbon reduction manager and was named president-elect of the Association of Energy Engineers.

YUN ZHOU is a Nuclear Security Fellow at the Belfer Center's Project on managing the atom and International Security Program at the John F. Kennedy School of Government, Harvard University. She received her PhD in nuclear engineering from the University of California, Berkeley, in 2006.